MACMILLAN DICTIONARY OF

ASTRONOMY

SECOND EDITION

VALERIE ILLINGWORTH

MACMILLAN PRESS
LONDON

© Laurence Urdang Associates Ltd 1979

Material for new edition
© Market House Books Ltd 1985

First published 1979; second edition 1985 by
THE MACMILLAN PRESS LTD
London and Basingstoke

Associated companies in Auckland, Delhi, Dublin,
Gaborone, Hamburg, Harare, Hong Kong,
Johannesburg, Kuala Lumpur, Lagos, Manzini,
Melbourne, Mexico City, Nairobi, New York,
Singapore, Tokyo

Paperback reprinted 1985

British Library Cataloguing in Publication Data

Macmillan dictionary of astronomy. – 2nd ed
1. Astronomy – Dictionaries
520'.3'21   QB14

ISBN 0-333-39062-8
ISBN 0-333-39243-4 Pbk

Prepared for automatic typesetting by
Market House Books Ltd Aylesbury

Printed and bound in Great Britain by
Anchor Brendon Ltd, Tiptree, Essex

THE MACMILLAN DICTIONARY OF ASTRONOMY

Contributors

Rosamund L. Briggs, BSc
Peter H. Cadogan, BSc, PhD
Heather Couper, BSc, FRAS
Charles A. Cross, MA, FRAS
John Daintith, BSc, PhD
Tony Dean, BSc, PhD, FRAS
Peter Duffett-Smith, BSc, PhD, FRAS
Professor Andrew Fabian, BSc, PhD, FRAS
Peter Gill, FRAS
Nigel Henbest, BSc, MSc, FRAS
David W. Hughes, BSc, PhD, FRAS
Garry E. Hunt, DSc, PhD, FIMA, FRAS, FMet.S, MBCS
Valerie Illingworth, BSc, MPhil
Professor Philip L. Marsden, BSc, PhD, FIP, FRAS
Alan Pickup, FRAS
James Pinfold, BSc, PhD
Professor Ken Pounds, BSc, PhD, FRS, CBE
Gareth Rees, BSc, PhD, FRAS
Allan J. Willis, BSc, PhD, FRAS

Preface

The tremendous advances in astronomy in the past five years, combined with the enthusiastic reception of the original edition, have encouraged the publishers to call for this second edition of the *Dictionary of Astronomy*. New and improved ground-based observational techniques and instruments and several highly successful astronomical satellites and planetary probes have produced a flood of information together with new and modified ideas about celestial objects. These advances in astronomy are described in the new dictionary: entries in the first edition have been updated, often extensively, and over 250 entries have been added, bringing the total to more than 2300.

The dictionary contains several long entries in which a word of primary importance is defined and discussed and in which closely associated words, printed in italics, are also defined. In these and in the shorter entries, an asterisk (*) before a word or group of words indicates additional entries that the reader could consult for further information. Synonyms and abbreviations of entries are placed in brackets immediately following the headword. Tables at the back of the dictionary contain general information while tables and illustrations in the text relate to specific entries.

I would like to express my thanks and appreciation to all those involved in the preparation of the second edition, especially the contributors who produced the new and updated entries and those at astronomical centres in the UK, Europe, and the USA who have readily supplied information on request.

Valerie Illingworth
November 1984

A

AAT. *Abbrev. for* Anglo-Australian telescope.

Abell catalogue. The standard catalogue of rich *clusters of galaxies visible from the northern hemisphere. It was published by George Abell in 1958.

aberration. 1. (aberration of starlight). The apparent displacement in the position of a star due to the finite speed of light and to the motion of the observer, which results primarily from the earth's orbital motion around the sun. It was discovered in 1729 by the English astronomer James Bradley. Light appears to approach the observer from a point that is displaced slightly in the direction of the earth's motion. The angular displacement, α, is given by the ratio v/c, where v is the earth's orbital velocity and c is the speed of light. Using the earth's mean orbital speed gives the *constant of aberration*, equal to 20.4955 arc seconds. Over the course of a year, the star appears to move in a small ellipse around its mean position; the ellipse becomes a circle for a star at the pole of the *ecliptic and a straight line for one on the ecliptic. The maximum displacement, i.e. the semimajor axis of the ellipse, is 20.5 arc seconds.

The aberration due to the earth's orbital motion is sometimes termed *annual aberration* to distinguish it from the very much smaller *diurnal aberration* that results from the earth's rotation on its axis. *Compare* annual parallax.

2. A defect in the image formed by a lens or curved mirror, seen as a blurring and possible false coloration in the image. Aberrations occur for all light rays lying off the optical axis and also for those falling at oblique angles on the lens or mirror surface. The four principal aberrations are *chromatic aberration (lenses only), *spherical aberration, *coma, and *astigmatism. *Curvature of field and *distortion are other aberrations. Chromatic aberration occurs when more than one wavelength is present in the incident light beam. For light of a single wavelength, only the latter five aberrations occur. These image defects may be reduced – but not completely eliminated – in an optical system by a suitable choice of optical materials, surface shape, and relative positions of optical elements and stops. *See also* achromatic lens; correcting plate.

3. A defect in the image produced by an electronic system using magnetic or electronic lenses.

absolute magnitude. *See* magnitude.

absolute temperature scale. *See* thermodynamic temperature scale.

absolute zero. The zero value of the *thermodynamic temperature scale, i.e. 0 K ($-273.15\,°C$). Absolute zero is the lowest temperature theoretically possible. At absolute zero molecular motion almost ceases.

absorption lines, bands. *See* absorption spectrum.

absorption nebula. *See* dark nebula.

absorption spectrum. A *spectrum that is produced when *electromagnetic radiation has been absorbed by matter. Typically, absorption spectra are produced when radiation from an incandescent source, i.e. radiation with a *continuous spectrum, passes through cooler matter. Radiation is absorbed (i.e. its intensity is diminished) at selective wavelengths so that a pattern of very narrow dips or of wider troughs – i.e. *absorption lines* or *bands* – are superimposed on the continuous spectrum.

The wavelengths at which absorption occurs correspond to the energies required to cause transitions of the absorbing atoms or molecules from lower *energy levels to higher levels. In the hydrogen atom, for example, absorption of a photon with the required energy results in a 'jump' of the electron from its normal orbit to one of higher energy (*see* hydrogen spectrum).

The absorption lines (or bands) of a star are produced when elements (or compounds) in the outermost layers of the star absorb radiation from a continuous distribution of wavelengths generated at a lower level in the star. Basically the same elements occur in stars. Since each element has a characteristic pattern of absorption lines for any particular temperature (and pressure) range, there are several different types of stellar spectra depending on the surface temperature of the star. *See* spectral types. *See also* emission spectrum.

abundance. *See* cosmic abundance.

Acamar. *See* Eridanus.

acceleration. 1. (linear acceleration). Symbol: a. The rate of increase of velocity with time, measured in metres per second per second (m s^{-2}), etc.

2. (angular acceleration). Symbol: α. The rate of increase of *angular velocity with time, usually measured in radians per second per second.

acceleration of gravity. Symbol: g. The acceleration to the centre of a planet (or other massive body such as a natural satellite) of an object falling freely without air resistance, i.e. acceleration due to downward motion

in a gravitational field. It is equal to GM/R^2, where G is the gravitational constant and M and R are the mass and radius of the planet. The acceleration is thus independent of the mass of the accelerated object, i.e. it is the same for all bodies (neglecting air resistance) falling at the same point on the surface of the planet, satellite, etc.

On earth the standard value of the acceleration of gravity is 9.806 65 metres per second per second. The value varies from place to place on the earth's surface because of different distances to the earth's centre and greater acceleration towards the equator. In addition, it is affected by local deposits of light or heavy materials.

accretion (aggregation). The increase in mass of a body by the addition of smaller bodies that collide and stick to it. The relative velocity of any two colliding bodies must be low enough for them to coalesce on impact rather than fly apart. Once a large enough body forms, its gravitational attraction accelerates the accretion process. Accretion is now assumed to have had an important role in the formation of the planets from swarms of dust grains. *See* solar system, origin.

accretion disc. *See* black hole; mass transfer.

Achernar (α Eri). A conspicuous bluish-white star that is the brightest in the constellation Eridanus. m_v: 0.5; M_v: −2.2; spectral type: B5 V; distance: 35 pc.

Achilles. The first member of the *Trojan group of *minor planets to be discovered, in 1906. It lies east of Jupiter and precedes the latter in its orbit of the sun. *See* Table 3, backmatter.

achondrite. Any of a class of *stony meteorites that lack the characteristic *chondrules of the *chondrites. They are usually more coarsely crystallized than the chondritic stones and are more similar chemically and mineralogically to some terrestrial rocks. They contain very little iron and nickel.

achromatic lens (achromat). A two-element lens – a doublet – that greatly reduces *chromatic aberration in an optical system. The components, one converging and the other diverging in action, are of different types of glass (i.e. they have different *refractive indices); the combination focuses two selected colours, say red and blue, at a common image plane with a small spread in focal length for other colours. The difference in optical power (reciprocal of focal length) for the two colours in one element must cancel that in the other element. By a suitable choice of glasses and surface curvatures, the doublet can be *aplanatic as well as achromatic, so that three major aberrations are minimized.

Residual colour effects in an achromat can be further reduced by using a compound lens of three or more elements – an *apochromatic lens*; each element has an appropriate shape and dispersive power so that three or more colours can be focused in the same image plane.

acronical rising (or setting). The rising (or setting) of a star at or just after sunset. When the rising is acronical the setting is *cosmic* and vice versa.

Acrux. *See* Alpha Crucis.

active galaxy. A galaxy that is emitting unusually large amounts of energy from a very compact central source. (A separate category, *starburst galaxy, is now employed for a galaxy where a high infrared luminosity arises from intense star formation.) This central powerhouse may be observed directly, as in *Seyfert galaxies, *N galaxies, *BL Lac objects, or *quasars; in *radio galaxies

it is the radio-emitting lobes created by *beams emanating from the powerhouse that are observed. There is mounting evidence that quasars and Seyferts are associated with spiral galactic structure while radio galaxies, BL Lac objects, and possibly N galaxies might represent activity in elliptical galaxies.

Observations of the motions of stars and gas in galaxies such as M87 (*see* Virgo A) and NGC 4151 (*see* Seyfert galaxy), in addition to theoretical arguments (*see* quasar), strongly suggest that the energy output is derived from the gravitational potential of a supermassive *black hole: the energy would arise from an accretion disc of gases spiralling into the black hole. This gas could come from the interstellar medium of a spiral galaxy (especially when perturbed by gravitational effects in *interacting galaxies), from the tidal disruption of stars near the black hole, or from flows of intergalactic gas on to the central galaxy of a *cluster of galaxies, as the gas cools (*see* Perseus cluster).

active optics. The techniques and technology by which any distortion, arising from atmospheric effects, of a *wavefront entering a telescope can be sensed and information fed back to a deformable primary mirror whose profile is continuously adjusted to compensate. This should in theory lead to diffraction-limited performance at all times (*see* Airy disc).

active prominence. *See* prominences.

active-region filament. *See* prominences.

active regions. Regions of intense localized magnetic field on the sun that extend from the *photosphere, through the *chromosphere, to the *corona. They may encompass a variety of phenomena, such as *sunspots, *faculae/*plages, *filaments (or *prominences), and *flares, and are characterized by enhanced emission of radiation at x-ray, ultraviolet, and radio wavelengths.

active sun. The term applied to the sun around the maximum of the *sunspot cycle, when the profusion of *active regions ensures a high level of *solar activity. *Compare* quiet sun.

Adhara (ε CMa). A very luminous remote blue-white giant that is the second brightest star in the constellation Canis Major. It is a visual double star with an 8th-magnitude companion at a fixed separation of 8″. m_v: 1.5; M_v: −5.0; spectral type: B2 II; distance: 200 pc.

adiabatic process. A process that takes place in a system with no heat transfer to or from the system. In general, a temperature change usually occurs in an adiabatic process.

adiabatic theory. *See* galaxies, formation and evolution.

Adonis. A member of the *Apollo group of *minor planets. Discovered in 1936 when it passed 0.015 AU from the earth, it was not observed again until it was recovered in 1977. Its *perihelion distance of 0.44 AU is the fourth smallest among the minor planets. *See* Table 3, backmatter.

advance of the perihelion. The gradual movement of the *perihelion of a planet's elliptical orbit in the same direction as that of the planet's orbital motion. This advance results from the slow rotation of the major axis of the planet's orbit due to gravitational disturbances by other planets and to the curvature of *spacetime around the sun. The small contribution from the curvature of spacetime was predicted by Einstein's general theory of *relativity.

The value of this relativistic contribution towards the advance of the perihelion of Mercury is about 43 arc seconds per century. This agrees almost exactly with the discrepancy between the experimentally determined value for the advance and that predicted by classical Newtonian mechanics. It was therefore an important confirmation of general relativity. Recent measurements of the advance of the perihelia of Venus and Earth have also been very close to Einstein's predicted values for those planets.

aeon. A period of one thousand million years: 10^9 years.

aerial. *See* antenna.

aerolite. *See* stony meteorite.

aeronomy. The physics and chemistry of the upper atmosphere of the earth, i.e. its temperature, density, motions, composition, chemical processes, reactions to solar and cosmic radiation, etc. The term has been extended to include the physics and chemistry of the atmospheres of the other planets.

aerospace. The earth's atmosphere and the space beyond.

Ae stars. Hot stars of *spectral type A that in addition to the normal absorption spectrum have bright emission lines of hydrogen. These lines are thought to arise in an expanding atmospheric shell of matter lost from the star. Like some *Be stars they are probably similar to *T Tauri stars except that they have larger masses.

afocal system. An optical system in which both the object and the secondary image are infinitely distant. It is the usual adjustment in a simple refracting telescope and is produced when the objective and eyepiece lenses are separated by the sum of their focal lengths: the lenses are then *confocal.*

AG catalogue. *See* AGK.

age of the earth. The oldest rocks found in the earth's crust, in SW Greenland, have been assigned ages of 3.8 thousand million years from radioisotope studies, but other properties of the rocks support the belief that the planet shares a common origin with the rest of the solar system, about 4.6 thousand million years ago. The cycling of the crustal material by plate tectonics (*see* earth) and the eroding effects of ice, water, and wind mean that none of the earliest rocks survive intact and that evidence of an early meteoritic bombardment, so apparent on the moon, Mercury, and Mars, is absent. *See* earth; solar system, origin.

age of the universe. The observed expansion and evolution of the universe suggest that it has a finite age, considered as the time since the *big bang. The inverse *Hubble constant, $1/H_0$, gives a measure of the age if the rate of expansion has always been constant. Since gravitation tends to diminish the expansion rate, H_0 can only give an upper limit. Using the value of H_0 of 55 km s^{-1} Mpc^{-1} gives an upper limit of 18 thousand million (10^9) years.

In the standard *cosmology (solutions of Einstein's field equations without a cosmological constant) with *deceleration parameter q_0, the age is given as one of the three alternatives:

$$H_0^{-1}q_0(2q_0-1)^{-3/2}[\cos^{-1}(q_0^{-1}-1)-q_0^{-1}(2q_0-1)^{1/2}]$$
$$2/3H_0^{-1}$$
$$H_0^{-1}q_0(1-2q_0)^{-3/2}[q_0^{-1}(1-2q_0)^{1/2}-\cosh^{-1}(q_0^{-1}-1)]$$

The choice depends on whether q_0 exceeds, equals, or is less than $\frac{1}{2}$ (and >0), i.e. on whether the universe is closed, flat, or open, respectively. Ages of 12 and 15 $\times$ 10^9 years thus correspond to values of q_0 of $\frac{1}{2}$ and 0.15, for H_0 equal to 55 km s^{-1} Mpc^{-1}.

Lower limits to the age of the universe, other than through measurements of H_0 and q_0, can be found from *radiometric dating of the earth and Galaxy and from studies of globular clusters. The relative abundances of radioactive elements, such as uranium, and their decay products yield an estimate of the time since formation of that body of material. This method gives an age for the universe of 14–16 $\times$ 10^9 years. The age of a globular cluster may be estimated by comparison of the observed main-sequence *turnoff point in the Hertzsprung–Russell diagram for the cluster with theoretical models. The oldest globular clusters in our Galaxy then work out to be 15–18 $\times$ 10^9 years old. Both of these ages should be less than the age of the universe and in particular they must be less than H_0^{-1}. This is the case for H_0 equal to 55 km s^{-1} Mpc^{-1}; but some observations suggest a value as high as 100 km s^{-1} Mpc^{-1} (*see* Hubble constant), in which case the upper limit on the age of the universe is 10 $\times$ 10^9 years, and there would be a serious (and unexplained) conflict with the direct measurements.

aggregation. *See* accretion.

AGK (AG catalogue). *Abbrev. for* Katalog der Astronomischen Gesellschaft. A general catalogue of star positions originally proposed by Friedrich Argelander in 1867. The first version, known as *AGK 1*, covered the northern sky and listed about 190 000 stars: by 1912 zones between *declinations 90° and −18° had been published. A revision, known as *AGK 2*, was based on photographic measurements begun in the 1920s and was published 1951–58. A further set of photographic plates was produced in the 1950s to determine *proper motions, which appeared in the *AGK 3* version; this was distributed on magnetic tape from 1969.

airglow. The faint everpresent glow arising in the earth's atmosphere that is light emitted during the recombination of ionized atoms and molecules following collisions with high-energy particles and radiation, mainly from the sun. This dim light, often termed *nightglow* when seen at night, interferes with observations of faint celestial bodies.

air shower (extensive air shower). *See* cosmic rays.

Airy disc. The bright disc-like image of a point source of light, such as a star, as seen in an optical system with a circular *aperture. The disc is formed by *diffraction effects in the instrument and is surrounded by faint diffraction rings that are only seen under perfect conditions (see illustration). About 87% of the total light intensity lies in the disc. The disc diameter, first calculated by George Airy in 1834, is the factor limiting the angular *resolution of the telescope.

AI Velorum stars. *See* dwarf Cepheids.

Alba Patera. Probably the largest central-vent volcanic structure on Mars. Located north of the *Tharsis Ridge, it has a low dome containing a central caldera and enclosed by a ring of fractures 600 km in diameter. Associated lava flows cover an area more than 1500 km across. *See* Mars, volcanoes.

albedo. The reflecting power of a nonluminous object, such as a planet or planetary surface feature. It is the ratio of the total amount of light reflected in all directions to the amount of incident light: an albedo of 1.0 indicates a perfectly reflecting surface while a value of 0.0 indicates a totally black surface that absorbs all incident light.

A comparison of albedo observations with measures of terrestrial substances may be made to infer the surface characteristics of planets, satellites, or minor planets. Volcanic rocks, whether on the earth or the moon, typically have very low albedos: the albedos of the moon and Mercury are 0.07 and 0.06 respectively. Clouds, such as those of Venus and the *giant planets, have albedos of 0.7 or more. *See also* Table 1, backmatter.

Albireo (β Cyg). An orange giant in the constellation Cygnus. It is a beautiful double star with a deep blue companion 35″ away. m$_v$: 3.2 (A), 5.5 (B); spectral type: K5 II (A), B8 V (B).

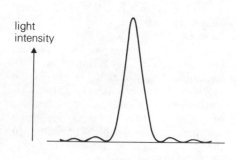
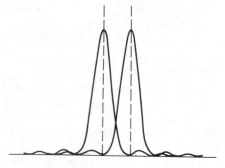

Light distribution in Airy disc image of single point source (left) and two just resolvable point sources

Alcaid (Benatnasch; η UMa). A blue-white star that is one of the seven brightest in the *Plough. m_v: 1.86; M_v: −2.3; spectral type: B3 V; distance: 70 pc.

Alcor (80 UMa). A white star that lies in the *Plough and forms an optical double with *Mizar. m_v: 4.1; spectral type: A5 V; distance: 27 pc.

Alcyone (η Tau). A bluish-white double, possibly multiple, star in the constellation Taurus that is the brightest star in the *Pleiades. m_v: 2.86; M_v: −2.7; spectral type: B; distance: 125 pc.

Aldebaran (α Tau). A conspicuous red giant that is the brightest star in the constellation Taurus and lies in the line of sight of but much nearer than the Hyades. It is a slow irregular variable. It is also both a visual binary with an 11th-magnitude red dwarf companion at 31″ distance and an optical double with an independent binary system 122″ away. m_v: 0.9; M_v: −0.8; spectral type: K5 III; radius (by interferometer): 45 times solar radius; distance: 21 pc.

Alfvén's theory. A theory proposed by the Swedish physicist Hannes Alfvén in 1942 for the formation of the planets out of material captured by the sun from an interstellar cloud of gas and dust. As atoms fall towards the sun, they become ionized and subject to the control of the sun's magnetic field. Ions are concentrated in the plane of the solar equator where the planets coalesce. The theory, in its original form, had difficulty in accounting for the inner planets, but was important in suggesting the role of *magnetohydrodynamics in the genesis of the solar system. *See* solar system, origin.

Algeiba (γ Leo). An orange giant that is the second-brightest star in the constellation Leo. It is a multiple star, forming a visual binary (separation 4″.4, period 619 years) with a 4th-magnitude yellow companion. m_v: 2.02 (A), 1.84 (AB); M_v: −0.5; spectral type: K0p III (A), G7 III (B); distance: 32 pc.

Algenib (γ Peg). A remote bluish-white subgiant that is one of the brighter stars of the constellation Pegasus and lies on the Great Square of Pegasus. m_v: 2.8; spectral type: B2 IV; distance: 143 pc.

Algol (Demon Star; Winking Demon; β Per). A white star that is the second-brightest one in the constellation Perseus. It was the first *eclipsing binary to be discovered, being the prototype of the *Algol variables, although its variations in brightness were known to early astronomers. The theory of a darker companion periodically cutting off the light of the brighter star (Goodricke, 1782, Pickering, 1880) was confirmed spectroscopically in 1889. The brighter star (Algol A) is about three times the sun's diameter; the fainter orange 3rd-magnitude companion (Algol B) is about 20% larger. The two stars revolve about one another in a period of 68.8 hours, and the eclipses cause the magnitude to drop from 2.2 to 3.47 (see illustration). Algol A is about 3.7 times as massive as the sun while Algol B has a mass of only 0.8 solar masses. According to *stellar evolution theory, a more massive star evolves more rapidly; yet in the Algol system the more massive Algol A is still a main-sequence star while Algol B has evolved to become a subgiant. This is the *Algol paradox*, which is explained by *mass transfer from Algol B to Algol A (*see* Algol variables).

There is also a third more distant star, Algol C, which orbits Algol A and B in 1.86 years. Slight changes in the time of Algol's eclipses have suggested a fourth star, but these changes are now believed to be due to *apsidal motion and to continuing

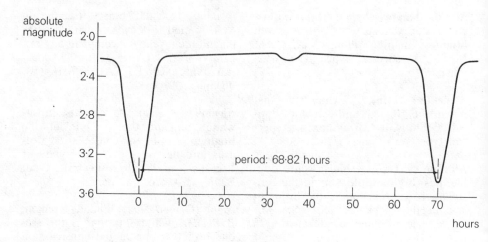

absolute magnitude

period: 68·82 hours

hours

Light curve of Algol

slow mass transfer from Algol B to Algol A. These streams of gas make Algol an erratic radio and x-ray source. m_v: 2.1 (A), 3.5 (B); M_v: −0.2 (A), 1.2 (B); spectral type: B8 V (A), G7 IV (B); distance: 29 pc.

Algol paradox. *See* Algol variables; Algol.

Algol variables (Beta Persei stars). A subclass of *eclipsing binary stars, named after *Algol, where the brighter and more massive star is still on the *main sequence while the less massive companion has evolved more and has become a *subgiant. This seemingly contradicts the theory of stellar evolution, which predicts that more massive stars evolve more rapidly, and is known as the *Algol paradox*. It is explained (Crawford 1955) as a result of extensive *mass transfer: the now less-massive star originally contained most of the system's mass, and it evolved rapidly beyond the main sequence. As it expanded, this star lost up to 85% of its mass to the companion (*see* W Serpentis star) to end up as a faint low-mass subgiant, while the companion became a massive hot and brilliant star, still on the main sequence. Mass transfer con-

tinues at a very slow rate in Algol systems, causing variations in the orbital period and feeble radio and x-ray emission.

As a result of the mass transfer, the two stars have the unusual property of being roughly the same size (several times larger than the sun) but having very different luminosities. They thus have a *light curve characterized by deep primary minima when the dim subgiant eclipses the bright main-sequence star, alternating with scarsely detectable secondary minima when the subgiant is eclipsed.

Algonquin Radio Observatory. The observatory at Lake Traverse, Ontario, Canada run by the National Research Council. Its principal instrument is a 46 metre (150 foot) fully steerable radio dish, in operation since 1966.

Alhena. *See* Gemini.

Alioth (ε UMa). A white *spectroscopic binary, period 4.15 years, that is the brightest star in the constellation Ursa Major. It is a *spectrum variable star. m_v: 1.78; M_v: −0.02; spectral type: A0p; distance: 25 pc. *See also* Plough.

Alkaid. *See* Alcaid.

Allende. *See* meteorite.

Almach (γ And). An orange giant that is the third-brightest star in the constellation Andromeda. It is a triple star, forming a very fine visual binary with its apparently greenish companion (B), 10″ distant, with which it has a common proper motion; B is a spectroscopic binary, period 56 years. m_v: 2.28 (A), 5.1 (B), 2.16 (AB); M_v: −2.3; spectral type: K3 II (A), A0p (B); distance: 80 pc.

Almagest. (Arabic: the Greatest). An astronomical work compiled by Ptolemy of Alexandria in about AD 140. It was translated into Arabic in the ninth century and became known in Europe when it was translated into Latin in the late 12th century. Its 13 volumes cover the whole of astronomy as conceived in ancient times, with a detailed description of the *Ptolemaic system of the solar system. It also included a star catalogue giving positions and *magnitudes (from 1 to 6) of 1022 stars.

almucantar. 1. A *small circle on the *celestial sphere parallel to the horizon.
  2. An instrument for measuring altitude and azimuth.

Alnilam. *See* Orion.

Alnitak. *See* Orion.

alpha (α). The first letter of the Greek alphabet, used in *stellar nomenclature usually to designate the brightest star in a constellation or sometimes to indicate a star's position in a group.

Alpha Capricornids. *See* Capricornids.

Alpha Centauri (Rigil Kent; α Cen). A triple star that is the brightest star in the constellation Centaurus, the third-brightest star in the sky, and the nearest star to the sun. The two brighter components, A and B, form a yellow-orange *visual binary (separation 17″.7, period 80.1 years) and are similar in mass and size to the sun. *Proxima Centauri is the third component and the nearest to the sun. m_v: 0.0 (A), 1.38 (B), −0.27 (AB); M_v: 4.4 (A), 5.8 (B), 4.2 (AB); spectral type: G2 V (A), K1 V (B); mass: 1.09 (A), 0.89 (B) times solar mass; distance: 1.31 pc.

Alpha Crucis (Acrux; α Cru). A conspicuous white subgiant that is the brightest star in the constellation Crux. It is a visual binary (separation 6″), both components being spectroscopic binaries. Alpha and Gamma Crucis point towards the south celestial pole. m_v: 1.6 (A), 2.1 (B), 0.83 (AB); M_v: −3.7; spectral type: B1 IV (A), B1 V (B); distance: 80 pc.

alpha particle (α particle). The nucleus of a helium atom, i.e. a positively charged particle consisting of two protons and two neutrons. It is thus a fully ionized helium atom. Alpha particles are very stable. They are often ejected in nuclear reactions, including *alpha decay* in which a parent nucleus disintegrates – or breaks up – into an alpha particle and a lighter daughter nucleus.

Alphard (α Hya). An orange giant that is the brightest star in the constellation Hydra and lies in a fairly isolated position. m_v: 2.05; M_v: −0.7; spectral type: K4 III; distance: 35 pc.

Alphecca or **Alphekka (Gemma; α CrB).** A blue-white star that is the brightest one in the constellation Corona Borealis. It is an *eclipsing binary (separation 42″, period 17.36 days). m_v: 2.23; M_v: 0.5; spectral type: A0 V (A), G6 (B); distance: 22 pc.

Alpheratz (α And). A bluish-white giant that is the brightest star in the constellation Andromeda. It is both a spectroscopic binary (period 96.7 days) and a *spectrum variable. It originally lay in the constellation Pegasus, as Delta Pegasus, and is still considered part of the Great Square of Pegasus. m_v: 2.07; M_v: −0.5; spectral type: B9p III; distance: 31 pc.

ALSEP. *Abbrev. for* Apollo Lunar Surface Experiments Package. Any of the experimental packages carried to the moon by *Apollos 11–17, set up by the astronauts, and left there to transmit information back to earth. The content of each package differed, becoming more sophisticated as the programme progressed. In the Apollo 17 ALSEP, a central station and thermal generator provided the main power; the experiments included an analysis of any residual atmosphere, detection and measurement of any lunar ejecta and meteorite impacts, measurement of any gravitational anomalies, lunar surface studies, and seismic measurements.

Altair (α Aql). A nearby conspicuous white star that is the brightest star in the constellation Aquila. It is an optical double. m_v: 0.77; M_v: 2.3; spectral type: A7 IV–V; distance: 5.1 pc.

altazimuth mounting (azimuthal mounting). A *mounting in which the telescope swings in azimuth about a vertical axis and in *altitude about a horizontal axis (see illustration). It is easy to make and use and needs no counterpoise weights to balance the telescope. It provides a very firm support and is well adapted for terrestrial observations and for following rapidly moving objects such as artificial satellites. Its great disadvantage is the need to adjust both altitude and azimuth simultaneously and at different rates to follow the diurnal motion of a heavenly body. The application of very precise computer-controlled drive mechanisms to altazimuth mountings has lead to their wider use in large telescopes. The *Multiple Mirror Telescope, the new *William Herschel Telescope, and the Soviet six metre telescope at *Zelenchukskaya have altazimuth mountings, as does the radio telescope at *Jodrell Bank. The altazimuth design allows a smaller and less costly *dome to be used on an optical telescope. *Compare* equatorial mounting.

altitude. The angular distance of a point or celestial object above or below the *horizon, or of an object, such as an artificial satellite, above mean sea level. Altitude and azimuth are coordinates in the *horizontal coordinate system.

aluminizing. A process whereby a very thin but perfectly uniform coating of aluminium is deposited by evaporation on a suitable base. It is used in astronomy to produce the reflective layer of a mirror. The aluminizing is

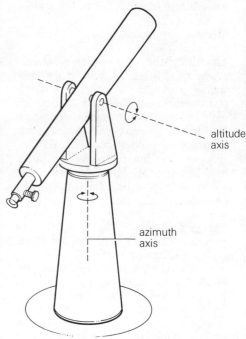

altitude axis

azimuth axis

Altazimuth mounting

done in an evacuated chamber. The aluminium layer is usually protected by a silicon coating that is applied in the same way and that oxidizes to silica when exposed to air. The surface is harder and more stable than silver and is also able to reflect shorter wavelengths than silver.

Amalthea. One of the inner group of *Jupiter's satellites that was discovered in 1892 and until 1979 was thought to be the innermost satellite. It has been greatly distorted by the gravitational pull of Jupiter, becoming markedly elongated in shape (155 × 165 × 270 km); its long axis points towards Jupiter. Photographs from *Voyager I show it to have crater-like features and a low surface reflectivity (i.e. albedo). *See* Table 2, backmatter.

Amor group. The *minor planets that come within the *perihelion distance of Mars but not within the orbit of the earth. The group is named after the minor planet Amor, discovered in 1932. *Compare* Apollo group. *See* Table 3, backmatter.

amplitude. 1. Symbol: Δm. The difference between the maximum and minimum *magnitudes of a *variable star, i.e. the range in magnitude of the star. The amplitude of pulsating variables is related to the logarithm of the period.
   2. The maximum instantaneous deviation of an oscillating quantity from its average value.

AMPTE. *Abbrev. for* active magnetospheric particle tracer explorer. Any of three satellites built by West Germany, the UK, and the USA that were launched from a Thor Delta rocket into different orbits on 9 Aug. 1984 for a series of interactive experiments to study the earth's *magnetosphere. The German and UK satellites were put in highly eccentric orbits into the *solar wind while the US craft was in a lower orbit. Lithi-

um and later barium atoms released into the magnetosphere from the German satellite formed positive ions. The US satellite used these as tracers to study how solar energy, carried by the solar wind, is intercepted and stored in the magnetosphere as charged particles.

Am stars. *Peculiar main-sequence stars of spectral types from A0 to F0 in which there is an over-abundance of heavier elements and rare earths and (less so) of iron, and an apparent under-abundance of calcium. They rotate slower than normal A stars and are almost all short-period spectroscopic binaries. Current theory suggests that tidal effects slow the A star's rotation, leading to an unusually stable atmosphere where heavy elements can diffuse up from the interior. *Compare* Ap stars.

Ananke. A small satellite of Jupiter, discovered in 1951. *See* Jupiter's satellites; Table 2, backmatter.

Andromeda. A constellation in the northern hemisphere close to *Pegasus, the brightest stars being *Alpheratz (α), *Mirach (β), and the fine visual binary *Almach (γ). It contains the *Mira stars R and T Andromedae, the spiral *Andromeda galaxy, and the bright planetary nebula NGC 7662. Abbrev.: And; genitive form: Andromedae; approx. position: RA 1h, dec +40°; area: 722 sq deg.

Andromeda galaxy (M31; NGC 224). The largest of the nearby galaxies, visible to the unaided eye as a faint oval patch of light in the constellation Andromeda. It is an intermediate (Sb) spiral (*see* Hubble classification), orientated at an angle of about 15° from the edge-on position, and has a bright elliptical-shaped nucleus. Its distance is currently estimated as 660 kiloparsecs (2.15 million light years). With a total mass of about 3×10^{11} solar masses (roughly double that of our own galaxy) and an overall diameter of approximately 45 000 parsecs, M31 is the largest of the established members of the *Local Group. It has two dwarf elliptical satellites: NGC 205 and NGC 221 (M32).

Andromedids. *Another name for* Bielids.

Anglo-Australian Observatory. The observatory sited on Siding Spring Mountain in New South Wales, Australia. The chief instruments are the 3.9 metre *Anglo-Australian Telescope and the 1.2 metre *UK Schmidt Telescope.

Anglo-Australian Telescope (AAT). The 3.9 metre (154 inch) reflecting telescope of the *Anglo-Australian Observatory, sited at an altitude of 1165 metres. It was funded jointly by the governments of Australia and the UK, each country having a half share in observing time. It began regular operation in 1975 and is now equipped for both optical and infrared observations. It has *Ritchey–Chrétien optics and a *Cer-Vit mirror and works at a *focal ratio of f/3.5 at the prime focus, f/8 and f/15 at the Cassegrain foci, and f/35 at the coudé focus. It has a limiting magnitude of 23–25 (depending on wavelength and technique used), the maximum field of view covering about one square degree of sky. The telescope is computer controlled so that the pointing and tracking are exceptionally accurate.

angstrom. Symbol: Å. A unit of length equal to 10^{-10} metres. It is mainly used to specify the wavelength of light and ultraviolet radiation.

angular acceleration. *See* angular velocity.

angular diameter. *See* apparent diameter.

angular distance. *See* apparent distance.

angular measure. Units of angle or length given in terms of degrees (°), arc minutes ('), and *arc seconds ("):

$$60'' = 1'$$
$$60' = 1°$$
$$360° = 2\pi \text{ radians (a full circle)}$$

angular momentum (moment of momentum). Symbol: L. A property of any rotating or revolving system whose value depends on the distribution of mass and velocity about the axis of rotation or revolution. It is a vector directed along the axis and for a body orbiting about a point it is given by the vector product of the body's linear *momentum and *position vector r, i.e. it is the product m $(v \times r)$. It is also expressed as the product of the *moment of inertia (I) and *angular velocity (ω) of the body.

In a closed system, such as the solar system or an isolated star, there is always *conservation of momentum.

angular resolution. *See* resolution.

angular separation. *See* apparent distance.

angular velocity. Symbol: ω. The rate at which a body or particle moves about a fixed axis, i.e. the rate of change of angular displacement. It is expressed in radians per second and is a vector quantity. For a point of *position vector r, an angular velocity ω about the origin imparts to the point a velocity $\omega \times r$ at right angles to the position vector. *Angular acceleration* is the rate of change of angular velocity.

anisotropy. A dependence of physical properties upon direction; a lack of *isotropy. A lack of *homogeneity implies anisotropy but the reverse is not necessarily true.

annihilation. A reaction between an *elementary particle and its *antiparticle. An electron and a positron, for example, interact and produce two gamma-ray photons. Hadrons, such as the proton and the antiproton, also undergo annihilation, as do *quarks and antiquarks.

annual equation (annual inequality). The periodic *inequality in the moon's motion that arises from variations in solar attraction due to the *eccentricity of the earth's orbit. Its period is the *anomalistic year and the maximum displacement in longitude is 11'8".9. *See also* evection; variation; parallactic inequality.

annual inequality. *See* annual equation.

annual parallax (heliocentric parallax). The *parallax of a celestial body that results from the change in the position of an observational point during the earth's annual revolution around the sun: nearby stars are seen to be displaced in position relative to the more remote background stars. The instantaneous parallax of a given star is the angle formed by the radius of the earth's orbit at the star; it varies through the year as the orbital radius varies. If the positions of a star are determined during one year they are found to describe an ellipse – a *parallactic ellipse* – on the celestial sphere. The annual parallax, π, of the star is equal to and can be measured from the semimajor axis of the ellipse.

Annual parallax is the maximum parallactic displacement of the star and occurs when the angle earth-sun-star is 90° (see illustration). It is thus expressed by the relation $\sin \pi = a/d$, where a is the semimajor axis of the earth's orbit and d is the star's distance from earth. Since π is extremely small ($<1''$), this reduces to $\pi = a/d$. If a is assumed to be unity (i.e. one *astronomical unit) then a measurement of π in arc seconds will be

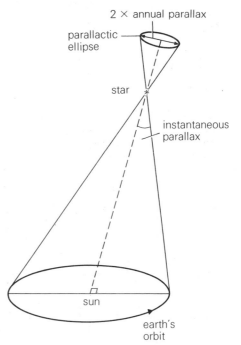

Annual parallax

equal to the reciprocal ($1/d$) of the distance in *parsecs.

The determination of distance from measurements of annual parallax is termed *trigonometric parallax*. Due to its tiny value annual parallax can only be measured (photographically) with reasonable accuracy for stars within about 30 parsecs ($\pi > 0''.03$). Friedrich Bessel published the first account of a successful distance measurement from parallax – that of 61 Cygni – in 1838. The distances of over 6000 stars have since been measured thus, the nearest star – Proxima Centauri – having the greatest value ($0''.76$) of annual parallax.

annular eclipse. *See* eclipse.

anomalistic month. The time interval of 27.554 55 days between two successive passage of the moon through the *perigee of its orbit.

anomalistic year. The time interval of 365.259 64 days between two successive passages of the earth through the *perihelion of its orbit. The anomalistic year is longer than the *sidereal and *tropical years because of the *advance of the perihelion caused (mainly) by planetary perturbations.

anomaly. Any of three related angles by means of which the position, at a particular time, of a body moving in an elliptical orbit can be calculated. For a body S moving around the focus, F, of an orbit (see illustration), the *true anomaly* is the angle v made by the body, the focus, and the point, P, of nearest approach – the pericentre. For a body orbiting the sun, P is the perihelion. The angle is measured in the direction of motion of S. If an auxiliary circle is drawn centred on the midpoint, C, of the major axis of the elliptical orbit, then the *eccentric anomaly* is the angle E between CS′ and CP, where S′ lies on the circle and is vertically above S. The *mean anomaly* is the angle M between P, F, and a hypothetical body moving at a constant angular speed equal to the *mean motion of S. It is thus the product of the mean motion and the time interval since S passed pericentre.

The eccentric and mean anomalies are related by *Kepler's equation*:

$$E - e\sin E = M$$

where e is the *eccentricity of the orbit. The coordinates (x, y) of the body S can be found from the equations

$$x = a(\cos E - e)$$
$$y = a\sin E\sqrt{(1 - e^2)}$$

where a is the semimajor axis of the orbit. *See also* orbital elements; equation of centre.

ANS. *Abbrev. for* Astronomical Netherlands Satellite.

ansae. The parts of *Saturn's rings that are visible on either side of the planet as viewed from the earth.

antapex (solar antapex). *See* apex.

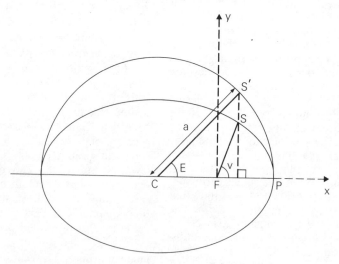

Anomaly

Antares (α Sco). A huge remote but conspicuous red supergiant that is the brightest star in the constellation Scorpius. It is in an advanced stage of evolution. It is a visual binary (separation 3″), its 7th-magnitude B-type companion with which it has a common proper motion appearing green by contrast. The companion is a peculiar hot radio source. m_v: 0.94; M_v: −4.7; spectral type: M1 Ib; diameter: about 285 times solar diameter; distance: 130 pc.

antenna (aerial). A device used for the transmission or reception of radio waves. When connected to a transmitter, the oscillating electric currents induced in the antenna launch radio waves into space. When connected to a *receiver, incoming radio waves from a distant source generate oscillating electric currents in the antenna, which are detected by the receiver. A practical antenna neither radiates nor receives equally in all directions, which may be demonstrated by its *antenna pattern* – a plot of relative *gain as a function of direction. The plot may be made in terms of the antenna's voltage response or power response, giving a *voltage pattern* or a *power pattern*. By the principle of reciprocity the antenna pattern for transmission is identical with that for reception.

A number of distinct *lobes* may often be identified in the antenna pattern. The lobe corresponding to the direction of best transmission or reception is the *main lobe* or *main beam*; all the others are called *side lobes* and are usually unwanted. The ratio of the power received in the main beam to the total power received by all the lobes is called the *beam efficiency*. The angle between the two directions in the main beam at which the power response has fallen to half its maximum value is called the *beamwidth*; it is a measure of the antenna's *directivity. If the angular separation of two cosmic ra-

dio sources is less than the beamwidth, the sources will not be resolved but will be observed as a single source. Some highly directional antennas have *pencil beams*, which are narrow main beams of circular cross section. Others have *fan beams* where the cross section of the main beam is greatly extended in one direction. *See also* array; dish; radiation resistance.

antenna pattern (field pattern; polar diagram). *See* antenna.

antenna temperature. The apparent temperature of the *radiation resistance of an antenna: for a lossless antenna (no energy dissipation) this is not the same as its physical temperature but is determined by the temperature of the distant emitting regions in the antenna's beam. It is thus a measure of the strength of a signal from a *radio source. A resistance at temperature T generates a noise power that, in bandwidth B, is given by $P = kTB$, where k is the Boltzmann constant. The antenna temperature, T_A, is equal to the (weighted) average *brightness temperature of the distant emitting regions so that the radiation resistance generates a noise power $P = kT_AB$; this is passed to the *receiver. A radio source whose angular size is smaller than the antenna beam and whose *flux density is S generates an antenna temperature equal to $SA_e/2k$, where A_e is the effective area (*see* array) of the antenna.

*Synchrotron emission from *cosmic rays in the Galaxy produces a diffuse radio emission centred broadly on the galactic plane. Any practical observation of a radio source has to be made against this background emission, which at low frequencies limits the *sensitivity of the *radio telescope. The antenna temperature of the diffuse emission is called the *background temperature*; the electrical *noise it produces in radio receivers is often called *cosmic noise*.

anthropic principle. A principle that was put forward in the 1960s by R. Dicke and maintains that the presence of life in the universe places constraints on the ways in which the very early universe evolved: the possible initial conditions are limited to those that give rise to an inhabited universe, i.e. what we observe must be restricted by the conditions necessary for our presence as observers.

antimatter. Matter composed entirely of *antiparticles. Although individual antiparticles are produced in cosmic-ray showers and in high-energy particle accelerators, the search for antimatter in the universe has so far proved unsuccessful.

antiparticles. A pair of *elementary particles, such as the electron plus positron or the proton plus antiproton, that have an identical rest mass but a charge, strangeness, and other fundamental properties of equal magnitude but opposite sign (positive rather than negative and vice versa). Thus, the electron has a negative charge and the positron an equal positive charge. A reaction of a particle with its antiparticle is called *annihilation. *See also* pair production.

Antlia (Air Pump, Pump). A small inconspicuous constellation in the southern hemisphere near Centaurus, the brightest stars being of 4th magnitude. Abbrev.: Ant; genitive form: Antliae; approx. position: RA 10h, dec −35°; area: 239 sq deg.

Antoniadi scale. *See* seeing.

apastron. The point in any orbit around a star that is farthest from the star. The nearest point is the *periastron*. The terms are mainly used in connection with the greatest separation or closest approach of the components of a binary star.

aperture. The clear diameter of the objective lens in a refracting telescope or of the primary mirror in a reflector. As the aperture is increased the telescope gathers more light, and so will discern fainter objects: the light-gathering power depends on area, i.e. on the square of the aperture. A larger aperture also produces a smaller *Airy disc and so has greater angular *resolution.

aperture ratio (relative aperture). The ratio d/f of the effective diameter (i.e. *aperture), d, of a lens or mirror to its *focal length, f. The ratio f/d is the *focal ratio.

aperture synthesis. A technique of radio astronomy devised by Martin Ryle in the late 1950s. It is used to obtain the high *resolution of a very large aperture with a number of small *antennas that are moved to cover the large aperture. The antennas are connected as interferometer pairs (*see* radio telescope) and the *amplitudes and *phases of the signals from a *radio source are recorded at all the antenna spacings necessary to cover the required area. The information is then processed in a computer where an inverse *Fourier transform is carried out to yield a radio map of the source. In practice, the rotation of the earth is often used to move the antennas in space; the technique is then called *earth-rotation synthesis*.

In the *Five-Kilometre Telescope* at Cambridge, England, for example, four moveable and four fixed *dishes are mounted on an east-west line to give 16 simultaneous interferometer pairs, each moveable dish with each fixed dish. The dishes are set to track a radio source for 12 hours during which time the earth, and hence the *baseline of the instrument, moves round by half a revolution. The data collected from each interferometer pair corresponds to a strip of an elliptical aperture and by moving the four moveable dishes each day half the aperture can be filled. The information for the other half of the aperture is the same as that already collected (except for a reversal of phase) so that a 12-hour run each day suffices. The major axis of the synthesized aperture equals the maximum baseline of the instrument, five km, while the minor axis depends on the sine of the *declination of the source; this limits the resolution that may be achieved for sources near the equator. The Five Kilometre is part of the Mullard Radio Astronomy Observatory and became operational in 1972.

The *Very Large Array* (*VLA*) near Socorro, New Mexico, uses 27 dishes mounted in a **Y** formation to offset this difficulty. The dishes, each 25 metres across, are arranged around a central control room. The VLA can synthesize a maximum aperture of 27 km diameter and produces radio maps, relatively quickly, with a very high resolution. The VLA is part of the *National Radio Astronomy Observatory and became fully operational in 1981.

The *Australia Telescope, which should be operating by 1988, combines the principles of aperture synthesis and *very long baseline interferometry (VLBI).

In the computer processing, the data from each interferometer pair are not necessarily given equal weight but are scaled according to a *grading function*: this reduces the levels of the side-lobe responses in the synthesized antenna beam. Some loss of resolution results from this process and in practice a compromise is achieved between high resolution and low side-lobe levels. It is often not necessary to fill in the entire aperture, a few equally spaced strips being sufficient. The final radio map will, however, then have *grating rings* on it – concentric elliptical responses at equal intervals from the middle of the observed area of sky. Their presence does not matter provided that all of the structure in a particular radio source is within the first grating ring.

apex (solar apex). The point on the celestial sphere towards which the sun and the solar system are moving relative to the stars in our vicinity. The apex lies in the constellation Hercules at a position RA 18h, dec $+30°$, close to the star Vega. The point diametrically opposite to the direction of this relative *solar motion* is the *antapex* (or *solar antapex*); it lies in the constellation Columba. The position of the apex can be determined from an analysis of the proper motions, radial velocities, and parallaxes of stars, based on the assumption that solar motion is equal and opposite to the group motion of stars. The velocity of this motion has been calculated as about 19.5 km s^{-1} with respect to the *local standard of rest, when all available information on proper motion and radial velocity is used. The value is lower when only the stars in the sun's immediate neighbourhood are considered. Solar motion produces the *secular parallax of stars.

aphelion. Symbol: Q. The point in the orbit of a planet, comet, or artificial satellite in solar orbit that is farthest from the sun. The earth is at aphelion on July 3.

Aphrodite Terra. *See* Venus.

aplanatic system. An optical system that is able to produce an image free from *spherical aberration and *coma.

apoapsis. *See* apsides.

apochromatic lens. *See* achromatic lens.

apocynthion. The point in the lunar orbit of a satellite launched from earth that is farthest from the moon. The equivalent point in the lunar orbit of a satellite launched from the moon is the *apolune*. *Compare* peri-cynthion.

apogalacticon. The point in a star's orbit around the Galaxy that is far-thest from the galactic centre. The nearest point is the *perigalacticon*.

apogee. 1. The point in the orbit of the moon or an artificial earth satel-lite that is farthest from the earth and at which the body's velocity is at a minimum. Strictly the distance to the apogee is taken from the earth's cen-tre. The distance to the moon's apo-gee varies; on average it is about 405 500 km. When a spacecraft or satellite reaches apogee, its *apogee rocket* can be fired to boost it into a more circular orbit or to allow it to escape from earth orbit. *See also* peri-gee.
2. The highest point above the earth's surface attained by a rocket, missile, etc.

apojove. The point in the orbit of a satellite of Jupiter that is farthest from Jupiter. The nearest point is *per-ijove*. Corresponding terms for satel-lites of Mars are *apomartian* and *perimartian* and for satellites of Sat-urn *aposaturnian* and *perisaturnian*.

Apollo. The American space pro-gramme for landing men on the *moon and returning them safely to earth. A manned lunar landing, to be achieved before 1970, was proposed to Congress in 1961 by J. F. Kennedy following the first manned flight by the USSR. The first six Apollos were used in flights designed to test the Saturn 1B and Saturn V launch vehi-cles, the *command and service module (CSM)*, and the *lunar module (LM)*. After three astronauts were killed in January 1967 during ground tests, the programme was delayed for 18 months while the *command module (CM)* was redesigned.

The method selected for the flight programme was *lunar orbit rendezvous*.

The astronauts were to travel to and from the moon in the command mod-ule, which contained the controls and instruments. Rocket engines and fuel supplies were housed in a separate *service module (SM)*. On entering lu-nar orbit the command module pilot would remain in the CM while the commander and lunar-module pilot made the landing in the LM. On completion of the mission on the moon's surface, the descent stage was to remain on the moon: the ascent stage of the LM was to carry the as-tronauts into lunar orbit and rendez-vous with the command and service modules. The craft would then em-bark on its return journey, the LM being jettisoned. The SM was to be jettisoned just prior to re-entering the earth's atmosphere.

Details of the Apollo flights are given in the table. Apollo 7 was the first manned test flight of the CSM. Apollo 8 was the first to be launched by Saturn V and to enter lunar orbit. The LM was tested in earth orbit by Apollo 9, in lunar orbit by Apollo 10, and landed on the moon for the first time by Apollo 11 on 20 July 1969. Apollo 12 landed with higher accura-cy close to *Surveyor 3, parts of which were returned to earth. Both missions returned basalts from the *maria. Apollo 13 was aborted safely after an explosion in the service mod-ule and its target was taken over by Apollo 14. This was to collect rocks from Cone crater (25 million years old) in the *Fra Mauro formation. Mobility was increased by the use of a *Modular Equipment Transporter but the terrain was unexpectedly rug-ged. Complex non-mare basalt brec-cias were returned.

The last three Apollos visited sites with multiple objectives, collected deep (two metre) drill cores (in addi-tion to wide cores, soils, pebbles, and larger rocks), and used a *Lunar Rov-ing Vehicle. Apollo 15 returned anor-thosite and green glass from the Apennine *mountains and mare ba-

Apollo Missions

Mission	Astronauts** (Commander) (LM pilot) (CM pilot)	Date (of landing)	Landing site coordinates (accomplishment)	EVA time (hr)	Distance traversed (km)	Sample weight (kg)
Apollo 7	Schirra Eisele Cunningham	October 1968	(first test of CSM in earth orbit)	–	–	–
Apollo 8	Borman Lovell Anders	December 1968	(first men in lunar orbit)	–	–	–
Apollo 9	McDivitt Scott Schweickart	March 1969	(test of LM in earth orbit)	–	–	–
Apollo 10	Stafford Young Cernan	May 1969	(full dress rehearsal for lunar landing)	–	–	–
Apollo 11	Armstrong* Aldrin* Collins	(20) July 1969	M. Tranquillitatis 0°67′N 23°49′E (first landing)	2.2	0.5	21.7
Apollo 12	Conrad* Bean* Gordon	(19) November 1969	Oc. Procellarum 3°12′S 23°23′W (Surveyor 3)	7.7	1.3	34.4
Apollo 13	Lovell Swigert Haise	April 1970	(Mission aborted: explosion in service module)	–	–	–
Apollo 14	Shepard* Mitchell* Roosa	(31) January 1971	Fra Mauro formation 3°40′S 17°28′E (Cone crater)	9.2	3.4	42.9
Apollo 15	Scott* Irwin* Worden	(30) July 1971	Hadley Rille 26°06′N 3°39′E (first test of LRV)	18.3	27.9	76.8
Apollo 16	Young* Duke* Mattingley	(21) April 1972	Cayley-Descartes 8°59′S 15°31′E (lunar highlands)	20.1	27.0	94.7
Apollo 17	Cernan* Schmitt* Evans	(11) December 1972	Taurus-Littrow 20°10′N 30°46′E (first scientist)	22.0	30.0	110.5

**Astronauts marked with an asterisk landed on the moon.

salts from *Hadley Rille. Apollo 16 landed in the vicinity of South Ray and North Ray craters (2 and 50 million years old respectively) in the *highlands to sample the *Cayley and *Descartes formations, but returned only anorthositic breccias rather than volcanic rocks. Apollo 17 included a scientist-astronaut for the first time. It landed in a dark and *light mantled mare-filled valley between high massifs on the borders of Mare *Serenitatis. Several boulders were sampled and unusual orange glass was found near a 30 million year old *dark halo crater, Shorty.

Experiments performed on the moon by Apollo included *solar-wind, lunar-atmosphere, *cosmic-ray and neutron detection, heat-flow and magnetic-field measurements, active and passive seismometry, and laser ranging. Apollos 15, 16, and 17 were equipped with experiments in the service module, including metric and panoramic photography, laser altimetry, radar sounding, magnetometry, and gamma-ray, ultraviolet, and alpha-particle spectrometry. Subsatellites were ejected from the SM for particle and field measurements. Spent boosters were crashed as seismic sources. After Apollo 17, the remaining hardware of the programme was used in *Skylab and the *Apollo-Soyuz test project.

See also Mercury project; Gemini project; Ranger; Lunar Orbiter probes; Luna probes; Zond probes; moonrocks.

Apollo-Amor objects. *See* Apollo group.

Apollo group. The *minor planets, including *Hermes and *Icarus, that have *perihelia inside the orbit of the earth. Like members of the *Amor group, most are very small bodies and can be observed only when close to the earth. The group is named after the one km diameter minor planet *Apollo*, which was discovered when it approached within 0.07 AU of the earth in 1932 but was lost because of uncertainties in its orbit; it was recovered in 1973. Many of their spectra closely resemble that of *chrondrite meteorites. Members of the Apollo and Amor groups are often classed collectively as *Apollo-Amor objects*, the perihelia of such bodies usually being taken as smaller than 1.3 AU. *See* Table 3, backmatter.

Apollo-Soyuz test project (ASTP). The first international manned space flight, finally agreed to in 1972 and achieved by the US and the USSR in 1975. An American *Apollo spacecraft – Apollo 18 – and a Soviet *Soyuz craft – Soyuz 19 – were launched into earth orbit on July 15, rendezvoused on July 17 at an altitude of 225 km, and successfully docked. The crews of two cosmonauts (Leonov and Kubasov) and three astronauts (Stafford, Slayton, and Brand) visited each other's craft and conducted joint experiments and surveys. The mission involved major design modifications in both spacecraft.

Apollo Telescope Mount (ATM). The complex telescope mount on NASA's orbiting space laboratory *Skylab. It was sited at the centre of the cross-shaped solar panels, being about 4.5 metres in overall length and about 3.5 metres in diameter; the experimental rack was almost 3 metres long. The instruments that it carried were larger and more complex than those previously shot into space. Free from the distortion and absorption effects of the earth's atmosphere they were able to make observations of very high accuracy and resolution.

The six principal instruments operated in the x-ray and/or ultraviolet spectral regions, not accessible to ground-based instruments. Two x-ray telescopes (0.3–6, 0.6–3.3 nanometres (nm)) and a white-light coronagraph (350–700 nm) were used in studying the solar *corona. An XUV spectroheliograph (15–62 nm) a UV spectroheliometer (30–140 nm), and a UV spectrograph (97–394 nm) provided data on the *chromosphere. There were also two hydrogen-alpha telescopes, operating at 656.3 nm, which observed the low chromosphere. The observations, made between May 1973 and Feb. 1974, led to great advancements in solar astronomy.

apolune. *See* apocynthion.

apomartian. *See* apojove.

aposaturnian. *See* apojove.

apparent. Denoting a property of a star or other celestial body, such as altitude or brightness, as seen or measured by an observer. Corrections must be made to obtain the true value.

apparent diameter (angular diameter). The observed diameter of a celestial body, expressed in degrees, minutes, and/or seconds of arc. *See also* stellar diameter.

apparent distance (angular distance; angular separation). The observed distance between two celestial bodies, expressed in degrees, minutes, and/or seconds of arc.

apparent magnitude. *See* magnitude.

apparent noon. The time at which the sun's centre is seen to cross the meridian. *See* apparent solar time.

apparent place. The position on a celestial sphere centred on the earth, determined by removing from the directly observed position of a celestial body the effects depending on the observer's location on the earth's surface – *atmospheric refraction, diurnal *aberration, and diurnal *parallax. It is the position at which the object would be seen from the earth's centre, displaced by *planetary aberration and referred to the *true equator and equinox.

apparent sidereal time. *See* sidereal time.

apparent solar time. Time measured by reference to the observed (apparent) motion of the sun, i.e time as measured by a sundial. The *apparent solar day* is the interval between two successive *apparent noons. The time on any day is given by the *hour angle of the sun plus 12 hours. Because the sun's motion is nonuniform (*see* mean sun) apparent solar time does not have a constant rate of change

and cannot be used for accurate timekeeping. The difference between apparent time and *mean solar time is the *equation of time.

apparition. The period of time during which a particular planet, star, etc., can be observed.

appulse. 1. The apparent close approach of one celestial body to another. Often it refers to the close approach of a planet or minor planet to a star, without the occurrence of an *occultation.
2. A penumbral *eclipse of the moon.

April Lyrids. *See* Lyrids.

apsidal motion. Rotation of the line of *apsides (i.e. the major axis) of the elliptical orbit of a single or a binary star so that the orbit is not closed and there is a gradual shift in the position of periapsis. It can occur if the gravitational field of a third body can produce a significant effect on a two-body situation, as with the earth, moon, and sun, or if one or both stars of a binary system is appreciably distorted from a spherical shape.

apsides (*sing.* apsis; apse). The two points in an orbit that lie closest to – *periapsis* – and farthest from – *apoapsis* – the centre of gravitational attraction. The prefixes *peri-* and *apo-* are used in describing the apsides of specific orbits. The velocity of an orbiting body at the apsides is either maximal (at periapsis) or minimal (at apoapsis). The *line of apsides* is the straight line connecting the two apsides and is the major axis of an elliptical orbit.

Ap stars. *Peculiar main-sequence stars of spectral types from B5 to F5 in which the spectral lines of certain elements (mainly Mn, Si, Eu, Cr, and Sr) are selectively enhanced and are sometimes of varying intensity (*see* spectrum variables). The enhancement and its variation is apparently associated with strong and often variable magnetic fields. Like the *Am stars, Ap stars are generally slow rotators, but they differ in being single stars rather than spectroscopic binaries. The relative enhancement may be due to diffusion in the stable atmosphere caused by slow rotation, modified by the effect of strong magnetism. The spectral peculiarities could also relate to stellar temperature, the *manganese stars* being hotter than the *silicon stars*, which in turn are hotter than the *europium-chromium-strontium stars*. *See also* magnetic stars.

Apus (Bird of Paradise). A small inconspicuous constellation in the southern hemisphere near Crux, the brightest stars being of 4th magnitude. Abbrev.: Aps; genitive form: Apodis; approx. position: RA 16h, dec −75°; area: 206 sq deg.

Aquarids. Either of two active *meteor showers: the *Eta Aquarids*, radiant: RA 336°, dec 0°, maximize on May 4, having a peak duration of 10 days and a *zenithal hourly rate of 20; the *Delta Aquarids*, double radiant: RA 339°, dec −17° and 0°, maximize on July 28, have a peak duration of about 20 days, and a ZHR of about 30. The Eta Aquarids have an orbit that is closely aligned with the post-perihelion path of *Halley's comet.

Aquarius (Water Bearer). An extensive zodiac constellation in the southern hemisphere near Pegasus, the brightest stars being of 3rd magnitude. It contains the planetary nebulae NGC 7293 (the *Helix nebula) and NGC 7009 (the *Saturn nebula*), and the globular clusters M2 and M72. Abbrev.: Aqr; genitive form: Aquarii; approx. position: RA 22h, dec −10°; area: 980 sq deg.

Aquila (Eagle). An equatorial constellation near Cygnus, lying in the Milky Way, the brightest stars being the 1st-magnitude *Altair and some of 3rd magnitude. Eta Aquilae is a bright *Cepheid variable. Abbrev.: Aql; genitive form: Aquilae; approx. position: RA 19.5h, dec +3°; area: 652 sq deg.

Ara (Altar). A small constellation in the southern hemisphere near Centaurus, lying in the Milky Way, the brightest stars being of 3rd magnitude. Abbrev.: Ara; genitive form: Arae; approx. position: RA 17h, dec −55°; area: 237 sq deg.

archaeoastronomy. The astronomy of early or nonliterate cultures, of interest to astronomers, archaeologists, and anthropologists. Many megalithic sites in the UK, Europe, and North and South America are thought to have been used for astronomical measurements and predictions.

arch filament. *See* prominences.

arc second (arc sec; second of arc). Symbol: ″. A unit of angular measure equal to 1/3600 of a degree. It is used, for example, to give the apparent diameter of objects, the angular separation of binary stars, and the proper motion and annual parallax of stars. An *arc minute*, symbol: ′, is 60 arc seconds, i.e. 1/60 of a degree.

Arcturus (α Boo). A conspicuous red giant that is the brightest star in the constellation Bootes and also in the northern sky. The three stars Alioth, Mizar, and Alcaid in the handle of the Plough point in its direction. It is a high-velocity star and shows spectral peculiarities. m_v: −0.05; M_v: −0.3; spectral type: K2 IIIp; radius (by interferometer): 28 times solar radius; distance: 11 pc.

areal velocity. The velocity at which the *radius vector sweeps out unit area. It is constant for motion in an elliptical orbit. *See* Kepler's laws.

Arecibo radio telescope. The radio telescope, 305 metres (1000 ft) in diameter and spherical in shape, sited in a natural hollow at Arecibo, Puerto Rico. It is the world's largest single-dish telescope. It is operated by Cornell University and has been in use since 1963. The original metallic mesh lining was replaced in the early 1970s by tens of thousands of reflecting panels in order to improve reception of lower wavelengths, including the 21 cm line of hydrogen. Although not steerable some directionality can be obtained by moving the feed antennas along the north-south girder on which they are carried, 130 m (425 ft) above the dish. The Arecibo dish, with its huge collecting area, can also be used for *radar astronomy.

Arend–Roland comet (1957 III). A *comet discovered by S. Arend and G. Roland in the course of routine work on *minor planets. It was unusual because it had a sunward spike: many of the solid particles ejected from the *nucleus near *perihelion remained in the orbital plane of the comet; as the earth passed through the orbital plane the reflection from the particles combined to give a luminous spike.

areography. The geography of *Mars. Surface coordinates are defined by *aerographic longitude* and *latitude*. Longitude is measured in degrees westwards from a prime meridian chosen to pass through the dark-albedo feature Sinus Meridiani near the Martian equator. The exact position of the meridian is now fixed by a 0.5 km crater pit, Airy–0, within the crater Airy, 300 km south of the equator.

Argo or **Argo Navis (Ship).** A former very extensive unwieldy constellation in the southern hemisphere subdivided in the mid-18th century into the constellations *Carina, *Puppis, *Vela, and *Pyxis. Its stars are still identi-

fied by a single set of Bayer letters (*see* stellar nomenclature), the brightest being *Canopus or Alpha Carinae.

argument of the perihelion. *See* orbital elements.

Argyre Planitia. A 1000 km diameter basin in Mars' southern hemisphere, visible as a bright area to earth-based observers. Like *Hellas Planitia it is probably an impact feature. *See* Mars, surface features.

Ariane. The three-stage *launch vehicle built for the *European Space Agency (ESA). The building programme was officially approved in 1973, financed mainly by France. The original version, Ariane-1, was declared operational in Jan. 1982 following three successful test flights out of four; these occurred on 24 Dec. 1979, 18 June 1981, and 20 Dec. 1981. The first series of operational flights was carried out under ESA responsibility; the first successful flight took place on 16 June 1983, an earlier launch in Sept. 1982 ending in failure. Following this promotional series, responsibility for production and marketing of Ariane was transferred to the private-law company *Arianespace*, set up in 1980 to finance, produce, market, and launch Ariane; its first launch occurred in May 1984. The launch site is near the equator, at Kourou, French Guiana. The rocket is optimized for launching craft into geostationary orbit (35 800 km) for communications and meteorological purpose, into lower (200 km) circular earth orbits for scientific missions, and into sun-synchronous earth orbits for terrestrial observations. Ariane is expendable, however, unlike NASA's reusable space shuttle.

Ariane-1 can carry a payload of up to 1.7 tonnes (1700 kg) into geostationary orbit. Improvements to the rocket engines have led to the more powerful versions Ariane-2 and Ariane-3, Ariane-3 having the advantage of two solid-fuel boosters in addition to the three liquid-fuel stages. Ariane-3, first launched in June 1984, can carry 2.58 tonnes into geosynchronous transfer orbit and can launch two satellites simultaneously. Ariane-4 is under construction for the ESA and should fly by 1986. Even more powerful (new first stage and up to four solid boosters), it should launch 4.3 tonnes into orbit. A further version, Ariane-5, is planned.

Ariel. A satellite of Uranus, discovered in 1851. *See* Table 2, backmatter.

Ariel V. A highly successful British *x-ray astronomy satellite launched into a 550 km equatorial orbit in Oct. 1974 from the San Marco platform off the Kenyan coast. Its six experiments (five UK, one US) were designed to detect, accurately locate, and measure spectral and temporal features of cosmic x-ray sources. Ariel V was spin-stabilized with the spin axis and period controlled by gas jets. Tracking, command, and data reception were carried out from NASA ground stations at Quito and on Ascension Island. The UK control station at the Rutherford Appleton Laboratory had data links to the experiment groups at the universities of Birmingham, Leicester, and London. The satellite and all experiments remained fully operational until mid-1977 when the control gas was exhausted. Subsequent operation with magnetometer attitude control was extended up to re-entry in Mar. 1980.

Ariel V's major scientific achievements were the production of a new atlas of *x-ray sources including a number of short-lived transients, the establishment of *Seyfert galaxies as powerful x-ray emitters, and the discovery of slow x-ray pulsators (believed to be *neutron stars spinning with periods of minutes) and of abundant iron (by detection of its charac-

teristic x-radiation) in *supernova remnants and in *clusters of galaxies. Technically it was also the first small scientific satellite to be operated in near real-time.

Ariel VI. A British satellite that was launched June 1979 and was the final small satellite in the Ariel series. Its three experiments were designed to measure the flux of heavy primary cosmic rays, i.e. those above iron in the periodic table, to produce a soft x-ray map of the sky (at energies 0.1–1.5 keV), and to determine the time variability and spectra of individual x-ray sources (many of which were discovered by *Ariel V) by means of a *proportional counter pointed for several hours towards each source. Each experiment operated successfully for most of the satellite's life (to early 1982) but the scientific returns were limited by a series of spacecraft problems, including interference from powerful ground radars that caused random switching of both payload and spacecraft elements.

Aries (Ram). A zodiac constellation in the northern hemisphere near Auriga, the brightest stars being the 2nd-magnitude *Hamal (α) and 3rd-magnitude Sheratan (β). Gamma Arietis is a wide visual binary with components equal in magnitude (4.4) and both A stars. Abbrev.: Ari; genitive form: Arietis; approx. position: RA 2.5h, dec +21°; area: 441 sq deg. *See also* first point of Aries.

Aries, first point of. *See* first point of Aries.

Arizona meteorite crater. The archetypal impact explosion crater, 1280 metres across and 180 m deep, formed about 6000 years ago by a 250 000 tonne *meteorite, 70 m in diameter, impacting at about 16 km s^{-1}. The rim is composed of folded and inverted rock layers that have been pushed out from a central point. The meteorite almost vaporized completely, only a few hundred tonnes being scattered over 80 km² around the crater in the form of tiny iron *spherules.

armillary sphere. A device, dating back to antiquity, composed of a set of graduated rings representing circles on the *celestial sphere, such as the ecliptic, celestial equator, and colures. The whole globe often revolved about an axis – the polar axis – within horizon and meridian circles. Movable sighting adjustments enabled a star to be observed and its coordinates to be read off the relevant circles.

arm population. Young *population I stars occurring in the spiral arms of the Galaxy.

Arneb. *See* Lepus.

array. A system used in radio astronomy in which a number of discrete elements are connected together to form a composite *antenna. An element may for example be a single *dipole or a *dish. The ratio of the aperture area occupied by all the elements to the area that would be occupied by a continuous aperture of the same overall length is called the *filling factor*; it is a measure of the fraction of the aperture that is actually there. The *effective area* for an array operating at a wavelength λ with a *directivity D is given by

$$A_e = D\lambda^2/4\pi$$

It is usually less than the physical area, even when the filling factor is equal to unity, because of small irregularities in the aperture distribution. For a perfect array with a uniform aperture distribution the effective and physical areas are indeed equal.

An array may produce a *grating response* in one or more directions away from the main lobe (*see* antenna) which, if the elements of the array were all nondirectional, would be as strong as the main lobe. However, the *antenna patterns of the elements together with geometrical effects usually reduce the power in the grating response to an acceptable level. *See also* Butler matrix; feeder; aperture synthesis.

Arsia Mons. A 19 km high 400 km wide volcano at the southern edge of the *Tharsis Ridge of Mars. It appears to be older than the other Tharsis volcanoes and has the largest summit caldera, 140 km across. *See* Mars, volcanoes.

artificial satellite. *See* satellite, artificial.

ascending node. *See* nodes.

Ascraeus Mons. A 19 km high volcano on the *Tharsis Ridge of Mars. It is 400 km wide at its base and has a summit caldera 50 km across. *See* Mars, volcanoes.

ashen light. 1. The faint glow claimed to be occasionally observed on the unlit area of Venus in its crescent phase. Its cause is unknown but it might result from bombardment of atmospheric atoms and molecules by energetic particles and radiation, as with terrestrial *airglow.

2. *See* earthshine.

aspect (configuration). The apparent position of any of the planets or the moon relative to the sun, as seen from earth. Specific aspects include *conjunction, *opposition, and *quadrature, which differ in the *elongation of the object concerned.

aspheric surface. Any surface that does not form part of a sphere. A paraboloid reflector, used in both optical and radio telescopes, is aspheric. Although production costs are high, aspheric surfaces can greatly improve lens and mirror performance.

association. A loose group of young stars of similar *spectral type. *OB-associations* are groups of massive and highly luminous main-sequence stars of spectral types O and B. They occur in regions rich in gas and dust in the spiral arms of the Galaxy. They have dimensions ranging from a few parsecs to several hundred parsecs. Often an *open cluster is found near the centre of an association, e.g. the Zeta Persei association surrounds *h and Chi Persei. *R-associations* are groups of bright young stars of slightly lower mass (3–10 solar masses) that illuminate *reflection nebulae. *T-associations* are groups of *T Tauri stars, i.e. young stars of about the sun's mass. Most contain less than 30 stars though some contain as many as 400. R- and T-associations are often found in the vicinity of young open clusters.

Associations are generally too sparsely populated to be gravitationally bound systems and there is strong evidence that they represent the aftereffects of comparatively recent multiple star births. In some cases the stars appear to be expanding from a common centre: by extrapolating back their present velocities an estimate of the age of the system can be derived. For instance, the association II Persei shows an expansion age of slightly over one million years.

A stars. Stars of *spectral type A. They are blue-white or white and have a surface temperature (T_{eff}) of about 7500 to 10 000 kelvin. Balmer absorption lines of hydrogen dominate the spectrum, reaching maximum strength for A0 to A3 stars; lines of ionized and neutral metals slowly strengthen with the H and K lines of ionized calcium appearing in later subdivisions. A few A stars (*Am and *Ap stars) have anomalously high abundances of metals and rare earth elements. *Colour indices for A0 stars are zero. Vega, Sirius, Deneb, and Altair are A stars.

asteroid. *See* minor planets.

asteroid belt. The zone near to the plane of the *ecliptic and between the orbits of Mars and Jupiter that contains the majority of asteroids, i.e. *minor planets. The inner edge is well-defined and the orbits strongly concentrated in the plane of the ecliptic. The orbital *inclination becomes more scattered on moving through the belt. The outer edge is rather diffuse. All the bodies are in *direct motion. *See also* Kirkwood gaps.

astigmatism. An *aberration of a lens or mirror system that occurs when light falls obliquely on the system and is focused not as a single point image but as two perpendicular and separated lines. In the reflecting system shown in the illustration rays from points A and B on the mirror converge to the vertical line image ab; rays from C and D converge to the horizontal line image cd. The pencil of reflected rays, elliptical in cross section, cannot produce a sharp image anywhere along its path; the plane of optimum focus occurs between ab and cd where the pencil has its smallest cross section. Astigmatism is not as severe an aberration as *coma.

ASTP. *Abbrev. for* Apollo-Soyuz test project.

Astraea. The fifth *minor planet to be discovered, in 1845. *See* Table 3, backmatter.

astration. The continual cycling of material from the *interstellar medium into stars and back again, which steadily enriches the galaxies with elements heavier than hydrogen.

ASTRO-B. Pre-launch name of the Japanese satellite *Tenma.

astroblemes. Scars, usually ancient, left in the earth's crust by meteorite impact. Criteria used for recognition

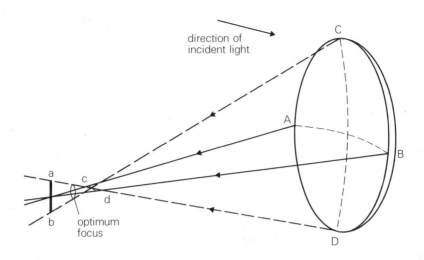

direction of incident light

Astigmatism due to concave mirror

are shatter cones, high-pressure silica polymorphs, shock microdeformations of quartz, impactite glasses, traces of nickel, and nickel-iron associations. All are of a geologically youthful age, none older than 10 million years. The largest are Vredefort, South Africa (40 km diameter), Nordlinger Ries, Germany (25 km), Deep Bay, Saskatchewan, Canada (14 km), Lake Bosumtwi, Ghana (10 km), and Serpent Mound, Ohio (6.4 km).

The velocity of a large meteorite, i.e. one over thousands of tonnes, is dissipated very little by atmospheric braking effects and it hits the ground at its cosmic velocity of 11 to 74 km s^{-1} and explodes. Most of the kinetic energy goes to accelerate and excavate the material from the volume that eventually becomes the crater. Usually only an exceedingly small fraction ($<0.01\%$) of the projectile survives. The resulting crater has a diameter between 20 and 60 times that of the impacting body, this factor depending on the incident velocity.

ASTRO–C. A 500 kg Japanese x-ray astronomy satellite that is due for launch in Feb. 1987 and will further extend the scientific observations of the satellites *Hakucho and *Tenma. The major payload element will be a 5000 cm^2 array of proportional counter detectors capable of detailed spectral and variability studies of a wide range of galactic and extragalactic sources in the 1–20 keV band. AS-TRO–C will be the first collaborative Japan–UK space mission, UK groups at Leicester University and the Rutherford Appleton Laboratory providing the detectors and integration and scientific operations support.

astrochemistry. The study of the chemistry of celestial bodies and of the intervening regions of space. It involves the detection and identification, principally by spectroscopy, of the inorganic and organic chemicals present and the study of the reactions by which these neutral and charged atoms and molecules could have been produced and of future chemical processes. *See also* molecular-line radio astronomy.

astrodynamics. The application of *celestial mechanics, the ballistics of high-speed solids through gases, propulsion theory, and other allied fields to the planning and production of the trajectories of spacecraft.

Astrographic Catalogue. *See* Carte du Ciel.

astrolabe. An instrument, dating back to antiquity, used to measure the altitude of a celestial body and to solve problems of spherical astronomy. From the 15th century it was employed by mariners to determine latitude, until replaced by the sextant. There were various types of astrolabes. One simple form consisted of a graduated disc that could be suspended by a ring to hang in a vertical plane. A movable sighting device – the alidade – pivoted at the centre of the disc. Modern versions are still used to determine stellar positions and hence local time and latitude. *See also* prismatic astrolabe.

astrometric binary. *See* binary star.

astrometry (positional astronomy; spherical astronomy). The branch of astronomy concerned with the measurement of the positions of celestial bodies on the *celestial sphere, with the conditions, such as *precession, *nutation, and *proper motion, that cause the positions to change with time, and with the concepts and observational methods involved in making the measurements. In *photographic astrometry* the positions of stars, planets, etc., are determined in relation to reference stars on photographic plates. In *meridian astrometry* the positions of celestial bodies are determined relative to the motions of the earth. *See also* radio astrometry; long-focus photographic astrometry.

Astron. An astronomical observatory launched in Mar. 1983 by the USSR into a highly elliptical earth orbit.

The main instrument is a Soviet–French 0.8 metre ultraviolet telescope, spectral range 110–360 nm. There are also x-ray spectrometers on board.

astronautics. The science and technology of spaceflight.

Astronomers Royal

John Flamsteed	1675–1719
Edmund Halley	1720–42
James Bradley	1742–62
Nathaniel Bliss	1762–64
Nevil Maskelyne	1765–1811
John Pond	1811–35
Sir George Biddell Airy	1835–81
Sir William H.M. Christie	1881–1910
Sir Frank Watson Dyson	1910–33
Sir Harold Spencer Jones	1933–55
Sir Richard Woolley	1955–71
Sir Martin Ryle	1972–82
Prof. F. Graham Smith	1982–

Astronomer Royal. An eminent British astronomer appointed to the post of Astronomer Royal of England and, until the retirement of Sir Richard Woolley in 1971, Director of the Royal Greenwich Observatory. The two titles were separated in 1972 when Sir Martin Ryle became Astronomer Royal. Astronomers Royal are listed in the table. The first Astronomer Royal, John Flamsteed, was appointed in 1675 following the foundation of the Royal Greenwich Observatory. The office of *Astronomer Royal for Scotland* was created by Royal Warrant in 1834 and is held by the Director of the Royal Observatory, Edinburgh. Malcolm Longair succeeded Vincent Reddish to the title in 1980.

Astronomical Netherlands Satellite (ANS). A joint Netherlands/US satellite that was launched in Aug. 1974 and operated until 1976. It contained a 20-cm ultraviolet telescope providing broad-band (about 15 nm) photometric data on selected astronomical sources at the five wavelengths of 155 nm, 180 nm, 220 nm, 250 nm, and 350 nm. A catalogue of such observations for about 3500 stars has been completed. In addition the satellite carried two x-ray telescopes that were mainly used for the study of x-ray transient sources.

astronomical triangle. A *spherical triangle on the *celestial sphere formed by the intersection of the great circles joining a celestial body, S, the observer's zenith, Z, and the north celestial pole, P (see illustration). The relationships between the angles and sides of a spherical triangle are used for transformation between *equatorial and *horizontal coordinate systems: the angle at S (q) is the *parallactic angle*; that at P (t) is the *hour angle of S; that at Z (A) is the azimuth of S. The sides are equal to the *zenith distance (ζ), the complement of the terrestrial latitude ($90° - \phi$), and the complement of the declination ($90° - \delta$). The parallactic angle, for example, is given by the sine rule:

$$\sin q = (\cos \phi \sin t)/\sin \zeta$$
$$= (\cos \phi \sin A)/\cos \delta$$

astronomical twilight. *See* twilight.

astronomical unit (AU or au). A unit of length that is used for distances, especially within the solar system. It was originally the length of the *semimajor axis of the earth's orbit, but is now defined dynamically by Kepler's third law:

$$n^2a^3 = k^2(1 + m)$$

a is the semimajor axis of an elliptical orbit (in AU), n is the sidereal *mean motion (in radians per day), m is the mass (in *solar masses), and k is the *Gaussian gravitational constant. The semimajor axis of the earth's orbit is thus

1.000 000 031 AU

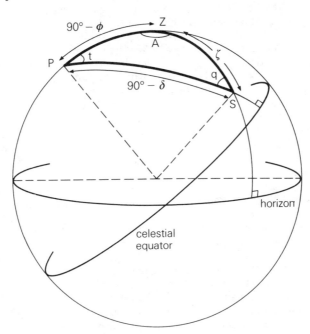

Astronomical triangle

One astronomical unit is equal to 149 597 870 kilometres or 499.005 light-seconds.

astronomy. The observational and theoretical study of celestial bodies, of the intervening regions of space, and of the universe as a whole. Astronomy is one of the oldest sciences and has developed in step with advances in instrumentation and other technology and with advances in physics, chemistry, and mathematics. The main branches are *astrometry, *celestial mechanics, and the major modern field of *astrophysics with its subsections of *cosmology, *radio astronomy, and *x-ray, *gamma-ray, *infrared, and *ultraviolet astronomy. *Stellar statistics is another modern field.

astrophysics. The study of the properties, constitution, and evolution of celestial bodies and the intervening regions of space. It is concerned in particular with the production and expenditure of energy in systems such as the stars and galaxies and in the universe as a whole and with how this affects the evolution of such systems. Astrophysics developed in the 19th century and is closely related to particle physics, plasma physics, thermodynamics, spectroscopy, solid-state physics, and relativity. *Cosmology, *radio astronomy, *x-ray, *gamma-ray, *infrared, and *ultraviolet astronomy are usually considered subsections of astrophysics.

asymptotic giant branch. See giant.

ataxite. See iron meteorite.

Àten. A small member of the *Apollo group of minor planets. Discovered in 1976, it was the first minor planet to be found with a semimajor axis of less than one astronomical unit and therefore an orbital period of less than one year. See Table 3, backmatter.

atmosphere. The gaseous envelope that surrounds a star, planet, or some other celestial body. The ability of a planet or planetary satellite to maintain an appreciable atmosphere depends on the *escape velocity at its surface and on its temperature. Small bodies, such as the moon, have low escape velocities that may be exceeded comparatively easily by gas molecules travelling with the thermal speeds appropriate to their temperature and mass. The speed of a gas molecule increases with temperature and decreases with molecular weight; the lighter molecules, such as hydrogen, helium, methane, and ammonia, can therefore escape into space more readily than heavier ones such as nitrogen, oxygen, and carbon dioxide. Some bodies, such as Pluto, may be so cold that most potential atmospheric gases lie frozen on their surfaces. A body may gain an atmosphere by any of several processes, including a temporary acquirement by collision with an object containing frozen gases, such as a cometary nucleus.

The study of planetary atmospheres – their contents, meteorology, and evolution – has developed rapidly with the advent of planetary space probes in the 1960s. The planets *Earth, *Mars, *Venus, *Jupiter, and *Saturn, and Saturn's satellite *Titan, have been examined at close range by probes and much has been learned about their atmospheres. (*Mercury has been found to have only the most tenuous of atmospheres.) The deep hydrogen, helium, methane, and ammonia atmospheres of Jupiter, Saturn, and the other *giant planets probably derive directly from the nebula and icy planetesimals from which the planets are now thought to have coalesced by accretion (see nebular hypothesis). The *terrestrial planets, however, were too small to capture the warm gases of the early nebula; the present atmospheres of Venus, Earth, and Mars are thought to have escaped from volcanoes after being liberated from chemical combination in sub-surface rocks by radioactive heating.

Exploration of the solar system has provided a unique opportunity to examine weather systems of planetary atmospheres. Venus: slowly rotating cloud over the planet with a huge *greenhouse effect. Mars: a thin atmosphere where the weather systems are strongly influenced by the large topography of the surface. Jupiter and Saturn: large fluid atmospheres, rapidly rotating, with internal heating to assist in driving the weather phenomena. Each one of these examples is a natural laboratory for geophysical fluid dynamics that provides insight into general meteorological processes.

See also corona; chromosphere; solar wind.

atmosphere, composition. The percentage composition of clean dry air in the earth's atmosphere is given in the table. The concentrations of CO_2, CO, SO_2, NO_2, and NO are greater in industrial areas. The amount of ozone increases in the ozonosphere (*see* atmospheric layers) and depends on meteorological and geographical variations. The fraction of water vapour in normal air varies from 0.1–2.8% by volume and 0.06–1.7% by mass, with meteorological and geographical variations and a decrease with altitude.

atmospheric extinction. *See* extinction.

atmospheric layers. The gaseous layers into which the earth's atmosphere is divided, mainly on the basis of the variation of temperature with altitude. The *troposphere* is the lowest layer, extending from sea level to the *tropopause* at an altitude of between 8 km at the poles and 18 km within the tropics. It contains three-quarters of the atmosphere by mass and is the layer of clouds and weather systems. It is a region, heated from the ground by infrared radiation and convection, in which the temperature falls with increasing height to reach a minimum of approximately −55°C at the mid-latitude and polar tropopause and maybe −80°C at the equatorial tropopause.

The *stratosphere*, lying above the tropopause, is an atmospheric layer in which the temperature is at first steady with altitude before increasing to a maximum of 0°C at about 50 km, which marks the *stratopause*. This temperature variation inhibits vertical

Composition of earth's lower atmosphere

gas		percentage composition of clean dry air by volume	by mass
nitrogen	N_2	78.084	75.523
oxygen	O_2	20.947	23.142
argon	Ar	0.934	1.290
carbon dioxide	CO_2	0.032	0.050
neon	Ne	0.0018	0.0013
helium	He	0.0005	0.0001
methane	CH_4	0.0002	0.0001
krypton	Kr	0.0001	0.0003
carbon monoxide	CO	0.0001	0.0001
sulphur dioxide	SO_2	0.0001	0.0002

With smaller traces of hydrogen (H_2), nitrous oxide (N_2O), ozone (O_3), xenon (Xe), nitrogen dioxide (NO_2), radon (Rn), and nitric oxide (NO).

air movements and makes the stratosphere stable. The heating arises from the absorption of high-energy solar ultraviolet radiation by ozone (O_3) molecules. These molecules, themselves formed by the action of ultraviolet radiation on oxygen molecules, constitute the *ozonosphere*: this atmospheric region lies between about 25 and 40 km altitude.

Above the stratopause is the *mesosphere*, in which the temperature falls with height to reach about −90°C at the *mesopause* at an altitude of about 85 km. Within the mesosphere the heat contribution from ultraviolet absorption by ozone decreases as the ozone becomes less plentiful. The *thermosphere*, above the mesopause, has a temperature that rises with height as the sun's far-ultraviolet radiation is absorbed by oxygen and nitrogen. This process gives rise to ionized atoms and molecules and leads to the formation of the *ionosphere layers above altitudes of about 60 km. This is also the domain of *meteors and *aurorae.

Although the temperature climbs to about 500°C at 150 km and about 1300°C at 500 km, these are the kinetic temperatures (i.e. measures of the average random motions) of the few atoms and molecules found at such heights: their combined heating ability is negligible. The atmospheric density decreases to about 0.25 of its sea-level value at the tropopause, 0.0009 at the stratopause, and 0.000 007 at the mesopause. The density at 500 km averages less than one million millionth of its sea level value but is prone to large variations resulting from solar heating and disturbances during periods of enhanced *solar activity.

The *exosphere*, above 500 km, is the region in which atmospheric constituents lose collisional contact with each other because of their rarity and can leak away into space. It contains the *Van Allen radiation belts and the *geocorona and extends to the magnetopause where it meets the interplanetary medium.

atmospheric pressure. The force per unit area (i.e. pressure) exerted by a column of atmosphere extending vertically upwards to the limit of the atmosphere. It is usually expressed in millibars (mb), atmospheres (atm), or newtons per square metre (N m^{-2}). The standard atmospheric pressure at the earth's surface at sea level is 1013 mb or 101 325 N m^{-2}. On the surface of Venus and Mars the atmospheric pressure is about 90 000 mb and 7.5 mb, respectively.

atmospheric refraction. *Refraction of light passing obliquely through a body's atmosphere. Light entering the earth's atmosphere is bent towards the earth, with the result that stars and other celestial bodies appear to be displaced very slightly towards the *zenith. The amount of displacement, i.e. the resulting increase in altitude, depends on the body's distance from the observer's zenith and is gratest (about 35 arc minutes) for objects on the horizon. *See also* zenith distance.

atmospheric windows. Regions of the *electromagnetic spectrum that can be transmitted through the earth's atmosphere without significant absorption or reflection by atmospheric constituents. All spectral regions are at least partially absorbed but there are two nearly transparent ranges: the *radio window* and the *optical window* (see illustration). The radio window extends from a wavelength of about a millimetre to about 30 metres (i.e. it includes frequencies from roughly 300 gigahertz to 10 megahertz); it thus allows high-energy radio waves from celestial *radio sources to be received by ground-based radio telescopes. Lower energies are reflected by the *ionosphere while above 100 GHz molecular absorption increases.

The optical window spans wavelengths of about 300 to 900 nanome-

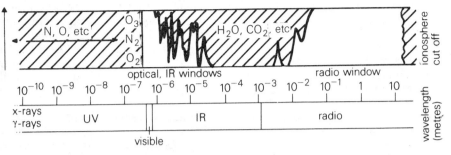

Atmospheric windows

tres (nm) and thus includes some of the near-ultraviolet and near-infrared regions as well as the whole visible spectrum. In addition there are several narrow-band *infrared windows* at micrometre (μm) wavelengths. The photometric designations of these windows are J (1.25 μm), H (1.6 μm), K (2.2 μm), L (3.6 μm), M (5.0 μm), N (10.0 μm), and Q (21 μm). There are also small but useable windows at 350 and 460 μm. The atmospheric constituents responsible for the dominant absorptions are water vapour, carbon dioxide, nitrous oxide, and ozone. Observatories studying infrared sources are therefore best sited in very dry or mountainous regions where the effect of overlying water vapour is reduced and/or the atmosphere is thinner.

Beyond the low-wavelength end of the optical window no radiation can penetrate the atmosphere: ultraviolet radiation with wavelengths between about 230–300 nm is absorbed by atmospheric ozone; shorter wavelengths down to about 100 nm are blocked by oxygen and nitrogen molecules; the shortest wavelengths are absorbed by atoms of oxygen, nitrogen, etc., in the upper atmosphere. The spectral regions from space rendered inaccessible by atmospheric absorption must be studied by means of instruments carried in satellites, rockets, etc. Even then, high-energy ultraviolet and low-energy x-rays from distant stars can be absorbed by *geocoronal hydrogen and helium.

atom. The smallest part of a chemical *element that can retain its chemical identity, i.e. can take part in a chemical reaction without being destroyed. An atom can however be transformed into one or more other atoms by nuclear reactions. Nearly all the mass of an atom is concentrated in the small central *nucleus, which is made up of protons and neutrons. Electrons occupy the remaining space in the atom and may be thought of as orbiting the nucleus. The number of protons in a particular nucleus, i.e. the atomic number of the atom, is equal to the number of electrons in a neutral (non-ionized) atom. The arrangement and behaviour of these electrons determine the interactions of the atom with other atoms and thus govern chemical properties and most physical properties of matter. *See also* energy levels; isotopes; ion.

atomic clock. The most accurate type of clock to date, in which the periodic process used as the timing mechanism is an atomic or molecular event with which is associated a constant and very accurately known *frequency. Various atoms and molecules have been employed in atomic clocks, the frequency involved in the timing mechanism being a microwave frequency; such frequencies can be produced and manipulated by sophisticated electronic techniques. The accuracy is at present better than one part in one thousand million million. *See also* atomic time.

atomic number. Symbol: Z. The number of protons in the nucleus of an atom. This is equal to the number of electrons orbiting the nucleus in a neutral atom. The *isotopes of an element have the same atomic number but different *mass numbers.

atomic time. A time system that operates at a constant rate and that is measured by means of an *atomic clock. The definition of the *second has been in terms of atomic time since 1967. At the time of definition the SI atomic second was equal to the ephemeris second. *See also* International Atomic Time.

attenuation. The reduction in magnitude of a quantity as a function of distance from source or of some other parameter. The attenuation of a beam of radiation, i.e. the reduction in intensity or flux density, results from absorption and/or scattering of the waves or particles as they pass through matter.

AU or **au.** *Abbrev. for* astronomical unit.

audiofrequency. Any frequency at which a sound wave would be audible, i.e. a frequency in the range 15 hertz to 20 000 hertz.

Auger shower. *Another name for* air shower. *See* cosmic rays.

augmentation. The difference at a particular time between the topocentric and geocentric semidiameter of a celestial body. *See also* diurnal libration.

Auriga (Charioteer). A conspicuous constellation in the northern hemisphere near Orion, lying in the Milky Way, the brightest stars being *Capella (α) and the 2nd-magnitude *Algol

variable Menkalinan (β). *Epsilon Aurigae and *Zeta Aurigae are eclipsing binaries. The area contains the bright open clusters M36, M37, and M38. Abbrev.: Aur; genitive form: Aurigae; approx. position: RA 5.5h, dec +40°; area: 657 sq deg.

aurora. A display of diffuse changing coloured light seen high in the earth's atmosphere mainly in polar regions. Aurorae show varying colours from whitish green to deep red and often have spectacular formations of streamers or drapery. They are formed usually within about 20° of the magnetic poles and occur at altitudes of approximately 100 km. They are caused by charged particles from the *solar wind or from *flares that become trapped by the earth's magnetic field and interact with gases in the upper atmosphere as they spiral around the magnetic field lines between the poles.

Australian National Radio Astronomy Observatory. The Australian CSIRO observatory near Parkes, New South Wales, at which the chief instrument is a fully steerable 64 metre (210 ft) radio dish.

Australia Telescope. A new radio telescope combining the principles of *aperture synthesis and *very long baseline interferometry (VLBI). Planned by Australia's CSIRO, it is due to begin operation in 1988. It will consist of a 6 kilometre array of five 22 metre radio dishes at Culgoora, connected by radio links to an identical dish at Siding Spring and the existing 64 metre dish at Parkes, all in New South Wales, Australia. The telescope will have a maximum *baseline of about 300 km.

australite. *See* tektites.

autocorrelation function. A mathematical function that for a real function f (t) is given by

$$\int_{-\infty}^{\infty} f(u)f(u-t)du$$

It describes, for example, the time or distance over which a signal is *coherent. The *cross-correlation function* describes how two different signals compare as they are displaced relative to one another; for two real functions $f(t)$ and $g(t)$, it is given by

$$\int_{-\infty}^{\infty} g(u-t)f(u)du$$

automation. The use of electronic equipment, especially computer systems, for the automatic control of instruments and processes and for automatically acquiring and processing information. The applications in astronomy include positioning and tracking of telescopes, direct control of instruments associated with telescopes, the storage of information and its reduction by sampling or simple analysis, and the processing of the information. The information in its 'raw' form is collected by electronic detectors and fed through cables to a control room. The astronomer in the control room can see on display screens both the telescope's field of view and the incoming data. He can send commands to the telescope's drive computer to move the telescope if necessary, or to another computer to control the telescope's instrumentation. The information can then be processed as required. *See also* imaging; remote operation.

AXAF. The Advanced X-ray Astrophysics Facility, intended by NASA to be a long-lived x-ray observatory for the 1990s, with in-orbit servicing from the space shuttle. A successor to the Einstein Observatory, AXAF will carry an array of six large (1.2 m outer diameter) coaxial *grazing incidence mirrors in a nested configuration. The optics will provide 0.5 arc second imaging and operate from 0.1 to 10 keV. Focal-plane instrumentation is expected to include a variety of x-ray cameras (*microchannel plate, *CCD) and spectrometers, and the mission is seen as the x-ray

equivalent of the Space Telescope (optical) and the VLA (radio) as major long-term facilities in astronomy.

axial period. The period of time during which a body makes one complete *rotation on its axis. For planets it is usually referred to the direction of a fixed star and is thus equivalent to the *sidereal day.

axial tilt of the earth. *See* obliquity of the ecliptic.

axis. 1. An imaginary line that usually passes through the centre of a body or system and about which the body is often symmetrical or has some form of symmetry. It is the imaginary line about which a rotating body turns or about which an object, such as the celestial sphere, appears to rotate.
2. A reference line on a graph.

azimuth. *See* horizontal coordinate system.

azimuthal mounting. *See* altazimuth mounting.

B

Baade's star. *Another name for* Crab pulsar. *See* Crab nebula.

background radiation. *See* cosmic background radiation; microwave background radiation.

background temperature. *See* antenna temperature.

Baily's beads. A string of brilliant points of sunlight sometimes seen, very briefly, at the moon's edge just before or just after totality in a solar *eclipse. They are the result of sunlight shining through the valleys on the moon's limb as the sun just disappears or emerges from behind the moon. They were first described by Francis Baily after the eclipse of 1836. *See also* diamond-ring effect.

Balmer series, lines. *See* hydrogen spectrum.

balun. *See* feeder.

band. 1. *See* waveband.
  2. *See* band spectrum.

band spectrum. A *spectrum consisting of one or more bands of emitted or absorbed radiation. The bands consist of numbers of closely spaced lines and result from transitions between *energy levels in molecules, allowing different levels of vibration or rotation.

bandwidth. 1. The range of frequencies over which an instrument, such as a *radio telescope, makes an appreciable response to input signals.
  2. (coherence bandwidth). The range of frequencies over which electromagnetic or other emissions from a source maintain *coherence.

Barlow lens. An *achromatic diverging lens placed behind the eyepiece of a telescope, just inside the primary focal plane, in order to increase the effective focal length of the objective or primary mirror. This increases the magnification of the telescope so that a long-focus eyepiece may be used to give the higher powers needed to separate optical doubles, or for planetary observation under the best seeing conditions.

Barnard's loop or **ring.** *See* Orion.

Barnard's star. A red dwarf star in the constellation Ophiuchus that was discovered in 1916 by E.E. Barnard. The fourth-nearest star to the sun, it has the largest known proper motion ($10''.3$ per year) and thus moves a distance equivalent to the moon's di-

ameter in 180 years. Observations of the star's position over many years show slight oscillating variations in right ascension and declination. It is thought that this wobbling motion could be due to the presence of one or more planets, of about Jupiter's mass, orbiting the star. m_v: 9.53; M_v: 13.21; spectral type: M5 V; distance: 1.83 pc.

barred spiral galaxy. *See* galaxies; Hubble classification.

barycentre. The *centre of mass of a system of bodies, e.g. the earth-moon system.

barycentric dynamical time (TDB). *See* dynamical time.

baryons. A class of *elementary particles, including the proton and neutron, that take part in strong interactions (*see* fundamental forces). Baryons are composed of a triplet of *quarks. Antibaryons, i.e. the *antiparticles of baryons, consist of a triplet of antiquarks. The *baryon number* is the total number of baryons minus antibaryons in a system. It was thought that in any particle interaction the baryon number before and after remains unchanged, i.e. baryon number is a conserved quantity. This is now being questioned.

baseline. The straight line or arc of a great circle between two observational points, as between the elements of an interferometer (*see* radio telescope). The longer the baseline between two radio telescopes the finer the detail that can be resolved in a radio source. *See also* very long baseline interferometry.

basin. A huge *crater. Lunar basins are multiringed structures that are several hundred kilometres in diameter and were all produced during the first 700 million years of lunar history by the impact of asteroidal or cometary bodies. The youngest basins (*Orientale, *Imbrium, Crisium, Nectaris, and Serenitatis) may have been excavated during a cataclysmic period around 3900 million years ago. The *ejecta blankets of basins are extensive and provide the basis for *highland stratigraphy. Basins are so defined by their systems of concentric ring structures, most of which are now in the form of ridges and mountain arcs. Basins on the moon's nearside were filled with lava at some time up to 3000 million years ago to produce *maria: the resulting mare has the same name as the basin (e.g. Mare Imbrium). Several unfilled basins (*thassaloids*) exist on the farside, the largest of which was revealed as a vast depression by the *Apollo 15 laser altimeter. At least 28 basins are now known to exist. *See also* rille; mascon.

Bayer letters. *See* stellar nomenclature.

BC. *Abbrev. for* bolometric correction. *See* magnitude.

BD. *See* Bonner Durchmusterung.

beam. 1. *See* antenna.
2. A well-defined elongated region of space down which energy is passing. In the beam model of double radio sources (*see* radio source structure) the central galaxy shoots out two beams in opposite directions in which the energy may be transported by *relativistic electrons, low-frequency waves, or some other mechanism. Hot spots are formed where the beams impinge on the surrounding *intergalactic medium.

beam efficiency. *See* antenna.

beamwidth. *See* antenna.

Becklin-Neugebauer object. *See* BN object.

Beehive. *See* Praesepe.

Bellatrix (γ Ori). A remote luminous blue-white giant that is the third brightest star in the constellation *Orion. m_v: 1.63; M_v: -4.1; spectral type: B2 III; distance: 140 pc.

Benatnasch. *See* Alcaid.

bent double radio source. *See* radio source structure.

Besselian year. A period of one complete revolution in *right ascension of the *mean sun. The beginning of a Besselian year was used as a standard *epoch until replaced in 1984 by the *Julian year. A Besselian year started when the mean sun reached an RA of 18h 40m, some time on Jan. 1; the precise time varied from year to year. The beginning of the Besselian year 1980, say, is now written B1980.0 rather than just 1980.0.

Be stars. *Irregular variables of spectral type B in which bright emission lines of hydrogen are superimposed on the normal absorption spectrum. They are now known to be identical to *shell stars. Some Be stars are young stars, rather more massive than *Ae stars: together they are sometimes classed as *Herbig emission-line stars*, and are heavier versions of the *T Tauri stars. These Be stars are rotating very rapidly and are slowly losing mass to an expanding shell that surrounds the star and is drawn into a disc around the equator due to the rotation. Other Be stars are in close binary systems, with shells that consist of gas being accreted from an evolved companion star (*see* mass transfer).

beta (β). The second letter of the Greek alphabet used in *stellar nomenclature usually to designate the second-brightest star in a constellation or sometimes to indicate a star's position in a group.

Beta Canis Majoris stars. *See* Beta Cephei stars.

Beta Centauri (Hadar; β Cen). A luminous remote but still conspicuous blue-white giant that is the second-brightest star in the constellation Centaurus. It is a visual binary, separation $1''.2$. m_v: 0.63; M_v: -4.8; spectral type: B1 III; distance: 120 pc.

Beta Cephei stars (Beta Canis Majoris stars). A small group of *pulsating variables, the prototypes being β Cep and β CMa. They are hot massive luminous stars of spectral types O9 to B3 with short periods of rotation (about 3–7 hours) and a very small variation in visual brightness (about 0.01–0.25 magnitudes), although the range is much greater at ultraviolet wavelengths. Some of the stars have two or even more periods of *radial-velocity variation. The pulsation mechanism is still uncertain. Their position on the Hertzsprung-Russell diagram for *pulsating variables is just above the main sequence, beginning at type B3.

Beta Crucis (Mimosa; β Cru). A remote luminous blue-white giant that is the second-brightest star in the constellation Crux. It is a variable star (period 0.25 days) and may also be a double star. m_v: 1.29; M_v: -4.3; spectral type: B0 III; distance: 130 pc.

beta decay. A type of *radioactive decay in which an atomic nucleus spontaneously transforms into a daughter nucleus and either an electron plus antineutrino or a positron plus *neutrino. The daughter nucleus has the same *mass number as the parent nucleus but differs in *atomic number by one. The electrons or positrons ejected by beta decay have a spread of energies, extra energy being taken up by the antineutrinos or neutrinos, respectively.

Beta Lyrae, Beta Lyrae star. *See* W Serpentis star.

beta particle. An energetic electron or positron ejected by *beta decay.

Beta Persei stars. *See* Algol variables.

Betelgeuse (α Ori). A remote luminous red supergiant that is the second-brightest star in the constellation *Orion. It is a semiregular variable with a period of about 5.8 years; the normal magnitude range is 0.3 to 0.9 but the magnitude has reached 0.15 and been as low as 1.3. It has a 14th-magnitude companion at a separation of $40''$ and is a strong source of infra-red radiation. *IRAS has found that the long-wavelength infrared is emitted from three concentric shells, the largest with a radius of 1.5 parsecs, ejected within the past 100 000 years. Images of the surface of Betelgeuse have been produced by means of *speckle interferometry. M_v: -5.7; spectral type: M2 Iab; diameter: about 800 times solar diameter; distance: 120 pc.

Bethe cycle, Bethe-Weizsäcker cycle. *Other names for* carbon cycle.

B²FH. The collaboration between Geoffrey and Margaret Burbidge, William Fowler, and Fred Hoyle (hence the initials) that led to the theory of the synthesis of the elements in stars, published in 1956 and 1957. The four collaborators worked out the nuclear reactions by which almost all naturally occurring nuclear species could be synthesized in close to their observed abundances in successive generations of stars. *See also* nucleosynthesis.

Biela's comet (1852 III). A comet, discovered in 1826, having a period of 6.62 years and a perihelion distance of 0.86 AU. At the 1846 return it appeared to be distinctly elongated into a pearshaped form and actually divided into two separate comets some 10 days later, these comets travelling in practically the same orbit, one preceding the other by 280 000 km. The brightness of the two parts fluctuated drastically. At the next return in 1852 their separation had increased eightfold, the two periods differing by about 15 days. Neither comet has been seen since. Other comets have been observed to break up and to have portions detatch from the main *nucleus; these include Brooks' comet (1889 V), Swift 1860 III, and 1882 III.

Bielids (Andromedids). A *meteor shower with its *radiant in Andromeda. It is named after *Biela's comet (1852 III), which passed within 30 000 km of the earth's orbit in 1832 and broke up in the 1850s. The associated *meteor stream is in the same orbit as the comet and produced a meteor storm on earth on Nov. 27, 1872 and 1885, the latter having an observed rate of over 75 000 meteors per hour. Due to the rapid regression of the *nodes of the orbit very little has been seen since.

big bang theory. A *cosmological model in which all matter and radiation in the universe originated in an explosion at a finite time in the past. This theory has been remarkably successful in explaining the expansion of the universe, the *microwave background radiation, and the cosmic abundance of *helium. An evolving universe was first discussed in the 1920s by Aleksandr Friedmann, Georges Lemaître, and others. A *hot big bang,* in which the temperature of matter and radiation decreased with time, was suggested by George Gamow in the 1940s in an attempt to explain the observed abundances of the elements by cosmological *nucleosynthesis. A neglected prediction of primeval-fireball radiation from the hot early phases following the big bang was verified by the discovery of

the cosmic microwave background in 1965.

The universe came into existence some 10 to 20 thousand million years ago (*see* age of the universe) and has since been expanding and cooling from an intial state of extreme density and temperature. The uncertainty principle of quantum mechanics prevents our speculating on times shorter than 10^{-43} seconds after the big bang. Current theories of *elementary particles allow some calculation of the conditions from 10^{-35} seconds onwards. At this time the universe would have contained roughly equal amounts of matter and antimatter in the form of particles and their corresponding *antiparticles: *electrons and positrons; *neutrinos and antineutrinos; *quarks and antiquarks (quarks are the particles that make up *protons and *neutrons). The asymmetrical decay of hypothetical particles (called X-bosons) at this time into matter rather than antimatter may explain why the universe is today apparently composed entirely of matter. The time around 10^{-35} seconds may have been the critical period of a sudden and enormous expansion of the universe, according to the *inflationary universe theory.

The temperature dropped as the universe aged. By 10^{-4} seconds after the big bang, quarks and antiquarks had annihilated (*see* annihilation) and the small excess of quarks over antiquarks (by 1 in 10^9) survived to combine into protons and neutrons. Electrons and positrons annihilated when the universe was about 10–100 seconds old, leaving a small residue of electrons. The *photons of radiation produced by matter-antimatter annihilation then dominated the energy density, i.e. energy per unit volume, ε, of the early universe (see illustration). This was the *radiation era.*

Radiation loses energy in the expansion – by the *cosmological redshift – and together with the decrease

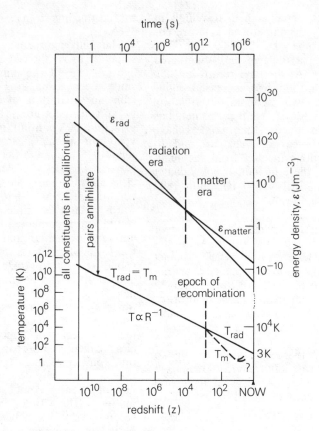

Big - bang theory of evolution of universe

in photon number density this causes the radiation energy density, ε_{rad}, to vary with the *cosmic scale factor, R, as R^{-4}. The matter density, ε_{matter}, decreases less slowly, as R^{-3}, and thus dominated the total energy density after about 10 000 years, when the temperature was about 10 000 K. This marked the beginning of the *matter era*. The matter was composed primarily of an ionized gas of electrons, protons, and helium nuclei. At about 3000 K electrons can recombine with protons to form neutral hydrogen. Prior to that epoch the scattering of photons on the electrons coupled matter and radiation so that they shared a common temperature. The possibility of recombination marked the beginning of the *decoupling era*

some 300 000 years after the big bang. Since recombination the radiation has cooled from about 3000 K down to the observed 3 K of the microwave background at a rate proportional to R. The matter, no longer coupled to the radiation, has interacted to form stars and galaxies (*see* galaxies, formation and evolution).

Deuterium and helium were synthesized in the dense hot phase at about 100 seconds after the big bang when the temperature was about 10^9 kelvin: primordial *neutrons combined with protons to form deuterium before the neutrons, with a half-life of 12 minutes, were able to decay. The deuterium nuclei then combined to give helium. Most of the helium in the universe today was formed at this time. The lack of any stable nuclei with *atomic numbers of 5 and 8 prevented formation of heavier elements, except lithium. (The heavier elements are generated by nuclear reactions in the stars – See nucleosynthesis.) Calculations yield estimates of approximately 25–30% by mass for the helium abundance if an expansion rate consistent with observations is used.

Big Bear Solar Observatory. A solar observatory located on an artificial island in Big Bear Lake, S. California. The altitude is 2070 metres. It is operated as part of the *Hale Observatories. The chief instrument is a 65 cm reflector. See also solar telescope.

Big Dipper. US name for the Plough.

billion. In the US, one thousand million. In the UK, one million million. American usage is becoming current in the UK.

binary galaxy. Another name for physical double galaxy. See double galaxy.

binary pulsar. A *pulsar that is in orbit about another star, and is detected by its intrinsic emission of radiation (usually radio waves) rather than by radiation resulting from *mass transfer (see x-ray pulsators). The orbital motion is inferred from apparent changes in the pulse period as the pulsar orbits its companion; in the four binary pulsars known to date (1984) the companion star has not been detected directly but is believed to be a neutron star or a white dwarf. The properties of the systems vary widely: the pulse period ranges from 0.006 seconds in PSR 1953+29 (the second *millisecond pulsar discovered) to 0.87 seconds for PSR 0820+02; the orbital period ranges from 4.7 years for the latter, down to only 7.75 hours for PSR 1913+16 (see below).

PSR 1913+16 was the first binary pulsar to be discovered (in 1974), and is still sometimes called 'the binary pulsar'. It has a pulse period of 59 milliseconds. Its short period and highly eccentric $e = 0.617$) orbit have led to accurate determinations of the system's parameters. The masses of the pulsar and the unseen companion are identical, and at 1.4 solar masses are equal to the *Chandrasekhar limit for collapse to a neutron star. The orbital period is decreasing at a rate of 75 microseconds per year, implying that the stars are spiralling together. The only mechanism for losing orbital energy in this system is *gravitational waves, and the analysis of PSR 1913+16 gives a rate of energy loss that is precisely in accord with the theory of general *relativity.

binary star (binary). A pair of stars that are revolving about a common *centre of mass under the influence of their mutual gravitational attraction. In most cases the stars may be considered to be moving in elliptical orbits described by *Kepler's laws. Binary and *multiple stars are very common in the Galaxy: in a recent survey of 123 nearby sunlike stars over half (57%) were found to have one or more companions. Studies of the orbital motions in binaries are important because they provide the only direct means of obtaining stellar masses. Optical determination of orbits is only possible if the components are sufficiently far apart to be distinguished (see visual binary). In an astrometric binary one component is too faint to be observed directly; the presence of this unseen component is inferred from perturbations in the motion of the visible component.

Orbital motion in *spectroscopic binaries is revealed by variations in radial velocity. Most spectroscopic binaries are close binaries, in which the components are too close together to be seen separately. Stars in a close binary are often distorted into nonspherical shapes by mutual *tidal forces. If the two stars are physically separate, it is a *detached* binary system; in a *semidetached* system gas is drawn off one star on to the other; a *contact* binary consists of two stars sharing gas. In the latter two cases the gas flow from one star to the other (*see* equipotential surfaces; mass transfer) profoundly alters the evolution of the stars (*see* W Ursae Majoris stars; W Serpentis stars; Algol variables). When one star is a compact *white dwarf or *neutron star, the infalling gas powers *novae outbursts and *x-ray binary systems.

The orbital planes of binaries appear to be randomly oriented and only a minority of systems are *eclipsing binaries, most of which are also spectroscopic binaries. *See also* cataclysmic variable; common envelope star; symbiotic star; RS Canum Venaticorum star.

bipolar group. *See* sunspots.

bipolar nebula. A small nebula that appears hourglass-shaped and that represents outflow of gas in opposite directions from a central star that is often obscured by dust. A bipolar nebula can occur at two very different stages of a star's evolution. It can represent the early stages of a star's life, when it is the visible sign of the *bipolar outflow of *T Tauri wind from a young star. It can also result from the directional loss of matter from an evolved *giant star as it begins to eject mass to become a *planetary nebula.

bipolar outflow. A flow of gas from a very young star, forming two oppositely directed beams; it is a recently discovered but apparently common feature of *stellar birth. In the youngest stars the bipolar flow is detected by the radio emission from molecules like *carbon monoxide; when stars have dispersed most of the surrounding gas and dust, the outflow is visible optically as a *bipolar nebula or as 'jets' extending away from a *T Tauri star for thousands of astronomical units. Both kinds of outflow can reach speeds of several hundred kilometres per second. The outflow from T Tauri stars can produce a line of *Herbig-Haro objects. These flows are too powerful to be driven by the pressure of the star's radiation. They are probably *T Tauri winds that are channelled outwards from the star's poles, a disc of dense gas around the equator preventing them from escaping isotropically. Such discs are indeed often detected in the centres of the different kinds of bipolar outflow at the correct orientation to account for the observed outflow of less-dense gas.

birefringent filter (Lyot filter; Lyot-Öhman filter). A type of *interference filter for studying the solar corona (*see* coronagraph). It was devised independently by the French astronomer Bernard Lyot in 1933 and the Swedish astronomer Yngve Öhman in 1938. Its action depends on the interference of the polarized light occurring in birefringent (double refracting) crystalline laminae: a stack of such laminae, of suitable thickness, transmits well-separated and very narrow wavebands of about 0.07 nanometres.

black body. A body that absorbs all the radiation falling on it, i.e. a body that has no reflecting power. It is also a perfect emitter of radiation. The concept of a black body is a hypothetical ideal. The radiation from stars, and their *effective and *colour temperatures, can however be described by assuming that they are black bodies.

Black-body radiation is the *thermal radiation that would be emitted from a black body at a particular temperature. It has a continuous distribution of wavelengths. The graph of the energy, or intensity, of the radiation has a characteristic shape (see illustration) with a maximum value at a given wavelength, λ_{max}. At lower tempera-

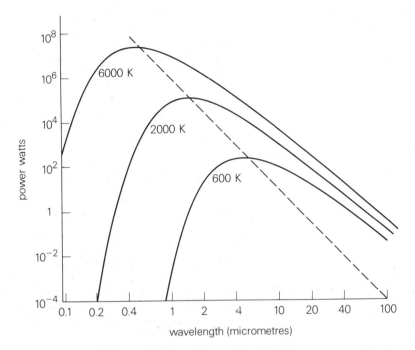

Black body radiation curves for different temperatures

tures the black-body radiation is mainly in the infrared region of the spectrum. As the temperature increases the maximum of the curve moves to shorter wavelengths. Curves at different temperatures follow the relationship

$$\lambda_{max}T = 2.9 \times 10^3$$

where the wavelength is measured in micrometres. This is known as the *Wien displacement law*. The radiation from a hot black body is greater at every wavelength than the radiation from a cooler black body. The total radiation flux thus increases rapidly with increasing temperature, as described by *Stefan's law.

Stefan's law gives the total energy emitted over all wavelengths per second per unit area of a black body. The energy emitted at a particular wavelength by a black body can be predicted from *Planck's radiation law*. This law was derived by Max Planck from his *quantum theory, propounded in 1900. If $E(\lambda,T)$ is the energy emitted per unit wavelength interval at wavelength λ, per second, per unit area, into unit solid angle, by a body at thermodynamic temperature T, then

$$E(\lambda,T) = (2hc^2/\lambda^5)[e^{hc/\lambda kT} - 1]^{-1}$$

h is the Planck constant, k the Boltzmann constant, and c the speed of light. Stefan's law and the Wien displacement law can both be derived from Planck's law.

Planck's radiation law only describes the continuous spectrum emitted by a black body. The continuous radiation from a star, as opposed to a black body, does not usually follow Planck's law exactly, although over a broad region of the spectrum the law is a good approximation.

black drop. An optical phenomenon seen near the beginning and end of *transits of Venus and Mercury across the sun's disc. Caused by irradiation and refraction of sunlight in the planet's atmosphere, it takes the form of an apparent link between the sun's limb and the silhouette of the planet against the solar surface. It reduced the accuracy of early attempts to determine the *solar parallax by observing transits from different parts of the earth.

black dwarf. *See* white dwarf.

black hole. An object so collapsed that its *escape velocity exceeds the speed of light. Although a black hole has yet to be unambiguously detected, there are strong theoretical grounds for believing that they exist. A collapsing object becomes a black hole when its radius has shrunk to its *Schwarzschild radius and light can no longer escape. Although gravity will make the object shrink beyond this limit, the 'surface' having this critical value of radius – the *event horizon – marks the boundary inside which all information is trapped. Calculations, however, indicate that space and time (*see* spacetime) become highly distorted inside the event horizon, and that the collapsing object's ultimate fate is to be compressed to an infinitely dense *singularity at the centre of the black hole. It has also been shown that the distortion of spacetime just outside the Schwarzschild radius causes the production of particles and radiation that gradually rob the black hole of energy and thus slowly diminish its mass. This *Hawking radiation* (Stephen Hawking, 1974) depends inversely on the black hole's mass, so the most massive black holes 'evaporate' most slowly. A black hole of the sun's mass would last 10^{66} years.

Once matter (or antimatter) has disappeared into a black hole, only three of its original properties can be ascertained: the total mass, the net electric charge, and the total angular momentum. The proof that a black hole has only these three observable properties is summed up in the 'No Hair' theorem: a black hole has no hair. Since all black holes must have mass, there are four possible types of black hole: a *Schwarzschild black hole* (1916) has no charge and no angular momentum; a *Reissner-Nordstrom black hole* (1918) has charge but no angular momentum; a *Kerr black hole* (1963) has angular momentum but no charge; a *Kerr-Newman black hole* (1965) has charge and angular momentum. (The dates in brackets indicate when the named mathematician (s) solved the equations of general relativity for these particular cases.) In astrophysics the simple Schwarzschild solution is often used, but real black holes are almost certainly rotating and have very little electric charge so that the Kerr solution should be the most applicable.

The most promising candidate for black holes are massive stars that explode as *supernovae, leaving a core in excess of 3 solar masses. This core must undergo complete *gravitational collapse because it is above the stable limit for both *white dwarfs and *neutron stars. Once formed, a black hole can only be detected by its gravity. Finding black holes only a few kilometres across (the size of the event horizon for a single-star black hole) is exceedingly difficult, but chances are increased if the black hole is a member of a close binary system. If the components of a binary system are close enough, mass transfer can occur between the primary star and its more compact companion (*see* equipotential surfaces). Matter will not fall directly on to the companion, however, for it has too much angular momentum; instead it forms a rapidly spinning disc – an *accretion disc* – around the compact object. If the latter is a black hole considerable energy can be produced, predominantly at x-ray wavelengths, as matter in the accretion disc loses momentum and spirals in.

Candidates for black holes in a binary system have been detected. The first to be found was the *x-ray binary *Cygnus X-1, comprising a 20 solar-mass B0 supergiant accompanied by an invisible companion with a mass 10 times that of the sun. This massive nonluminous object is probably a black hole emitting x-rays from its accretion disc. Even stronger is the evidence in two x-ray binaries in the Large Magellanic Cloud, LMC X-1 and LMC X-2, where the visible star is about 6 solar masses and the unseen component again 10 solar masses; unless the latter is a black hole, it should be much brighter than the visible star.

Supermassive black holes of 10^6 to 10^9 solar masses probably lie in the centres of some galaxies and give rise to the *quasar phenomenon and the phenomena of other active galaxies. If a huge black hole is able to form and to capture sufficient gas and/or stars from its surroundings, the rest-mass energy of infalling material can be converted into radiation or energetic particles.

At the other mass extreme are the more speculative *mini black holes*, weighing only 10^{11} kg and with radii of 10^{-10} metres. These could have formed in the highly turbulent conditions existing after the big bang. They create such intense localized gravitational fields that their Hawking radiation makes them explode within the lifetime of the universe; their final burst of gamma rays and microwaves should be detectable but has not yet been found.

Blazhko effect. *See* RR Lyrae stars.

blink comparator. *See* comparator.

BL Lac objects. Extremely compact violently variable extragalactic objects that resemble *quasars but lack both emission and absorption lines in their spectra. Several dozen are known, the first to be identified being BL Lacertae, which was mistakenly classified as a variable star and much later (1968) shown to be the optical counterpart of a peculiar radio source. BL Lac objects are most easily identified from radio surveys, so most of those known are strong radio sources, but, like quasars, the peak of their emission lies in the infrared.

Light and radio waves from these objects are strongly polarized, indicating intense magnetic fields, and there are rapid variations in both strength and direction. BL Lac objects are also violently variable in luminosity at all wavelengths, sometimes flaring up five magnitudes (a factor of 100) in only a few weeks. This indicates that the size of their energy-producing regions cannot be more than a few light-weeks across. Distance measurements to BL Lac objects have been hampered by their absence of spectral lines but some (BL Lac, AP Lib) are embedded in a faint 'fuzz', believed to be the surrounding galaxy. This contains weak absorption lines, from which it is possible to measure a *redshift. A few BL Lac objects (AO 0235 + 164) show 21 cm absorption due to hydrogen in an intervening galaxy, and the redshift of this absorption sets a lower limit to the object's distance. Such measurements place the BL Lac objects at distances typical of quasars, and with a comparable energy output. The BL Lac objects appear to lie in elliptical galaxies, however, while there is mounting evidence that quasars occur in spiral galaxies.

bloomed lens. *See* coating of lenses.

blueshift. An overall shift of the spectral lines in a spectrum towards shorter wavelengths. It is observed in the spectra of celestial objects approaching the earth. *See* redshift.

BN object. *Short for* Becklin-Neugebauer object. A point-like infrared source in the *Orion molecular cloud (OMC-1). It is thought to be a young B0 or B1 star, surrounded by a compact *HII region and an expanding dust envelope characterized by a strong *stellar wind. The infrared luminosity of the dust is about a thousand times the sun's luminosity at a temperature of ~ 600 K. The object is still above the *main sequence and evolving towards it, and may represent the earliest phase of stellar evolution that has so far been identified.

Bode's law (Titius–Bode law). A relationship between the distances of the planets from the sun. Take the sequence 0, 3, 6, 12, 24,..., where each number (except the 3) is twice the previous one, add 4 to each, and divide by 10. The resulting sequence (0.4, 0.7, 1.0, 1.6, 2.8, 5.2,...) is in good agreement with the actual distances in *astronomical units (AU) of most planets, provided that the minor planets are included and considered as one entity at a mean distance of 2.8 AU. The law fails to predict the correct distances for Neptune and Pluto, but may have some significance with respect to the formation of the solar system. Named after Johann Bode, who published it in 1772, it was formulated by Johann Titius of Wittenberg in 1766.

Bok globules. Small dark cool (10 K) clouds of gas and dust seen as near-circular objects against a background of stars or of an *HII region; they are named after the American astronomer Bart J. Bok. They are believed to represent a phase in the contraction of some *protostars when the material has become sufficiently dense to be opaque. The main sites for star formation are now known to be giant *molecular clouds, but Bok globules give rise to some of the Galaxy's lower-mass stars. Most globules have diameters between 0.2 and 0.6 parsecs, and absorb between one and five magnitudes of light. Their mass varies from 20 to 200 masses. During the globule stage it is thought that the protostars collapse more slowly than before but the internal tempera-

ture and pressure build up to become sufficient for nuclear reactions to commence. *IRAS and other infrared telescopes have located protostars within some Bok globules; *bipolar outflows from these produce *Herbig–Haro objects seen at the globule's edge.

bolide. A brilliant *meteor that appears to explode, i.e. a detonating *fireball. The brighter ones are caused by ablating *meteorites that subsequently fall to earth. About 5000 bolides occur in the earth's atmosphere each year.

bolometer. A radiation-sensing instrument that is used in astronomy to measure the total energy flux of radiation entering the earth's atmosphere, and is especially useful in the infrared and microwave regions of the spectrum. It is essentially a small resistive element capable of absorbing electromagnetic radiation. The resulting temperature rise is a measure of the power absorbed. There is a large variety of bolometers, including semiconductor and supercooled devices.

bolometric correction (BC). *See* magnitude.

bolometric magnitude. *See* magnitude.

Boltzmann constant. Symbol: k. A constant given by the ratio of the universal gas constant to the Avogadro number and equal to $1.380\,622 \times 10^{-23}$ joules per kelvin. It is the constant by which the mean kinetic energy of a gas particle can be related to its thermodynamic temperature.

Boltzmann equation. An equation derived by the Austrian physicist Ludwig Boltzmann in the 1870s that shows how the distribution of molecules, atoms, or ions in their various *energy levels depends on the temperature of the system; the system is in thermal equilibrium, with *excitation balanced by de-excitation. The equation gives the ratio of the number density (number per unit volume) of molecules, atoms, or ions, N_2, in one energy level to the number density, N_1, in another lower energy level as

$$N_2/N_1 = (g_2/g_1)e^{-E/kT}$$

g_1 and g_2 are the degeneracies of the two levels, i.e. the multiplicity of energy levels with the same energy, E is the energy required to excite the molecule, atom, or ion from the lower to the higher energy level, and T is the thermodynamic temperature. Thus as T increases a greater number of species will become excited.

The Boltzmann equation together with the *Saha ionization equation are widely used to interpret the absorption and emission spectra of stars and to determine stellar temperatures and densities.

Bonner Durchmusterung (BD). A general star catalogue prepared by the Prussian astronomer Friedrich Argelander and published in Bonn 1859–62. The original catalogue contained 324 189 stars – i.e. all those visible in the three inch Bonn refractor – lying between *declinations of +90° to −2°. The limiting magnitude was about 9.5. Accompanying star charts were published in 1863. It was extended by E. Schönfeld in 1886 to a declination of −23° and then included about 458 000 stars.

Cataloguing was extended to the south polar regions through an analogous work, *The Córdoba Durchmusterung (CD)*, compiled at the Córdoba Observatory, Argentina, and finally completed in 1930. This catalogue contains about 614 000 stars, brighter than 10th magnitude, from declinations −23° to −90°.

Bootes (Herdsman). A large constellation in the northern hemisphere near Ursa Major. The brightest stars are *Arcturus (the brightest star in the northern sky) and the 2nd-magnitude yellow-blue visual binary Epsilon Bootis, one of several double stars. The area also contains one of the most distant *radio galaxies, 3C–295. Abbrev.: Boo; genitive form: Bootis; approx. position: RA 14.5h, dec +30°; area: 907 sq deg.

bosons. A class of *elementary particles with integer values (positive, negative, or zero) of the basic *quantum of *spin; the *photon is an example. More than one boson can exist with an identical set of quantum numbers (numbers assigned to the various quantities that describe a particle). Bosons do not therefore satisfy the *Pauli exclusion principle. *See also* fermions; fundamental forces.

Boss General Catalogue. A whole-sky star catalogue listing the positions and *proper motions of 33 342 stars for the epoch 1950.0, down to a limiting magnitude of 7. It was prepared initially by the American astronomer Lewis Boss, with a preliminary publication in 1910, and was completed by his son Benjamin Boss in 1937.

bound-free absorption. A process of absorption that occurs when an electron bound to an atom or ion is given sufficient energy by an absorbed photon of radiation to free itself from the atom. If however the photon energy can only excite the electron to a higher energy level so that it remains bound to the atom, *bound-bound absorption* takes place. *See also* opacity.

bow shock. *See* interplanetary medium.

Brackett series. *See* hydrogen spectrum.

Brans–Dicke theory. A relativistic theory of gravitation put forward in the 1960s by Carl Brans and Robert Dicke as a variant of Einstein's general theory of *relativity. It is considered by many astronomers to be the most serious alternative to general relativity. It is a scalar-tensor theory, i.e. a theory in which the gravitational force on an object is due partly to the interaction with a scalar field and partly to a tensor interaction. Newton's gravitational constant is replaced by a slowly varying scalar field. The effect is to allow the strength of gravity to decrease with time. In the limit that this variation is zero, the various Brans–Dicke theories of gravitation that now exist reduce to Einstein's general relativity. Current observations limit the variation of Newton's gravitational constant to be less than one part in 10^{10} per year. This means that for local applications of a noncosmological nature Brans–Dicke theory is indistinguishable from general relativity.

bremsstrahlung. (literally: 'braking radiation'). Electromagnetic radiation arising from the rapid deceleration of electrons. It has a continuous spectrum. *See* thermal emission; synchrotron emission.

bridge. *See* radio source structure.

brightest stars. The stars with the greatest apparent visual *magnitude. Many are very much more luminous than the sun and others appear bright because of their proximity. The most luminous lie in the supergiant and giant regions of the *Hertzsprung–Russell diagram. *See* Table 6, backmatter.

bright nebula. An *emission nebula or *reflection nebula as opposed to a *dark nebula. *See also* nebula.

brightness. The intensity of light or other radiation emitted from – absolute or intrinsic brightness – or received from – apparent brightness – a celestial body, the latter decreasing as the distance from the body increases. Intrinsic brightness is directly related to *luminosity. Apparent brightness is considered in terms of

apparent *magnitude: a star one magnitude less than another is about 2.5 times brighter. If two stars belong to the *main sequence then the brighter star is the hotter of the two. *See also* jansky.

brightness temperature. Symbol: T_b. The apparent temperature of radiation from a distant emitting region, i.e. the temperature that a *blackbody – a perfect radiator – would have in order to radiate the same observed *flux density per unit solid angle. In the case of an interstellar cloud, it may be equal to the physical temperature of the cloud if the radiation is by *thermal emission and the cloud is optically thick (*see* optical depth). If the cloud is optically thin, the brightness temperature is reduced. *See also* antenna temperature.

Brooks 2. *See* Jupiter's comet family.

brown dwarf. A theoretical 'star' formed by the contraction of a lump of gas with a mass too small for nuclear reactions to begin in the core. This limit on *stellar mass is uncertain but is thought to be about 0.08 solar masses. An object below this limit will shine for only 100 million years as a result of gravitational contraction on the *Hayashi track, and then cool off. Its interior consists of *degenerate matter. Some of the least luminous red *dwarf stars may actually be brown dwarfs. The best candidate is LHS2924; its mass is uncertain but it has an absolute magnitude of +20 (making it the least luminous star currently known) and an effective temperature of only 1950 K.

B stars. Stars of *spectral type B that are massive hot blue stars with a surface temperature (T_{eff}) of about 10 000 to 28 000 kelvin. Absorption lines of neutral helium (He I) dominate the spectrum, reaching maximum intensity for B2 stars. Balmer lines of hydrogen strengthen from B0 to B9; lines of ionized magnesium and silicon are also present. A few B stars – the *Be stars – also have emission lines from a circumstellar shell of gas. B0, B1, and B2 stars are found in OB *associations in the spiral arms of galaxies. Spica, Rigel, Bellatrix, and Alpha Crucis are B stars.

burst. 1. A brief flux of intense radiation with a sudden onset and rapid decay, as is observed from Jupiter and from the sun at radio wavelengths. *See also* x-ray burst sources; gamma-ray bursts.
 2. (flare). *See* train.

burst source (burster). *See* x-ray burst sources.

Butler matrix. An electronic phasing device by which the signals from 2^n separate elements of an *array are combined together with their correct relative *phases to produce 2^n outputs, each of which corresponds to a main lobe (*see* antenna) pointing in a different direction.

butterfly diagram. *See* sunspot cycle.

B-V. *See* colour index.

Bw stars. Blue stars that, unlike normal *B stars, have only weak helium lines.

C

Caelum (Chisel). A small inconspicuous constellation in the southern hemisphere near Orion, the brightest stars being of 4th magnitude. Abbrev.: Cae; genitive form: Caeli; approx. position: RA 4.5h, dec −40°; area: 125 sq deg.

Calar Alto. Mountain site of the *German–Spanish Astronomical Centre.

calendar. Any of various present-day, past, or proposed systems for the reckoning of time over extended periods: days are grouped into various periods that are suitable for regulating civil life, fixing religious observances, and meeting scientific needs. All calendars are based either on the motion of the sun (solar calendar), or of the moon (lunar calendar), or both (lunisolar calendar) so that the length of the year corresponds approximately to either the *tropical (solar) year or the *lunar year. Their complexity results primarily from the incommensurability of the natural periods of *day, *month, and *year: the month is not a simple fraction of the year and the day is not a simple fraction of the month or the year. Present-day calendars include the *Gregorian calendar, which is in use throughout most of the world, and the Moslem, Jewish, and Chinese calendars.

calendar year. The interval of time that is the basis of a *calendar. The *Gregorian calendar year contains an average of 365.2425 days.

calibration. A procedure carried out on a measuring instrument, such as a *radio telescope, by means of which the magnitude of its response is determined as a function of the magnitude of the input signal. The calibration of a radio telescope provides an absolute scale of the output deflection against *antenna temperature.

California nebula (IC 1499). An *emission nebula – an *HII region – that lies in the constellation Perseus and is ionized by the star Zeta Persei.

Callisto. The faintest of the four giant Galilean satellites of Jupiter. It is a heavily cratered object with a radius of 2400 km and a density of 1.6 g cm^{-3}. There are a number of ray systems, i.e. craters from which bright streaks radiate. Some major systems of concentric ring mountains are prominent, particularly in the neighbourhood of the huge basin *Valhalla*, which is located slightly north of the equator. The basin is 600 km in diameter and the outermost rings are 3000 km across. This basin is comparable with *Orientale Basin on the moon, the Caloris Basin on *Mercury, and *Hellas Planitia on Mars. Callisto is thought to have a thick crust of ice and rock extending to a depth of 200 to 300 km, beneath which is thought to be a mantle of 1000 km of convecting water or soft ice. Callisto, the outermost of the Galilean satellites, is beyond the major charged particle environment of Jupiter. *See also* Jupiter's satellites; Table 2, backmatter.

Caloris Basin. *See* Mercury.

Camelopardalis (Giraffe). A large inconspicuous constellation in the northern hemisphere near Ursa Major, the brightest stars being of 4th magnitude. It contains the prototype of the Z Camelopardalis *dwarf novae stars. Abbrev.: Cam; genitive form: Camelopardalis; approx. position: RA 6h, dec +70°; area: 757 sq deg.

Canada–France–Hawaii Telescope (CFHT). *See* Mauna Kea.

canals. Linear markings on *Mars that were first observed by Giovanni Schiaparelli in 1877 but later charted by many observers, most notably by Percival Lowell who advocated that they were irrigation ditches dug by Martians to distribute their planet's scarce water resources. *Mariner and *Viking spacecraft observations show that the canals do not exist.

Cancer (Crab). An inconspicuous zodiac constellation in the northern hemisphere near Ursa Major, the brightest stars being of 4th magnitude. There are many double and variable stars, including the multiple

star Zeta Cancri. The area also contains the open clusters *Praesepe and the fainter M67 and the strong radio source NGC 2623. Abbrev.: Cnc; genitive form: Cancri; approx. position: RA 9h, dec +20°.

Canes Venatici (Hunting Dogs). An inconspicuous constellation in the northern hemisphere near Ursa Major, the brightest star being the 3rd-magnitude binary *Cor Caroli. It contains the bright globular cluster M3 and the *Whirlpool galaxy, M51. Abbrev.: CVn; genitive form: Canum Venaticorum; approx. position: RA 13h, dec +40°; area: 465 sq deg.

Canis Major (Great Dog). A conspicuous constellation in the southern hemisphere, lying partly in the Milky Way. The brightest stars are *Sirius (the brightest and one of the nearest stars in the sky), the 1st-magnitude giant *Adhara (ε), and the giant Mirzam (β) and supergiant Wezen (δ), both of 2nd magnitude and very remote and luminous. The area contains the open cluster M41. Abbrev.: CMa; genitive form: Canis Majoris; approx. position: RA 7h, dec −20°; area: 380 sq deg.

Canis Minor (Little Dog). A small constellation in the northern hemisphere near Orion, lying partly in the Milky Way. The brightest star is the visual binary *Procyon. Abbrev.: CMi; genitive form: Canis Minoris; approx. position: RA 7.5h, dec +5°; area: 183 sq deg.

cannibalism. 1. (galactic cannibalism) The swallowing of one galaxy by a larger galaxy. It occurs especially in the centre of a *cluster of galaxies, where successive acts of cannibalism have probably produced cD galaxies, the most massive galaxies known. The largest cD galaxies (such as NGC 6166 in the rich cluster Abell 2199) appear to have several cores, each thought to be the core of a smaller galaxy that has been incorporated into the cD.

2. (stellar cannibalism) The process by which a *giant star in a close binary system can swallow its companion. *See* common envelope star.

Canopus (α Car). A conspicuous and luminous cream supergiant that is the brightest star in the constellation Carina and the second-brightest star in the sky. m_v: -0.72; M_v: -4.6; spectral type: F0 II; distance: 60 pc.

canyon. *See* valley, lunar; Mars, surface features.

Cape Canaveral. *See* Kennedy Space Center.

Capella (α Aur). A conspicuous yellow giant that is the brightest star in the constellation Auriga. It is a spectroscopic triple star (period 104 days) and has a high lithium content. m_v: 0.09; M_v: -0.6; spectral type: G8 III, G5 III, M5 V; distance: 13.7 pc.

Cape Photographic Durchmusterung. A general star catalogue of the southern sky compiled by J.C. Kapteyn from photographic plates produced by David Gill in Cape Town. It was published 1896–1900. It lists positions and *magnitudes of some 455 000 stars, brighter than 10th magnitude, from declinations $-19°$ to $-90°$.

Caph. *See* Cassiopeia.

Capricornids. A minor *meteor shower, radiant: RA 315°, dec $-15°$, that maximizes on July 25. Another minor shower, the *Alpha Capricornids*, radiant: RA 309°, dec $-10°$, maximizes on Aug. 1.

Capricornus (Sea Goat). A zodiac constellation in the southern hemisphere near Sagittarius, the brightest stars being of 3rd magnitude. Alpha Capricorni is a naked-eye double star

(magnitudes 3.7, 4.5): the brighter component is a binary and the fainter component an optical double. The area contains the globular cluster M30. Abbrev.: Cap; genitive form: Capricorni; approx. position: RA 21h, dec $-20°$; area: 414 sq deg.

captured rotation. *See* synchronous rotation.

capture theory. *See* encounter theories; moon.

carbonaceous chondrite. An uncommon class of *meteorites but very important because of their mineralogical and chemical composition, especially as regards the presence of hydrated minerals and organic (carbon) compounds. They are very easily crumbled and contain water-soluble compounds and must therefore be collected soon after fall. Although all meteorites were formed very early in the solar system's history, carbonaceous chondrites are possibly the most primitive form of matter in the solar system. *See also* chondrite.

carbon cycle. A chain of *nuclear fusion reactions by which energy may be generated in stars. The overall effect of the cycle is the transformation of hydrogen nuclei into helium nuclei with emission of gamma-ray photons (γ), positrons (e^+), and neutrinos (ν). The major sequence of reactions is as follows:

$$^{12}C + {}^1H \rightarrow {}^{13}N + \gamma$$
$$^{13}N \rightarrow {}^{13}C + e^+ + \nu$$
$$^1H + {}^{13}C \rightarrow {}^{14}N + \gamma$$
$$^1H + {}^{14}N \rightarrow {}^{15}O + \gamma$$
$$^{15}O \rightarrow {}^{15}N + e^+ + \nu$$
$$^1H + {}^{15}N \rightarrow {}^{12}C + {}^4He$$

The carbon nucleus, ^{12}C, reappears at the end of the cycle and can be regarded as a catalyst for the reaction:

$$4{}^1H \rightarrow {}^4He + 2e^+ + 2\nu + 3\gamma$$

Because nitrogen (N) and oxygen (O) intermediates are involved, the cycle is often termed the *carbon-nitrogen cycle* or *CNO cycle*.

The carbon cycle is very strongly temperature dependent and becomes the dominant energy-producing mechanism at core temperatures exceeding about 1.6×10^7 kelvin. It is therefore thought to be the major source of energy in hot massive stars of spectral types O, B, and A. *See also* proton-proton chain reaction.

carbon monoxide. A molecule, CO, consisting of an atom of oxygen bound to a carbon atom. It is commonly found in giant *molecular clouds, where there is one CO molecule to about 10 000 hydrogen molecules. Collisions with hydrogen and other molecules excite the CO molecules and cause them to emit characteristic radio waves at wavelengths of 2.6 mm and 1.33 mm. Carbon monoxide is thus used in radio astronomy as the best tracer of molecular gas over wide areas: the molecular hydrogen in the cool clouds has no emission at radio wavelengths. The CO emission lines are analysed to determine the density, velocity, and temperature of the molecules in the clouds.

carbon–nitrogen cycle. *See* carbon cycle.

carbon stars (C stars). Rare red giant stars of low temperature that have an over-abundance of carbon relative to oxygen and also an unusually high abundance of lithium. WZ Cassiopeia is an example. The spectra show strong bands of carbon compounds, including C_2, CN, and CH. In the earlier Harvard classification (*see* spectral types) these stars were divided into *R stars* and *N stars*: N stars are similar to *M stars, being cooler and much redder than the K-type R stars. *See also* S stars.

cardinal points. The four principal points on the *horizon (see illustration). The *north point* (n) lies at the intersection of the horizon with the *celestial meridian nearest the north

*celestial pole; the *south point* (s), diametrically opposite, is the equivalent intersection point closest to the south celestial pole. The *east point* (e) and *west point* (w) lie at the intersections of the horizon with the *celestial equator, the east point being 90° and the west point 270° clockwise from the north point.

Carina (Keel). A constellation in the southern hemisphere near Crux, lying partly in the Milky Way. It was once part of the constellation *Argo. The brightest stars are the supergiant *Canopus and three 2nd-magnitude stars. The irregular variable *Eta Carina is associated with extensive nebulosity. The area also contains the large globular cluster NGC 2808. Abbrev.: Car; genitive form: Carinae; approx. position: RA 9h, dec −60°; area: 494 sq deg.

Carina arm. *See* Galaxy.

Carlsberg Automatic Transit Circle. *See* Roque de los Muchachos Observatory.

Carme. A small satellite of Jupiter, discovered in 1938. *See* Jupiter's satellites; Table 2, backmatter.

Carte du Ciel. A whole-sky photographic atlas proposed in Paris in 1887 and planned to include stars of magnitude 14 or brighter. This ambitious scheme involved 18 observatories in photographing areas of the sky only 2° square using specially designed 33 cm refracting telescopes. The survey has never been fully completed. Technical advances have since made it possible to photograph much greater areas with very much less work. The associated *Astrographic Catalogue,* listing the positions of stars down to 12th magnitude, was essentially complete by 1958.

cascade. *See* cosmic rays.

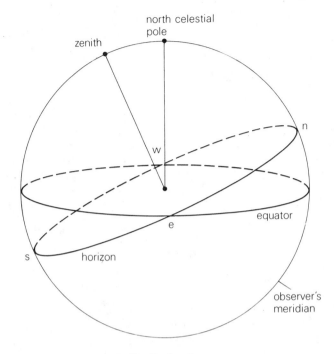

Cardinal points

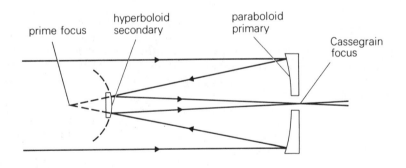

Ray path in Cassegrain configuration

Cassegrain configuration. One of the configurations in which a large reflecting telescope can be used. A small secondary mirror with a convex hyperboloid shape intercepts the light before it reaches the prime focus of the *primary mirror (see illustration). The secondary reflects the light back down the tube, through a central hole in the primary mirror, to a focal point just behind it. This is the *Cassegrain focus*. It is more accessible than the prime focus.

Cassegrain focus. See Cassegrain configuration.

Cassegrain telescope. A compound *reflecting telescope designed by the Frenchman Guillaume Cassegrain in 1672. It is similar to the *Gregorian telescope but has a secondary mirror with a convex hyperboloid shape mounted inside the focal plane of the primary mirror. The telescope is compact, portable, and easily mounted and thus very popular. Most large modern reflectors include this facility. *See* Cassegrain configuration.

Cassini Division. *See* Saturn's rings.

Cassiopeia. A conspicuous constellation in the northern hemisphere, lying partly in the Milky Way. The five brightest stars form a W-shape, which includes the variable *Schedar (α), the blue-white visual binary Cih (γ), which is an irregular variable, and the cream-coloured Caph (β), all of 2nd magnitude. Eta, close to Alpha, is also a binary star. The area contains some fine open clusters (M52, M103, and NGC 457) and globular clusters and the remnants of two recent supernovae – *Tycho's star and the radio source *Cassiopeia A. Abbrev.: Cas; genitive form: Cassiopeiae; approx. position: RA 1h, dec +60°; area: 598 sq deg.

Cassiopeia A. An intense *radio source about three kiloparsecs distant in the constellation Cassiopeia. It has the largest *flux density at low frequencies of all the discrete radio sources (apart from the sun), the value being 11 000 *jansky at 178 megahertz. It is a *supernova remnant – probably of a supernova explosion (unrecorded) in the late 17th century – and has a ring-like structure about four arc minutes in diameter. The radio radiation is by *synchrotron emission. It is also an extended source of soft x-rays.

Castor (α Gem). A white star that is the second-brightest one in the constellation Gemini. It is a visual triple star, the two brighter components having an orbital period of about 380 years; the third component, a flare star, is located some distance away.

All three are spectroscopic binaries. m_v: 1.56; M_v: 0.8; spectral type: A1 V (A), A5 (B); distance: 14 pc.

cataclysmic variable. A close binary-star system where one member is a *white dwarf, and mass transfer on to the latter causes sudden large and unpredictable changes in brightness. The main classes of cataclysmic variable are *novae, *recurrent novae, *dwarf novae, and *symbiotic stars. The outbursts of cataclysmic variables are also detected at ultraviolet and x-ray wavelengths; their x-ray output is however much less than that of *x-ray binaries, which are similar systems where the compact star is a neutron star rather than a white dwarf.

The small separation of the two stars means that the two cores almost certainly shared a *common envelope at the previous stage of evolution. Initially the system was probably a *W Ursae Majoris binary consisting of two main-sequence stars in contact.

catadioptric telescope. A telescope that uses both refraction and reflection to form the image at the prime focus. By introducing a full-aperture *correcting plate in front of the primary mirror, the telescope designer can correct many *aberrations over an exceptionally wide field of view, in a way that combines the best features of both *refracting and *reflecting telescopes. *See* Schmidt and Maksutov telescopes.

catalogue equinox. The intersection of the *hour circle of zero right ascension of a star catalogue with the celestial equator. It is thus the origin of the catalogue. It is usually an approximation to the *dynamical equinox of some date, differing from it as a result of the limits of observational accuracy. There is also a difference, over a period of time, due to the motion of the dynamical equinox; this time-dependent difference is called *equinox motion.*

catastrophic theories. *See* encounter theories.

catena. A chain of craters. The word is used in the approved name of such a surface feature on a planet or satellite.

catoptric system. An optical system in which the principal optical elements are mirrors. *Compare* catadioptric telescope; dioptric system.

cavus. A steep-sided depression. The word is used in the approved name of such a surface feature on a planet or satellite.

Cayley formation. *See* highland light plains.

CCD. *Abbrev. for* charge-coupled device. A light-sensitive silicon chip originally developed for use in TV cameras and now frequently used in highly sensitive electronic equipment associated with telescopes to detect light or ultraviolet radiation from space. Recently CCD devices operating at x-ray energies up to about 6 keV have been demonstrated. The radiation is directed on to the CCD and an electrical signal is produced. This signal can rapidly generate on a TV screen a pictorial representation of a small area of sky; the signal can also be fed into a computer for storage and further analysis. A CCD is about 30 times as sensitive as the best photographic emulsion so that exposure times can be reduced from hours to minutes.

cD galaxy. *See* clusters of galaxies; cannibalism.

Celescope experiment. *See* Orbiting Astronomical Observatory.

celestial axis. The line joining the north and south *celestial poles and passing through the centre of the *celestial sphere; the extension of the earth's axis to the celestial sphere.

celestial equator. The great circle in which the extension of the earth's equatorial plane cuts the *celestial sphere. The plane of the celestial (and hence terrestrial) equator is perpendicular to the celestial axis and is the reference plane for the equatorial coordinates *right ascension and *declination. The orientation of the celestial equator is slowly changing as a result of the *precession of the earth's axis.

celestial latitude (ecliptic latitude). *See* ecliptic coordinate system.

celestial longitude (ecliptic longitude). *See* ecliptic coordinate system.

celestial maser. *See* maser source.

celestial mechanics. The study of the motions and equilibria of celestial bodies subjected to mutual gravitational forces, usually by the application of Newton's law of *gravitation and the general laws of mechanics, based on *Newton's laws of motion. Satellite and planetary motions, tides, precession of the earth's axis, and lunar libration are all described by these laws within the limits of accuracy of measurements. Newtonian mechanics is much simpler to use than the more accurate general theory of *relativity. Despite fundamental differences, the equations of relativistic mechanics, based on the concepts of relativity, reduce in a first approximation to those of Newtonian mechanics, observational and predicted Newtonian values normally being very close.

celestial meridian. *See* meridian.

celestial poles. The two points at which the extension of the earth's axis of rotation cuts the *celestial sphere. As a result of the *precession of the earth's axis the positions of the poles are not fixed but trace out a circle on the celestial sphere in a period of about 25 800 years. The *north celestial pole* revolves about a point in the constellation Draco and at present lies in Ursa Minor close to the star *Polaris. The *south celestial pole* lies in the constellation Octans. *See also* pole star.

celestial sphere. An imaginary sphere of indeterminate but immense radius that provides a convenient surface on which to draw and study the directions of celestial bodies and other points in the sky. As circumstances require, the celestial sphere may be centred on the observer, at the earth's centre, or at some other location. Celestial bodies, etc., may thus be thought of as points projected onto the surface of the celestial sphere. The earth's west-to-east rotation causes an apparent rotation of the celestial sphere about the same (extended) axis once every 24 hours but in an east-to-west direction.

The measurements that may be made with the celestial sphere, using the four main *coordinate systems of positional astronomy, concern not the distances but only the directions of celestial bodies. Angular relationships are established by the methods of spherical trigonometry (*see* spherical triangle). The principal reference circles on the celestial sphere are the *celestial equator, the *ecliptic, and an observer's *horizon, which are all great circles. The principal points are

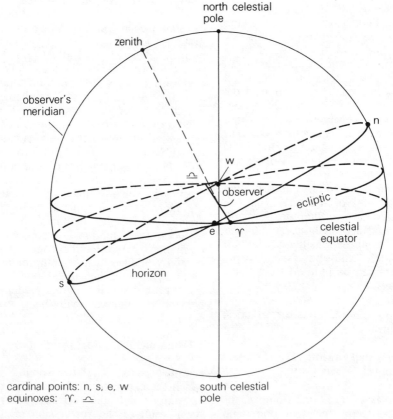

Celestial sphere

the poles of these circles and the points of intersection – the *equinoxes and *cardinal points – of the circles (see illustration).

cell (mirror cell). The enclosure that holds the *primary mirror in a reflecting telescope. It must hold the mirror so that the *collimation of the optical elements is maintained as the direction of observation changes. It must also support the mirror so that it does not sag to an unacceptable degree under its own weight. Without carefully designed support over the whole rear face the mirror of the 5-metre Hale telescope would sag by 0.0625 mm, which is about 500 times the allowable quarter wave limit. The cell is usually provided with a closure or lid that protects the optical surface when the telescope is not in use.

Centaurus (Centaur). An extensive conspicuous constellation in the southern hemisphere almost surrounding Crux and lying partly in the Milky Way. The brightest stars are the triple star *Alpha Centauri, the remote 1st-magnitude *Beta Centauri, and several stars of 2nd and 3rd magnitude. *Proxima Centauri is the nearest star to the sun. The area also contains the fine naked-eye globular cluster Omega Centauri, the strong radio source *Centaurus A, and the x-ray binary *Centaurus X-3. Abbrev.: Cen; genitive form: Centauri; approx. position: RA 13h, dec −50°; area: 1060 sq deg.

Centaurus A. An intense *radio and *x-ray source in the southern constellation Centaurus and a source also of infrared radiation and gamma rays. It is identified with the galaxy NGC 5128 lying at a distance of only 5 megaparsecs from the solar system. This is an elliptical star mass, 100 kpc in diameter, cut across by broad belts of gas and dust. The belts are rotating steadily, unlike the elliptical

mass, which is rotating very slowly if at all. A complex elongated radio structure emerges from the centre of the gas and dust belts, approximately along their axis of rotation and extending about 400 kpc in each direction. The structure consists broadly of two large lobes more or less symmetrically disposed about a central nucleus, from which a *jet extends towards one of the lobes. The jet is broken up into a number of *knots. This huge radio galaxy has a flux density at 86 megahertz of 8700 *jansky by *synchrotron emission.

Centaurus A is also one of the brightest hard x-ray sources, its spectrum being measured to about 500 kiloelectronvolts. It is also variable on timescales down to a few days, suggesting that most of the x-ray emission probably arises in the nucleus. An Einstein Observatory image showed not only the bright nucleus but in addition a line of sources, i.e. an x-ray jet, significantly along the axis of the radio lobes. The x-ray emission follows closely the radio jet but extends beyond it into the radio lobe.

Centaurus X-3. An *x-ray binary, discovered by the *Uhuru satellite in 1971. The binary nature was determined from the observation of regular x-ray eclipses with a period of 2.09 days; in 1974 the optical companion star was identified as a 13th magnitude supergiant of spectral type O6.5. Rapid pulsation of the x-ray source at a 4.8 second period is widely interpreted as being associated with the rotation of a magnetized *neutron star. It may thus be regarded as an x-ray *pulsar.

central force. A force on a moving body that is directed towards a fixed point or towards a point moving according to known laws. The gravitational attraction between the sun and a planet is an example.

centre of gravity. The point in a material body at which a single force, which is the resultant of all external forces on the body, may be considered to act. The centre of gravity of a body in a uniform gravitational field coincides with the body's *centre of mass. A body does not possess a centre of gravity however if the external forces are not equivalent to a single resultant force or if the resultant does not always act on the same point of the system. This situation occurs in a nonuniform gravitational field. The moon has strictly no centre of gravity although the resultant force of the earth's gravitational attraction always passes within a few metres of the centre of mass.

centre of inertia. *See* centre of mass.

centre of mass (centre of inertia). The point in a material system at which the total mass of the system may be regarded as being concentrated and that moves as if all external forces on the system could be reduced to a single force acting at that point. When a body has a centre of gravity, which it does in a uniform gravitational field, this point coincides with the centre of mass.

Two bodies, moving under the influence of their mutual gravitation, will orbit around their centre of mass, which lies on the line between them. The distances, r_1 and r_2, of the bodies from this point depend on their masses, m_1 and m_2, the more massive body lying closer. For circular orbits:

$$r_1/r_2 = m_2/m_1$$

centrifugal force. *See* centripetal force.

centripetal acceleration. *See* centripetal force.

centripetal force. A force, such as gravitation, that causes a body to deviate from motion in a straight line to motion along a curved path, the force being directed towards the centre of

curvature of the body's motion. The force reacting against this constraint, i.e. the force equal in magnitude but opposite in direction, is the *centrifugal force*. The centrifugal force results from the inertia of all solid bodies, i.e. their resistance to acceleration, and unlike gravitational or electrical forces, cannot be considered a real force. The centripetal force is equal to the product of the mass of the body and its *centripetal acceleration*. The latter is the acceleration towards the centre, and for a body moving in a circle at a constant angular velocity ω it is given by $\omega^2 r$, where r is the radius.

Cepheid. *Short for* Cepheid variable.

Cepheid instability strip. *See* pulsating variables.

Cepheid variables. A large and important group of very luminous yellow giants or supergiants that are *pulsating variables with periods ranging from about 1–70 days. Over 700 are known in our galaxy and several thousand in the Local Group. There are two categories: *classical Cepheids* (also known as *Type I Cepheids*) are massive young population I objects found in spiral arms on the galactic plane; *W Virginis stars* (or *Type II Cepheids*) are much older and less massive population II objects found in the galactic centre and halo, especially in *globular clusters, and are thus similar in distribution to *RR Lyrae stars. The classical Cepheids are about 1.5–2 magnitudes more luminous than W Virginis stars of the same period. The luminosity variations of both categories are continuous and extremely regular so that the periods can be measured very accurately. Characteristic periods are 5–10 days (classical) and 12–30 days (W Virginis stars); the *amplitudes are typically 0.5–1 in magnitude.

The prototype of the classical Cepheids is *Delta Cephei*, discovered

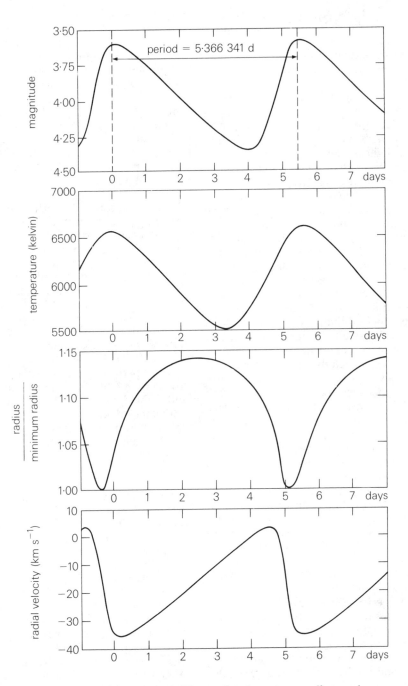

Variation in brightness, temperature, radius, and
radial velocity of expanding and contracting
layers of Delta Cephei

1784. The changes in brightness were found in the 1890s to be accompanied by and principally caused by changes in stellar temperature and also by changes in radius (see illustration). This was later explained in terms of pulsations in the outer layers of the stars that appear at a late evolutionary state (*see* pulsating variables). The relation between period of pulsation and light variation of Cepheid variables was discovered during 1908–12 by Henrietta Leavitt. Miss Leavitt could only measure the apparent rather than the absolute magnitude of the Cepheids under study. An independent determination of the distance of a Cepheid of known period would lead to the graphical relation between period and luminosity, or absolute magnitude (*see* distance modulus). It was quickly realized by Shapley that this *period-luminosity relation was an invaluable tool for measurements of distance out to the nearest galaxies and thus for studying the structure of our own galaxy and of the universe (*see* distance determination). The *Hubble Space Telescope will be able to study classical Cepheids in galaxies as far away as the Virgo cluster (about 15 megaparsecs). Much work has been done to establish the graph of period versus absolute magnitude, mainly involving independent measurements of the distances or luminosities of Cepheids. Baade and Kukarkin were consequently able to demonstrate in the 1950s the existence of the two distinct categories of Cepheids, having separate but parallel period-luminosity relations.

A *period-colour relation* was also discovered whereby the *colour index of a Cepheid increases, i.e. becomes more positive, as the period increases: a relatively long period implies a redder and cooler star. It was thus also possible to derive a *period-spectrum relation* between *spectral type and period. Both Cepheid categories follow these relations.

Cepheus. A constellation in the northern hemisphere near Cassiopeia, lying partly in the Milky Way, the brightest star, Alderamin (α), being of 2nd magnitude. It contains several pulsating variable stars, including the prototype *Cepheid variable Delta Cephei, the red irregular variable Mu Cephei or *Garnet star, and the prototype of the *Beta Cephei stars. Abbrev.: Cep.; genitive form: Cephei; approx. position: RA 22h, dec +70°; area: 588 sq deg.

Cerenkov counter. An instrument for detecting and measuring the velocity of very energetic particles such as cosmic rays: light – *Cerenkov radiation* – is emitted by the particles as they pass through a transparent nonconducting (dielectric) material at a speed greater than the speed of light in that material. The radiation is emitted at a fixed angle, θ, to the direction of motion of the particle, where

$$\cos \theta = c/nv$$

v is the particle velocity, n the refractive index of the transparent medium, and c the speed of light in a vacuum. This cone of light is focused on to a *photomultiplier so that an amplified electric pulse is registered. The velocity of the particle is determined from the angle of the cone.

Ceres. The first *minor planet to be discovered, found by Piazzi in 1801 following the predicted existence by *Bode's law of a planet at about 2.8 AU from the sun. It is the largest minor planet (diameter 1003 km) and has a mass of 10^{21} kg; it may contain as much as half the total mass of the minor-planet system. It has a low *albedo of 0.05 and a spectrum similar to that of the *carbonaceous chondrite meteorites. Unlike the rather smaller minor planet *Vesta it can never be seen by the unaided eye. *See* Table 3, backmatter.

Cerro Las Campanas Observatory. *See* Las Campanas Observatory.

Cerro Tololo Interamerican Observatory. An observatory near La Serena, Chile, at an altitude of 2200 metres. It is run by a group of American universities – the Association of Universities for Research in Astronomy. Its principal instruments are a 4 metre (158 inch) reflecting telescope, which began operating in 1974, and a 1.5 metre (60 inch) reflector.

Cer-Vit. *Trademark* A glass-ceramic material that expands and contracts very little when its temperature changes and is thus used in telescope optics.

Cetus (Whale). An extensive equatorial constellation near Orion, the brightest star being the 2nd-magnitude orange giant Diphda (β). It contains several *Mira stars including the prototype *Mira Ceti, the *flare star UV Ceti, and the *Seyfert galaxy M77. Abbrev.: Cet; genitive form: Ceti; approx. position: RA 1.5h, dec −10°; area: 1231 sq deg.

chain craters. *See* crater chain.

Chameleon. A small inconspicuous constellation in the southern hemisphere near Crux, the brightest stars being of 4th magnitude. Abbrev.: Cha; genitive form: Chameleontis; approx. position: RA 11h, dec −80°; area: 132 sq deg.

Chandler wobble. A small continuous variation in the location of the geographic poles on the earth's surface. It leads to a *variation of latitude* of points on the earth since latitude is measured from the equator midway between the poles. The variation in polar location is resolved into two almost circular components, one (diameter: 6 metres; period: 12 months) resulting from seasonal changes in ice, snow, and atmospheric mass distribution; the second (diameter: 3–15 metres; period: about 14 months) is believed to arise from movements of material within the earth.

Chandrasekhar limit. The limiting mass for a nonrotating *white dwarf. It depends slightly on the star's composition, being 1.44 solar masses for a helium white dwarf, dropping to 1.40 solar masses for a carbon composition and 1.11 solar masses for an iron composition. The limit is raised substantially if the white dwarf has a rapidly rotating core. A star whose mass exceeds this limit will be forced to undergo further *gravitational collapse to become a *neutron star or even a *black hole, because its material will be unable to support itself against the force of gravity. *Compare* Schönberg–Chandrasekhar limit.

channels. Meandering valleys, some over 1000 km long, that look like dried watercourses on the Martian surface. They are not related to the mythical Martian canals. *See* Mars, surface features.

chaos. Broken terrain. The word is used in the approved name of such a surface feature on a planet or satellite.

charge. A property of certain *elementary particles that causes them to attract or repulse each other. Charged particles have associated electric and magnetic fields that allow them to interact with each other and with external electric and magnetic fields. Charge is conventionally 'negative' or 'positive': like charges repel, unlike charges attract. The *electron possesses the natural unit of negative charge, equal to 1.6022×10^{-19} coulombs. The *proton carries a positive charge of the same magnitude. If matter is charged, it is due to an excess or deficit of electrons with respect to protons.

charge-coupled device. *See* CCD.

Charon. The satellite of *Pluto.

chasma (plural: **chasmata**). A deep valley with steep sides. The word is used in the approved name of such a surface feature on a planet or satellite.

Chi Persei. *See* h and Chi Persei.

Chiron. *Minor planet number 2060 that was discovered in 1977 but was subsequently identified on photographs taken as early as 1895. It has a 50.68 year orbit that differs from orbits of all other known minor planets in lying almost entirely beyond that of Saturn. With an uncertain diameter of 300 to 400 km, it is possible that Chiron is one of the brighter members of a distant swarm of minor planets. *See* Table 3, backmatter.

chondrite. A type of *stony meteorite that contains *chondrules. They are the most abundant class of meteorite in the solar system (about 86%). *Compare* achondrite.

chondrules. Near-spherical bodies composed chiefly of silicates, with sizes between 0–2 mm and 4 mm, found embedded in *chondrites. They are usually aggregates of olivine, $(Mg,Fe)_2SiO_4$, and pyroxene $(Mg,Fe)SiO_3$. They may also be single crystals, wholly glass, or crystal and glass, in a wide range of proportions. They appear to have been free fluid drops made spherical by surface tension and then solidified and crystallized. There is a possibility that these silicate drops were produced by lightning discharges in the dusty primitive solar nebula or that they are crystallized droplets of impact melt produced when two minor planets collided. Chondrule-like bodies have been found in lunar soils.

christmas tree. *See* feeder.

chromatic aberration. An *aberration of a lens – but not a mirror – whereby light composed of different wavelengths, i.e. ordinary white light, is brought to a focus at different distances from the lens (see illustration). It arises from the variation with wavelength of the *refractive index of the lens material: red light is refracted (bent) less than blue light (*see* dispersion). False colours therefore arise in the image. Chromatic aberration can be reduced by using an *achromatic lens. Before the introduction of achromats, objective lenses of very long focal length were used in telescopes to reduce the aberration; this led to very cumbersome instruments.

chromosphere. The stratum of a star's atmosphere immediately above the *photosphere and below the *corona. The chromosphere is considerably less dense than the photosphere, and its gases are characterized by an emission rather than an absorption spectrum. The chromosphere of the sun is naturally the best studied.

In the solar chromosphere the temperature rises over a few thousand kilometres from 4000 kelvin at the temperature minimum to around 50 000 kelvin where it reaches the transition region (*see* sun). The rise in temperature (which continues in the transition region and inner corona) was thought to be the result of ascending shock waves, but this mechanism does not tally with detailed observations of the coronae of the sun and other stars. It is now proposed that magnetic activity is responsible (*see* corona).

The sun's chromosphere is visible under natural circumstances only when the photosphere is totally eclipsed by the moon (*see* eclipse). It is then seen in profile at the sun's limb. It may, however, be observed at times other than totality with the aid of a *sppresumably magnetic activity akin to the solar cycle. Chromospher-

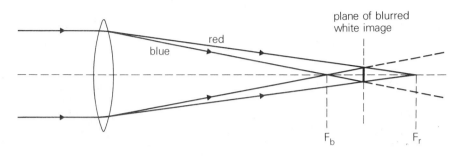

Chromatic aberration of convex lens

ic brightness is related to speed of rotation, being higher for stars that rotate rapidly (either because they are young or because of the effect of a companion – *see* RS Canum Venaticorum star).

chromospheric network. A large-scale cellular pattern in the solar *chromosphere. It is most clearly defined at the centre of the sun's disc in the light of the K *Fraunhofer line of singly ionized calcium, when it appears as a mosaic of cells outlined (at least partially) by bright mottles. Its appearance is similar though less pronounced in the core of the Hα line of neutral hydrogen, but in the wings (i.e. on either side of the core) the mottles appear dark instead of bright. Both bright and dark mottles have been tentatively identified with *spicules, visible beyond the sun's limb. They seem to differ only in their height above the *photosphere, bright mottles being situated in the lower chromosphere, dark mottles at heights of up to about 10 000 km.

The chromospheric network coincides with the underlying *supergranulation, the horizontal flow within each supergranular cell concentrating magnetic flux at its boundary; at the boundary an enhanced vertical magnetic field is produced, which is responsible for the localized excess

heating that manifests itself as the bright mottles.

Chryse Planitia. A relatively smooth Martian plain, 1600 km across and 2.5 km below the average planetary surface. It appears to have suffered water erosion in the past and was chosen as the site of the *Viking 1 landing in 1976. *See* Mars, surface features.

Cih. *See* Cassiopeia.

Circinus (Compasses). A small constellation in the southern hemisphere near Centaurus, lying in the Milky Way. The brightest star is of 3rd magnitude. Abbrev.: Cir; genitive form: Circini; approx. position: RA 15h, dec −60°; area: 93 sq deg.

circle of longitude. Any great circle that passes through the poles of the *ecliptic and is thus at right angles to the plane of the ecliptic. *Circles of latitude* lie parallel to the ecliptic plane.

circle of position. An imaginary small circle on the earth's surface on which an observer is located and that is centred on the *substellar point of a particular star. Its radius is equal to the *zenith distance of the star. The star will thus have the same zenith distance from every point on the circle of position.

circulator. A three- or four-port electronic device in which signals entering port number 1 appear only at port number 2, while signals entering port number 2 emerge only at port number 3, and so on. It is used, for example, in a *radio telescope employing a negative-resistance preamplifier, such as a *parametric amplifier or a *maser amplifier; the circulator isolates the preamplifier from the *antenna, which, if connected directly, might oscillate.

circumpolar stars. Stars that are permanently above the horizon from a given observational point on earth, i.e. they never set. For stars to be circumpolar from a given geographical latitude ϕ, their declination (angular distance above or below celestial equator) must be greater than $90° - \phi$. For an observer at the north pole all stars in the northern hemisphere of the celestial sphere will be circumpolar while at the equator there are no circumpolar stars although almost all stars will be seen at some time during the year.

cislunar. Of or in the region of space between the earth and the moon.

civil twilight. *See* twilight.

civil year. The length of the year as reckoned for ordinary purposes. Any civil year must contain an exact number of days, which in the *Gregorian calendar usually amounts to 365 but is equal to 366 in *leap years.

classical Cepheids. *See* Cepheid variables.

clock star. A bright star, usually situated near the celestial equator, whose position and proper motion are very accurately known so that it can be used for the exact determination of

time, for the determination of the error of observatory clocks, and for correction of the positional observations of other stars.

close binary. *See* binary star.

closed universe. *See* universe.

cluster. A group of stars whose members are sufficiently close to each other to be physically associated. Clusters range from dense congregations of many thousands of stars to loose groups of only a few stars. Current theories of stellar evolution suggest that stars form in *associations, whose densest groupings remain bound as clusters while the other stars disperse; this view is supported by the relatively well-defined *Hertzsprung-Russell diagrams of clusters, which indicate that cluster members are of essentially the same composition and age (*see also* turnoff point). Since the stars in a cluster are all at approximately the same distance, the observed differences in appearance are believed to be due mainly to differences in mass. An estimate of distance is obtained from the technique of *main-sequence fitting. *See* globular cluster; open cluster.

clusters of galaxies. Groups of galaxies that may contain up to a few thousand members. The majority of galaxies appear to occur in clusters or in smaller groups such as doubles or triples. Our own Galaxy is a member of a small irregular cluster – the *Local Group. The nearest large cluster is the *Virgo cluster. The densest clusters, which typically contain a thousand or more members, are apparently roughly spherical and consist almost entirely of elliptical and S0 galaxies. Irregularly shaped clusters may be large, as with the Virgo cluster, or small but tend to be less dense and to contain all types of galaxies. Adjacent clusters are loosely grouped into larger *superclusters.

Large clusters with an unusually high concentration of galaxies in the centre are called *rich clusters*. Several thousand are listed in the *Abell catalogue, examples being the *Coma cluster and *Perseus cluster. It has been shown that the mass required to keep the galaxies in rich clusters gravitationally bound is, on average, about 10 times greater than the mass observed in the main body of the galaxies (*see* missing mass; dark halo). X-ray observations, particularly from the *Einstein Observatory, show that clusters contain a large amount of hot (up to 10^8 K) gas; the mass of the gas is not, however, sufficient to explain the missing mass. In the irregularly shaped clusters the gas is associated with individual galaxies, but in the regular clusters it forms a large pool between the galaxies. This provides evidence that the regular clusters are more 'evolved': their galaxies have interacted so often that their gases are stripped off into a common pool. Another feature typical of most rich clusters is a centrally located supergiant elliptical galaxy that has a greatly extended halo of faint stars and is usually a strong radio source: these are termed *cD galaxies*. They may have grown to this size by galactic *cannibalism.

C-M diagram. *Abbrev. for* colour-magnitude diagram. *See* Hertzsprung-Russell diagram.

CNO cycle. *See* carbon cycle.

coalesced star. *See* common envelope star.

Coalsack (Southern Coalsack). A prominent *dark nebula about 170 parsecs away in the southern constellation Crux. It has an angular diameter of 4° and can be seen with the naked eye, projected against the Milky Way.

coaltitude. *See* zenith distance.

coating of lenses. The deposition of one or more thin uniform transparent coatings on lens surfaces, usually by vacuum evaporation or other electronic techniques; this reduces the amount of scattered light resulting from reflection at each refracting surface, say the interfaces between air and lens and between lens and air. A *bloomed lens* is coated with a single film of suitable refractive index and thickness: destructive *interference can then occur between light reflected from the coating-air surface and that reflected from the lens-coating surface. *Multilayer coating* depends on thin-film interference. Using up to 30 layers of controlled thickness, extremely low levels of reflected light can be achieved. Multilayer coating can also be used to produce very high levels of reflection from a *mirror.

coded mask telescope. A recently developed telescope system that enables high-quality sky images to be made using nonfocusing optics; it is a development of the simple pinhole camera. The technique is particularly applicable at gamma-ray wavelengths and greatly improves the accuracy with which cosmic γ-ray sources can be located within every point in the telescope's field of view, typically 10° square. The system consists of a crossword-like mask of γ-ray opaque lead or tungsten elements positioned over, but separated from, a flat γ-ray detector. The detector is capable of measuring the point at which a γ-ray photon from a cosmic source makes contact, as well as its energy. A patterned shadow of the mask will be cast on the detector plane by photons not reaching the detector. The location of the γ-ray source can be determined uniquely from the position of the shadow and the known mask-detector separation.

coelostat. A flat mirror that can be driven by a clock mechanism so as to rotate from east to west about an axis parallel to the earth's rotational axis, thus compensating for the apparent west to east rotation of the celestial sphere. The mirror may thus continuously reflect light from the same area of the sky into the field of view of an instrument that is fixed in position, usually by means of an additional optical system. *See also* siderostat; heliostat; solar telescope.

coherence. The degree to which an oscillating quantity maintains a constant *phase and *amplitude relationship at points displaced in space or time. Hence the *coherence length* of a train of waves describes the distance over which the oscillations are appreciably correlated while the *coherence time* defines the time for which the character of the wavetrain remains more or less unchanged. *See also* autocorrelation function.

coherence bandwidth. *See* bandwidth.

colatitude. The complement of the celestial latitude (β) of a celestial body, i.e. the angle $(90° - \beta)$. *See* ecliptic coordinate system.

collimation. The alignment of the optical elements in a telescope. For a simple refractor the only adjustment required is to position the object lens at right angles to the optical axis by adjusting the screws on its supporting cell. In a simple Newtonian reflector the orientation of both the primary and diagonal mirrors must be adjusted, and the latter positioned correctly opposite the *draw tube. Final small adjustments are then made to obtain bright star images as free as possible from *coma.

collimator. A device used to produce a parallel or near parallel beam of light or other radiation in an instrument. One example, used in spectroscopes, is a converging lens or mirror at whose focal point is a narrow slit

upon which light is focused from behind.

colour. 1. The colour of a star is measured and quoted in terms of its *colour index but basically depends on surface temperature: the star will radiate predominantly blue, white, yellow, orange, or red light in descending order of temperature (*see* spectral types). Only the brightest stars have recognizable colours. Two stars of the same colour may have very different *luminosities.
2. *Short for* colour index.

colour coding. *See* false-colour imagery.

colour difference photography. A method for enhancing subtle colour differences on the moon by combining photographs exposed through infrared and ultraviolet filters.

colour excess. *See* colour index.

colour index. The difference between the apparent *magnitude of a star measured at one standard wavelength and the apparent magnitude at another (always longer) standard wavelength. Its value depends on the spectral distribution of the starlight, i.e. whether it is predominantly blue, red, etc., and it is therefore an indication of the colour (i.e. temperature) of the star. It is independent of distance. Prior to the UBV system the *international colour index* was mainly used; this is the difference between photographic and photovisual *magnitudes $(m_{pg} - m_{pv})$. In the now widely used UBV system (*see* magnitude) colour index is usually expressed as the difference $B-V$, where B and V are the magnitudes measured with blue starlight (at a wavelength of 440 nanometres) and greenish-yellow starlight (550 nm), respectively. The colour index $U-B$ is also used, where U is the apparent magnitude measured with ultraviolet radiation (365 nm) from a star.

Stars are classified into *spectral types, which are further subdivided into luminosity classes; each has a characteristic *intrinsic colour index*, given as $(B-V)_0$ and $(U-B)_0$. These two indices are defined as zero for A0 main-sequence stars and are therefore negative for hotter stars, i.e. those emitting more ultraviolet (O and B stars), and positive for cooler ones (A1 to M stars). Since colour index is easily measured, it is usually used on graphs in preference to spectral type or temperature. Any excess of the measured value of colour index of a star over the expected intrinsic value indicates that the starlight has become reddened by passage through interstellar dust (*see* extinction). The difference between the values is the *colour excess*, E, of the star:

$$E = (B-V) - (B-V)_0$$

The value of E gives the amount of reddening.

There are also colour indices relating to *magnitudes measured at red and infrared wavelengths. For example, in the indices $V-R$ and $V-I$, I and R are the magnitudes measured at 0.7 μm and 0.9 μm.

colour-luminosity diagram. *See* Hertzsprung-Russell diagram.

colour-magnitude diagram (C-M diagram). *See* Hertzsprung-Russell diagram.

colour temperature. Symbol: T_c. The surface temperature of a star expressed as the temperature of a *black body (i.e. a perfect radiator) whose energy distribution over a range of wavelengths corresponds to that of the star. It can thus be found by matching the energy distribution in the star's continuous spectrum to that of a black body (given by Planck's radiation law). With increasing temperature the star emits a higher proportion of blue and ultra-

violet radiation and the position (wavelength) of maximum radiated energy on the energy distribution curve shifts accordingly; the basic shape of the curve remains unchanged (see illustration). Colour temperature is related to *colour index $(B-V)$ by an approximation of Planck's radiation law:

$$T_c = 7300/[(B-V) + 0.6] \text{ kelvin}$$

As a star's spectrum is not precisely that of a black body, the colour temperature and *effective temperature are not equal: the former is always higher, the difference being greatest for hot (O and B) stars. Although the colour temperature is not as closely related to the star's surface temperature as is the effective temperature, it has the advantage of being found very precisely by measurements of the star's colour index. The sun's colour temperature is 6500 kelvin.

Columba (Dove). A constellation in the southern hemisphere near Canis Major, the two brightest stars being of 3rd magnitude. Abbrev.: Col; genitive form: Columbae; approx. position: RA 6h, dec −35°; area: 270 sq deg.

colures. The two great circles passing through the celestial poles and intersecting the ecliptic at either the *equinoxes (*equinoctial colure*) or the *solstices (*solsticial colure*).

coma. 1. An *aberration of a lens or mirror that occurs in a telescope when the optical elements are misaligned so that light falls on the objective or primary mirror at an oblique angle. The light is not imaged as a point in the focal plane but as a fan-shaped area: each zone of the lens or mirror produces an off-axis image in the form of a circular patch of light, the diameter and position of the circle centre varying steadily from zone to zone (see illustration). The dimensions of the resulting combination of zone images, i.e. the fan-

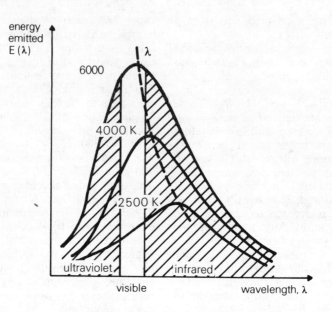

Planck radiation curves for various colour temperatures

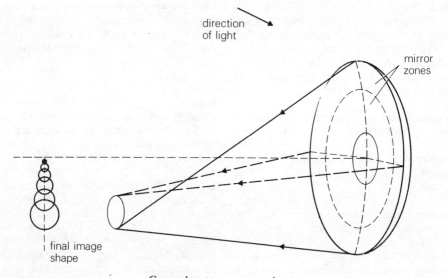

Coma due to concave mirror

shaped image, depends on the obliquity of the light falling on the lens or mirror.

2. The principal part of most *comets, consisting of a diffuse luminous nebulous cloud of gas and dust that surrounds the *nucleus. The visual boundary of the coma merges with the sky background. The size varies with the comet's distance from the sun and with comet age but can grow to typically 10^4 to 10^5 km in diameter.

The gas and dust originate in the nucleus; the gas density at the surface of the nucleus is typically 10^{13} molecules per cm^3. The luminosity is mainly produced by *fluorescence from a variety of carbon, nitrogen, hydrogen, and oxygen radicals and reflection of sunlight by dust particles. The shape seems roughly circular but fan-shaped comas have been reported. It is transparent: even faint stars shine through. *See also* head.

Coma Berenices (Berenice's Hair). An inconspicuous constellation in the northern hemisphere near Bootes, the brightest stars being of 4th magnitude. The north galactic pole lies in this constellation, which contains a huge cluster of faint galaxies, the *Coma cluster. Abbrev.: Com; genitive form: Comae Berenices; approx. position: RA 12.5h, dec $+20°$; area: 386 sq deg.

Coma cluster. A spherical rich *cluster of galaxies, near the north galactic pole, in the constellation Coma Berenices that contains at least 1000 bright elliptical and S0 galaxies. It is about 6 megaparsecs in diameter. The core is dominated by two large galaxies that are both thought to be radio sources: an elliptical, NGC 4889, and an S0 galaxy, NGC 4874. The cluster is about 90 megaparsecs distant and is moving away from us at approximately 6700 km s^{-1}.

Comas Solá. A *comet that had its orbit severely disturbed when it approached within 0.19 AU of Jupiter in 1912. The major changes were in period (from 9.43 years to 8.53 years), *perihelion distance (from 2.15 AU to 1.77 AU), and *inclination (from 18.1° to 13.7°).

combined magnitude. The apparent brightness of two or more stars so close that they are observed as a single star. It is not equal to the sum of the individual magnitudes but is a logarithmic function of the sum of the individual brightnesses (*see* magnitude).

comet. A minor member of the solar system that travels around the sun in an orbit that generally is much more eccentric than the orbits of planets. Typical comets have three parts: the *nucleus, *coma, and *comet tail. The gas and dust tails of a comet only appear when the comet is near the sun and always point away from the sun. The nucleus is the permanent solid portion of a comet and is thought by the majority of astrophysicists to be a kilometric-sized *dirty snowball*; this model was first suggested by Fred L. Whipple. The solar system contains only a few observable comets that have nuclei in the tens of kilometres size range. The numbers of actual (as opposed to observed) comets increase enormously as one goes to smaller and smaller sizes, but these get much less easy to see. The satellite missions to *Halley's comet and *Giacobini–Zinner should reveal much about cometary composition.

Cometary orbits fall into two classes. *Short-period comets* have periods of less than 150 years and orbits lying completely or almost completely inside the planetary system. They seem to have been captured into the inner solar system by the gravitational attraction of the major planets. At any epoch about 50 of these are bright enough to be detected as they come to *perihelion. They have a mean *eccentricity of 0.56 and a mean *inclination of 11°. Nearly all of them (Halley's comet is an exception) are moving in direct orbits, i.e. in the same direction as the planets. The second class have near-parabolic orbits with periods in excess of 150 years. These orbits are orientated at random, indicating that these comets do not come from any specific direction in space. Many of them have been seen only once (for example

Kohoutek has a period in excess of 70 000 years).

The majority of comets observed from earth have perihelions near the earth. This is simply due to observational selection. The brightness of a comet is proportional to $1/r^n\Delta^2$, where r is the distance between the comet and the sun and Δ is the comet-earth distance. The power n is on average 4.2 but can vary widely around this value. On average a comet passes perihelion about 1000 times before decaying away. Cometary decay produces *meteor streams.

Comets are named after their discoverers. In any year, say 1985, comets are designated 1985a, 1985b, etc., in order of discovery. Permanent designations 1985 I, 1985 II, etc., are given later, these being in order of date of perihelion passage.

There are two prevalent theories of cometary origin. In the first theory comets were produced out of icy *planetesimals, at the same time as the origin of the solar system, and were stored in the *Oort cloud. In the second they are produced by gravitational accretion every time the sun passes through an interstellar dust cloud. This latter theory predicts that the solar system's cometary reservoir is periodically being topped up. The first theory predicts a cometary population that decreases with time.

cometary nebula. A fan-shaped *reflection nebula whose illuminating star lies at the vertex of the fan. The *Hubble nebula is an example.

comet family. A distinct group of ·comets, the members of which have *aphelion distances that coincide approximately with points on the orbit of a particular major planet: this grouping is shown on a plot of comet numbers as a function of aphelion distance, in which the distribution peaks at the orbits of the planets Jupiter, Saturn, Uranus, and Neptune. The planet has captured the comet

from a long-period orbit. Two frequency concentrations at 53 AU and more strongly at 83 AU have been found, indicating possible transplutonian planets. *See* Jupiter's comet family.

comet group. A group of comets, occurring in the solar system, which, apart from the time of perihelion passage, have very nearly the same orbital elements. 1668, 1843 I, 1880 I, 1882 II, 1887 I and 1948 VII is a typical group.

comet tails. Tails of gas and dust that point in the antisolar direction away from the cometary *nucleus and only appear when a *comet is near the sun, i.e. closer than about 2 AU (see illustration). Not all comets have tails. The luminosity is due to both molecular and atomic emission and to reflection of sunlight. The gas and dust is forced away by *radiation pressure and by *solar-wind interactions. The *ion tail* is usually long and narrow, showing many diverse visible and rapidly changing structures. It consists exclusively of ionized molecules, moving at velocities between 10 and 100 km s^{-1}, the repulsive force being 20 to 100 times greater than gravity. The tail can be a million km wide and lengths of 10 million km are not uncommon. *Dust tails* are more strongly curved than ion tails. They contain solid particles, which simply reflect sunlight. Particles are ejected from the *nucleus along curved paths known as *syndynames*. The tail is an envelope over the reflection regions of particles of similar mass. The closer the dust tail is to the ion tail in curvature the smaller are the responsible particles.

command and service module (CSM). *See* Apollo.

command module (CM). *See* Apollo.

common envelope star. A hypothetical type of *giant star that consists of

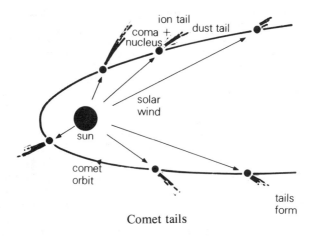

Comet tails

two stellar cores orbiting within a large common envelope of gas. Since the two cores are hidden within the envelope the star should superficially resemble a normal giant, and as a result no common envelope stars have yet been unambiguously identified. Their existence is strongly attested, however, by evidence for systems just before and just after this stage of evolution. Contact binaries of the *W Ursae Majoris type should become common envelope stars when one member expands as a red giant. Again, a close binary system where one member is a white dwarf must have had a common envelope at a previous evolutionary stage when the star now seen as a white dwarf went through its red giant phase: these systems are observed either as *cataclysmic variables or in the centres of some *planetary nebulae; the nebula's gas is probably the original common envelope, ejected as a result of the orbital motion of the two star cores within it. In some common envelope stars, the two cores may not eject the envelope gas but instead spiral together and eventually fuse into a single core. Angular momentum from the orbital motion would spin up the distended star; some rapidly spinning giant stars, like FK Comae, may be examples of such *coalesced stars*.

common proper motion stars. *See* visual binary.

compact galaxy. A member of a class of *peculiar galaxies, first catalogued by Fritz Zwicky in the 1960s. Their brightness is so concentrated that they are almost indistinguishable from stars on normal photographs. Detailed examination has shown that they are either *active galaxies, *starburst galaxies, or *extragalactic HII regions, and the term compact galaxy is now rarely used.

comparator. An instrument by means of which two photographs of the same area of the sky may be compared and any difference in the brightness or position of an object quickly detected without resorting to exact measurements. The movement of a comet, the proper motion of a star, or the presence of a variable star may thus be revealed. In the *blink comparator* the two photographs are viewed in rapid succession through an eyepiece: changes in position produce an apparent movement of an image while variable stars are detected as a pulsation in brightness. In the *stereo-comparator* the two photographs are viewed simultaneously by binocular vision: objects that have changed in position or brightness (i.e. size) appear to stand out of the plane of the picture.

comparison spectrum. The spectrum of one or more known substances under terrestrial conditions that is used in astronomy as a standard of comparison for investigating spectra of celestial objects; it provides, for example, a zero doppler shift.

components. Stars, or possibly planets, that are constituent members of a *binary star, *triple star, or *multiple-star system.

Compton scattering (Compton effect). An interaction between a *photon of electromagnetic radiation and a charged particle, such as an electron, in which some of the photon's energy is given to the particle. The photon is therefore reradiated at a lower frequency (i.e. with a lower energy) and the particle's energy is increased. In *inverse Compton emission* the reverse process takes place: photons of low frequency are scattered by moving charged particles and reradiated at a higher frequency.

Compton telescope. A gamma-ray detector design based on the detection of a *Compton scattering interaction of a gamma ray in a first detector, usually a plastic scintillator, followed by the detection of the Compton scattered photon in a second detector. This method allows the simultaneous determination of photon direction and energy to be made.

concave. Curving inwards. A concave mirror converges light. A biconcave or planoconcave lens, thinner in the middle than at the edges, has a diverging action.

concentration. The diameter of the telescopic image of a star, etc. The image is increased from point size by unavoidable optical (diffraction) effects in the telescope (*see* Airy disc),

and for ground-based telescopes is further increased by distortion due to atmospheric conditions at the time of observation, i.e. by the *seeing.

configuration. *See* aspect.

confocal lenses. *See* afocal system.

confusion. The running together of the traces from different *radio sources in the output of a *radio telescope. Confusion becomes important at the value of *flux density when there is, on average, more than about one source in the *beam at once. The radio telescope becomes *confusion limited* when this flux density is appreciably higher than its *sensitivity. Further increase in sensitivity will not then result in fainter sources being detected.

conic sections (conics). A family of curves that are the locus of a point that moves so that its distance from a fixed point (the focus) is a constant fraction of its distance from a fixed line (the directrix). The fraction, e, is the eccentricity of the conic. The value of the eccentricity determines the form of the conic. If e is less than 1 the conic is an ellipse. A circle is a special case of this with $e = 0$. If $e = 1$ the conic is a parabola and if e exceeds 1 the conic is a hyperbola.

These curves are known as conic sections because they can be obtained by taking sections of a right circular cone at different angles: a horizontal section gives a circle, an inclined one an ellipse, one parallel to the slope of the cone is a parabola, and one with an even greater inclination is a hyperbola (see illustration). Conics are important in astronomy since they represent the paths of bodies that move in a gravitational field. *See also* orbit.

conjunction. The alignment of two bodies in the solar system so that they have the same longitude as seen from the earth (see illustration at

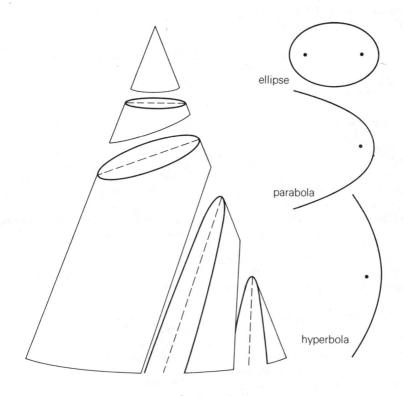

ellipse

parabola

hyperbola

Conic sections

elongation). It is also the time at which such an alignment occurs. Often the term refers to a conjunction between a planet and the sun. An *inferior planet is in *superior conjunction* when in line with the sun and the earth and on the far side of its orbit with respect to the latter; when it lies between the sun and the earth it is said to be in *inferior conjunction*. Conjunction may also occur between two planets, or between the moon and a planet or the moon and the sun – the latter occurring at every new moon.

conservation of mass-energy. The principle that in any system of interacting bodies the total mass plus total energy is constant, where energy and mass are to be considered interconvertible according to the equation $E = mc^2$. The principle is a consequence of the theory of special *relativity and is a generalization of the classical laws of *conservation of energy* and *conservation of mass*. These state that in a closed system energy (or matter) cannot appear or disappear spontaneously and cannot be lost or gained. Energy can only be transformed into different forms but the total energy must remain constant. Similarly the total mass in the system must be constant.

conservation of momentum. The principle that in any system of interacting bodies the total linear *momentum in a fixed direction is constant provided that there is no external force acting on the system. The *angular momentum of a system of rotating and/or revolving bodies is also conserved provided that no external torque is applied.

constellation. Any of the 88 areas into which the whole of the northern and southern hemisphere of the sky (or *celestial sphere) is now divided. Every star, galaxy, or other celestial body lies within, or sometimes overlaps, the boundaries of one of the constellations. These boundaries were established unambiguously by the International Astronomical Union in 1930 along arcs of *right ascension and *declination for the epoch 1875, Jan. 1.

Originally the constellations had no fixed limits but were groups of stars considered by early civilizations to lie within the imagined outlines of mythological heros, creatures, and other forms. In the *Almagest, Ptolemy, in about AD 140, listed 48 constellations that were visible from the Mediterranean region. During the 16th, 17th, and 18th centuries many new groupings were formed (and several later discarded), especially in the previously uncharted regions of the southern hemisphere. Since then further additions have proved unacceptable.

The constellations seen on any clear night depend on the latitude of the observer, and change with the time of year and time of night. If, for a particular latitude, ϕ, the stars are *circumpolar (never set), they will be visible on every night of the year. Other stars never rise above the horizon: these have *declinations $\phi - 90°$ or $\phi + 90°$ for the northern and southern hemispheres, respectively. The remaining stars can be seen only when they are above the horizon during the night, a star rising earlier, on average, by about two hours per month.

Each constellation bears a Latin name, as with Ursa Major, the genitive form of which is used with the appropriate letter or number for the star name (*see* stellar nomenclature), as with Alpha Ursae Majoris. For simplification the three-letter abbreviations for the constellations are more

usually used, as with UMa. *See* Table 5, backmatter.

contact. *See* eclipse.

contact binary. *See* binary star; equipotential surfaces.

continental drift. *See* earth.

continuous spectrum. A *spectrum consisting of a continuous region of emitted or absorbed radiation in which no discrete lines are resolvable. The emission from hot fairly dense matter will produce a continuous spectrum, as with the continuous emission of ultraviolet, visible, and infrared radiation from the sun's *photosphere. *Synchrotron emission is another example of continuous emission.

continuum. The *continuous spectrum that would be measured for a body if no absorption or emission lines were present.

contour map. *See* radio source structure.

convection. A process of heat transfer in which energy is transported from one region of a fluid to another by the flow of hot matter in bulk into cooler regions. *See also* energy transport; convective zone.

convective zone. A zone in a star where *convection is the main mode of *energy transport. This occurs throughout a protostar on the *Hayashi track, and beneath the surface of a *main-sequence star, where the temperature is sufficiently low to enable the nuclei of hydrogen and heavier elements to recombine with free electrons to form atoms and negative ions. The presence of these atoms and ions and their ability to absorb photons increases the *opacity of the medium to the passage of radiation from below and results in a steeper temperature gradient, thereby triggering turbulent convection. Stars rather more massive than the sun, fusing hydrogen by the *carbon cycle, have a convective zone at the core, which mixes the nuclear fusion region.

The surface convective zone of a late-type star like the sun is almost certainly involved in the production of its magnetic field, and hence in the cycle of *solar activity and in the heating of the *chromosphere and *corona, but the details are still uncertain. *See also* stellar structure; granulation.

convex. Bulging outwards. A convex mirror diverges light. A biconvex or planoconvex lens, fatter in the centre than at the edges, has a converging action.

convolution. A mathematical operation that is performed on two functions and expresses how the shape of one is 'smeared' by the other. Mathematically, the convolution of the functions f (x) and $g(x)$ is given by
$$\int_{-\infty}^{\infty} f(u)g(x-u)du$$
It finds wide application in physics; it describes, for example, how the transfer function of an instrument affects the response to an input signal. *See also* radio-source structure.

coordinated universal time (UTC). *See* universal time.

coordinate system. A system by which the direction of a celestial body or a point in the sky can be defined and determined by two spherical coordinates, referred to a fundamental great circle lying on the *celestial sphere and a point on the fundamental circle (see illustration). One coordinate (a) is the angular distance of the celestial body measured perpendicular to the fundamental circle along an auxiliary great circle passing through the body and the poles of the fundamental circle. The other coordinate (b) is the angular distance measured along the

fundamental circle from a selected zero point to the intersection of the auxiliary circle.

There are four main coordinate systems: the *equatorial, *horizontal, *ecliptic, and *galactic coordinate systems (see table). They are all centred on the earth. Transformations can be made from one system to another by means of the relationships between the angles and sides of the relevant *spherical triangles. The *astronomical triangle, for example, relates equatorial and horizontal coordinates; the triangle formed by the celestial body and the poles of the equator and ecliptic relates equatorial and ecliptic coordinates. *See also* heliocentric coordinate system.

Copernican system. A *heliocentric system of the solar system that was proposed by Nicolaus Copernicus and eventually published in 1543 in his book *De Revolutionibus*. It uses some of the basic ideas of the *Ptolemaic system, including circular orbits and epicycles, and was no more accurate in its predictions. Copernicus, however, maintained that the planets move around the sun (in the relative positions accepted today), the sun's position being offset from the centre of the orbits. The apparent motions of celestial bodies such as the sun were explained in terms of the rotation of the earth about its axis and also the earth's orbital motion.

The planetary motion can be represented by two uniform circular motions: one is an epicyclic motion of the planet about a point D on the circular orbit; the other, unlike that of the Ptolemaic system, is a uniform circular motion of D about the centre, C, of the orbit. This requires that the rate of motion of D about C is exactly half that of the epicyclic rate of motion with respect to a fixed direction.

There was a strong and prolonged reaction – especially by the Church – to the Copernican system, which ef-

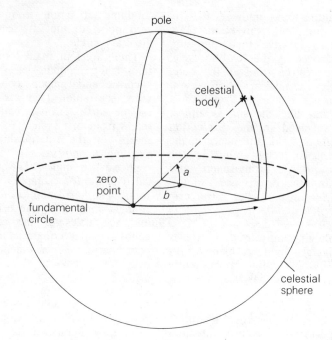

Coordinate system

coordinate system	fundamental circle	zero point	coordinates
equatorial	celestial equator	vernal equinox	right ascension declination
ecliptic	ecliptic	vernal equinox	celestial longitude celestial latitude
horizontal	horizon	north point	azimuth altitude
galactic	galactic equator	galactic centre	galactic longitude galactic latitude

fectively displaced the earth as the centre of the universe. There was also a sudden revival in astronomical observation in order to test the theory, notably by Tycho Brahe. Tycho's detailed observations, which showed the inadequacies of the Copernican system, were used in the formulation of *Kepler's laws of planetary motion.

Copernican System. *See* Copernicus.

Copernicus. 1. A young lunar crater to the south of Mare Imbrium that is 90 km in diameter and has an extensive system of *rays and *secondary craters. Possible ray material was collected by the crew of Apollo 12 and dated at 850 million years; this provides a calibration point for the *Copernican System*, which is the most recent era in lunar history.

2. NASA's *Orbiting Astronomical Satellite OAO-3, launched in Aug.

1972 and operated until Dec. 1980. It carried a 0.9 metre telescope and a grating spectrometer used primarily for measurements of ultraviolet radiation, but also carried x-ray equipment. *See* ultraviolet astronomy; x-ray satellites.

Coprates canyon. *See* Valles Marineris.

Cor Caroli (α CVn). A bluish-white star that is the brightest one in the constellation Canes Venatici and was named in honour of Charles II. It is a visual binary (separation 20″), the brighter component (A or α^2 CVn) being the prototype of the *spectrum variable stars. It has a period of 5.469 days and its spectrum shows strong profuse lines of rare earths, iron-peak elements, and silicon. m_v: 2.89 (A), 5.60 (B); spectral type: B9.5 pv (A), F0 V (B).

Córdoba Durchmusterung. *See* Bonner Durchmusterung.

core. 1. The central region of a star, such as the sun, in which energy is generated by thermonuclear reactions. **2.** The central region of a differentiated planet or satellite, such as the *earth or *moon.

corona. 1. The outermost part of a star's atmosphere. It is large, tenuous, and has a temperature of over a million kelvin. The sun's corona is the most easily observed.

Optically, the solar corona has two main components: the *K-corona (or inner corona) consists of rapidly moving free electrons, exhibits a *continuous spectrum, and attains a temperature of around 2 000 000 K at a height of about 75 000 km; the *F-corona (or outer corona) consists of relatively slow-moving particles of interplanetary dust, exhibits an *absorption spectrum, and extends for many million kilometres into the *interplanetary medium. A third component,

the *E-corona, consists of relatively slow-moving ions and exhibits an *emission spectrum superimposed on the *continuum of the K-corona.

The *white-light corona* comprises the overlapping K-corona and F-corona. It is visible under natural circumstances only in profile beyond the sun's limb on the rare occasions when the *photosphere is totally eclipsed by the moon. It may however be observed out to distances of several solar radii at times other than totality, with the aid of a balloon- or satellite-borne externally occulted *coronagraph. Similarly the E-corona may be observed at specific wavelengths with the aid of the Lyot coronagraph, used at certain high-altitude observatories.

The corona may also be observed against the sun's disc at extreme-ultraviolet and x-ray wavelengths, using rocket- or satellite-borne instrumentation. X-ray observations have revealed the structure of the solar corona: x-ray telescopes on *Skylab and the *Solar Maximum Mission have played a major role in this. Strong x-ray emission is associated with *active regions, and an absence of x-ray emission with *coronal holes. There is little or no evidence for a uniform corona – its structure is determined by the strength and configuration of the localized magnetic fields.

The overall shape of the solar corona changes with the phase of the *sunspot cycle. At sunspot minimum it is roughly symmetrical, with long equatorial *streamers* and *plumes* orientated in the direction of the sun's polar magnetic field. At sunspot maximum it is less symmetrical, although more evenly distributed about the sun's disc as a whole. Its changing shape is due principally to the presence of individual streamers above active regions, the mean heliographic latitude of which progresses towards the equator as the sunspot cycle proceeds. Solar radio emission at metre wavelengths originates in the corona and may exhibit an intense burst or series of bursts at the time of a large *flare.

It has been found that in general the coronae of normal stars are sources of x-rays: the x-ray telescope on the *Einstein Observatory was able to detect all types of stars apart from red supergiants. The idea that coronae are heated by shock waves rising through the chromosphere from the photosphere (acoustic heating) is now considered untenable since such heating would only occur in late-type stars like the sun. Instead, the coronal output is thought to be linked to the star's rotation. In certain binary systems (*RS CVn stars and some *flare stars) the orbital motion forces a star to spin rapidly, and these stars are unusually strong x-ray sources. Young stars, like those in Orion or the Pleiades, rotate faster than older stars of the same spectral type; they too have powerful x-ray emitting coronae. The link is almost certainly that fast rotation causes a stronger magnetic field, and this provides more heat input to the corona; the details are still obscure.

2. The *dark halo of our Galaxy.

Corona Australids. A minor *meteor shower that is observable in the southern hemisphere, radiant: RA 245°, dec −48°, and that maximizes on March 16.

Corona Australis (Southern Crown). A small inconspicuous constellation in the southern hemisphere near Scorpius, lying partly in the Milky Way. The brightest stars are of 4th magnitude. It contains the just-visible globular cluster NGC 6541. Abbrev.: CrA; genitive form: Coronae Australiş; approx. position: RA 19h, dec −40°; area: 128 sq deg.

Corona Borealis (Northern Crown). A small constellation in the northern hemisphere near Ursa Major in which there is a semicircle of stars, the brightest being the 2nd-magnitude

*Alphecca. It also contains the prototype of the variable *R Coronae Borealis stars and the *recurrent nova T Coronae Borealis. Abbrev.: CrB; genitive form: Coronae Borealis; approx. position: RA 15.5h, dec +30°; area: 179 sq deg.

coronagraph. An optical instrument designed in 1930 by the French astronomer Bernard Lyot for observing and photographing the sun's *corona, *prominences, etc., at times other than at solar eclipses. The image of the bright solar surface is artificially eclipsed by means of a blackened occulting disc set at the centre of the focal plane of the imaging lens of a telescope. A lens is used since scattering of light from its surface is much less than with a mirror. Strong sunlight scattered from the elements of the instrument must be eliminated so that only the faint coronal light is focused on the photographic plate or other detector. To reduce atmospheric scattering, high-altitude sites are used. Narrow-band optical filters are usually placed in front of the detectors so that the emission lines of the *E-corona can be studied.

coronal holes. Regions of exceptionally low density and temperature compared with the surrounding solar *corona. They overlie areas of the sun's disc that are characterized by relatively weak divergent (and therefore primarily unipolar) magnetic fields. They are thought to be the primary source of the long-lived high-speed streams in the *solar wind.

coronal lines. Bright emission lines observed in the spectrum of the sun's *E-corona.

coronal rain. *See* prominences.

correcting plate (correcting lens; corrector). The lens placed at the front of a *catadioptric telescope to correct the aberration of its spherical primary mirror. In this way a very wide field can be obtained, sensibly free from aberrations, even at *focal ratios well below one. See also Schmidt and Maksutov telescopes.

correlation receiver. A radio-astronomy *receiver in which either the *noise power from a single *antenna is split and multiplied with itself or the noise power from the two arms of an interferometer (see radio telescope) are multiplied together. Signals that are uncorrelated (completely unrelated) give an output whose average value is zero when multiplied together. Correlated signals, however, produce components near zero frequency that give a steady deflection in the output of the multiplier. Correlation receivers are used to reduce the effect of man-made *interference and of changes in the *gain and other parameters of the system. See also phase-switching interferometer.

Corvus (Crow). A small fairly conspicuous constellation in the southern hemisphere near the star Spica, the four brightest stars being of 2nd and 3rd magnitude. It contains the associated radio sources NGC 4038 and 4039. Abbrev.: CrV; genitive form: Corvi; approx. position: RA 12.5h, dec −20°; area: 184 sq deg.

COS-B. A satellite of the European Space Agency that was launched in Aug. 1975 to investigate celestial objects emitting gamma rays and to measure time variations of such emission. It operated over the energy range 70 to 5000 MeV and performed so efficiently that the mission was twice extended, operations ceasing in 1982. See also gamma-ray astronomy.

cosine rule. See spherical triangle.

cosmic abundance. The relative proportion of each *element found in the universe. The standard cosmic abundance is based on that of the solar

Cosmic abundance of elements,
using solar system as standard

element	atomic number	approx. composition by mass
hydrogen (H)	1	73%
helium (He)	2	25%
oxygen (O)	8	0.8%
carbon (C)	6	0.3%
neon (Ne)	10	0.1%
nitrogen (N)	7	0.1%
silicon (Si)	14	0.07%
magnesium (Mg)	12	0.05%
sulphur (S)	16	0.04%

system, determined from observations of the relative line strengths in the spectrum of solar radiation, from geological surveys of the earth, and from analysis of meteorites (see table). The solar abundance in terms of numbers of atoms gives 94% hydrogen, 5.9% helium, about 0.06% oxygen, 0.04% carbon, and trace numbers of other atoms. Abundances deduced for most stars from their spectra usually agree quite well with the standard, although old stars tend to have rather less of the heavier elements (see population I, population II) and there are a number of odd stars with abundances quite unlike the rest. The abundances deduced from the absorption and emission lines of the *interstellar medium do not agree so well, principally in that there is an apparent shortage of refractory elements such as iron. It is likely that these elements are present, however, but are bound up with the *cosmic dust.

cosmic background radiation. Unresolved radiation from space. The cumulative effect of many unresolved – and individually weak – discrete sources provides the background at radio, x-ray, and possibly other wavelengths. One important radiation is the *microwave background radiation, which peaks at about 1 mm wavelength (i.e. a frequency of 3×10^{10} hertz). This is considered to be due to the hot *big bang. *See also* x-ray astronomy; gamma-ray astronomy; infrared background radiation; radio source.

cosmic censorship, principle of. *See* singularity.

cosmic dust. Small particles or grains of matter found in many regions of space. Their size ranges from about 10 μm to less than 0.01 μm. They are thought to be composed of carbon, silicate, or iron material, which may in some cases have icy mantles. Within the solar system they are associated with the *zodiacal light. In the *interstellar medium they are found in *molecular clouds and *dark nebulae, causing *interstellar extinction. The grains are also found in circumstellar shells causing the *infrared excess seen in the spectrum of many stars.

cosmic noise (Jansky noise). *See* antenna temperature.

cosmic rays. Highly energetic particles that move through space at close to the speed of light and that continuously bombard the earth's atmosphere from all directions. They were discovered by V.F. Hess during a balloon flight in 1912. The energies of the individual particles are immense, ranging from about 10^8 to over 10^{20} electronvolts (eV). The intensity (or particle flux) of cosmic rays is very low; it is greatest at low energies – several thousand particles per square metre per second – dropping with increasing energy but apparently levelling off at high energies. Cosmic rays are primarily nuclei of the most abundant elements (atomic weights up to 56) although all known nuclei are represented. *Protons (hydrogen nuclei) form the highest proportion; heavy nuclei are very rare. Also present are a small number of *electrons, *posi-

trons, antiprotons, *neutrinos, and *gamma-ray photons. All these particles are known collectively as *primary cosmic rays.

On entering the earth's atmosphere the great majority collide violently with atomic nuclei in the atmosphere producing *secondary cosmic rays*; these consist mainly of *elementary particles. The large number of particles produced from one such collision form an *extensive air shower* (or simply an *air shower*), in which there is a very rapid, highly complex, but well-defined sequence of reactions. The initial products are principally charged and neutral *pions. The neutral pions decay (disintegrate) almost immediately into gamma rays, which give rise to *cascades* of electrons and positrons by *pair production and to more photons by *bremsstrahlung radiation from the decelerating electrons and positrons. The charged pions decay into *muons, which subsequently decay into neutrinos and electrons. The original primary cosmic ray nucleus, and many of the secondary cosmic rays, undergo more collisions to create further generations of particles. Following the initial collision the number of particles in the shower increases to a maximum at some point high in the atmosphere, where the creation of new particles is balanced by absorption; the number then decreases. Very few primary cosmic rays reach the earth's surface, the final products including electrons, muons, neutrinos, gamma rays, and some initial products of the primary collision. The maximum number of particles in an air shower depends on the energy of the primary cosmic ray: half a million particles can be produced at maximum development by a primary nucleus of 10^{15} eV. It is however a major problem to distinguish air showers produced by different primary nuclei of the same energy.

Lower-energy primary cosmic rays ($<10^{13}$ eV), i.e. those with the largest intensity, can be studied directly by sending particle detectors, such as scintillation counters, above the earth's atmosphere in satellites, spaceprobes, rockets, and balloons. High-energy particles ($>10^{16}$ eV) have too low a flux for direct measurements and can only be studied through their extensive air showers using arrays of detectors at ground level; particles with energies exceeding 10^{20} eV have been detected with giant arrays of instruments covering areas greater than one km^2. Intermediate energies require sensitive detector arrays on mountaintops or large detectors in satellites.

Most of the lighter primary cosmic-ray nuclei are considered products of collisions of heavier nuclei that occur during their journey from source. These parent nuclei, with other heavy nuclei and most of the primary electrons, are created in some as yet unknown *nucleosynthesis process of catastrophic origin. Medium and lower energy particles probably originate in *supernova explosions: both the Crab nebula and the supernova remnant of the Vela pulsar are thought to be sources. The sun produces very low energy cosmic rays, the strongest solar *flares generating energies up to 10^{10} eV. The extraordinary acceleration processes that could produce the highest energies are not yet identified. There is no evidence as yet for a maximum energy for cosmic rays, although this has been predicted.

It is generally believed that almost all cosmic rays with energies less than about 10^{18} eV are generated by sources within the Galaxy. It is also thought that these particles are confined to the Galaxy for probably tens of millions of years by the complex and very weak *galactic magnetic field. Cosmic-ray particles, being charged, are affected by a magnetic field: they are deflected from their initial paths and become trapped in the galactic field, the lowest energies being most deflected. During their long confinement their directions of travel become almost uniformly scattered. As cosmic rays enter the solar system they can be further scattered by magnetic irregularities in interplanetary space: the total intensity at the earth's orbit is twice as great at sunspot minimum than at sunspot maximum. Although lower-energy cosmic rays are confined very effectively, higher-energy particles tend to 'leak out' of the Galaxy; above a critical energy at which the effect of the magnetic field becomes negligible, they can all escape. Thus the cosmic-ray intensity should decrease with energy.

The intensity however levels off at high energies and this has been taken as evidence of other sources of cosmic rays outside the Galaxy. Since high-energy particles travel in approximately straight lines (being unaffected by magnetic fields) their arrival direction should indicate their source direction: measurements made with the giant detector arrays point to sources well beyond the galactic plane, i.e. to possible extragalactic sources. Gamma rays can arise in space from the interaction of primary cosmic rays with interstellar matter. It is hoped that measurements of the distribution of gamma rays in space, especially those of high energies ($>10^8$ eV), will reflect the distribution of cosmic rays and thus reveal the galactic or extragalactic nature of their sources.

cosmic rising (or setting). *See* acronical rising (or setting).

cosmic scale factor. Symbol: R. A measure of the size of the universe as a function of time. R is related to the *redshift parameter, z, by
$$1 + z = R(t_0)/R(t_1)$$
where t_0 is the present time and t_1 is the time at emission of the radiation. The quantity $(1 + z)$ thus gives the factor by which the universe has expanded in size between the time t_1 and the present, t_0. The *proper distance* between the origin and a point at a comoving coordinate r (i.e. in a

system expanding with the universe) is given by the product of $R(t)$ and $\sin^{-1}r$, r, or $\sinh^{-1}r$ depending on whether the *universe is closed, Euclidean, or open, respectively. R is also related to the *Hubble constant H_0:

$$H_0 = (dR/dt)/R$$

cosmic year. The period of revolution of the sun about the centre of the Galaxy, equal to about 220 million years.

cosmogony. The study of the origin and evolution of cosmic objects and in particular the solar system. *See* solar system, origin; galaxies, formation and evolution; stellar evolution.

cosmological constant. *See* cosmological models; static universe.

cosmological models (world models). Possible representations of the universe in simple terms. Models are an essential link between observation and theory and act as the basis for prediction. Complications are added only when necessary (Occam's razor). A simple model for a two-dimensional universe is the surface of an ex-

panding balloon, on which *Hubble's law and the isotropy of the *microwave background radiation may be demonstrated.

Most standard cosmological models of the universe are mathematical and are based on the *Friedmann universe*, derived by Aleksandr Friedmann in 1922 and independently by Georges Lemaître in 1927. They assume the *homogeneity and *isotropy of an expanding (or contracting) universe in which the only force that need be considered is *gravitation. The *big bang theory is such a model. These models result from considerations of Einstein's field equations of general *relativity. When the pressure is negligible the equations reduce to

$$(dR/dt)^2/R^2 + kc^2/R^2 = (8\pi/3)G\rho$$

for energy conservation – this is known as the *Friedmann equation* – and

$$\rho R^3 = \text{constant}$$

for mass conservation. R is the *cosmic scale factor, ρ the *mean density of matter, G the gravitational constant, and c the speed of light; k is the curvature index of space of value $+1$ (closed *universe), -1 (open universe), or 0 (flat or *Einstein–de Sitter universe*). See illustration.

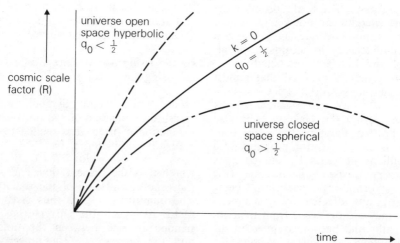

Models of universe with different deceleration parameters, q_0

Other models involving the *cosmological constant*, Λ, have been proposed, such as the *de Sitter model*, in which no mass is present, the *Lemaître model*, which exhibits a coasting phase during which R is roughly constant, the *steady-state theory, and those in which the gravitational constant, G, varies with time (*see* Brans-Dicke theory). The cosmological constant is an arbitrary constant. Although it is possible for it to have any value that does not conflict with observation, it is highly probable that it is close to zero. Cosmological models involving Λ are considered nonstandard. In the standard (Friedmann) models Λ = 0. *See also* static universe.

cosmological principle. A principle stating that there are no preferred places in the universe. This is the basis for much of modern *cosmology and is an extreme case of the Copernican view that the earth is not the centre of the universe. Neglecting local irregularities, measurements of the limited regions of the universe available to earth-based observers are then valid samples of the whole universe.

cosmological redshift. The *redshift resulting from the expansion of the universe rather than from gravitational effects of intervening matter (*see* gravitational redshift) of from the motion of an intragalactic object away from the solar system. *See also* cosmic scale factor.

cosmology. The study of the origin, evolution, and large-scale structure of the *universe. *Cosmological models describing the behaviour of the *cosmic scale factor with time are constructed from gravitational theory. These are then tested by comparison with, for example, *source counts, the *Hubble diagram, and the cosmic *helium abundance. The angular size of cosmic objects does not necessarily vary linearly with *redshift and can

be used as a further test – the *angular-size redshift test*. The nonlinearity arises because such objects were closer to us in the past and because radiation is bent by the gravitational attraction of intervening matter.

Cosmos. A series of over 1600 Soviet satellites, the first of which was launched in March 1962 and the 1000th in March 1978. The many and varied applications have included astronomical, ionospheric, atmospheric, geomagnetic, geodetic, and biological studies. The satellites have also apparently been used for navigation, reconnaissance, ocean surveillance, and military communications.

COSMOS. A system of highly versatile automated measuring equipment that together with another system, GALAXY, was developed by the *Royal Observatory, Edinburgh, in collaboration with industry, to determine and prepare all the data required from the sky photographs of the *UK Schmidt Telescope.

coudé telescope. Any telescope in which the light emerges along the *polar axis so that its direction remains fixed while the object observed goes through its normal diurnal motion. This arrangement is essential when the light is subjected to analysis by bulky and delicate instruments, such as the *spectrograph, which can then be mounted in a fixed position at the *coudé focus*.

There are two main ways in which the fixed coudé focus can be secured. With an equatorially mounted telescope, one or more extra mirrors can be placed so as to reflect the emerging light along the polar axis (see illustration *a*). This is the practice in large reflecting telescopes where the coudé focus is provided in addition to the prime and Cassegrain foci. It is also the basis for the amateurs' *Springfield mount* (see illustration *b*) in which the diagonal of a Newtonian

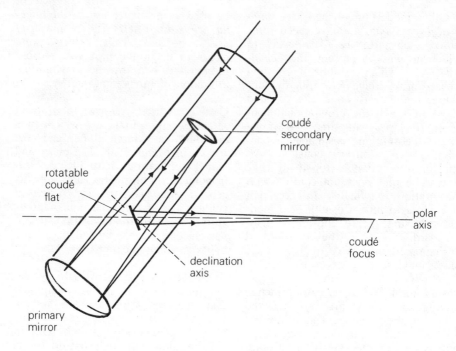

a Light path in telescope with coudé facility

telescope sends the light along the declination axis to an additional diagonal, which reflects it up the polar axis to the observer's eye.

In the other alternative the whole telescope is mounted in a fixed position along the polar axis; a large moveable plane mirror is driven to reflect the observed object into the telescope. This is the basis of many *solar telescopes.

counterglow. *See* gegenschein.

Crab nebula (M1; NGC 1952). A turbulent expanding mass of gas and dust with luminous twisting filaments of ionized gas, lying about 2000 parsecs away in the constellation Taurus. It is the *supernova remnant of a *supernova (probably of Type II) that was almost certainly observed by Chinese and Japanese astronomers in 1054 and was sufficiently bright (magnitude about −5) to be visible in daylight for over three weeks. At pre-

sent it is about four parsecs in diameter. The Crab nebula emits *synchrotron radiation of all wavelengths (with the possible exception of gamma rays) and is a particularly strong source of x-rays (Taurus X-1) and radio waves (*Taurus A).

The star whose explosion produced the Crab nebula is now a young *optical pulsar (the *Crab pulsar* NP 0532) identified as such in 1967. Its pulsations are also observed at radio, infrared, x-ray, and gamma-ray wavelengths. These pulsations have a period of only 0.0331 seconds. The pulsar is the power house for the Crab nebula: energy is being lost by the pulsar in the form of highly energetic electrons, causing the pulsar to slow down very gradually; the electrons interact with the intense magnetic field extending into the nebula and radiate synchrotron emission. The energy loss of the pulsar equals the total energy radiated by the nebula (10^{30} to 10^{31} J s^{-1}). The ultraviolet

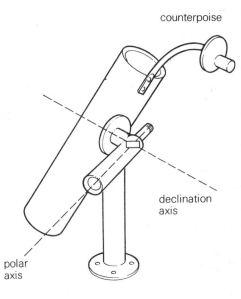
counterpoise

declination
axis

polar
axis

b Coudé telescope: equatorial Springfield

component of this radiation ionizes
the gas in the filaments, causing the
atoms to fluoresce. The Crab nebula
can thus be considered an *emission
nebula.

Crab pulsar. *See* Crab nebula.

Crater (Cup). An inconspicuous constellation in the southern hemisphere
near Leo, the brightest stars being of
4th magnitude. Abbrev.: Crt; genitive
form: Crateris; approx. position: RA
11.5h, dec −15°; area: 282 sq deg.

crater chain (chain craters). Any linear
association of *craters. Some lunar
crater chains are incipient sinuous
*rilles, i.e. lava tubes (in *maria) that
have suffered only partial collapse.
Others are not confined to the maria
and consist of overlapping *secondary
craters. Chains of volcanic craters (as
in fissure eruptions) have not been
unequivocally recognized but *see*
wrinkle ridges.

crater cluster. A group of *secondary
craters. Crater clusters were visited by
the crews of Apollo 15 and 17.

crater counting. *See* craters.

craters. Circular depressions on the
surface of the moon and several other
bodies in the solar system. They have
diameters ranging from less than 1
metre to more than 1000 km, although the larger structures are more
usually termed *basins or *maria. The
largest lunar craters are considered to
be the 200 km *walled plains*. Some of
the more interesting lunar craters are
listed in the table. Craters appear all
over the moon's surface and are most
prominent when near the terminator.
Small craters are everywhere more numerous than larger ones and crater-
size distribution curves provide the
basis for the estimation of relative
ages by *crater counting*. The entire
surface of *Mercury is heavily
cratered while on Mars the craters
predominate in the southern hemisphere (*see* Mars, surface features).
Many planetary satellites are now
known to be cratered, some very
heavily.

The vast majority of craters are
now believed to have originated by
meteoritic impact, although in some
cases their present appearance results
from later modification by erosion or
volcanic processes. A few craters in
the lunar maria, particularly those associated with sinuous *rilles, may be
volcanic in origin. Some, such as
those at the summits of the volcanoes
of the earth or Mars, are solely volcanic. Other small craters result from
subsidence or surface collapse.

In profile, a typical crater has a
floor slightly below the level of the
surrounding surface and a raised rim
that slopes gently outwards but falls
more steeply towards the crater floor.
The inner slope of the rims of large
craters may be *terraced*, while the terrain beyond the rim may be hummocky and broken with numerous

Lunar Craters

Crater	Characteristics*	Coordinates	
Aitken	mare-filled, farside (L)	173°E	17°S
Alphonsus	dark halo crater, rilles (L)	3°W	13°S
Archimedes	flat-floored, Mare Imbrium (M)	4°W	29°N
Aristarchus	young, bright, TLP source (S)	47°W	24°N
Aristillus	young, close to Apollo 15 site (S)	2°E	34°N
Arzarchel	central peak, S of Alphonsus (M)	2°W	18°S
Autolycus	young, close to Apollo 15 site (S)	2°E	31°N
Bailly	extremely large (L)	68°W	66°S
Clavius	very large (L)	14°W	58°S
Cleomedes	large, N of Crisium (L)	56°E	27°N
Copernicus	young, rayed, secondaries (M)	20°W	10°N
Cyrillus	overlaps with Theophilus (L)	24°E	3°S
Eratosthenes	type crater for Eratosthenian System (M)	12°W	14°N
Gassendi	N Mare Humorum, TLP source (L)	39°W	17°S
Giordano Bruno	extremely young, historical? (S)	119°E	46°N
Grimaldi	very dark floored, TLP source (L)	68°W	5°S
Jules Verne	farside crater (L)	148°E	35°S
Kepler	very young, rayed, Oc. Procellarum (S)	38°W	8°N
Langrenus	large, E Mare Fecunditatis (L)	62°E	9°S
Plato	very dark floored, N Mare Imbrium (L)	9°W	52°N
Proclus	partially rayed, W Mare Crisium (S)	47°E	16°N
Ptolemaeus	large, flat-floored (L)	2°W	9°S
Reiner	very deep, Oc. Procellarum (S)	55°W	7°N
Schickard	large, S Mare Humorum (L)	55°W	44°S
Schiller	elongated, secondary? (M)	39°W	52°S
Stevinus	young, bright, S Mare Fecunditatis (M)	54°E	33°S
Theophilus	young, unrayed, N Mare Nectaris (L)	27°E	2°S
Tsiolkovsky	mare-filled, farside (L)	130°E	21°S
Tycho	very young, prominent ray system (M)	11°W	43°S
Wargentin	flooded with Orientale ejecta (M)	60°W	49°S
Van de Graaff	double, farside, magnetic anomaly (L)	174°E	27°S

*diameter ranges: L, greater than 100 km; M, 50–100 km; S, smaller than 50 km

very small craters. Some large craters also have mountainous *central peaks*.

The impact events that created most of the surviving craters on solar-system bodies are thought to have occurred between 3000 and 4000 million years ago as the bodies collided with numerous *planetesimals. Possible the planetesimals were themselves the debris from collisions between larger objects that accreted at the same time as the planets, about 4600 million years ago, between the orbits of Mars and Jupiter. Most of the present-day survivors of the planetesimals, the mi-

nor planets, orbit between Mars and Jupiter but the *Apollo and *Amor objects do stray within the orbits of the earth or Mars and, together with cometary nuclei, still produce occasional impacts, though at a much reduced rate.

During a crater-forming impact, the kinetic energy of the impacting body, which may be several kilometres across and moving at many kilometres per second, goes towards vaporizing and compressing much of itself and the rock on which it impinges. It is the subsequent explosion of these gases that excavates the crater, throwing material (*ejecta) away from the crater centre to form the rim and the crater's hummocky *ejecta blanket. Chunks within the ejecta give rise to secondary impact craters nearby. In a large crater the steep inner rim can then slump to form the observed terraces, while the central peaks may arise from a rebound of the rocks immediately below the site of the explosion. Craters may also initially have *ray systems, i.e. bright streaks radiating outwards from the crater.

Later impacts may destroy the crater completely, form a new overlapping crater, or partially fill the crater with ejecta. Lava may well up through cracks in the floor of the largest impact features to produce lava plains, such as the lunar maria. There may be rim obliteration and erosion and the occurrence of isostatic uplift. As a result the depth-to-diameter ratio, which for lunar craters may be as high as 1:5, decreases. On the earth, erosion by wind, water, ice, and tectonic forces has removed nearly all the evidence that our planet was subject to the same ancient bombardment as the moon; only a few recent impact craters, such as the *Arizona meteorite crater, remain relatively intact.

Crepe ring. *See* Saturn's rings.

Crimean Astrophysical Observatory. An observatory of the Academy of Sciences of the USSR, and an outstation of the *Pulkovo Observatory. It is sited at Simeiz in the Ukraine at an altitude of 680 metres. The chief instruments are a 2.6 metre (102 inch) reflecting telescope, which has been operating since 1961, and a solar tower, commissioned in 1954. A 22 metre radio dish for millimetre observations, first used in 1967, is located nearby on the Black Sea. The observatory was restored after its destruction in the Second World War.

Crisium, Mare. A lava-filled *basin close to the moon's eastern limb from which soil was returned by *Luna 24.

critical density. *See* mean density of matter.

Crommelin's comet. An inconspicuous comet with a period of 27.7 years, first observed in 1818. Its apparition in March 1984 was observed internationally: the results were used to test techniques to be used in the observation of *Halley's comet, perihelion Feb. 1986, the observing conditions for the two comets being very similar.

cross-correlation function. *See* autocorrelation function.

crust. *See* earth.

Crux or **Crux Australis (Southern Cross).** A very small but conspicuous constellation in the southern hemisphere, lying in the Milky Way. The four brightest stars, *Alpha, *Beta, Gamma, and Delta Crucis form a cross (see illustration). The 1st-magnitude Alpha and the 2nd-magnitude red giant Gamma form the longer arm that points approximately to the south pole. The 1st-magnitude Beta and 3rd magnitude Delta form the transverse arm. The 4th-magnitude Epsilon, between Alpha and Delta, interferes with the figure's regularity, making it more kite-shaped. The area also contains the brilliant cluster the *Jewel Box and the dark *Coalsack nebula. Abbrev.: Cru; genitive form: Crucis; approx. position: RA 12.5h, dec −60°; area: 68 sq deg.

C stars. *See* carbon stars.

Culgoora. Site of the CSIRO Solar Radio Observatory in New South Wales, Australia. The principal instrument is a large *radio heliograph, which provides a near instantaneous moving display of radio activity in the sun's atmosphere: 96 dishes, 14

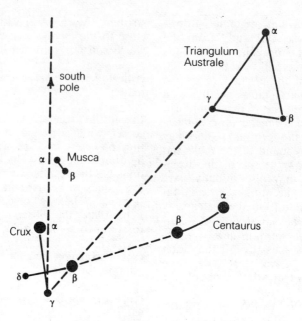

Crux and nearby bright constellations

metres in diameter, are equally spaced around the circumference of a 3 km diameter circle. The resolving power is thus that of a 3 km dish. The received signals are transmitted to a central receiver and processed electronically to produce a television image of the radio sun.

Culgoora will also be the site for the five-dish aperture synthesis array planned as part of the new *Australia Telescope.

culmination (meridian passage). The passage of a celestial body across an observer's *meridian. As a result of the earth's rotation this occurs twice daily although both culminations can be observed only for *circumpolar stars. *Upper culmination* (or *transit*) is the crossing closer to the observer's *zenith; for circumpolar stars and the moon this is also called *culmination above pole*. *Lower culmination* (or *culmination below pole* for circumpolar stars and the moon) is the crossing further from the zenith (see illustration).

curvature. *See* radius of curvature.

curvature of field. A minor *aberration of a lens or mirror whereby an object lying in a plane perpendicular to the optical axis produces an image lying not on a plane but on a curved surface. If there is any *astigmatism present there will be two curved focal surfaces.

curvature of spacetime. *See* spacetime.

cusp. Either of the tapering points of the crescent phase of the moon, Venus, or Mercury.

Cyclops. An immense *array of radio dishes that has been proposed as a means for detecting signals from extraterrestrial civilizations. Although an extremely costly project, the array could also be used for radio astronomy, for radar studies in the solar system, and for tracking space vehicles to very great distances. *See also* SETI.

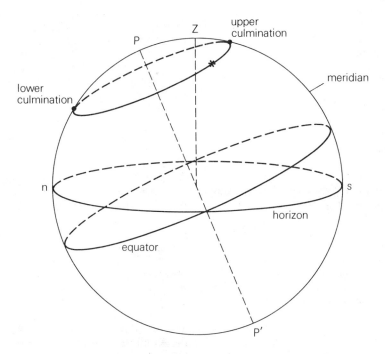

Culminations of a circumpolar star

cyclotron radiation. *See* synchrotron emission.

61 Cygni. A faint star in the constellation Cygnus that is one of the nearest stars to the sun. Since it had the largest known proper motion at the time (5″.25 per year) it was the first star to have its trigonometric parallax measured (F. Bessel, 1838). The value obtained, close to the presently accepted 0.29″, gives the distance. It is a visual binary (separation 24″.6, period 720 years) with one of the components itself a binary (period 5 years). Perturbations found in the motion of the visual-binary components have been interpreted as due to the influence of one or more massive planets. m_v: 5.20 (A), 6.03 (B); M_v: 7.53 (A), 8.36 (B); spectral type: K5 V (A), K7 V (B); mass: 0.59 (A), 0.50 (B) times solar mass; distance: 3.42 pc.

Cygnids. A minor *meteor shower, radiant: RA 290°, dec +55° (near Kappa Cygnus), that maximizes on Aug. 20.

Cygnus (Swan). A large conspicuous constellation in the northern hemisphere, lying in the Milky Way. The five brightest stars – Alpha (1st mag.), Gamma and Epsilon (2nd mag.), and Beta and Delta (3rd mag.) – form the *Northern Cross:* α, γ, β comprise the long arm and ε, γ, δ the transverse arm. Alpha is the extremely luminous supergiant *Deneb and Beta the beautiful binary *Albireo. There are several other noticeable binaries and some variables, including the *Mira star Chi Cygni, the brightest *dwarf nova SS Cygni, and the prototype of the *P Cygni stars. There is a large open cluster, M39, some complex emission and dark nebulosity, and the supernova remnant the *Cygnus Loop, comprising several bright nebulae. The constellation also contains the star 61 *Cygni, the powerful radio source *Cygnus A, and the x-ray binaries *Cygnus X-1 and *Cygnus X-3. Abbrev.: Cyg; genitive form: Cygni; approx position: RA 20.5h, dec +40°; area: 804 sq deg.

Cygnus A. An intense *radio source in Cygnus that has a *flux density at 178 megahertz of 8100 *jansky. It is a typical double radio source, a *radio galaxy with a *redshift of 0.057 indicating a distance from the solar system of around 230 megaparsecs. The bright outer lobes are about 60 kpc from the central 16-magnitude galaxy with which the radio source is identified and over which there is some weak radio emission. Cygnus A is also a source of low-energy x-rays.

Cygnus Loop. An *emission nebula in the constellation Cygnus. The *Veil nebula* forms part of the Cygnus Loop. It is the *supernova remnant of an old supernova that exploded probably some 50 000 years ago. The remnant now has an area of about 3 square degrees and is about 800 parsecs away. Its rate of expansion has gradually slowed to its present 100 km s^{-1} as the expanding shell collides with and sweeps up any surrounding interstellar matter. This creates a shock wave that compresses, heats, and ionizes the gas atoms, which then become visible as the emission nebula.

Cygnus X-1. An intense galactic x-ray source considered to represent the best observational evidence for the existence of a *black hole of stellar mass. Identification of Cygnus X-1 with the luminous supergiant star HDE 226868 was followed by the discovery of a 5.6 day binary period. Analysis of the optical observations showed the mass of the 'unseen' companion star (the x-ray source) to lie in the range 6–15 solar masses, well above the limit allowed for a *neutron star (≤3 solar masses) or *white

dwarf (≤1–4 solar masses). Support for the view that Cygnus X-1 involves gas accretion from the optical supergiant onto a black hole is provided by the rapid (millisecond) and irregular variability of the x-ray emission.

Cygnus X-3. A luminous x-ray binary source with an unusually short (4.8 hour) binary period. No optical identification exists but the short period suggests that the system consists of a *neutron star and a low-mass companion. The sinusoidal x-ray *light curve is interpreted as being due to scattering of x-rays from the (unseen) central source by a hot dense wind driven off the companion star. Recent observations of variable radio emission and very high energy (10^{15} eV) gamma rays, both modulated at the 4.8 hour period, make Cygnus X-3 a unique source. It is postulated that these unusual properties may arise from the unseen neutron star being very rapidly rotating (a young pulsar).

D

dark halo. The large region around our Galaxy and other galaxies that is believed to contain distributed *dark matter.

dark halo crater. A small lunar crater, originally believed to be volcanic (e.g. a cinder cone or fumarole) but now considered to be produced when an impacting meteorite excavates dark surface materials.

dark matter. Matter that probably comprises 90% (or more) of the mass of the universe but is undetectable except by its gravitational effects. Dark matter was first suspected in *clusters of galaxies when the galaxies were found to move with too high a speed to be retained in the cluster by their gravitational influence on each other.

The term *missing mass* was coined for the necessary invisible matter in the cluster, amounting to 10 to 100 times the amount of visible matter in the galaxies and in the *intergalactic medium. In the 1970s investigations of the rotation of spiral galaxies indicated that these galaxies have a *dark halo containing some 10 times as much matter as the visible galaxy. Simultaneously statistical analyses of the motions of galaxies on a larger scale than clusters suggested that the universe in general has a ratio of perhaps 30:1 of dark matter to detectable matter (except within galaxies, where ordinary matter dominates).

The nature of the dark matter is highly elusive, and there is no certainty even that the dark haloes of spiral galaxies consist of the same matter as the missing mass of clusters and the universe at large. A proportion may consist of objects composed of normal matter, but according to standard *big-bang models the whole of the missing mass cannot consist of matter containing *baryons (protons and neutrons) since that amount of baryonic matter would have produced a higher abundance of helium than is observed. The bulk of the dark matter does not therefore consist of dim stars or smaller chunks of solid matter, nor of *black holes that have formed from stars. Candidates currently (1984) under serious consideration are certain *elementary particles – *neutrinos, if these have a mass greater than zero, or hypothetical particles called photinos, gravitinos, and axions.

dark nebula (absorption nebula). A cloud of interstellar gas and dust that is sufficiently dense to obscure partially or completely the light from stars and other objects lying behind it and sufficiently large and suitably located to produce a noticeable effect. These *nebulae can be observed as dark extrusions in front of bright (*emission or *reflection) nebulae or

as blank regions or regions with a greatly diminished number of stars in an otherwise bright area of sky. They are situated along the spiral arms of the Galaxy. Although the absorption is caused by *cosmic dust, the dark nebulae are composed predominantly of molecular hydrogen. Small dark nebulae, called *Bok globules, can sometimes be seen in large numbers superimposed on bright nebulae. Although having no optical features dark nebulae can be studied through their radio and infrared emissions. The *Coalsack, *Horsehead nebula, and the *Great Rift are dark nebulae.

Davy Chain. A prominent *crater chain on the moon that was produced by secondary *ejecta.

Dawes limit. *See* resolution.

day. The period of the earth's rotation on its axis, equal to 86 400 seconds (24 hours) unless otherwise specified. The 4-minute difference between the *sidereal day and the 24-hour day results from the superimposition of the earth's orbital motion round the sun on the rotational motion.

day numbers. Quantities that depend on the position and motion of the earth, and are used in reducing mean place of a celestial object to apparent place. *See* true place.

dec. *Short for* declination.

deceleration parameter. Symbol: q_0. The rate at which the expansion of the universe changes with time as a result of self-gravitation. It is dimensionless and can be given in terms of the *cosmic scale factor, R:
$$q_0 = -R(d^2R/dt^2)/(dR/dt)^2$$
Values of q_0 less than 0.5 are associated with an ever-expanding open universe; values exceeding 0.5 imply that the universe is closed and will collapse at a finite time in the future. If q_0 is equal to 0.5 then the universe is

infinite and space is Euclidean. *See also* mean density of matter; universe.

declination (dec). Symbol: δ. A coordinate used with *right ascension in the *equatorial coordinate system. The declination of a celestial body, etc., is its angular distance (from 0° to 90°) north (counted positive) or south (counted negative) of the celestial equator. It is measured along the hour circle passing through the body.

declination axis. The axis at right angles to the *polar axis about which a telescope on an *equatorial mounting is turned in order to make adjustments in declination. The optical axis then follows a particular hour circle across the sky.

declination circle. A *setting circle on the declination axis that enables an equatorially mounted telescope to be set at the declination of the object to be observed. Its scales are graduated from 0°, when the telescope is aligned with the celestial equator, to $\pm 90°$, when it is aligned with the north or south poles.

decoupling era. *See* big bang theory.

deferent. *See* Ptolemaic system.

deflection of light. *See* relativity, general theory.

degeneracy pressure. *See* degenerate matter.

degenerate matter. Matter in a highly dense form that can exert a pressure as a result of quantum mechanical effects. Degenerate matter occurs in *white dwarfs and *neutron stars. During the *gravitational collapse of a dying star, the electrons are stripped from their atomic nuclei, and nuclei and electrons exist in a closely packed, highly dense mass. As the density increases, the number of electrons per unit volume increases to a point when the electrons can exert a considerable pressure, called *degeneracy pressure*. This pressure is a result of the uncertainty principle of quantum mechanics. Unlike normal pressure, degeneracy pressure is essentially independent of temperature, depending primarily on density. At the immense densities typical of white dwarfs (10^7 kg m^{-3} or more) it becomes sufficiently large to counteract the gravitational force and thus prevents the star from collapsing further. The gross properties of a white dwarf are therefore described in terms of a gas of degenerate electrons.

Above a certain stellar mass (the *Chandrasekhar limit) equilibrium cannot be attained by a balance of gravitational force and electron degeneracy pressure. The star must collapse further to become a neutron star. It is then the degeneracy pressure exerted by the tightly packed neutrons that balances the gravitational force. *See also* black hole.

degenerate star. A star composed primarily of degenerate matter, i.e. a *white dwarf or a *neutron star.

Deimos. A satellite of Mars, lying further from the planet than the slightly larger satellite *Phobos. It is an irregular body measuring $11 \times 12 \times 15$ km, and orbits above Mars' equator at a distance of 23 460 km from the centre of the planet in a period of 1.26 days. Like Phobos, it may be a captured *minor planet of a type similar to the *carbonaceous chondrite meteorites. However, it appears to have none of the grooves and crater chains found on Phobos while many of its craters are less distinct, probably because of infilling by crater ejecta that may lie tens of metres thick on the surface. The larger craters, such as *Voltaire* and *Swift,* measure only about 2 km across. *See* Table 2, backmatter.

Delphinus (Dolphin). A small constellation in the northern hemisphere near Cygnus, the brightest stars being of 3rd magnitude. Gamma Delphinus is a 4th-magnitude double. Abbrev.: Del; genitive form: Delphini; approx. position: RA 20.5h, dec $+10°$; area: 189 sq deg.

delta (δ). The fourth letter of the Greek alphabet, used in *stellar nomenclature usually to designate the fourth-brightest star in a constellation or sometimes to indicate a star's position in a group.

Delta Aquarids. *See* Aquarids.

Delta Cephei. *See* Cepheid variables.

Delta Scuti stars. A class of relatively young *pulsating variables, typified by Delta Scuti, that are about 1.5–3 solar masses and range in spectral type from about A3 to F6. They were originally grouped with the *dwarf Cepheids. They are numerous but tend to be inconspicuous because of the very small variation in brightness, usually about 0.01–0.4 magnitudes. Like dwarf Cepheids they have very short periods ranging from about 0.8–5 hours. A typical feature is a variation in the amplitude and shape of the light curve; irregularities in the light-curve contour result from the superimposition of two or more harmonic modes in which the star is pulsating.

delta T. Symbol: ΔT. From 1984, the increment to be added to *universal time, UT1, to give terrestrial *dynamical time, TDT. Prior to 1984 it was the increment to be applied to UT1 to give *ephemeris time. ΔT is not constant but at present is increasing in an irregular manner: between 1970–85 it has changed from about $+40$ to $+54$ seconds.

Demon Star. *See* Algol.

Deneb (α Cyg). A very luminous remote white supergiant that is the brightest star in the constellation Cygnus and lies at one end of the long arm of the Northern Cross. It has very low proper motion ($0''.003$ per year). m_v: 1.25; M_v: -7.2; spectral type: A2 Ia; distance: about 500 pc.

Denebola (β Leo). A white star that is the third-brightest in the constellation Leo. m_v: 2.13; M_v: 1.6; spectral type: A3 V; distance: 13 pc.

density. **1.** Symbol: ρ. The mass per unit volume of a body or material. The mean density of a celestial body is its total mass divided by its total volume. A wide variation in densities is found in the universe, ranging from about 10^{-20} kg m^{-3} for interstellar gas to over 10^{17} kg m^{-3} for neutron stars. The *mean density of matter in the universe is of the order of 10^{-27} kg m^{-3}.
  2. The number of electrons, ions, or other particles per unit volume.

density parameter. *See* mean density of matter.

density-wave theory. A theory that attempts to explain how the structure of a spiral galaxy is sustained by the flow of material in and out of the regions of the spiral arms under the influence of gravity. It was proposed in 1964 by C.C. Lin and F.H. Shu and based on work by B. Lindblad. The spiral structure is not permanent: it is regarded as a wave pattern in which the spiral arms are regions of low gravitational potential where matter concentrates. The spiral arms therefore comprise the loci of wave crests of a density wave. The wave pattern is not tied to the matter but moves through it at an angular velocity that can differ appreciably from that of the matter; the relative motion is on average about 30 km s^{-1}. It is this motion that somehow maintains the overall appearance of the galaxy.

Stars move into the travelling gravitational troughs of the spiral arms and can remain there for a considerable time until their orbits take them out again. Interstellar gas also comes under the influence of the density wave. It has been suggested that when the gas encounters the moving wave it undergoes a sudden compression, which in turn could be a possible trigger for interstellar cloud collapse and star formation. Young stars should then be found preferentially in spiral arms, which is indeed the case.

Although the density-wave theory explains the spiral shape of galaxies and the presence of young stars in spiral arms, it does not show how the density wave originated. In addition it predicts the dissipation of the spiral arms after only a few galactic rotations, i.e. after several hundred million years, which is much shorter than the lifetime of galaxies. The theory is widely believed to apply to galaxies suffering gravitational perturbations from a passing neighbour or from the central bar of a barred spiral galaxy: these galaxies have sharply defined arms. The patchier arms of many isolated galaxies may not be due to density waves but to *self-propagating star formation.

Descartes formation. A range of lunar hills to the NW of Mare Nectaris that have an unusually high *albedo. They were once thought to consist of acid volcanics but are now considered to have been impact-generated.

descending node. *See* nodes.

de Sitter model. *See* cosmological models.

detached binary. *See* binary star; equipotential surfaces.

deuterium. *See* hydrogen.

diagonal mirror. *See* Newtonian telescope.

Diamond of Virgo. *See* Spica.

diamond-ring effect. A bright flash of sunlight from a conspicuous *Baily's bead sometimes very briefly observed, together with the sun's inner corona, at the moon's limb just before or after totality in a solar eclipse.

diaphragm. *See* stop.

dichotomy. The moment of exact half-phase of the moon, Mercury, or Venus.

Dicke-switched receiver. A radio-astronomy *receiver in which the input is continuously switched between the *antenna and a reference noise source of constant noise power. The switching, at a rate of between 10 and 1000 hertz, is carried out by a *Dicke switch*, which is typically a semiconductor-diode switch or a ferrite switch. The switching reduces the fluctuations in the output of the receiver that arise from *gain changes in the amplification stages and seriously limit the performance of total-power *receivers. The amplitude of the component at the switching frequency in the output of the first detector is proportional to the difference in powers between the noise source and the antenna; it is detected using a *phase-sensitive detector. The effect of a change in gain is much reduced provided that the change takes place much more slowly than in one switch cycle.

differential rotation. Rotation of different parts of a system at different speeds. It occurs in a system, such as a star, composed primarily of gas. A solid body, like the earth, rotates at a constant rate for all positions. The *sun and the giant planets, such as Jupiter, rotate differentially such that equatorial regions move somewhat quicker than those closer to the poles. *See also* galactic rotation.

differentiation. The process by which layers of different density are formed from originally homogeneous molten rock. It is generally accepted that melting and differentiation occurred very early in the history of the earth and resulted from the rapid increase in the young planet's temperature; the heat was generated by decay of radioactive elements in the planetary material, from the impact of infalling material, and from contraction of the planet. The widespread melting led to massive differentiation of the earth into the iron-rich core and the lighter silicate (rocky) regions of the mantle and crust (*see* earth). Differentiation also occurred early in the moon's history. It is uncertain whether other terrestrial planets have differentiated.

diffraction. The apparent bending of light or other electromagnetic radiation at the edges of obstacles, with the resulting formation of light and dark bands or rings (diffraction patterns) at the edges of the shadow. An example of a diffraction pattern is the *Airy disc observed in telescope images of distant celestial objects. A small aperture or slit produces similar annular or banded patterns. Diffraction results from the wave nature of electromagnetic radiation. *See also* interference.

diffraction grating. An optical device that is used for the production and study of *spectra, usually by means of a *spectrograph. Its action depends on the *diffraction of light or other radiation by a very large number of very close and exactly equidistant linear slits. The slits are produced by ruling very fine closely spaced scratches on glass or polished metal, forming either a *transmission* or *reflection grating*. Reflection gratings can be plane or concave; the latter can act as a focusing element for incident light.

The diffracted light, once focused, produces a series of sharp spectral lines for each resolvable wavelength present in the incident light. For a plane wave of a single wavelength λ, incident on a transmission grating at angle i, the successive wave trains passing through the slits will travel different pathlengths. If the path difference between two adjacent slits is a whole number of wavelengths, the wavetrains will be brought to a focus as a bright image of the radiation source at a particular angle of refraction, d (see illustration), where

$$\sin d = n\lambda/s - \sin i$$

s is the spacing between adjacent slits and n is an integer. Bright images will in fact be produced for each of the angles d corresponding to $n = 1, 2, 3$ ... These numbers denote the *orders* of the image. If, for a particular order, d is made equal to i, then

$$\sin d = n\lambda/2s$$

If the incident light is composed of various wavelengths, the image of any particular order, $n = 1, 2, 3 \ldots$, will appear at different points since d varies with wavelength. Hence a spectrum is produced. The angular dispersion of the spectral lines, i.e. $dd/d\lambda$, will be high when the slits are very fine and closely spaced (i.e. s is small). For the spectral resolution – separation of images of very nearly equal wavelength – to be high, the total number of slits must be large: several thousand slits per centimetre are common for the visible and ultraviolet regions of the spectrum.

diffraction pattern. The pattern of alternating light and dark rings or bands produced when light is diffracted by a circular opening or object (rings) or a rectangular slit or an edge (bands). *See also* Airy disc.

Dione. A satellite of Saturn, discovered in 1684. It has a diameter of 1120 km and density of 1.4 g cm^{-3}, and is slightly larger than the satellite Tethys. An important characteristic of Dione is the nonuniformity of its brightness. The trailing hemisphere is

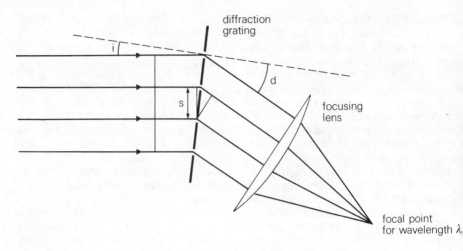

Diffraction by a plane transmission grating

dark, with an albedo of approximately 0.3, while the brightest features of the leading hemisphere have an albedo of approximately 0.6. Only Iapetus, of the Saturn system, displays a greater variation of brightness between the hemispheres. The surface shows evidence of a number of craters of 30–40 km, with a few large craters 165 km in diameter. There are some broad ridges in the southern part of the heavily cratered plains with a long linear valley more than 500 km in length near the south pole. Dione's relatively high density compared with the neighbouring satellites indicates a higher rock content of the interior. The bright streaks on the surface may relate to outpouring of water from the interior. A minor satellite has been discovered to share the orbit of Dione. *See also* Table 2, backmatter.

dioptric system. An optical system, such as a refracting telescope, in which all the optical elements are lenses.

dipole. A simple linear balanced *antenna that is fed from its centre. It usually consists of two equal-length straight opposing elements, though these may sometimes be folded back on themselves and their far ends joined together in the middle to form a *folded dipole*. The elements may be of any length but the overall length of the dipole is usually a multiple of half a wavelength of the received (or transmitted) radio wave; the reactive component of the impedance seen at its feed point is then small.

A *half-wave dipole* has two elements of length one quarter of a wavelength each and has a *radiation resistance of about 70 ohms. A *full-wave dipole* is twice as long as a half-wave dipole and has a radiation resistance of several thousand ohms. The *gain of a half-wave dipole over an isotropic radiator is about two decibels.

See also dish; array; radio telescope.

directivity. A measure of the ability of an *antenna to transmit or receive radiation in one direction only. It is defined as the ratio of the maximum radiation intensity in the wanted direction to the average radiation intensity over all directions. If an antenna is operating at a wavelength λ and has an effective area A_e (*see* array), then the directivity, D, is given by

$$D = 4\pi A_e/\lambda^2$$

See also gain.

direct motion. 1. The apparent west to east motion of a planet or other object as seen from earth against the background of stars. *Retrograde motion* is an apparent motion from east to west. When a superior planet near *opposition is overtaken by the earth, moving with higher relative velocity, its normal direct motion seems to become temporarily retrograde and it appears to undergo a loop or zigzag in the sky (see illustration); the turning points between these motions, when the planet appears motionless in the sky, are known as *stationary points*.

2. The anticlockwise orbital motion, as seen from the north pole of the ecliptic, of a planet or comet around the sun or of a satellite around its primary. A clockwise orbital motion is *retrograde*. The planets all have direct motion.

3. The anticlockwise rotation of a planet, as seen from its north pole. Venus and Uranus have a clockwise *retrograde* rotation.

dirty-snowball theory. *See* comet; nucleus.

disc. 1. The two-dimensional projection of the surface of a star or planet. **2.** *See* Galaxy.

disc population stars. *See* population I, population II.

dish. A type of *antenna used in radio telescopes and consisting of a large spherical or parabolic metal reflector, usually circular in outline, by means of which radio waves are brought to a focus above the dish centre. The waves are collected at the focus by a secondary antenna, called a *feed*. At low radio frequencies the feed may be a *dipole mounted in front of a small reflector; at high frequencies it is usually a horn antenna (*see* waveguide). The signals picked up by the feed are transferred by means of the *feeder to the *receiver for amplification and analysis. Both the angular *resolution and the *sensitivity of the telescope increase with the area of the dish.

The dish is often mounted so that it can be steered to point in different directions and can be made to track a moving object. Some dishes however,

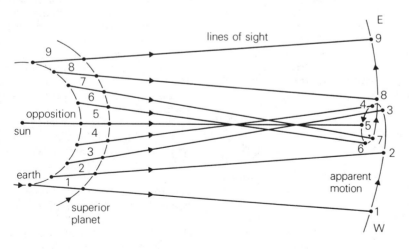

Direct and retrograde motion of superior planet

like the 300-foot dish at Green Bank, USA, can only move in one coordinate (*see* coordinate system) and rely on the earth's rotation to provide coverage in the other coordinate. The 1000-foot dish mounted in a natural hollow in the ground at Arecibo, Puerto Rico, achieves partial beam steering by moving the feed.

dispersion. 1. The separation of a beam of light into its component colours, i.e. into its component wavelengths, so that a spectrum is formed. It arises in a lens or prism because of the variation of the *refractive index of the transmitting medium with wavelength. Normally the refractive index of a transparent substance increases as the wavelength decreases: blue light is refracted (bent) more than red light. Dispersion is the cause of *chromatic aberration. *See also* spectrograph.

2. The retardation of radio waves that occurs when they pass through ionized gas in the interstellar medium, the speed of propagation depending on frequency: the lower the frequency the greater the delay. This delay is not observable in a continuous radio signal but is detectable in a pulsed one, such as that from a pulsar. The amount of dispersion gives an indication of distance to a pulsar. *See also* dispersion measure.

dispersion measure. (DM). The number of electrons per unit volume, n_e, integrated along the line of sight to a source:
$$DM = \int n_e dl$$
where dl is an element of the path. It is the factor that determines, for example, the time delay (i.e. *dispersion) between the arrival of radio pulses from a *pulsar at two different frequencies.

distance determination. The distances of celestial objects can be determined directly by means of radar or laser ranging and from measurements of *parallax, and indirectly from photometric and spectroscopic methods. Radar and laser measurements can only be made for comparatively tiny distances, i.e. those within the solar system: the earth-moon distance has been measured to a few cm by laser beaming; radar measurements of the distances of Venus, Mercury, and Mars have been used in determining planetary distances from the sun by *Kepler's laws. The distance from earth to sun is thus now known very accurately.

This earth-sun distance is used as the baseline in the determination of the *annual parallax of stars, from which stellar distances out to about 30 parsecs can be determined trigonometrically. At greater distances the error in measuring parallax becomes as large as the parallax itself. These distances have therefore to be found indirectly from measurements of *luminosity, *magnitude, and other stellar properties: distances are determined from relationships connecting these properties, including *distance modulus, *moving-cluster parallax, and the *period-luminosity relation for *Cepheid variables.

A distance scale can be established whereby the distances found for one set of distance indicators, such as open clusters (using *main-sequence fitting, for example), can be used to calibrate the next most distant indicators – Cepheid variables – occurring in such clusters. Cepheids, along with novae and the most luminous stars, are in turn used to measure the distances to galaxies in the *Local Group and in other nearby groups of galaxies. Out to this distance (a few megaparsecs), the distance scale is believed to be accurate to 10% or better.

In the case of more distant galaxies, however, determination of distance currently differ by up to a factor two. Traditional indicators of these larger distances include: the size of a galaxy's *HII regions; the brightness of its *globular clusters; the maximum brightness of its Type I *supernovae; its total luminosity as inferred, for a luminous spiral, from its detailed appearance and as a standard brightness for cD galaxies (the brightest galaxies) in *clusters of galaxies. The best technique at present is the *Tully-Fisher method, which relies on a relation between a galaxy's absolute magnitude and the spread of its rotation velocities.

These methods provide distances out to at least 100 megaparsecs, where the motions of galaxies are dominated by the *expansion of the universe. A galaxy's recession velocity is determined from its *redshift. The ratio of recession velocity to measured distance for the more distant galaxies gives a value for the *Hubble constant; the Hubble constant is then used to derive distances to farther galaxies and quasars from their measured redshifts. The uncertainty of a factor two in the distance scale leads to a corresponding uncertainty in the value of the Hubble constant. These should both be reduced considerably with the advent of the *Hubble Space Telescope, which should be able to discern individual Cepheid variables out to the distance of the *Virgo cluster (15 megaparsecs) and thus provide more accurate distances for galaxies in the Hubble flow (see Hubble's law).

distance modulus. The difference between the apparent magnitude, m, and the absolute magnitude, M, of a star and therefore a measure of distance (see magnitude):
$$m - M = 5 \log(d/10) = 5 \log d - 5$$
where d is the distance in parsecs. Distance modulus is used to determine the distances of stars and stellar clusters, especially those over about 30 parsecs away.

distortion. An *aberration of a lens or mirror in which the image has a distorted shape as a result of nonuniform lateral magnification over the field of view. In *barrel distortion* the magnification decreases towards the edge of the field; in *pincushion distortion* there is greater magnification at the edge.

diurnal circle. *See* diurnal motion.

diurnal libration. Daily geometrical *libration brought about by the change in the position of the observer with respect to the earth-moon line as the earth rotates on its axis. This libration is most significant near the equator, where it amounts to $57'2''.6$; this is equal to the moon's mean equatorial horizontal parallax (see diurnal parallax). An observer closer to the poles will however experience an equivalent geometrical libration because of his displacement from the *ecliptic. *See also* physical libration; augmentation.

diurnal motion. The apparent daily motion of celestial bodies across the sky from east to west. It results from the earth's rotation from west to east, the axis of the diurnal motion coinciding with the earth's axis. The apparent path of a celestial body arising from diurnal motion is a circle parallel to the celestial equator and is termed a *diurnal circle.*

diurnal parallax (geocentric parallax). The *parallax of a celestial body that results from the change in position of an observational point during the earth's daily rotation. It is the angle measured at a given celestial body between the direction to the earth's centre and that to an observer on the earth's surface. It is thus equal to the difference between the topocentric and geocentric *zenith distances, η and η_0, respectively (see illustration). It is only appreciable for members of the solar system. As the observer's position changes during the day, the diurnal parallax of the body varies from a minimum, when the body is

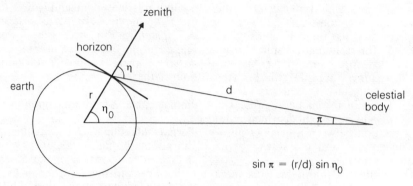

$$\sin \pi = (r/d) \sin \eta_0$$

Diurnal parallax

on the observer's meridian, to a maximum, when it is on the horizon. This maximum value is the object's *horizontal parallax*, π. It is given by:
$$\sin \pi = r/d$$
where r is the earth's radius and d is the distance to the celestial object.

The earth's nonspherical shape causes the length of the baseline to vary with latitude. If this variation is significant, as it is with nearby bodies like the moon, then for the horizontal parallax to be a maximum it must be measured from the equator. This gives the *equatorial horizontal parallax:* the average value for the moon – the *lunar parallax* – is 3422".45; the mean value for the sun – the *solar parallax* – is 8".794

D-layer (D-region). *See* ionosphere.

D lines. A close prominent doublet of lines in the yellow region of the visible spectrum of sodium. They occur at wavelengths of 589.6 nm and 589.0 nm and, being easily recognizable, are useful as standards in spectroscopy. The D lines are strong *Fraunhofer lines.

Dobsonian. A telescope with a simple but highly stable *altazimuth mounting that can be constructed and used by an amateur astronomer. It swings very smoothly between side bearings and is mounted on a simple sturdy

rotating platform. Relatively large telescopes can be used.

docking. The locking together of two spacecraft in space following their rendezvous.

Dog Star. *See* Sirius.

dome. 1. The roof of an observatory, which protects the telescope and associated equipment from bad weather and from the heating effect of the sun. The telescope has access to the sky through a *shutter* in the dome, which is closed when the telescope is not in use. Where the telescope has an *equatorial mounting the dome is movable and hemispherical. Rotation of the dome on a circular track allows different areas of the sky to be brought into view. It is too complex to relate the dome's rotation with the movement of the telescope about its inclined axis. The telescope must thus be able to swing freely inside the structure.

For a telescope with an *altazimuth mounting the rotation rates of dome and telescope are the same. The dome needs only to be slightly larger than the telescope, and can be square or oblong. The cost of the dome is thus greatly reduced.

2. A low lunar hill with a gentle convex-upward profile, of probable volcanic origin. Domes are confined to the *maria and are generally less than 15 km in diameter.

Dominion Astrophysical Observatory. An observatory at Victoria, British Columbia, Canada at an altitude of 220 metres. The chief instruments are a 1.8 metre (72 inch) reflector, which went into use in 1918 as the world's largest telescope and is now equipped with a new Cer-Vit mirror, and a 1.2 m (47″) reflector, in operation since 1962. They are used mainly for spectroscopy.

Donati's comet (1858 VI). A spectacular *comet famous for its *coma with multiple *haloes*: parabolic envelopes with vertices towards the sun and foci near the apparent *nucleus. These haloes were regular and sharp and are thought to have been produced by repetitive ejection of material from a single active area exposed successively to solar radiation as the solid cometary nucleus rotated every 4.6 hours.

doppler broadening. Broadening of spectral lines due to the random motion of emitting or absorbing atoms. As a result of the *doppler effect, atoms moving away from the observer show lines with a slight shift to longer wavelengths; atoms moving towards the observer show a slight shift to shorter wavelengths. The overall effect is that the line is broader than the natural width (determined by quantum mechanical uncertainty).

The motion of the atoms may be due to thermal motion, in which case the effect is larger for lighter atoms. Turbulence of stellar material, rapid rotation of a star, or an expanding stellar atmosphere can also produce doppler broadening. *See also* line broadening.

doppler effect. A change in apparent frequency – and hence wavelength – of a wave motion as a result of relative motion of the source and observer. For *electromagnetic radiation emitted from a moving source, the magnitude of this change is known as the *doppler shift*. For a source moving away from the observer, the observed wavelength is longer than it would be if source and observer had no relative motion along the line joining them. This change to longer wavelengths (i.e. towards the red end of the visible spectrum for visible wavelengths) is called a *redshift. Conversely, if the source is moving towards the observer, there is a change to shorter wavelengths, i.e. a *blueshift.

When there is no relative motion between source and observer, the wavelength of a spectral line can be given by λ. If the relative velocity along the line of sight of source and observer is v, the change in wavelength ($\Delta\lambda$) of the spectral line is given by

$$\Delta\lambda/\lambda = v/c$$

where c is the speed of light. For values of v comparable with c, a relativistic expression is used:

$$\Delta\lambda/\lambda = [(c + v)/(c - v)]^{1/2} - 1$$

v is taken to be positive for a receding object and negative for an approaching one. Doppler shifts can be observed in all regions of the electromagnetic spectrum.

The doppler effect was first suggested by the Austrian physicist C.J. Doppler in 1842. It has proved invaluable in astronomy for measuring the *radial velocities of celestial objects and is also used for determining the orbital motions (and in some cases the masses) of *spectroscopic binaries and the rotational motions of bodies. *See also* doppler broadening.

doppler shift. *See* doppler effect.

Dorado (Swordfish). A small constellation in the southern hemisphere, the brightest star being of 3rd magnitude. It contains a large portion of the Large Magellanic Cloud in which lies the extremely luminous star *S Doradus and the complex Tarantula nebula, 30 *Doradus. Abbrev.: Dor;

genitive form: Doradus; approx. position: RA 5.5h, dec −60°; area: 179 sq deg.

30 Doradus (Tarantula nebula; NGC 2070). A very extensive very luminous *emission nebula in the Large *Magellanic Cloud (LMC) that is a grouping of *HII regions characterized by rapid complex motions. It is the brightest object (absolute magnitude −19) in the LMC and is visible to the naked eye. It is also the strongest radio source in the LMC. The most luminous object in the nebula, both at optical and ultraviolet wavelengths, lies near the centre and is called R136. It is surrounded by dozens of fainter yet still very bright stars. R136 has at least three components, designated a, b, and c. R136a is an intense UV source with a powerful *stellar wind. Observations indicate that it could be an extremely dense cluster of massive stars or it might be or contain a single *supermassive star of maybe 1000 solar masses or more.

dorsum. A ridge. The word is used in the approved name of such a surface feature on a planet or satellite.

double cluster. A pair of *clusters that lie relatively close together. A well-known example is the pair *h and Chi Persei, in which the individual clusters appear to be of similar age.

Double Cluster in Perseus. *See* h and Chi Persei.

double galaxy. Two galaxies observed in close proximity. *Physical doubles* are spatially close pairs assumed to be in orbit about a common centre of mass. *Optical doubles* are widely separated pairs that only appear close because of a chance alignment along the line of sight. Although orbital motions cannot be directly observed, the relative radial velocities of galaxies in a gravitationally bound pair can be used to estimate their total mass.

double-lined spectroscopic binary. *See* spectroscopic binary.

double quasar. *See* gravitational lens.

double radio source. *See* radio source structure.

double star. A pair of stars that appear close together in the sky. There are two types: *optical double stars* and *physical double stars*. Stars in an optical double, such as *Deneb, appear close only because they happen to lie in very nearly the same direction from earth; they are actually too far apart to be members of a dynamically linked system. Stars in a physical double are sufficiently close together for their motions to be affected by their mutual gravitational attraction. They are more usually known as *visual binaries.

Draco (Dragon). An extensive straggling constellation in the northern hemisphere near Ursa Major. The brightest stars are the 2nd-magnitude yellowish giant Eltanin (γ) and several of third magnitude including *Thuban (α), the *pole star in ancient times. The area contains several binaries, such as Psi Draconis, and the bright planetary nebula NGC 6543. Abbrev.: Dra; genitive form: Draconis; approx. position: RA 9.5 − 20.5h, dec 50° − 85°; area: 1083 sq deg.

draconic month (nodical month). The time interval of 27.2122 days between two successive passages of the moon through the ascending *node (or descending node) of its orbit.

Draco system. A member of the *Local Group of galaxies.

Draper classification. *Another name for* Harvard classification. *See* spectral types.

draw tube. A simple tube of 31.75 mm bore into which fits the telescopic *eyepiece in use. It is equipped with a mechanism, such as a rack and pinion, allowing fine adjustment of the eyepiece position in order to obtain a sharp focus. Some draw tubes are threaded to accomodate eyepieces with an 'RAS thread' but these are not very common.

drift chamber. A position-sensitive detector of ionizing radiation. The point of formation of a cloud of electrons, caused by a particle ionizing a gas, is determined by timing the constant velocity drift of the cloud towards a fine wire anode, under the influence of àn electric field. Drift chambers are used in high-energy gamma-ray telescopes for detecting photons by measuring the tracks of the electron-positron pair after a *pair production absorption.

drift scan. A profile of a *radio source made with a radio telescope by aiming the *antenna at a point beside the source and allowing the rotation of the earth to move the source through the antenna beam.

drive clock. *See* equatorial mounting.

Dubhe (α UMa). An orange-yellow giant that is one of the seven brightest stars in the *Plough and is one of the two *Pointers. It is a close visual binary (period 44 years). m_v: 1.81; M_v: −0.6; spectral type: G9 III; distance: 30 pc.

Dumbbell nebula (M27; NGC 6853). A *planetary nebula about 220 parsecs away in the constellation Vulpecula. It has an hourglass shape with a large angular diameter (330 arc seconds) and low surface brightness.

durchmusterung. *German* A star catalogue.

dust. *See* cosmic dust.

dust tail. *See* comet tails.

dwarf. *Short for* dwarf star.

dwarf Cepheids (AI Velorum stars; RRs stars). A group of *pulsating variables that until recently were classed as *RR Lyrae stars but have been shown to have a lower luminosity in addition to a shorter period of about 1.3 to 5 hours. The variation in luminosity (i.e. *amplitude) ranges from 0.2 or 0.3 up to 1.2 but can be higher. At the lower amplitudes they are difficult to differentiate from *Delta Scuti stars but are apparently older. Instabilities of period and light-curve shape are characteristic of these stars and many exhibit the Blazhko effect found in some *RR Lyrae stars. The pulsations are probably mixtures of oscillations of a fundamental mode and those at higher harmonic frequencies (*see* pulsating variables).

dwarf galaxy. A *galaxy that is unusually faint either because of its very small size, its very low surface brightness, or both. Since galaxies exist in a continuous range of sizes from the giant ellipticals downwards, the dividing line between average and dwarf is somewhat arbitrary. Since no spiral or S0 galaxies have been observed with total magnitudes below − 16, this is often used as a convenient demarcation line. Dwarf galaxies can only be seen when comparatively nearby so that their true abundance and distribution are uncertain. Also the lower limit to galactic size is not known: dwarf galaxies in the *Local Group have diameters ranging down to 300 parsecs but there may well be smaller and fainter systems still to be detected.

dwarf novae. A small group of intrinsically faint stars that are characterized by sudden increases in brightness occurring at intervals of a few weeks or months, the maximum brightness lasting only a few days. The change in brightness (i.e. *amplitude) is between 2–6 magnitudes. The first to be discovered, U Geminorum, is typical of the majority, which are therefore classified as *U Geminorum stars*. This subgroup displays a fairly smooth decline in brightness from the maximum, unlike the much smaller subgroup of *Z Camelopardalis stars* that can undergo *standstills*, i.e. periods of nearly constant intermediate brightness, before dropping to minimum brightness. Both the occurrence of the standstill and its duration – a few days to many months – are quite unpredictable. There are also periods of erratic light variations. The brightest and most carefully observed dwarf nova is *SS Cygni.

Dwarf novae are a class of *cataclysmic variables, i.e. close *binary stars in which the primary is a *white dwarf. The secondary is a cooler main-sequence star, spectral type K or G. The components have similar masses (about 0.7 to 1.2 solar masses); the orbital periods are between about 3 to 15 hours. The secondary is undergoing irregular expansion and in doing so can fill one Roche lobe of the *equipotential surfaces of the system. Hydrogen-rich gas then streams from the secondary and takes up a disc-shaped orbit around the primary. The outbursts occur when this disc brightens up, probably as a result of instabilities when its hydrogen changes from opaque to transparent and back. It does not involve an explosion, and no significant amount of mass is ejected. The gas in the disc spirals down on to the white dwarf, where it may eventually cause a nova explosion.

dwarf star (dwarf). Any star lying on the *main sequence of the *Hertzsprung-Russell diagram. The term arises from the early dual classification of stars into giants (*g*) and dwarfs (*d*): the sun was classified as *d*G2. This system has been superseded by that of luminosity classes (*see*

spectral types), and these 'dwarf' stars are of luminosity class V. The term *main-sequence star* is now more common, especially to prevent confusion with *white dwarfs; late-type main-sequence stars, however, are still referred to as *red dwarfs*.

dynamical equinox. The point of intersection of the ecliptic with the celestial equator at which the sun's declination changes from south to north. It is thus the ascending *node of the earth's mean orbit on the earth's equatorial plane. This point moves slowly with time. *See also* equinox; catalogue equinox.

dynamical parallax. An iterative method used to determine the distance and the mass of a *visual binary star, usually from measurements of the orbital period (*P* years) and apparent mean orbital size (*l″*) and an estimate of the mass of the two stars ($M_1 + M_2$ solar masses). The approximate semimajor axis (*a* AU) of the orbit is calculated from *Kepler's third law:

$$(M_1 + M_2)P^2 = a^3$$

The ratio of *a/l* gives an estimate of the distance of the binary. If the apparent magnitude is known, the absolute magnitude can be deduced from the *distance modulus and a more accurate value of total mass is hence obtained using the *mass-luminosity relation. The process is repeated as necessary using the corrected mass estimates, an inaccurate estimate in mass leading to a much smaller error in distance.

dynamical time. The family of time scales introduced in 1984 to replace *ephemeris time, ET, as the independent argument of dynamical theories and ephemerides. Like ET these time scales are independent of the earth's rotation.

Terrestrial dynamical time (TDT) is a unique time scale independent of any theory. It is used for apparent geocentric ephemerides. For practical purposes

TDT = TAI + 32.184 seconds

where TAI is the *International Atomic Time. This takes advantage of the direct (broadcast) availability of coordinated *universal time, UTC, which is an integral number of seconds offset from TAI. Continuity with pre-1984 practices has been achieved by setting the difference between TDT and TAI to the 1984 estimate of the difference between ET and TAI. The increment to be applied to *universal time, UT1, to give TDT is called delta T (ΔT). The unit of TDT is the day, 86 400 seconds, 24 hours, at mean sea level.

Barycentric dynamical time (TDB) is used for ephemerides and equations of motion that are referred to the *barycentre of the solar system. TDB differs from TDT only by periodic variations.

dynamics. The study of the motion of material bodies under the influence of forces. Newton's laws of motion form the basis of classical dynamics. When speeds approach the speed of light the special theory of *relativity must be taken into account.

E

Eagle nebula (M16). A large *emission nebula – an HII region – that is located about 2.5 kiloparsecs away in the direction of the constellation Serpens. It contains some hot young stars that are part of an open cluster.

early-type stars. Hot stars of *spectral type O, B, and A. They were originally thought, wrongly, to be at an earlier stage of evolution than *late-type stars.

earth or **Earth.** The third planet of the solar system and the only one known to possess life. It is the largest of the inner planets (equatorial radius: 6378 km) and has one natural satellite, the *moon. A summary of the earth's orbital and physical characteristics are given in Table 1, backmatter. The earth has a substantial *atmosphere, mainly of nitrogen and oxygen, and a *magnetosphere linked to a magnetic field. (*See* atmosphere, composition; atmospheric layers; geomagnetism; Van Allen radiation belts.) Two-thirds of the planet is covered by water, ocean depth ranging from 2500 to 6500 metres; the average land elevation is 860 metres. *See also* geoid.

On average the earth's surface transfers to the atmosphere an amount of energy equal to that it absorbs. The earth's *mean surface temperature*, i.e. that needed to keep earth and atmosphere in thermal equilibrium, is 13°C. Temperatures in the atmosphere generally decrease from equator to poles and from low to high atmospheric altitudes. These temperature gradients drive the circulation of the atmosphere. On average the earth emits into space an amount of radiant energy equal to that absorbed by surface plus atmosphere. The earth's *effective temperature*, i.e. that needed to keep earth in thermal equilibrium with space, is −18°C; this is the earth's temperature as measured from space.

Studies of the refraction and reflection of seismic waves propagating through the earth show that it consists of three main internal layers: the crust, mantle, and core. The *crust* has a thickness of about 20–70 km under the continents and an average of 6 km under the oceans. It has a density of about 3 times that of water. It consists largely of sedimentary rocks, such as limestone and sandstone, resting on a base of igneous rocks, such as granite (under the continents) and basalt (under both the continents and oceans). The *mantle* extends to a depth of 2900 km where its density reaches 5.5 times that of water. Its composition is thought to include a high proportion of silicate rocks. The rocks of the *lower mantle*, which starts at a depth of about 650 km, have a higher density than those of the *upper mantle*. The *core* increases in density from 10 times that of water at its junction with the mantle to 13 times at the centre. It is composed predominently of iron with several, possibly many, additional components. The *inner core* has a radius of about 1200 km and is solid. The *outer core* is liquid and is regarded as the seat of the earth's magnetic field (*see* geomagnetism).

It is believed that the pressure at the earth's centre is about 400 gigapascals (4 million atmospheres), while the internal temperature rises from 1200 kelvin (K) at a depth of 100 km, to 3000 K at 1000 km, and may exceed 5000 K at the centre. The heat required to maintain these temperatures is derived from the natural radioactivity of the earth's constituent rocks, but would have been greatly augmented soon after formation by gravitational compression and the impact of meteoritic material. This led to the widespread melting and *differentation that produced the present layered structure.

After 4.6 thousand million years the earth's internal heat is still a source of mechanical power, producing earthquakes and volcanic eruptions, raising mountains, and moving continent-sized blocks about its surface. It is regarded as a convective heat engine. According to the theory of *plate tectonics*, the crust and part of the upper mantle form the *lithosphere*, which is broken up into fairly rigid *plates*. The plates, all about 100 km thick, float on the less rigid material below the lithosphere; there are six major plates and many smaller ones. Convection within the upper mantle causes the plates, with their associated continental and/or oceanic crust, to move relative to each other:

the phenomenon of *continental drift*. The relative motion amounts to a few cm per year. At *transform* boundaries between two plates the plates can slide past each other. At *divergent* boundaries two plates are moving apart. Where this occurs in mid-ocean, molten rock from the mantle is injected into the crust. In continental lithosphere divergence produces stretching and attenuation and faults develop; it can culminate in the separation of two continental bodies and formation of a new oceanic basin. At *convergent* boundaries two plates are moving towards each other and one plate generally plunges under the other in a process called *subduction*.

See also age of the earth; solar system; solar system, origin.

earthgrazers. Members of the *Amor and *Apollo groups of minor planets, such as *Adonis, *Aten, *Hermes, and *Icarus, all having orbits that can bring them within about 0.1 AU (15 million km) of the earth.

earth-rotation synthesis (super-synthesis). *See* aperture synthesis.

earthshine (ashen light). Sunlight reflected from earth. Close to new moon, earthshine reflected by the moon back to the earth enables the whole lunar disc to become visible – the old moon in the new moon's arms. An observer on the moon sees the earth illuminated by earthshine and at new moon can measure the earth's *albedo.

eastern elongation. *See* elongation.

east point. *See* cardinal points.

eccentric anomaly. *See* anomaly.

eccentricity. Symbol: e. A measure of the extent to which an elliptical orbit departs from circularity. It is given by the ratio $c/2a$ where c is the distance between the focal points of the ellipse

planet	eccentricity
Mercury	0.205 628
Venus	0.006 787
Earth	0.016 722
Mars	0.093 377
Jupiter	0.048 45
Saturn	0.055 65
Uranus	0.047 24
Neptune	0.008 58
Pluto	0.249

and $2a$ is the length of the major axis. For a circular orbit $e = 0$. The major planets and most of their satellites have an eccentricity range of 0–0.25 (see table). Many comets and some of the minor planets and planetary satellites have very eccentric orbits. The eccentricity of an orbit varies over a long period due to changing gravitational effects: that of the earth's orbit varies between ~0.005 to 0.06 in a period of ~100 000 years. *See also* conic sections.

echelle grating. A *diffraction grating with very fine lines ruled much further apart than in other gratings. Such a grating has very high resolution but only over a fairly narrow wavelength band. The resolving power at a given wavelength depends primarily on the total ruled width and the angles of incidence and refraction; a high resolution can be obtained with coarse ruling if the grooves are properly shaped to reflect light, etc., in a narrow bundle. The echelle grating of the *echelle spectrograph* is used at high-order diffraction to give larger angular dispersion. Since the high orders overlap, cross dispersion with a second grating or prism must be used so that the spectral images of the different orders can be separated and stacked above one another at the detector.

eclipse. The total or partial obscuration of light from a celestial body as it passes through the shadow of another body. A planetary satellite is eclipsed when it passes through the shadow of its primary or another satellite. An eclipse of the sun is strictly an *occultation.

An eclipse of the sun – a *solar eclipse* – or of the moon – a *lunar eclipse* – occurs when the sun, moon, and earth lie in or nearly in a straight line (see illustration, *a*). If the plane of the moon's orbit lay exactly in the plane of the *ecliptic a solar eclipse would take place at each new moon and a lunar eclipse at each full moon. The two planes are however inclined at an angle of about 5°, intersecting at the *nodes of the moon's orbit. Eclipses are only observed when the sun is at or near a node and the moon is near the same node (solar eclipse) or the opposite one (lunar eclipse). The *ecliptic limits* are the maximum angular distances of the new or full moon from its node for an eclipse to take place.

Although the moon is 400 times smaller than the sun, it is also about 400 times nearer the earth. As a result, sun and moon have almost exactly the same angular size (about $\frac{1}{2}$°), so that it is possible for the moon to obscure the sun. The earth and moon both cast shadows in sunlight, the shadow having a dark cone-shaped inner region – the *umbra – and an outer lighter penumbral region. A solar eclipse occurs, between sunrise and sunset at new moon, when the moon passes directly in front of the sun so that the earth lies in the moon's shadow (see illustration, *b*). When the moon is sufficiently close to earth so that its apparent diameter exceeds that of the sun, then the umbra of the moon's shadow can just reach the earth's surface. It moves in a general west to east trend over a very narrow curved zone of the surface, known as the *path of totality*, which can be up to 250 km wide but

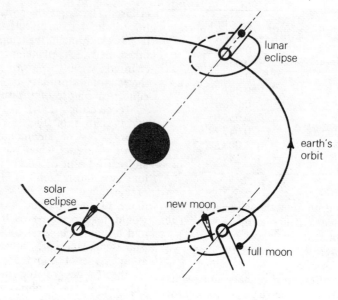

--- · --- · --- line of nodes

a Solar and lunar eclipses

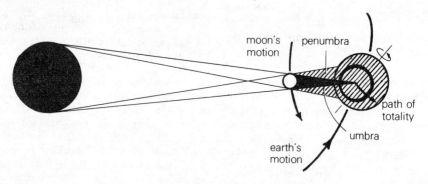

b Lunar shadow in solar eclipse

averages about 160 km. An observer at a point where only the penumbra will move past sees a *partial eclipse,* in which only part of the sun is obscured. An observer in the path of totality will experience a *total eclipse,* in which the sun is completely obscured. If the moon is far enough away to appear smaller than the sun, a rim (or annulus) of light will be seen around the eclipsed sun and an *annular eclipse* occurs. The period of annularity never exceeds 12.5 minutes and is normally much less.

In a total solar eclipse, *first contact* occurs when the moon just appears to touch the sun's western limb. As the moon gradually covers the sun, the landscape darkens and animals become disturbed. *Totality* begins at *sec-*

ond contact when the sun disappears from sight. The maximum duration of totality is 7m40s but is usually much less. Totality ends at *third contact*, just as the crescent sun emerges, and at *fourth contact* the whole disc of the sun is once more seen. The time between first and last contact can approach four hours. During totality the *chromosphere, *corona, and other phenomena can be observed and studied. There are between two and five solar eclipses each year. Total eclipses are however very rare at any particular place: the next UK occurrence is in 1999.

A lunar eclipse occurs, at full moon, when the moon passes into the shadow cone of the earth. It can be seen from any place at which the moon is visible above the horizon. A *total eclipse* occurs when the moon enters completely into the umbra of the earth's shadow. If only part of the moon enters the umbra the eclipse is *partial*. When the moon only enters the penumbral region, a *penumbral eclipse* takes place in which a slight, usually quite unappreciable darkening of the moon's surface occurs. The maximum duration of totality is 1h42m. The moon can usually be seen throughout totality, being illuminated by sunlight refracted by the earth's atmosphere into the shadow area. Since the bluer wavelengths are removed by scattering, the moon has a coppery-red colour. There are up to either two or three lunar eclipses each year. Up to seven eclipses can occur in one year, either five solar and two lunar or four solar and three lunar.

See also Saros.

eclipse year. The period of time between two successive passages of the sun through the same *node of the moon's orbit. It is equal to 346.620 days, being less than the *sidereal year because of *regression of the moon's nodes. There are almost exactly 19 eclipse years in one *Saros.

eclipsing binary. A *binary star whose orbital plane is orientated such that each component is totally or partially eclipsed by the other during each orbital period. The effect observed is a periodic decrease in the light from the system (see illustration). The deeper minimum corresponds to the eclipse of the brighter star. The light curve gives the period of revolution, and from the depths and shapes of the minima it is possible to estimate the inclination of the orbital plane. The duration of the eclipses compared to the time between eclipses indicates the radii of the stars in terms of the distance between them. If the system is also a double-lined *spectroscopic binary the individual masses and radii of the stars can be calculated.

Eclipsing binaries tend to be composed of large stars with small orbits and therefore the majority are close binaries (*see* binary star). In these systems *mass transfer affects the stars' evolution, leading to three main classes of eclipsing binaries. *W Ursae Majoris stars are in contact. In *W Serpentis stars one component is transferring mass rapidly to the other. In these two classes the stars are distorted into ellipsoids, an effect that shows in the light curve. *Algol variables represent a later stage; mass transfer is almost complete and the stars are nearly spherical.

ecliptic. The mean plane of the earth's orbit around the sun. Although the orbital plane is in fact defined by the motion of the centre of mass of the earth-moon system around the sun, the ecliptic refers to the earth alone, the resulting errors being negligible. The ecliptic may thus be taken as coincident with the sun's apparent annual path across the sky. The orbits of the moon and planets, apart from Pluto, lie very near the ecliptic.

The planes of the ecliptic and celestial equator are inclined at an angle equal to the tilt of the earth's axis. This angle, known as the *obliquity

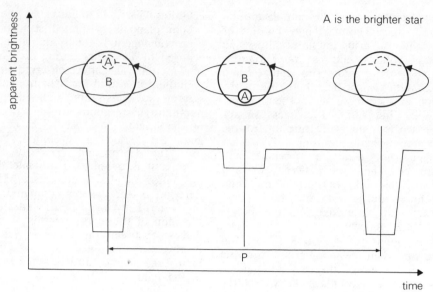

A is the brighter star

Schematic of orbits and light curve of eclipsing binary

of the ecliptic, is about 23°27′. The *equinoxes lie on the celestial sphere at the two points of intersection of ecliptic and celestial equator. The *poles of the ecliptic* lie at 90° from all points on the ecliptic at the positions RA 18h, dec +66° and RA 6h, dec −66°.5.

ecliptic coordinate system. A *coordinate system in which the fundamental reference circle is the *ecliptic and the zero point is the vernal (or dynamic) *equinox (♈). The coordinates are *celestial* (or *ecliptic*) *latitude* and *celestial* (or *ecliptic*) *longitude* (see illustration).

The celestial latitude (β) of a star, etc., is its angular distance (from 0° to 90°) north (counted positive) or south (counted negative) of the ecliptic; it is measured along the great circle through the body and the poles of the ecliptic. The celestial longitude (λ) of a body is its angular distance (from 0° to 360°) from the vernal equinox, measured eastwards along the ecliptic to the intersection of the body's circle of longitude; it is measured in the same direction as the

sun's apparent annual motion. Although observations are taken from the earth's surface the coordinates should strictly be geocentric and, tabulated as such, are universally applicable. A slight correction is therefore applied to convert surface (topocentric) observations to geocentric values.

The ecliptic system is older but less used than the *equatorial and *horizontal coordinate systems. It is sometimes used to give the positions of the sun, moon, and planets. *See also* heliocentric coordinate systems.

ecliptic limits. *See* eclipse.

E-corona. The part of the solar *corona that consists of relatively slow-moving ions and exhibits an *emission spectrum superimposed on the *continuum of the *K-corona. The emission lines have shown that the ions of the E-corona are very highly ionized, some species having lost up to 15 electrons. At visible wavelengths only *forbidden lines are present; the most prominent of these are the 'red' line at 637.4 nanometres (due to the

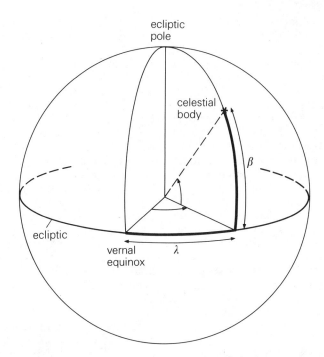

ecliptic
pole

celestial
body

β

ecliptic

vernal
equinox

λ

Ecliptic coordinate system

presence of Fe X), the 'green' line at 530.3 nm (due to Fe XIV), and the 'yellow' line at 569.4 nm (due to Ca XV).

ecosphere. The zone of space around the sun or some other star in which a planet would experience external conditions that are not incompatible with the existence of life. Prevailing conditions, as of atmospheric content or pressure, might however prove too hostile.

effective area of a radio telescope. *See* array.

effective temperature. Symbol T_{eff}, T_e. The surface temperature of a star, expressed as the temperature of a *black body (i.e. a perfect radiator) having the same radius as the star and radiating the same total amount of energy (E) per unit area per second. A star is thus being considered a

black body and its effective temperature is thus given by *Stefan's law:

$$T_{eff}^4 = E/\sigma = L/4\pi R^2$$

σ is Stefan's constant and L and R are the star's luminosity and radius. Although accurate determinations of effective temperature are difficult, each *spectral type has a characteristic range of T_{eff}; the total range is from about 2500 to 40 000 kelvin. The sun's effective temperature is 5800 K. The effective temperature is the best measure of the actual temperature of the gas in a star's outer layers.

Radiation temperature, T_r, is equivalent to the effective temperature measured not over the whole spectral range but over a narrow portion of the spectrum, for example at visible wavelengths, which gives the *optical temperature*.

Einstein–de Sitter universe. *See* cosmological models.

Einstein Observatory. NASA's second High-Energy Astrophysical Observatory, HEAO-2, launched in Nov. 1978 for x-ray observations and operated until loss of control gas in Apr. 1981. The unique capabilities of its 60 cm *grazing incidence telescope revolutionized x-ray astronomy, in the same way *Uhuru had done a decade before. High-resolution maps were obtained of many extended sources (clusters of galaxies, supernova remnants), the first high-resolution spectroscopy was achieved on a few bright sources, and the enhanced sensitivity led to the discovery of several thousand faint sources (down to ~ 10 nanojansky), mainly identified with normal galactic stars or active galaxies (quasars) at high redshifts.

Einstein shift. See gravitational redshift.

ejecta. Material excavated during the formation of a *crater or *basin. See also ejecta blanket; secondary craters.

ejecta blanket. A deposit of *ejecta from a *crater or *basin, the thickness of which decreases with distance from the crater or basin rim. The surface may exhibit transverse dunes, radial furrows, and *secondary craters.

Elara. A small satellite of Jupiter, discovered in 1905. See Jupiter's satellites; Table 2, backmatter.

E-layer (E-region). See ionosphere.

ELDO. Abbrev. for European Launcher Development Organization. See European Space Agency.

electromagnetic force. See fundamental forces.

electromagnetic radiation. A flow of energy that is produced when electrically charged bodies, such as electrons, are accelerated. Light is the most familiar form of such radiation.

It can equally be considered as a wave motion or as a stream of particles.

Electromagnetic waves consist of oscillating electric and magnetic fields that travel through space or through air, etc., carrying energy. The two fields vibrate in planes at right angles to each other and to the direction of wave motion, and are self-propagating. An electromagnetic wave thus requires no medium for its propagation (unlike sound waves). It travels through a vacuum at the *speed of light, c, which is a fundamental constant equal to about 3×10^5 km s^{-1}. The speed is slightly reduced on entering a medium, such as air or glass. Like other periodic waves, electromagnetic waves have a *wavelength λ and a *frequency ν, which are related by $\lambda \nu = c$.

Electromagnetic radiation can manifest itself as both waves and particles, or as either, and thus has a dual nature. Although *reflection, *interference, and *polarization can be explained in terms of wave motion, a particle-like nature must often be invoked, as when radiation interacts with atoms or molecules. A particulate nature was originally proposed by Newton but in its present form the concept is part of *quantum theory. Light and other kinds of radiation are considered as streams of particles travelling at the speed of light, c, and having zero mass; the particles are tiny packets of energy, called *photons. Energy is absorbed or emitted by matter in discrete amounts, i.e. in the form of photons. The photon is thus a quantum of electromagnetic radiation. The energy, E, of the photons is related to the frequency ν, of the radiation by $E = h\nu$, where h is the *Planck constant.

The range of frequencies (or wavelengths) of electromagnetic radiation is known as the electromagnetic spectrum (see illustration). The spectrum can be divided into various regions, which are not sharply delineated.

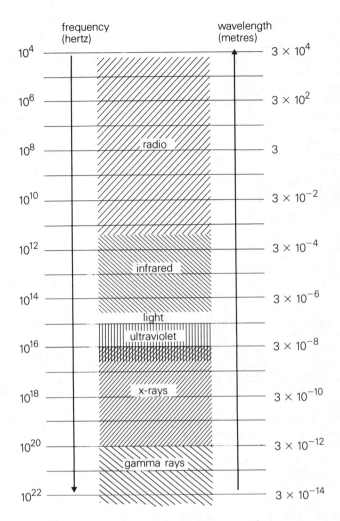

Electromagnetic spectrum

These regions range from low-frequency low-energy *radio waves through *infrared radiation, *light, *ultraviolet radiation, *x-rays, to high-frequency high-energy *gamma rays.

electromagnetic spectrum. *See* electromagnetic radiation.

electromagnetic wave. *See* electromagnetic radiation.

electron Symbol: e. A stable *elementary particle (a lepton) that has a negative *charge of 1.6022×10^{-19} coulomb, a mass of 9.1096×10^{-28} gram, and *spin ½. Electrons are constituents of all *atoms, moving around the central nucleus. They can also exist independently. The *antiparticle of the electron is the *positron*, e^+ or $\bar{e}$.

electronic imaging. *See* imaging.

electronographic camera. A highly sensitive electronic device that can be used for recording astronomical images. It is an *image tube in which the beam of electrons liberated from the photocathode is accelerated by an electric field and focused by a magnetic field on to a film coated with an *electronographic emulsion*. This is a very fine grain high-resolution 'nuclear' photographic emulsion in which electrons are recorded directly: every high-energy electron leaves a developable track so that an image can be produced by a process similar to that for a photograph. The resulting *electronograph* looks similar to a photograph with the important difference that the density at any point on the electronograph is proportional to the intensity at the corresponding point in the optical image through almost the whole intensity range. The device thus has a highly linear response. Photometric information can be obtained directly by measuring the density of the image. The device also has a high *quantum efficiency.

electronvolt. Symbol: eV. A unit of energy equal to the energy acquired by an electron falling freely through a potential difference of one volt. It is equal to 1.6022×10^{-19} joule. High-energy *electromagnetic radiation is usually referred to in terms of the energy of its photons: a photon energy of 100 eV is equivalent to a radiation frequency of 2.418×10^{16} hertz. The energies of *elementary particles are usually quoted in eV; their *rest masses are generally referred to in terms of their energies in eV.

element. Any of a large number of substances, including *hydrogen, *helium, carbon, nitrogen, oxygen, and iron, that consist entirely of *atoms of the same *atomic number, i.e. with the same number of protons in their nuclei. The atoms are not all identical however: *isotopes of an element can occur with different numbers of neutrons in their nuclei. When arranged in order of increasing atomic number along a series of horizontal rows, elements with similar chemical and physical properties fall into groups. The similarities in properties arise from similarities in the electron arrangements within the atom. This table of elements is called the periodic table. Over 100 elements are now known, more than 90 of which occur naturally on earth. The elements have stable isotopes and/or unstable (i.e. radioactive) isotopes.

The creation of the elements and the means by which they become distributed throughout the universe are major areas of study in astronomy (*see* nucleosynthesis). The abundance of each element in the universe, i.e. its *cosmic abundance, depends on the method of its synthesis – by nuclear reactions in stars, by *cosmic-ray collisions, etc. – and on its lifetime in its immediate surroundings and its long-term stability.

elementary particles. The basic building blocks of matter. At present the only truly elementary particles containing no internal subcomponents are believed to be *leptons and *quarks. Leptons include the *electron, *muon, the massive tau lepton, the *neutrinos, and their associated *antiparticles. Experiments show that if electrons possess internal constituents they must be smaller than 10^{-14} mm in size. Until recently the constituents of atomic nuclei (*neutrons and *protons) were thought to be elementary but it is now known that they possess internal structure in the form of triplets of quarks. Particles with quark constituents are called *hadrons*. Those, like the proton, composed of three quarks are called *baryons*. Particles known as *mesons* are produced by a variety of high-energy reactions and decay into stable particles. Mesons, which include the *pions, are composed of quark and antiquark pairs. Leptons participate in weak, electromagnetic, and gravitational interactions but not strong interactions (*see* fundamental forces). Hadrons participate in weak, electromagnetic, gravitational, and strong interactions.

elements, abundance. *See* cosmic abundance.

elements of an orbit. *See* orbital elements.

ellipse. A type of *conic section with an eccentricity less than one. The longest line that can be drawn through the centre of an ellipse is the major axis while the shortest line is the minor axis. The two axes are at right angles. There are two foci, which lie on the major axis and are symmetrically positioned on opposite sides of the centre. The sum of the distances from the foci to a point moving round the ellipse is constant and equal to the length of the major axis. An orbiting body moves in an ellipse with the primary at one of the foci. *See also* orbit; Kepler's laws.

ellipsoid. A surface or solid whose plane sections are circles or ellipses. An ellipse rotated about its major or minor axis is a particular type of ellipsoid, called a *prolate spheroid* (major axis) or an *oblate spheroid* (minor axis).

ellipsoidal reflector. A reflector whose surface is part of an ellipsoid. An object situated at one focus will be imaged, after reflection, at the other focus.

elliptical galaxy. *See* galaxies; Hubble classification.

ellipticity. *See* flattening.

El Nath (β Tau). A bluish-white giant that is the second-brightest star in the constellation Taurus. It was originally part of Auriga, as Gamma Aurigae.

m_v: 1.65; M_v: -2.9; spectral type: B7 III; distance: 80 pc.

elongation. The angular distance between the sun and a planet, i.e. the angle sun-earth-planet, measured from $0°$ to $180°$ east or west of the sun. It is also the angular distance between a planet and one of its satellites, i.e. the angle planet-earth-satellite, measured from $0°$ east or west of the planet. An elongation of $0°$ is called *conjunction, one of $180°$ is *opposition, and one of $90°$ is *quadrature (see illustration). When an inferior planet follows the sun in its daily motion, appearing east of the sun in the evening, it is in *eastern elongation*. When it precedes the sun, appearing west of the sun in the morning, it is in *western elongation*. The inferior planets, which cannot come to quadrature, reach positions of *greatest elongation (GE)*. The GE for both eastern and western elongation varies from $18°$ to $28°$ (Mercury) and from $45°$ to $47°$ (Venus).

Eltanin. *See* Draco.

Elysium Planitia. Mars' second main centre of volcanic activity, located on a bulge in the Martian crust about 3 km high and 2000 km in diameter. Its principal volcano is *Elysium Mons,* which is 250 km in base diameter and 15 km high. *See* Mars, volcanoes.

emersion. The reappearance of a celestial body following its eclipse or occultation.

emission lines. *See* emission spectrum.

emission measure. *See* thermal measure.

emission nebula. A region of hot interstellar gas and dust that shines by its own light. The gas atoms in the cloud are ionized – stripped of one or more electrons. The electrons liberated by the ionization process are free

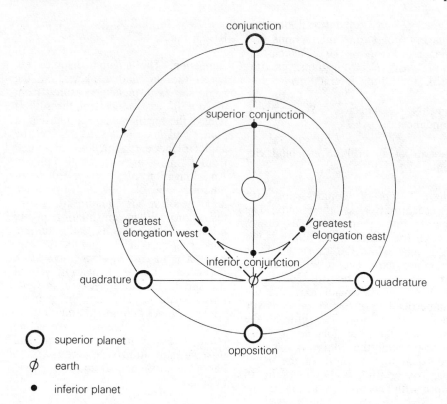

conjunction

superior conjunction

greatest
elongation west

greatest
elongation east

inferior conjunction

quadrature

quadrature

opposition

○ superior planet

∅ earth

● inferior planet

Elongations of superior and inferior planet

to move about the nebula. The light, and other radiation, emitted by the nebula results from subsequent interactions between ions and free electrons.

There are three classes of emission nebula, all of extremely low gas density. In *HII (ionized hydrogen) regions, such as the *Orion nebula, the gas is ionized by ultraviolet (UV) radiation from nearby hot young O and B stars; if there is sufficient gas close to one or more such stars a luminous nebula will form, surrounded by cooler neutral gas. The size is dependent on gas density and ultraviolet flux (*see* Strömgren sphere), ranging up to several parsecs. The temperature is always about 10 000 kelvin. In *planetary nebulae, typified by the *Ring nebula, the gas is ionized by UV radiation from the remnant of a dying

star lying within the nebula; the nebula comprises gas expelled from this star. In *supernova remnants the ionization mechanisms are very complex: the atoms can be ionized by ultraviolet *synchrotron radiation, as in the *Crab nebula, by interactions following the collision of an expanding remnant with surrounding interstellar gas atoms, as in the *Cygnus Loop, or possibly by other processes.

Emission nebulae contain primarily free electrons and hydrogen ions, together with some helium ions and trace amounts of oxygen, carbon, nitrogen, and other elements. These components can interact in various ways. Firstly, an electron can be recaptured by an ion, usually a hydrogen ion. This *recombination leads to the emission of radio, infrared, and optical radiation. Secondly, an electron can collide with an ion, such as ionized oxygen, without being captured. Instead, energy is transferred from electron to ion, which b'ecomes excited into an *energy level just above the ground state. The excited ion returns to its normal energy state by emitting light: at higher densities these transitions would be 'forbidden' (*see* forbidden lines). Thirdly, free electrons can be decelerated when they pass near an ion, energy being radiated away as weak *bremsstrahlung – mainly radio waves.

The spectacular colours, predominantly red and green, of an emission nebula are due to recombination and excitation. The most intense visible radiation from hydrogen is the red hydrogen alpha emission; red light is also emitted during a forbidden transition of singly ionized nitrogen. The strongest emission, however, results from a forbidden transition of doubly ionized oxygen (O^{++} or O III), the emission being green; the O III ions can only be produced by highly energetic UV photons. *See also* nebula.

emission spectrum. A *spectrum formed by emission of *electromagnetic radiation by matter. Energy has first to be supplied, either as heat (high temperature) or by some other mechanism such as absorption of electromagnetic radiation by the matter or by impact of electrons. The energy raises the atoms or molecules to higher *energy levels. In a simple example, the electron in a hydrogen atom may 'jump' from its normal orbit to one further from the nucleus (*see* hydrogen spectrum). The emission spectrum of hydrogen, or any other emitting material, is formed by transitions from such excited states to lower energy states, the excess energy appearing in the form of photons with characteristic frequencies.

The emission spectrum therefore consists of a specific pattern of narrow peaks – *emission lines* – which occur at these frequencies. These lines

may be superimposed on a *continuum, but not necessarily so. Emission lines can occur in stellar spectra if, for example, the star is surrounded by a hot shell of diffuse gas. *See also* excitation; absorption spectrum.

emissivity. Symbol: ε. A measure of a body's ability to radiate *electromagnetic radiation as compared to that of a perfect radiator – a *black body – at the same temperature.

Enceladus. A satellite of Saturn, discovered in 1789, that has now been found to have a varied surface of ancient and youthful geological developments. Its diameter is only 500 km and its density approximately 1.2 $g\,cm^{-3}$. The surface shows evidence of highly cratered regions (5–30 km diameter), smooth plains, and ridged plains. The valleys and ridges indicate crustal movements that may have been associated by faulting and the extrusion of fluids. Tidal effects from the nearby and larger satellite Dione may account for the production of the surface faulting and may allow the warmer fluid from the interior to resurface portions of the satellite. This situation also occurs on the Jovian satellite *Io. The E ring of Saturn shows a pronounced peak of brightness in the orbit of Enceladus and may consist of particles that have escaped from the satellite. *See also* Table 2, backmatter.

Encke Division. *See* Saturn's rings.

Encke's comet. The most observed comet, having been seen at nearly 50 apparitions. The comet has a period of 3.30 years – one of the shortest – and is thought to be the parent body of the Taurid *meteor stream. The period of Encke is decreasing by up to 2.7 hours per oribital revolution. This is caused by a jet effect from its rotating nucleus, which is not thermally symmetrical. The brightness of Encke has hardly decreased during the 165 years of observation. Since the orbit of Encke is extremely well known and a considerable bank of scientific data has been established, Encke is one of the prime targets for a spacecraft investigation of a comet.

encounter theories (catastrophic theories). Theories that propose that the planets formed from material ejected from the sun or a companion star during an encounter with another object. Although the first such theory, by G.L.L. de Buffon in 1745, postulated that a comet collided with the sun, most later theories have invoked an approach or collision involving another star.

The *planetesimal theory* (1901–05) of T.C. Chamberlin and F.R. Moulton suggested that a passing star caused the sun to eject filaments of material. These condensed into *planetesimals from which the planets formed by *accretion.

Tidal theories (1916–18) of Harold Jeffreys and James Jeans suggested that a star grazed the sun, drawing out into solar orbit a cigar-shaped filament of material that fragmented to form the planets: the larger planets – Jupiter and Saturn – condensed from the thicker central regions of the filament.

Variations on these ideas postulate that the encounter took place between a passing star and a binary companion star to the sun. This overcomes one objection to most encounter theories: that material ejected from the sun could not enter near-circular orbits at the observed planetary distances. However a remaining major problem with all these ideas is that most of the material drawn out from the sun or its postulated companion would have fallen back; any material ejected at stellar temperatures would expand violently and dissipate into space before it could form planets.

A modification of the Jeans-Jeffreys theory, the *capture theory* of M.M. Woolfson, attempts to overcome these problems. It proposes that the filament of matter was drawn off a relatively cool tenuous protostar moving past and tidally distorted by the condensed and more massive sun. Condensations of protoplanets formed from the fragmented filament, of which six were captured by the sun. A gradual rounding-off of the initial elliptical orbits and a major collision between the two innermost bodies led to the present distribution of solar-system bodies.

See also nebular hypothesis; solar system, origin.

energy. Symbol: *E*. A measure of the capacity of a body or system for doing work, i.e. for changing the state of another body or system. The SI unit of energy is the joule; the erg (10^{-7} joule) is also used. There are various forms of energy including heat, mechanical, electrical, nuclear, and radiant energy, all of which are interconvertible in the presence of matter. Mass is also regarded as a form of energy. In any closed system, the total energy, including mass, is always constant. *See also* conservation of mass-energy; nuclear fusion; nuclear fission.

energy level. One of a number of discrete energies that, according to quantum mechanics, is associated with an atomic or molecular system. For instance, a hydrogen atom, which consists of a proton and an orbiting electron, has an energy due to electrostatic interaction between the electron and proton and to the motion of the electron. This energy can only have certain fixed values, which correspond to different orbits of the electron, and these constitute the *electronic energy levels* of the atom. The state of lowest energy is the *ground state* of the atom and higher states (in which the electron is further from the nucleus) are known as *excited states*. Neither the hydrogen atom nor any other type of atom can take up energy continuously but can only 'jump' from one energy state to another: the energy is said to be *quantized*. This behaviour is quite general, applying to energies of vibration and rotation in molecules and to energy difference due to interaction of magnetic moments of the electrons. *See also* excitation; hydrogen spectrum.

energy transport. The flow of heat energy from a hot to a cooler region by means of radiation, convection, or conduction. In stars the energy flows from the hot interior to the surface, where it is radiated away into space. The most important means of energy transport in most stars is *radiative transport*, in which high-energy photons lose energy to the hot plasma through scattering, mainly by free electrons, and by absorption through photoionization of ions; this occurs in the *radiative zone* of the star. (The equation of radiative transport is given at *stellar structure*.) Convection is another mode of energy transport in which collections of hot fluid in the *convective zone of a star rise towards the surface, release their heat energy, and sink again to pick up more energy. Convection is a highly efficient process but the conditions must be right for its onset. Conduction is negligible in most stars but becomes important in very dense degenerate stars, such as white dwarfs.

The mode of energy transport has important effects on a star's evolution: radiation (and conduction) do not affect the shells of different composition within an evolved star, but convection mixes them to a uniform composition.

English mounting. *Another name for* yoke mounting. *See* equatorial mounting.

entropy. A measure of the amount of disorder in a physical system. It never decreases in any physical interaction of a closed system.

entropy per baryon. A cosmological parameter that measures the relative distribution of energy in disordered and ordered forms in the universe. It is determined by the number of photons per baryon (proton or neutron) in the universe today and has a value between 10^9 and 10^{10} according to present observations. *Big bang theories must explain the value of this ratio.

ephemeris (plural: **ephemerides**). **1.** A work published regularly, usually annually, in which are tabulated the daily predicted positions of the sun, moon, and planets for that period, together with information relating to certain stars, times of eclipses, etc. It is used in astronomical observation and navigation.
2. A series of predicted geocentric positions of a celestial object at constant intervals of time.

ephemeris meridian. The fictitious terrestrial meridian that rotates independently of earth at the uniform rate defined by terrestrial *dynamical time. It is
$$1.002\,738\,\Delta T$$
east of the Greenwich meridian (*see* delta T). *Ephemeris longitude* and *ephemeris hour angle* are measured relative to the ephemeris meridian.

ephemeris second. *See* ephemeris time.

ephemeris time (ET). A measure of time for which a constant rate was defined and that was used from 1958 to the end of 1983 as the independent variable in gravitational theories of the solar system. It was replaced in 1984 by *dynamical time. ET was reckoned from the instant, close to the beginning of 1900, when the sun's mean geometric longitude was 279° 41′ 48″.04, at which instant the measure of ephemeris time was 1900 Jan

0d 12h precisely. The primary unit of ET was the *tropical year at this epoch of 1900 Jan 0.5 ET, which contains a calculable number of seconds: the *ephemeris second* is the fraction

$$1/31\,556\,925.9747$$

of the tropical year 1900 Jan 0.5 ET. The ephemeris second was adopted in 1957 as the fundamental invariable unit of time until replaced in 1967 by the second of *atomic time. This SI-adopted second was defined so that it was equal to the ephemeris second within the error of measurement. The ephemeris day contains 86 400 ephemeris seconds. ET could be found by adding a correction, *delta T, to the *universal time UT1.

epicycle. See Ptolemaic system.

epoch. A precise instant that is used as a fixed reference datum, especially for stellar coordinates and *orbital elements. For example, the coordinates right ascension and declination are continuously changing, primarily as a result of the *precession of the equinoxes. Coordinates must therefore refer to a particular epoch, which can be the time of an observation, the beginning of the year in which a series of observations of an object was made, or the beginning of a half century. The *standard epoch* specifies the reference system to which coordinates are referred. Coordinates of star catalogues commonly referred to the *mean equator and equinox of the beginning of a *Besselian year. Since 1984 the *Julian year has been used: the new standard epoch designated J2000.0 is 2000 Jan. 1.5 TDB (barycentric *dynamical time). Epochs for the beginning of a year now differ from the standard epoch by multiples of the Julian year.

epsilon (ε). The fifth letter of the Greek alphabet, used in *stellar nomenclature usually to designate the fifth-brightest star in a constellation.

Epsilon Aurigae. An *eclipsing binary star about 600 parsecs distant in the constellation Auriga, with a period of 27.1 years. It consists of an extremely luminous F2 Ia supergiant about 15 times the mass of the sun with a very large companion of about the same mass. Once thought to be a very distended cool supergiant, the large occulting object is now believed to be a ring or shell of gases surrounding the true companion, a main-sequence B star. The gases arise from rapid *mass transfer from the F star (*see* W Serpentis stars).

Epsilon Eridani. One of the nearest stars to the sun. There is evidence of one or more associated planets. *See also* Table 7, backmatter; Ozma project.

Epsilon Indi. *See* Table 7 (nearest stars), backmatter.

equation of centre. The true *anomaly minus the mean *anomaly of a body moving in an elliptical orbit. It is the difference between the actual angular position of the body and the position it would have if its angular motion were uniform. It is a major *inequality in the moon's motion, being a direct result of the large *eccentricity (0.0549) of the moon's orbit and hence of its varying orbital velocity (*see* Kepler's laws).

equation of light. *See* light-time.

equation of the equinoxes. *See* sidereal time.

equation of time. The amount that must be added to the *mean solar time to obtain the *apparent time, i.e. it is the difference in time as measured by a sundial and by a clock. The equation of time varies continuously through the year; it has two maxima and two minima and is zero on four dates: April 15/16, June 14/15, Sept 1/2 and Dec 25/26. A positive value indicates that apparent time is ahead of mean time; the greatest positive value is 16.4 minutes, the greatest negative value being 14.2 minutes. The curve is the sum of two components, each reflecting a nonuniformity in the apparent motion of the sun: one component arises from the ellipticity of the earth's orbit, the other from the inclination of the ecliptic to the celestial equator (*see* mean sun).

equator. The great circle on the surface of a planet or star that lies in a plane – the *equatorial plane* – that passes through the centre of the body and is perpendicular to the axis of rotation.

equatorial coordinate system. The most widely used astronomical *coordinate system in which the fundamental reference circle is the celestial equator and the zero point is a *catalogue equinox (*see illustration at* coordinate system).

The coordinates are *right ascension (α) and *declination (δ), which are measured along directions equivalent to those of terrestrial longitude and latitude, respectively. *Sidereal hour angle is sometimes used instead of right ascension and *north polar distance instead of declination. Because of the slow westerly drift of the *dynamical equinox around the ecliptic the coordinates right ascension and declination are normally referred to the *mean equator and equinox for a standard epoch, which in present use is now 2000.0.

equatorial horizontal parallax. *See* diurnal parallax.

equatorial mounting. A telescope *mounting in which one axis (the polar axis) is parallel to the earth's axis, while the second axis (the declination axis) is at right angles to it. Its great advantage is that when the telescope is clamped in declination, and the po-

lar axis is driven to turn once in 24 hours in the opposite direction to the earth's rotation, then any star will remain stationary in the *field of view. Until recently almost every large telescope was equatorially mounted. The constantly changing stresses involved in swinging possibly hundreds of tonnes of asymmetrically shaped material about an inclined axis means that the mounting must be extremely strong and hence extremely costly. Many telescopes are now being constructed with a computer-controlled *altazimuth mounting.

There are several types of equatorial mountings, some being shown in the illustration. In the *fork mounting* the telescope swings in declination about an axis carried on two prongs of a fork; the fork itself rotates about a shaft that is the polar axis. In the *yoke* (or *English*) *mounting* the polar axis is in the form of a long frame with a bearing at each end. The telescope swings in declination about an axis between the sides of the frame. This mounting is simple, very rigid, and needs no counterpoise weights; the polar region is however inaccessible. The *horseshoe mounting* is a modification of the fork and yoke mountings: the upper end of the polar axis frame is made into a horseshoe shape to accommodate the telescope tube; the polar region may then be observed. Many of the giant reflectors use this exceptionally stable mounting. In the *German mounting* the declination axis is carried as a tee on the top end of the polar axis. The telescope is carried on one end of the declination axis with a counterpoise on the other end.

The drive mechanism for the telescope is usually called a *drive clock;* in modern telescopes it is often an electric motor controlled by a variable frequency generator. A telescope is usually driven, about its polar axis, at the *sidereal rate* so that one rotation is completed in 23h 56m 4s. Small corrections are required in drive rates

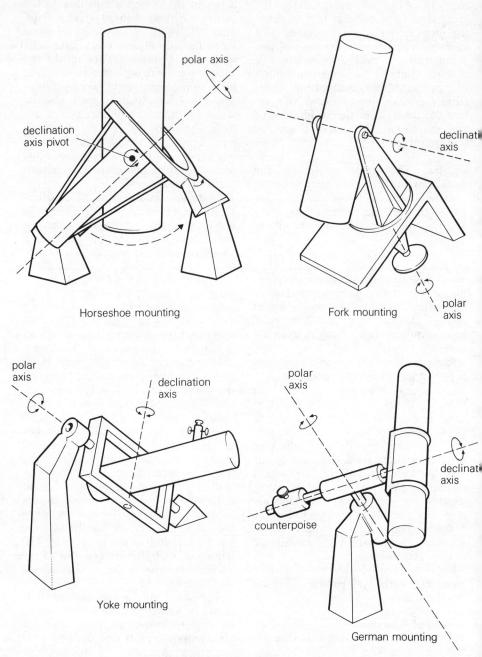

Horseshoe mounting

Fork mounting

polar axis

declination axis pivot

polar axis

declination axis

polar axis

declination axis

polar axis

declination axis

counterpoise

Yoke mounting

German mounting

Equatorial mountings

because of atmospheric refraction and when following the moon or planets, which move relative to the stars. These are obtained by varying the frequency supplied to the drive motor.

equatorial plane. *See* equator.

equator of illumination. The plane of symmetry of a lunar or planetary *phase.

equinoctial colure. *See* colures.

equinoctial points. *See* equinoxes.

equinoxes. 1. (equinoctial points). The two points on the celestial sphere at which the *ecliptic intersects the *celestial equator. They are thus the two points at which the sun in its apparent annual motion crosses the celestial equator. The sun crosses from south to north of the equator at the *dynamical equinox, originally called the *vernal equinox*, symbol: ♈, which lies at present in the constellation Pisces. The sun crosses from north to south of the equator at the *autumnal equinox,* symbol: ♎, which at present lies in Virgo. The dynamical or vernal equinox is the zero point for both the *equatorial and *ecliptic coordinate systems, although in star catalogues a *catalogue equinox is now used. The equinoxes are not fixed in position but are moving westwards around the ecliptic as a result of *precession of the earth's axis; the advance is about 50″ of arc per year. *See also* first point of Aries; true equator and equinox; mean equator and equinox.

2. The two instants at which the sun crosses these points, on about March 21 (dynamical equinox) and Sept. 23. On the days of the equinoxes the hours of daylight and of darkness are equal.
Compare solstices.

equinox motion. *See* catalogue equinox.

equipotential surfaces. Imaginary surfaces surrounding a celestial body or system over which the gravitational field is constant. For a single star the surfaces are spherical and may be considered as the contours of the potential well of the star. In a close *binary star the equipotential surfaces of the components interact to become hourglass-shaped (see illustration). The surfaces 'meet' at the *inner Lagrangian point*, L_1, where the net gravitational force of each star on a small body vanishes; the contour line through this point defines the two *Roche lobes*. When both components are contained well within their Roche lobes they form a *detached binary system*. If one star has expanded so as to fill its Roche lobe it can only continue to expand by the escape of matter through the inner Lagrangian point. This stream of gas will then enter an orbit about or collide with the smaller component. The system is then a *semidetached binary*: *dwarf novae, *W Serpentis stars, and some *Algol variables are examples. When both components fill their Roche lobes, as with *W Ursae Majoris stars, they form a *contact binary* sharing an outer layer of gas. Matter can then eventually spill into space through the *outer Lagrangian point*, L_2.

equivalence principle. The principle that, on a local scale, the physical effects of a uniform acceleration of some *frame of reference imitates completely the behaviour in a uniform gravitational field. This equivalence of the two frames of reference was introduced by Albert Einstein in his general theory of *relativity. It is a generalization of the observed direct proportionality between gravitational and inertial *mass.

equivalent focal length (e.f.l.). The property of a compound lens, such as an eyepiece, that can be used in the same way as the focal length of a sin-

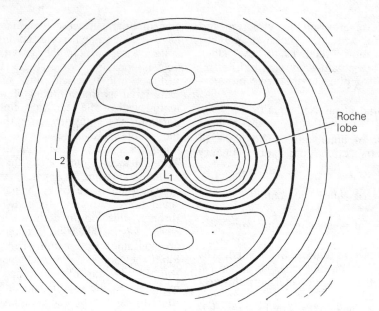

Equipotential surfaces of close binary star

gle thin lens in, say, the determination of magnifying power. For a two-lens system the e.f.l. is given by

$$f_1f_2/(f_1 + f_2 - d)$$

where f_1 and f_2 are the focal lengths of the components and d their separation.

Equuleus (Little Horse). A very small inconspicuous constellation in the northern hemisphere near Pegasus, the brightest stars being of 4th magnitude. Delta and Epsilon are triple stars. Abbrev.: Equ; genitive form: Equulei; approx. position: RA 21h, dec +7°; area: 72 sq deg.

Erfle eyepiece. *See* eyepiece.

Eridanus (River). A long straggling constellation in the southern hemisphere extending from Orion towards the southern polar region. Except for the 1st-magnitude *Achernar, the brightest stars are of 3rd and 4th magnitude. The 3rd-magnitude Acamar (Theta) is a fine double star, possibly once much brighter. Abbrev.: Eri; genitive form: Eridani; approx.

position: RA 1.5 to 5h, dec 0° to −58°; area: 1138 sq deg.

Eros. *Minor planet number 433 that was discovered in 1898 and is a member of the *Amor group. It passed within 23 million km of the earth in 1975. It appears to be a stony spindle-shaped object measuring 7 × 16 × 35 km. *See* Table 3, backmatter.

error box. The region of uncertainty in the position of a celestial source, usually a square or rectangular area of sky defined on a statistical basis. The term is often used in earlier sky surveys of x-ray, gamma-ray, and infrared sources.

eruptive centre. An active volcanic area on a planet or satellite.

ESA. *Abbrev. for* European Space Agency.

escape velocity. The minimum velocity required for an object to enter a parabolic trajectory around a massive body and hence escape from its vicinity. The object will keep moving away from the body although it travels at decreasingly lower speeds as the massive body's gravitational attraction continues to slow it down. If the object does not attain escape velocity it will enter an elliptical orbit around the body. If escape velocity is exceeded the object will follow a hyperbolic path. The escape velocity at a distance r from the centre of a massive body, mass m, is given by $\sqrt{(2Gm/r)}$, where G is the gravitational constant. The escape velocity from the earth's surface is about 11.2 km/s, from the moon 2.4 km/s, and from the sun 617.7 km/s. For a massive body to retain an atmosphere the average velocity of the gas molecules must be well below escape velocity.

ESO. *Abbrev. for* European Southern Observatory.

ESO/SERC Southern Sky Survey. *See* Southern Sky Survey.

ESRO. *Abbrev. for* European Space Research Organization. *See* European Space Agency.

ET. *Abbrev. for* ephemeris time.

Eta Aquarids. *See* Aquarids.

Eta Carinae (η Car). One of the most luminous and unstable stars in the Galaxy, lying at a distance of 2000 parsecs in the constellation Carina. It displays a large variation in magnitude at very irregular periods. From 1835–45 it outshone every star except Sirius, reaching a magnitude of −0.8 in 1843; it has now faded to 6.2. It is surrounded by a shell of cool dust that emits strongly at a wavelength of 20 μm. The decrease in light output observed after 1843 is assumed to have been produced by the ejection of the dust. The dust obscures the light output of the star (which may well have remained constant), converting the light to infrared radiation. Eta Carinae's total luminosity is about five million times that of the sun, its temperature is 29 000 K, and its mass is about 120 solar masses. It is probably related to the highly luminous *Hubble-Sandage variables seen in M31, but it is rather more evolved; it is surrounded by a small nebula of ejected gas that contains a high abundance of nitrogen, indicating that much *nucleosynthesis has taken place.

Eta Carinae nebula (NGC 3372). A large complex of dust and glowing gas in the southern Milky Way. The nebulosity is an extended *HII region in which there are clusters of young stars. Dark lanes of cool dust divide the nebula into islands of glowing gas. One of these dark and irregular dust lanes is known as the *Keyhole*, which is used to identify the bright emission *Keyhole nebula* in which the object *Eta Carinae is situated.

ether. An extremely elastic medium of negligible density that was thought in the 19th century to permeate all space and thus provided a medium through which light and other electromagnetic waves from celestial bodies could travel. (At that time all waves were considered to require a medium for their propagation). The speed of travel of the electromagnetic waves, i.e. the speed of light, was assumed to be fixed with respect to the ether. The earth, however, should move with respect to the ether and so, from considerations of relative motion, the speed of light should vary very slightly when measured in different directions.

Experiments were set up in the late 19th century to detect and measure the motion of the earth relative to the ether. The most notable experiment was carried out in 1887 by A.A. Michelson and E.W. Morley (the *Michelson-Morley experiment*). No such motion was found, indicating both that the ether did not exist and that electromagnetic waves did not require any medium for their propagation. The concept of the ether was thus discarded. The result of these experiments was later explained by Einstein's special theory of *relativity (1905) in which it was stated that the speed of light was independent of the velocity of the observer and was therefore invariant.

Eureca. *Short for* European retrievable carrier. A small reusable unmanned space platform that is being developed by the ESA. It will be jettisoned into earth orbit from a space shuttle and will be collected on a subsequent shuttle flight up to 6 months later. ESA plans to launch Eureca every 2 years starting in 1987. It will carry long-term scientific experiments that will be computer-controlled. Small thruster motors will control its position and solar panels will provide power. It is considerably less expensive than *Spacelab.

Europa. The smallest of the four giant Galilean satellites of Jupiter and the only one inferior in size and mass to the moon. It has a diameter of 3050 km and its density is 3.3 g cm^{-3}. There are very few craters on the surface of the satellite: only three have been identified with certainty and they are 18–25 km in diameter. The surface is covered in long linear features that may be 'cracks' in the surface through which material flows from the interior to cover up any recently formed impact crater. Radioactive heating and some internal tidal heating could be responsible for the internal heat. *See also* Jupiter's satellites; Table 2, backmatter.

European Southern Observatory (ESO). An observatory at La Silla, near La Serena, Chile, at an altitude of 2400 metres; even in normal conditions the *seeing is exceptional due to the clear dry stable climate. It is a European intergovernmental organiza-

tion (members: France, Germany, Netherlands, Sweden, Belgium, Denmark, Italy, and Switzerland). Its headquarters are at Garching, near Munich, where all activities have been concentrated since 1980. The chief instrument at the Observatory is the 3.6 metre (142 inch) reflecting telescope that commenced operation in 1976. It works at focal ratios of f/3, f/8, and f/30 at the prime, Cassegrain, and coudé foci, respectively. The limiting magnitude is 24 magnitudes or more. The observatory also has a 1 metre *Schmidt telescope, operational since 1968 and involved in the *Southern Sky Survey, a 2.2 metre telescope installed in 1983, and three 1.5 metre instruments. The 3.5 metre *New Technology Telescope should see first light in 1987. Studies of the projected Very Large Telescope continue.

European Space Agency (ESA). An organization for coordinating, promoting, and funding Europe's space programme. It was formed in 1975 from the merging of the *European Space Research Organization (ESRO)* and the *European Launcher Development Organization (ELDO)*. There are at present 11 member countries, including the UK. The head office is in Paris. The European Space Research and Technology Centre (ESTEC) is at Noordwijk in the Netherlands and is concerned with management of science and applications programmes. The European Space Operations Centre (ESOC) is at Darmstadt in Germany and is concerned with satellite operations and data acquisition and processing; it controls a network of ground stations for data reception from satellites. ESRIN (previously European Space Research Institute) at Frascati, Italy, houses a major information retrieval service and handles data from remote-sensing satellites.

ESA's scientific programme covers activities in infrared, visible, ultraviolet, x-ray, and γ-ray astronomy, solar physics, planetary science, studies of particles and magnetic fields (including the solar wind and earth's ionosphere and magnetosphere), and remote sensing of the earth. Economic support for the scientific programme is mandatory for all member countries. A group of members can agree to support a particular activity: the *Ariane launch vehicle (mainly France), *Spacelab (mainly Germany), communications satellites such as the Orbital Test Satellite (mainly UK).

EUV (extreme ultraviolet). *Another name for* XUV. *See* XUV astronomy.

eV. *Symbol for* electronvolt.

EVA. *Abbrev. for* extravehicular activity. Operations, including repair and maintenance, undertaken by an astronaut or cosmonaut outside his spacecraft. EVA was first achieved in 1965 by the Russians. *See* Voskhod; Gemini project.

evection. The periodic (31.807 day) *inequality in the motion of the moon that may amount to a displacement in longitude of up to 1°16′20″.4. It arises from changes in the *eccentricity of the moon's orbit (0.0432 to 0.0666) that are brought about by solar attraction. Evection was discovered by Hipparchus. *See also* variation; parallactic inequality; equation of centre; annual equation.

evening star. An informal name for *Venus when it lies east of the sun and is a brilliant object in the western sky after sunset. Such periods of visibility follow superior *conjunction and precede inferior conjunction. The term is sometimes applied loosely to evening visibility of Mercury or other planets.

event horizon. The boundary of a *black hole: for a nonrotating black hole it is a spherical 'surface' having a radius equal to the *Schwarzschild radius for a body of a given mass.

The event horizon marks the critical limit where the *escape velocity of a collapsing body becomes equal to the speed of light and hence no information can reach an external observer. Calculations predict that a body shrinking within its event horizon – or indeed any body that falls inside the event horizon of another – inevitably contracts to a *singularity at the precise centre of the black hole.

Evershed effect. A radial flow of material in the penumbra of a *sunspot, outward from the edge of the umbra, with a velocity of about 2 km s^{-1}. It forms part of a pattern of circulation determined by the configuration of the intense localized magnetic field. The gas ascends at or just beyond the outer edge of the penumbra, doubles back on itself in the *chromosphere, and then descends into the umbra (for the cycle to be repeated).

evolved star. A star, such as a *giant or a *white dwarf, that has passed the main-sequence stage of its evolution. *See* stellar evolution.

excitation. A process in which an electron bound to an atom is given sufficient energy to transfer from a lower to a higher orbit but not to escape from the atom. The atom is then in an *excited state*. The atom may be excited in two ways. In collisional excitation a particle, such as a free electron, collides with the atom and transfers some of its energy to it. If this energy corresponds exactly to the energy difference between two *energy levels, the atom becomes excited. The atom rapidly returns to a lower energy level, usually its ground state, possibly passing through several intermediate levels on the way. Photons are emitted during these transitions, so that an emission line spectrum results. The photon energies are equal to the energy differences between the levels involved.

In radiative excitation, a photon of radiation is absorbed by the atom. The photon energy must correspond exactly to the energy difference between two energy levels. This process produces an absorption line in a spectrum. The atom rapidly returns to its ground state, emitting photons in the process (as before). Since the photons can be emitted in any direction and can have different (lower) energies than the absorbed photon, the absorption line dominates the spectrum. *See also* Boltzmann equation; ionization.

excited state. *See* excitation; energy level.

exclusion principle. *See* Pauli exclusion principle.

exit pupil (Ramsden disc; Ramsden circle). The image of the objective or primary mirror of a telescope formed on the eye side of the *eyepiece. It is seen as a bright disc when the telescope is pointed at any extended bright surface. A convenient way to determine the magnification is to divide the telescope *aperture by the diameter of the Ramsden disc.

exobiology. The theories concerning 'living' systems that may exist on other planets or their satellites in the solar system or on planetary systems of other stars, the methods of detecting them, and their possible study. The probability of this *extraterrestrial life* occurring at some time on some planetary system of some star depends on the fraction of the possible 10^{20} stars in the universe – 10^{11} in our Galaxy – that have planetary systems and on the fraction of these planetary systems that could support, on one or more members, these 'living' organisms.

The degree to which exobiology may differ from terrestrial biology is a matter of dispute. There are no universal laws of biology as there are assumed to be for physics; organisms could evolve on planets or satellites moderately different from earth, in an unrecognizable form. There is no absolute definition of life but it is believed that any living material is built up from the common elements carbon, hydrogen, oxygen, and nitrogen and that the presence of water is probably necessary for sustaining life. In addition the environment should have an atmosphere to act as a shield from high-energy cosmic and stellar radiation, and to aid respiration where necessary, and the temperature should be reasonably uniform.

The course and rate of evolution of living organisms probably depends on the type of star in the planetary system, and its stability, as well as the planetary environment. The conditions may not be appropriate for the origin and evolution of living organisms in this present era. If they can form, however, many scientists believe that these organisms will inevitably become more developed. The number of life forms that might at this moment be 'intelligent' in a human sense and also able to communicate this intelligence is very uncertain: one group of scientists has suggested that at present there could be as many as one million advanced civilizations in our Galaxy; other estimates have been very much lower. Communications, however, are severely limited by distance: signals from a star 100 light-years (30.7 parsecs) away take 100 years to reach earth.

There is to date no direct evidence for extraterrestrial life. Searches for microorganisms in the soils of the moon and Mars have been unsuccessful. Searches for extraterrestrial intelligence (*SETI) have also been made and are once more in progress.

EXOSAT. The first European Space Agency mission in x-ray astronomy, a sophisticated 400 kg satellite launched in May 1983 carrying a payload of two *grazing incidence telescopes, a large (1800 cm^2) array of *proportional counters, and a (90 cm^2) gas scintillation proportional counter. The unusual orbit of EXOSAT, initially 380 km perigee, 195 000 km apogee, gives it the unique capability of long ($\sim$80 hr) observations, uninterrupted by earth occultation. With apogee over N Europe, direct contact with the ground station at Villafranca (Madrid) furthermore allows EXOSAT to be operated directly by the astronomer, much like a ground-based telescope. Although problems occurred with two detectors early in the mission, rendering one telescope inoperable, the spacecraft systems, including on-board computer and 3-axis attitude control system, were working well after a year in orbit. Outstanding results were being obtained, particularly on a variety of *x-ray binaries, *x-ray burst sources, and *active galaxies. Observations are expected to continue up to re-entry in 1987.

exosphere. *See* atmospheric layers.

expanding universe. The observation that radiation from distant galaxies is redshifted leads to the conclusion that all galaxies (beyond the Local Group) are receding from us or, more precisely, that the distance between clusters of galaxies is continuously increasing. The universe as a whole is therefore expanding. This expansion was discovered by Edwin Hubble in 1929 although it had already been suggested by theoretical cosmologists. It is the observational basis of the *big bang theory. The universe may eventually contract if the *deceleration parameter exceeds 0.5. *See* redshift; Hubble constant; Hubble's law.

Explorers. A large group of US scientific satellites, the first of which – Explorer 1 – was America's first satellite and was launched on Jan. 31 1958; it weighed 14 kg. It was able to confirm the existence of one of the *Van Allen radiation belts (already predicted) by means of the shielded Geiger counter that formed part of its 8.2 kg payload. Explorers have been used in studies of the upper atmosphere and the interplanetary medium, and as radio, ultraviolet, x-ray, and gamma-ray astronomy satellites.

extinction. The reduction in the amount of light or other radiation received from a celestial body as a result of absorption and scattering of the radiation by intervening *cosmic dust and by dust grains in the earth's atmosphere. The extinction decreases with wavelength of the radiation and increases with the pathlength through the absorbing medium and with the density of the medium.

The starlight is also reddened since the extinction of blue light by dust is greater than that of red light. The reddening may be given in terms of the colour excess, E,

$$E = (B-V) - (B-V)_0$$

where $(B-V)$ and $(B-V)_0$ are the observed and intrinsic *colour indices of the star. Most stars are reddened by a few tenths of a magnitude although values of up to two magnitudes are not uncommon. Stars lying behind extremely dense matter might only be detectable at radio or infrared wavelengths. *See also* infrared sources.

extragalactic HII region. A cloud of hot ionized gas that is isolated in extragalactic space and is believed to consist of intergalactic gas that has begun to condense into stars and will probably become a *dwarf irregular galaxy. Extragalactic HII regions were identified in 1970. They have no old stars and a low abundance of *heavy elements indicates that some (like the *compact galaxy IZw18) are less than 10^8 years old. These are the only 'galaxies' significantly different in age from our Galaxy.

extragalactic nebula. *Obsolete term for* galaxy.

extraterrestrial life. *See* SETI; exobiology.

extravehicular activity. *See* EVA.

extreme ultraviolet (XUV; EUV). The region of the electromagnetic spectrum bridging the gap between ultraviolet radiation and x-rays. *See* XUV astronomy.

extrinsic variable. *See* variable star.

eyelens. *See* eyepiece.

eyepiece. The magnifying system of lenses through which a telescopic image is viewed. It normally comprises a *field lens,* which receives the light rays, a stop, which defines the *field of view, and an *eyelens,* which directs the light into the eye. The magnification of a telescope is normally varied by using a short focal length eyepiece

to give high powers, and one of long focal length for low powers.

Some eyepieces are shown in the illustration. The *Huygens eyepiece* consists of two planoconvex lenses with their flat faces towards the eye and a field stop between them. The field lens has two or three times the focal length of the eyelens and their separation is half the sum of their focal lengths. Although it has good *eye relief and little *chromatic aberration, the large *spherical aberration gives poor performance except with long-focus refractors. The *Ramsden eyepiece* has two identical planoconvex lenses mounted convex faces together and separated by 2/3 to 3/4 of their focal length. It has much less spherical aberration than the Huygens but suffers from troublesome chromatic defects. The *Kellner eyepiece* is a Ramsden with an achromatic eyelens. It is an excellent eyepiece, widely used in bin-

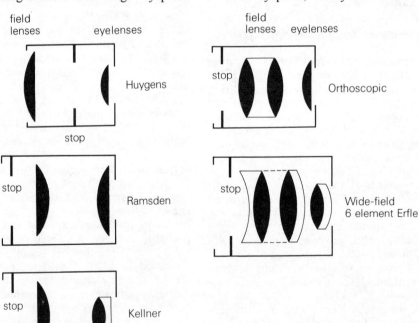

Types of eyepieces

oculars and field glasses. For best results the lenses must be coated to minimize ghost images formed by internal reflections. The *orthoscopic eyepiece,* in its commonest form, has a planoconvex eyelens and a cemented triplet field lens. Although expensive this eyepiece gives excellent results and has a long eye relief, which suits the bespectacled user. By the use of extra lens elements and specially figured (aspheric) lenses, *wide-field eyepieces* may be obtained; these have fields subtending up to 90° at the eye; the *Erfle* has a field of 68° and the *Bertele* up to 80°. Such eyepieces are useful for observation work requiring exceptionally wide fields, such as novae and comet searches, but they are very expensive and will not fit a standard *draw tube. *See also* Barlow lens.

eye relief. The distance between the eyeball and the eyelens of an *eyepiece at which the observer can see clearly the full *field of view. This is when the *exit pupil corresponds with the pupil of the eye. Low-power eyepieces usually have good eye relief and high powers very little. An observer who must wear spectacles may find it difficult to use the higher powers.

F

faculae. Bright patches in the upper part of the solar *photosphere that have a higher temperature than their surroundings and occur in areas where there is an enhancement of the relatively weak vertical magnetic field. With the exception of polar faculae, which consist of isolated granules and appear in high heliographic latitudes around the minimum of the *sunspot cycle, they are intimately related to *sunspots: they form shortly before the spots – in the same vicinity –

and persist for several weeks after their disappearance. Faculae are best seen when near the sun's limb, where *limb darkening renders them more readily visible. They are approximately coincident, albeit at a lower level, with the *plages visible in monochromatic light.

fall. *See* meteorite.

false-colour imagery. A process of colour coding in which colours are used in an image to indicate the distribution of temperature, intensity of emitted radiation, etc., of an object, such as the sun, or over a region of the earth, sun, sky, etc. The colours do not therefore occur in the original subject and are usually computer generated. Colour can also be used to indicate velocity since different *doppler shifts from a single observational wavelength can be calculated and converted into different colours. In addition, images taken at different wavelengths, e.g. a radio and an optical wavelength, and each allotted a colour can be superimposed to form a composite. *See also* imagery.

False Cross. A cross-shaped group of four 2nd-magnitude stars in the former constellation Argo. The group is now divided between Carina (Iota and Epsilon) and Vela (Kappa and Delta).

fan beam. *See* antenna.

Faraday rotation. The rotation of the plane of polarization experienced by a beam of plane-polarized radiation when it passes through a region where there are free electrons and a magnetic field. The plane of polarization is rotated through an angle equal to $\lambda^2 RM$, where λ denotes the radiation wavelength. RM is the *rotation measure* and is proportional to the integral along the line of sight of the product of the electron density, n_e, and the component of the magnetic flux density, *B*, parallel to the line of sight.

farside. *See* nearside.

fast ejection. *See* flares.

fast nova. *See* nova.

fault. A vertical or near vertical crustal displacement on the moon, Mars, or some other solid body. Lunar faults are usually double but single-sided faults also occur. *See also* wrinkle ridges; rille; Mars, surface features.

Faye's comet. A comet with a *coma that increased in size as it approached the sun. Most comets, Encke being a typical example, have comas that decrease as they near the sun. Between 1843 and 1955 Faye's brightness decreased from 4.2 to 11.1 magnitudes.

F-corona. The outer part of the solar *corona, responsible for the greater part of its intensity beyond a distance of about two solar radii. It consists of relatively slow-moving particles of interplanetary dust and exhibits an *absorption spectrum due to the scattering, by the particles, of light from the *photosphere. It extends for many million kilometres into the *interplanetary medium and merges, in the plane of the *ecliptic, with the particles that give rise to the *zodiacal light.

feed. *See* dish.

feeder. A system of wires or *waveguides used to connect a distant radio *antenna with its transmitter or *receiver so that radio-frequency power is transferred with minimum loss (energy dissipation). The wires are arranged to form a *transmission line*, which carries the power in the form of electromagnetic waves. When a transmission line is terminated by a resistance equal to its characteristic impedance all the power travelling down the line is absorbed by the load and none is reflected. The line is then said to be *matched* to its load. Transmission lines are of two types: balanced, as with the open-wire feeder, and unbalanced, as with coaxial cable. It is important that a balanced load, such as a *dipole, is fed by a balanced feeder. Where an unbalanced feeder is required to drive a balanced load, a balanced-to-unbalanced transformer, or *balun*, must be used.

In an *array many elements must be connected together to one receiver. One method of doing this is to join adjacent pairs of elements and then pairs of pairs and so on so that the total length of feeder from the receiver to any element is the same. Such an arrangement is sometimes called a *christmas tree*. *See also* Butler matrix.

fermions. A class of *elementary particles with half integer units of the basic *quantum of *spin; examples are *electrons, *protons, and *neutrinos. No two fermions can exist possessing an identical set of quantum numbers (numbers assigned to the various quantities that describe a particle). Fermions thus satisfy the *Pauli exclusion principle. *Compare* bosons.

fibrils. Predominantly horizontal strands of gas in the sun's lower *chromosphere – typically 11 000 km long and 725–2200 km wide – that are visible (in the monochromatic light of certain strong *Fraunhofer lines) as dark absorption features against the brighter disc. They form an ordered pattern in and around *active regions, where they tend to be aligned in the direction of the localized magnetic field. The overall configuration is stable over a period of several hours, though an individual fibril has a lifetime of only 10–20 minutes.

field. 1. The region in which a physical agency exerts an influence, measured in terms of the force experienced by an object in that region. Thus a massive body has an associated gravitational field within which an object will feel an attractive force. There are several types of field, including gravitational, magnetic, electric, and nuclear fields, each of which has its own characteristics and exerts its influence, strong or weak, over a particular range.
2. *See for* field of view.

field curvature. *See* curvature of field.

field equations. *See* relativity, general theory.

field galaxy. Any galaxy that is not a member of a *cluster of galaxies. Field galaxies are rare: the ratio of field galaxies to cluster galaxies is extremely uncertain, recent estimates ranging from 1:6 to essentially zero.

field lens. *See* eyepiece.

field of view (field). The area made visible by the optical system of an instrument such as a telescope at a particular setting.

field pattern. *Another name for* antenna pattern. *See* antenna.

figuring. The process of grinding and polishing the surface of a lens or mirror base to the exact shape desired.

filamentary structure. The structure of matter on the largest scales (greater than 100 megaparsecs), revealed by the results from large-scale *redshift surveys of galaxies. This structure consists of very long *filaments* of matter surrounding huge *voids* that are empty of matter. The largest *superclusters are elongated, and scattered *clusters of galaxies often link the ends of these superclusters into filaments that can be as much as 225

Mpc long (in the case of a filament found in Pisces and Cetus in 1983). These filaments surround empty voids, the first of which was discovered in Bootes in 1981 and is about 100 Mpc across. Most of the matter in the universe lies in the filaments, which occupy only 1% or 2% of the volume of space.

filaments. 1. (formerly sometimes termed **dark flocculi**). Clouds of gas in the sun's upper *chromosphere/inner *corona, with a higher density and at a lower temperature than their surroundings, that are visible (in the monochromatic light of certain strong *Fraunhofer lines) as dark absorption features against the brighter disc. When viewed beyond the limb they appear as bright projections and are termed prominences. *See* prominences.
2. *See* filamentary structure.

filar micrometer (position micrometer). A measuring device incorporated into the eyepiece of an instrument and used in astronomy to measure the angular separation and orientation of a visual binary star or to determine the position of an object relative to a nearby star whose coordinates are known. It consists essentially of two parallel wires that lie in the focal plane of the eyepiece and can be rotated about the optical axis of the telescope. A third wire lies perpendicular to the parallel wires. Two stars are aligned along this perpendicular wire. The position of one of the parallel wires can be controlled by turning a screw with a finely graduated micrometer head. The parallel wires are adjusted so as to coincide with the images of the two stars and the separation of the images can then be accurately measured.

filling factor. *See* array.

filter. A device that transmits part of a received signal and rejects the rest. The signal may be in the form of a beam of light or other radiation or may be an electrical signal. Optical, ultraviolet, and infrared filters are dyed plastic or gelatin, glass or glass-like substances, or confined liquids, all of which absorb incident radiation except for a relatively narrow band of wavelengths. Such filters are used in astronomical *photometry, especially in measurements of *magnitudes; transmitting bands typically 100 nanometres wide, they are termed broadband filters. Much narrower wavelength bands, of maybe 1 nm, can be obtained with *interference filters.

Electrical filters are devices whose *attenuation varies with frequency. Filters that allow low or high frequencies to pass without serious attenuation are called *low-pass* and *high-pass filters*, respectively. A filter that allows only a limited range of frequencies through is a *band-pass filter* while its converse is a *band-stop filter*. *See also* bandwidth.

filtergram. *See* spectroheliogram.

find. *See* meteorite.

finder. A low-power telescope with a wide *field of view that has its optical axis aligned with that of the main telescope. Since the field of view of the average amateur astronomer's telescope used at its lowest power is only about half a degree, some means of pointing it in the correct direction is needed. *Setting circles enable this to be done with a permanent *equatorial mounting but for a portable or a simple *altazimuth mounting a finder is essential. It should have a field of view of at least four to eight degrees and be provided with illuminated cross wires or a graticule.

fireball. A bright visual *meteor of *magnitude greater than -10. About 50 000 to 100 000 occur in the earth's atmosphere each year.

first point of Aries. Another name for the vernal *equinox, which over 2000 years ago, when Hipparchus first used the term, lay in the constellation Aries. Due to the westerly *precession of the equinoxes the vernal equinox now lies in Pisces and will subsequently move into Aquarius. Likewise the *first point of Libra*, or the autumnal equinox, once lay in Libra but is now located in Virgo.

first point of Libra. *See* first point of Aries.

first quarter. *See* phases, lunar.

fission. *See* nuclear fission.

FitzGerald contraction. *See* Lorentz-FitzGerald contraction.

Five-Kilometre Telescope. *See* aperture synthesis.

fixed stars. Originally stars in general, which, until the early 18th century and the discovery of *proper motion, were thought to have no relative motion and thus remained fixed in position in the sky. In comparison the planets were described as *wandering stars*.

FK4, FK5. *See* fundamental catalogue.

Flamsteed number. *See* stellar nomenclature.

flares. Sudden short-lived brightenings of small areas of the sun's upper *chromosphere/inner *corona that are optically visible usually only in the monochromatic light of certain strong *Fraunhofer lines. They represent an explosive release of energy – in the form of particles and radiation – that causes a temporary heating of the surrounding medium and may accelerate electrons, protons, and heavier ions to high velocities.

A typical flare attains its maximum brilliance in a few minutes and then slowly fades over a period of up to an hour. Although attempts have been made to classify them according to their structure, the most objective system considers the maximum area covered by the flare and to a lesser extent its intensity. The overwhelming majority of flares are small, covering areas of less than several hundred million square kilometres, but those that are larger are sometimes associated with a number of interesting phenomena.

During the 'flash phase' of a large flare, when it suddenly increases in brightness and rapidly expands to its maximum extent, material may be ejected in the form of a *surge, *spray, or (for a particularly energetic event) *fast ejection* – in which a compact portion of the flare is expelled without fragmentation, with a velocity of about 1000 km s^{-1}. Large flares may initiate a *Moreton wave*, which is a magnetohydrodynamic shock wave that spreads out from the centre of disturbance in a sector of about 90°, as it travels with a velocity of the order of 1000 km s^{-1} across the vertical magnetic fields of the inner corona. This causes a depression and subsequent relaxation of the underlying *chromospheric network and may induce distant *filaments, perhaps several hundred thousand kilometres from the flare, to undergo several damped vertical oscillations. Filaments in the vicinity of the flare may also be activated, before and/or during the flare, by changes in the configuration of the local magnetic field. This activation usually takes the form of increased internal motion or flow along the filament (small surges), sometimes accompanied by a gradual ascent of the filament itself. The aftermath of large flares may include coronal condensations of relatively dense material at temperatures of up to 4 000 000 kelvin.

The effects of energetic flares are by no means confined to the sun. Fade-out of short-wavelength radio signals is often experienced in the daylight hemisphere of the earth and is due to a temporary strengthening (by increased ionization) of the reflecting property of the D layer of the ionosphere (60–90 km altitude), which suppresses the passage of the signals to the higher layers where they are normally reflected. This is accompanied by a sudden increase in the electrical conductivity of the E layer (90–140 km altitude) and by disturbances of the earth's magnetic field. Ultraviolet radiation from the flare is responsible for these effects. X-rays may also be emitted and solar radio emission (from the inner corona) frequently exhibits an intense burst or series of bursts at centimetre or metre wavelengths. Occasionally, within about half an hour of the flare, low-energy *cosmic rays reach the earth, and within about 26 hours, on average, less energetic charged particles may also arrive. These latter particles spiral around the earth's magnetic field lines, causing *geomagnetic storms and their luminous counterpart, *auroral displays.

The nature of flares and the physical mechanism responsible for them are not completely understood. They invariably occur in *active regions, close to the line of inversion (*see* sunspots), where the gradient of the horizontal component of the magnetic field is steepest and therefore stresses are greatest. It is thought that energy is released when the stressed field reconnects to a lower potential energy configuration, and that this produces the flare; but much remains to be done before a quantitative picture will emerge.

A phenomenon similar to solar flares, but far more energetic, is thought to be responsible for the rapid brightening of *flare stars.

flare stars (UV Ceti stars). Intrinsically faint cool red main-sequence stars that undergo intense outbursts of energy from localized areas of the surface, causing transient but appreciable increases in the brightness of the star. The brightness can change by two magnitudes or more in several seconds, decreasing to its normal minimum in about 10 to 20 minutes. There are also radio and x-ray flares, not always coincident with optical flares. Flare stars are of *spectral type M or sometimes K with spectral emission lines of hydrogen and ionized calcium, i.e. they are *Me stars. They have unusually strong magnetic fields (typically 10^{-4} tesla), suggesting a similar mechanism to solar *flares. Most flare stars are either young (found in *associations) or are a component of a close *binary star: the fast rotation due to youth or to tidal effects, respectively, is probably responsible for the strong magnetism. UV Ceti (M6e V) is a typical flare star; it is the fainter component of the nearby star Luyten 726-8.

flash spectrum. The *emission spectrum of the solar *chromosphere and *E-corona. It is observable only for a few seconds during a total solar eclipse, at the beginning and end of totality. The majority of the spectral lines have the same wavelengths as the *Fraunhofer lines of the photospheric spectrum. Their presence in successive spectrograms taken during an eclipse provides information on the distribution of elements in the chomosphere.

flattening (oblateness; ellipticity). A parameter specifying the degree by which the form of a celestial body such as a planet differs from a sphere. It is the ratio
$$(a - b)/a$$
where a and b are the equatorial and polar radii respectively.

F-layer (F-region). *See* ionosphere.

flexus. A low curvilinear ridge with a scalloped pattern. The word is used in the approved name of such a surface feature on a planet or satellite.

flocculi. A term that has been applied to several different solar phenomena. In the older literature *plages and *filaments were often referred to as *bright* and *dark flocculi* respectively, while today the term is sometimes used to denote the bright mottles of the *chromospheric network.

fluoresecence. The transformation of *photons of relatively high energy (i.e. high frequencies, especially ultraviolet frequencies) to lower-energy photons through interactions with atoms. It is also the lower-energy radiation that is produced by the process.

flux. A measure of the energy, number of particles, etc., emitted from or passing through a surface per unit time.

flux collector. *See* infrared telescope.

flux density. A measure of the signal power to be expected from a discrete source of emission. The dimensions of flux density are energy per unit time per unit *bandwidth per unit area. The unit is the *jansky.

flyby. A trajectory that takes a spaceprobe close to a planet or satellite but does not permit it to enter an orbit about the body or land on it.

focal length. Symbol: f. The distance between the centre of a reflecting surface or refracting medium to the *focal point* or *focus* (see illustration). With a converging system, such as a *paraboloid surface, a concave mirror or thin convex lens, the focus, F, is the point to which a narrow beam of light, radio waves, etc., from a distant object, i.e. a parallel beam closely aligned to the axis, is brought to a sharply defined or focused image. In a convex mirror or thin concave lens it is the point from which a parallel beam,

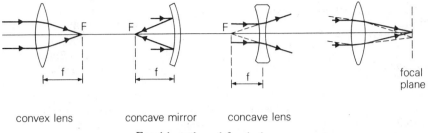

convex lens   concave mirror   concave lens

Focal length and focal plane

made divergent by the mirror or lens, appears to diverge. If the two surfaces of a lens do not have identical curvatures the lens will have two different focal lengths and focal points, depending on which surface the light falls.

The *focal plane* is the plane through the focus, at right angles to the optical axis, in which the image of a distant object will be formed. In some cases, as in the *Schmidt telescope, the image is focused on a curved surface – the *focal surface* – rather than a plane. *See also* equivalent focal length.

focal plane. *See* focal length.

focal point (focus). *See* focal length.

focal ratio. The ratio f/d of the *focal length, f, of a reflecting surface or refracting medium to its effective diameter, d, i.e. to its aperture. The numerical value of the ratio is usually written f/4 or f:4 for a ratio of 4, say. The reciprocal of the focal ratio (d/f) is the *aperture ratio. The *limiting magnitude, or the faintest detectable star, depends on the focal ratio of a telescope: for telescopes used under the same observing conditions, the larger the ratio the fainter the limit; if photographs are being taken, however, the larger the ratio the longer the necessary exposure time.

focal surface. *See* focal length.

focus (focal point). *See* focal length.

folded dipole. *See* dipole.

following spot (*f*-spot). *See* sunspots.

Fomalhaut (α PsA). A white star that is the brightest one in the constellation Piscis Austrinus. The infrared satellite IRAS has detected a shell of cool (about 50 K) material around this star that is interpreted as being due to solid material orbiting the star and may be protoplanetary system. *See also* Vega. m_v: 1.16; M_v: 1.9; spectral type: A3 V; distance: 7.0 pc.

forbidden lines. Lines that are not found in spectra under normal terrestrial conditions but are observed in certain astronomical spectra. In *emission nebulae, for example, atoms can be excited by impact of low-energy electrons. Under normal laboratory conditions such atoms would be de-excited by collisions with other atoms, etc., before they had time to radiate. Such collisions are very infrequent in nebulae and the '*forbidden*' *transitions* between the excited state and a state of lower energy can occur, producing the forbidden lines.

force. Symbol: *F*. Any physical agency that alters or attempts to alter a body's state of rest or of uniform motion. The force required to accelerate a body of mass m is given by ma, where a is the acceleration imparted.

There are many kinds of forces, including the gravitational force. The SI unit of force is the newton. *See also* fundamental forces.

fork mounting. *See* equatorial mounting.

Fornax (Furnace). An inconspicuous constellation in the southern hemisphere fairly near Orion, the brightest stars being of 4th magnitude. Abbrev.: For; genitive form: Fornacis; approx. position: RA 3h, dec −30°; area: 397 sq deg.

Fornax system. A member of the *Local Group of galaxies.

fossa. A long narrow shallow depression. The word is used in the approved name of such a surface feature on a planet or satellite.

Foucault's pendulum. A simple pendulum that demonstrates the rotation of the earth, first shown by J. B. L. Foucault in 1851. It consists of a massive ball that is suspended by a long wire and can swing freely with a minimum amount of friction. The pendulum swings steadily, tracing a straight line on the floor beneath it. The plane in which it swings, however, is observed to rotate during the day as a result of the earth's rotation. The period of rotation of the plane depends on the latitude of the place: at the poles the plane would appear to move through a complete circle in one sidereal day (23h 56m 4s) while on the equator it would not rotate at all. The pendulum is in fact swinging in a plane that is fixed relative to the stars.

Fourier transform. A mathematical operation by which a function expressed in terms of one variable, x, may be related to a function of a different variable, s, in a manner that finds wide application in physics. The Fourier transform, $F(s)$, of the function $f(x)$ is given by

$$F(s) = \int_{-\infty}^{\infty} f(x) \exp(-2\pi i x s)\, dx$$

and

$$f(x) = \int_{-\infty}^{\infty} F(s) \exp(2\pi i x s)\, ds$$

The variables x and s are often called *Fourier pairs*. Many such pairs are useful, for example, time and frequency: the Fourier transform of an electrical oscillation in time gives the spectrum, or the power contained in it at different frequencies.

Fra Mauro formation. The *ejecta blanket associated with the *Imbrium Basin. It was the target for Apollo 14.

frame of reference. A set of coordinate axes to which position, motion, etc., in a system may be referred and by which they can be measured.

Franklin-Adams charts. A whole-sky photographic star atlas of 206 plates prepared by John Franklin–Adams and published in 1914 as the *Photographic Chart of the Sky*. It has a limiting magnitude of 16.

Fraunhofer lines. Absorption lines in the solar (photospheric) spectrum, first studied in detail by Joseph von Fraunhofer in 1814. He catalogued over 500 lines and labelled the more striking ones with letters. Over 25 000 lines have now been identified. The most prominent lines at visible wavelengths are due to the presence of singly ionized calcium, neutral hydrogen, sodium, and magnesium; many weaker lines are due to iron. The strength of a particular line depends not only on the quantity of the element present but also on the degree of *ionization and level of *excitation of its atoms. By isolating an individual strong line it is possible, by virtue of its residual intensity due to the re-emission of radiation in the *chromosphere, to observe the chromosphere in the light of the element concerned.

See spectroheliogram. *See also* H and K lines; D lines; photosphere.

Fred Lawrence Whipple Observatory. An observatory on Mount Hopkins, Arizona. Formerly the *Mount Hopkins Observatory*, it was renamed in 1982. It is the largest field station of the Smithsonian Astrophysical Observatory. The chief instruments are the 4.5 metre *Multiple Mirror Telescope, a 1.5 metre reflector, and a 10 metre optical mosaic.

free-bound emission. The radiation emitted when a free electron is captured by an ion. *See* recombination-line emission.

free fall. Motion of a body under the influence of gravity alone, i.e. with no other forces acting. *See also* weightlessness.

free-free absorption. A process of absorption that occurs when a free electron (i.e. not bound to an atom or ion) absorbs a photon of radiation and moves to a higher energy state: the difference between final and initial electron energy is equal to the photon energy. *See also* opacity.

free-free emission. *See* thermal emission.

frequency. Symbol: f, ν. The number of oscillations per unit time of a vibrating system. The frequency of an electromagnetic wave is the number of wave crests passing a point per unit time. It is related to wavelength λ by $\nu = c/\lambda$, where c is the speed of light. The frequency of electromagnetic waves is measured in hertz.

Friedmann universe, equation. *See* cosmological models.

f-spot. *Short for* following spot. *See* sunspots.

F stars. Stars of *spectral type F that are white with a surface temperature (T_{eff}) of about 6000 to 7500 kelvin. The hydrogen lines in the spectrum weaken rapidly and the lines of ionized calcium (Ca II, H and K) strengthen from F0 to F9; there are many lines of neutral and singly ionized metals and heavy atoms. Canopus, Polaris, and Procyon are F stars.

full moon. *See* phases, lunar.

full-wave dipole. *See* dipole.

fundamental catalogue. A catalogue of *fundamental stars in which the precise positions and proper motions are listed for a given *epoch. Other star catalogues can be compiled by reference to a fundamental catalogue. The *Fourth Fundamental Catalogue (FK4)*, published 1963 in Heidelberg, contains data for 1535 stars to a limiting magnitude of 7 for the epoch 1950.0. *The Fifth Fundamental Catalogue (FK5)* is currently (1984) in preparation. It is based on the standard epoch J2000.0 (*see* Julian year) and takes into account recent theory and observations.

fundamental forces. The four basic forces of nature: the *gravitational force, the *electromagnetic force* (both long established), and the two nuclear forces, the *strong force* and the *weak force*. These forces act between *elementary particles, i.e. the basic building blocks of matter. In the case of the latter three forces, each is generated by the *interaction* between particles, which involves the exchange of an intermediate particle. These intermediates affectively 'carry' the force from one particle to another. Different intermediates (all *bosons) are exchanged in the three types of interaction (*see also* quantum gravitation).

The gravitational force occurs between all particles, the electromagnetic force between charged particles, e.g.

electrons and protons. The strong force arises between hadrons, e.g. protons and neutrons, and between the constituents of hadrons, i.e. *quarks, and is the force binding together particles in the *nuclei of atoms. The weak force occurs between both leptons and hadrons and is responsible for radioactivity. The four forces vary greatly in range: the gravitational and electromagnetic forces have an infinite range whereas the other two have an extremely short range. They also vary greatly in strength: the gravitational force is the weakest over very short distances ($\sim 10^{-39}$ times the strength of the strong force) but on a cosmic scale it dominates the others.

Attempts to construct a single theory unifying the four forces have been progressing. Steven Weinberg and Abdus Salem in the late 1960s found a mathematical description – the *electroweak theory* – that successfully unified the electromagnetic and weak forces so that they could be regarded as two aspects of the same phenomenon. There are various *grand unified theories (GUTs)* that aim to provide a mathematical framework in which the strong, electromagnetic, and weak forces emerge as parts of a single unified force. (The gravitational force has not yet been successfully incorporated.) A symmetry is said to relate one force to another. Since the forces are very different in strength and character this symmetry is broken in the present-day universe. GUTs predict that the symmetry holds only when the temperature is greater than about 10^{27} K, when particles would have extremely high energies, above 10^{24} electronvolts. Such extremes of temperature and energy occurred in the very young universe immediately after the *big bang. GUTs are thus of great importance to cosmology.

fundamental stars. A number of stars distributed over the whole celestial sphere whose positions and proper motions are known so accurately that they can be used to determine the positions of other stars.

FU Orionis. A very young star, thought to be a *T Tauri star, that is located in Orion in a cloud of gas and dust and is a strong source of infrared radiation. It flared up from about 16th to 10th magnitude in 1936 and has remained near that brightness. It has a high lithium content.

fusion. *See* nuclear fusion.

G

gain. 1. A measure of the amplification of an electronic device. If the power input to the device is P_1 and the power output is P_2, the gain expressed in decibels is given by
$$G = 10 \log_{10}(P_2/P_1)$$
Gains measured in this way can be added when amplifying stages are connected in cascade.

2. A measure of the directional advantage of using one radio *antenna as compared with another. It is usual to express the gain, G, of a particular antenna over an isotropic radiator. For a lossless antenna it is given by
$$G = 4\pi A_e/\lambda^2$$
where A_e is the effective area (*see* array) and λ is the wavelength; the gain is equal to the *directivity in this case. Sometimes the comparison is with a *dipole, which itself has a gain over an isotropic radiator of two decibels.

galactic centre. The innermost region of our *Galaxy, or its exact centre. *Interstellar extinction obscures this region at optical wavelengths and information on the very complex phenomena in the galactic centre has been derived mainly from radio, infrared, and x-ray observations. In the central region (radius 100 parsecs) there is strong 2 μm emission thought to be coming predominantly from *late-type stars. The peak of this infrared emission and the stellar density coincide with the radio emission from an HII region in *Sagitarrius A (West). This is generally considered to be the dynamic centre of the galaxy. Within the central one parsec region the late-type stars provide a luminosity of about 10^6 $L_\odot$ (i.e. a million times that of the sun) with a total mass of 3×10^6 $M_\odot$ (solar masses), implying the presence of about 10^6 stars. Also within this region are found clumps of plasma (i.e. ionized gas) typically of 1 $M_\odot$ each with a luminosity of about 10^5 $L_\odot$, probably arising from the *mass loss from the late-type stars.

The density of dust at the centre is very low but probably increases beyond a radius of 1 parsec to give rise to a double lobed structure, with a 100 μm luminosity of about 10^7 $L_\odot$. This may be heated by ultraviolet radiation from a central object located at the position of Sgr A West. The source of this heating radiation is an open question but could arise as a consequence of star formation in the galactic centre (but this appears improbable in the absence of dust) or possibly as emission from an exotic object such as a massive black hole. X-ray observations have been made with the *Einstein Observatory: images of a 1 × 1 degree field on the galactic centre show a group of weak x-ray sources together with a region of 'diffuse' emission, 25 × 15 arc minutes in extent. One of the point sources is apparently coincident with Sgr A West. The 'diffuse' emission may be due to many unresolved point sources or be truly diffuse. In the latter context it is interesting to note the similar shape and extent of the x-ray, radio, and 100 μm infrared radiation, suggesting a common connection with the distribution of gas and dust in the galactic centre region.

Further away from the centre, radio observations indicate that there is a thin rapidly rotating disc of hydrogen extending out to a radius of about 750 parsecs and also gas moving rapidly away from the centre. In particular there are two expanding arms of gas both roughly at a radius of 3 kiloparsecs. The arm on the sunward side of the centre is approaching us with a speed of 50 km s^{-1} and the one on the other side is receding at 135 km s^{-1}.

galactic circle. *See* galactic equator.

galactic cluster. 1. *See* open cluster.
   2. *See* clusters of galaxies.

galactic coordinate system. A *coordinate system used to study the structure, surroundings, and contents of the *Galaxy. The fundamental circle is the *galactic equator and the zero point lies in the direction of the galactic centre (in the constellation Sagittarius) as seen from earth (see illustration). The coordinates are *galactic latitude* and *longitude*.

The galactic latitude (b) of a celestial body is its angular distance (from 0° to 90°) north (counted positive) or south (counted negative) of the galactic equator; it is measured along the great circle passing through the body and the galactic poles. The galactic longitude (l) of a celestial body is its angular distance (from 0° to 360°) from the nominal galactic centre measured eastwards along the galactic equator to the intersection of the great circle passing through the body.

The position of zero galactic longitude, i.e. the nominal galactic centre, was agreed on (1959) by the International Astronomical Union (IAU); it lies at RA 17h 42.4m, dec −28°55′ (1950). More recent observations suggest the actual galactic centre coincides with a radio and infrared souce, *Sagittarius A West, a few arc minutes from this nominal position; the nominal centre is still used, however, as the zero point for galactic coordinates. (The true centre lies at

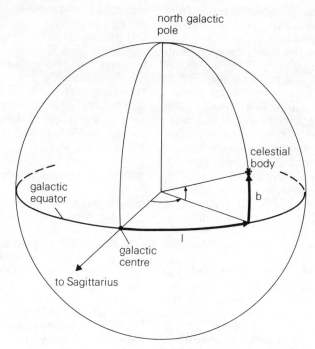

Galactic coordinate system

$l = -3'.34$, $b = -2'.75$.) The symbols l and b supersede the temporary designations l^{II} and b^{II} for the 1959 IAU system. (The old IAU system coordinates l^I, b^I referred to a slightly different galactic equator and a zero point defined by the ascending *node of galactic equator on celestial equator.)

galactic disc. *See* Galaxy.

galactic equator (galactic circle). The great circle on the *celestial sphere that represents the path of the *Galaxy. The nominal galactic centre lies on the galactic equator in the direction of the constellation Sagittarius and is the zero point for the *galactic coordinate system. The *galactic plane,* which contains the galactic equator, is the plane that passes most nearly through the central plane of the spiral disc of the Galaxy. The galactic plane

and celestial equator are inclined at an angle of about 62°.

galactic halo. *See* Galaxy. *See also* dark halo.

galactic latitude, longitude. *See* galactic coordinate system.

galactic magnetic field. A very weak and complex system of magnetic fields, known to exist in the interstellar regions of the Galaxy because of the *Faraday rotation of radiation from pulsars and other radio sources, from the *Zeeman splitting of the 21 cm hydrogen line arising in interstellar clouds, and from *interstellar polarization. The magnetic flux density lies in the range 10^{-10} to 10^{-9} tesla, with $3-6 \times 10^{-10}$ tesla being the best estimate at present. The direction of the field seems to be in the plane of the Galaxy, along the spiral arms. *See also* cosmic rays.

galactic plane. *See* galactic equator; Galaxy.

galactic poles. The two points on the celestial sphere that are 90° north and 90° south of every point on the *galactic equator. The *north galactic pole* lies in the constellation Coma Berenices at RA 12h 49.0m, dec +27°24' (1950). The *south galactic pole,* diametrically opposite, lies in Sculptor.

galactic rotation. The rotation of the *Galaxy about its centre, all its components sharing in this rotation to varying degrees. The predominant motion in the galactic disc is circular and parallel to the galactic plane. The orbital speed is determined by the mass within a star's orbit, not by the Galaxy's total mass, so the stars' rotation speeds do not follow *Kepler's laws. The rotation speed increases from near zero to 150 km s⁻¹ in the first kiloparsec of the Galaxy's radius, then increases more gradually to a peak near the sun's orbit (at a radius of 10 kpc). Further away it may fall off gradually or remain roughly constant. The rotation velocity of the *local standard of rest in the galactic plane is traditionally taken to be 250 km s⁻¹, but is only 220 km s⁻¹ if the Galaxy has a massive *dark halo. Objects in the galactic halo display much more random motions than those in the disc and the system as a whole has only a small residual rotation with respect to the galactic centre. *See also* Oort's fomulae.

galaxies. Giant assemblies of stars, gas, and dust into which most of the visible matter in the universe is concentrated. Each galaxy exists as a separate, though not always entirely independent, system held together and organized largely by the gravitational interactions between its various components. When capitalized, the term denotes specifically our own system, the Milky Way *Galaxy. Owing to their cloudlike appearance when viewed by eye through simple telescopes, galaxies other than our own were once known as 'nebulae'. This term is now strictly applied only to clouds of interstellar gas and dust.

The majority of the galaxies close enough to be observed in any detail can be divided into three broad categories: elliptical, spiral, and irregular. *Elliptical galaxies* are spheroidal systems with no clearly defined internal structures. They contain little or no interstellar matter and the stars within them (mainly *spectral types K and M) are predominantly old and well advanced in their evolution. Most ellipticals are highly concentrated objects, particularly near the centre. On photographs they look somewhat like fuzzy elliptical patches of light, usually diminishing smoothly in brightness from the centre outwards. The shape of the outline varies from almost circular through to narrow ellipses about three times as long as they are wide. Traditionally the latter were regarded as oblate (flattened) systems but recent analyses show that prolate (elongated) systems are also possible. Most elliptical galaxies are probably triaxial ellipsoids with their three axes of different lengths. The shape of the outline is the basis of *Hubble's classification.

Spiral galaxies are flattened, disc-shaped systems containing prominent *spiral arms* of interstellar matter and bright young stars that wind outwards from a dense central nucleus. Although two-armed spirals are the most common, systems with one arm or even, very occasionally, three arms have also been observed. In *normal spirals* the arms emerge directly from the nucleus, usually from opposite sides; in *barred spirals* the arms emanate from each end of a bright central bar that extends across the nucleus. Both types exist in a wide range of forms. At one extreme there are galaxies with large dominant nuclei and thin tightly wound spiral arms; at the other extreme there are galaxies with inconspicuous nuclei and prominent loosely wound spiral arms. The shape and structure of spirals and barred spirals is the basis of Hubble's classification.

Spiral galaxies are rich in gas and dust, most of which is distributed in clouds along the spiral arms. Stars in the nuclei of spirals appear to be predominantly well advanced in their evolution and the brightest individual stars observed there are *red giants of *population II. A fairly smooth axially symmetric distribution of old stars also extends beyond the nucleus. However, owing to the intrinsic faintness of most of its constituent stars, this system is much less conspicuous than the spiral arms embedded within it.

Although star formation is undoubtedly still taking place in the spiral arms, what triggers this process is still uncertain, as is the origin of the arms themselves. It has been proposed, in the *density-wave theory, that the spiral structure is maintained over a long period by gravitational effects. Alternatively, it has been suggested that the spiral structure arises and persists as a result of *self-propagating star formation involving supernovae. All spiral galaxies are in differential rotation, stars in the outer parts completing their orbits on average more slowly than those nearer the centre. As far as can be ascertained, the spiral arms always trail away from the direction of rotation.

Irregular galaxies have no discernable symmetry in shape or structure. They vary enormously in appearance but all are below average in size and contain large amounts of interstellar matter. In the early Hubble classification this term was applied to any galaxy that could not be fitted into the elliptical or spiral classes. Hubble's Irr II galaxies are now reclassified as *starburst, *active, and *interacting galaxies. The small or dwarf systems

presently labelled as irregular correspond to Hubble's Irr I galaxies.

In addition, astronomers now recognize a class of galaxies that appear to be intermediate between the spirals and ellipticals. Called *S0 galaxies*, these possess a central nucleus and the beginnings of a disc but show little or no evidence of spiral structure. The stars within them are predominantly old and very little gas is apparent, although dark dust clouds are sometimes seen in the disc. The term *lenticular galaxy* is used either to denote specifically these S0 galaxies, or to denote any galaxy that superficially resembles a biconvex lens.

The colours of galaxies vary according to the age of the stars responsible for most of the light. In general, ellipticals and S0s are the most red and irregulars, because of their high content of very young stars, are the most blue. Differences in the dominant *spectral type are also reflected in the mass to luminosity ratio, spirals and irregulars being, on average, brighter than ellipticals or S0s of similar mass.

The brightest (and also rarest) galaxies are the *cD galaxies*, i.e. *supergiant ellipticals*, with absolute magnitudes of about -22.5 and masses in excess of 10^{12} solar masses. Ellipticals also display the greatest variation in mass, ranging down to extreme dwarfs (about 10^6 solar masses) that are no brighter than the most luminous *globular clusters. Spirals appear to exist only as large or giant systems, with masses typically of the order of 10^{10} or 10^{11} times that of the sun. No irregulars are as bright as the giant spirals and some are extreme dwarfs.

Of the 1000 brightest galaxies, about 75% are spiral, 20% elliptical, and 5% irregular. However, when allowance is made for the many *dwarf galaxies, the true proportions turn out to be nearer to 30:60:10. Few galaxies exist in total isolation. Double and multiple systems are common and many galaxies are also members of larger groups known as *clusters of galaxies. Clusters can in turn form loosely bound aggregates called *superclusters.

How and why galaxies have evolved to their present shapes is still uncertain although it appears that spirals contain much more *angular momentum than ellipticals. Turbulence and vorticity in the early universe may play a role. However, it is generally conceded now that the principal types of galaxy represent separate species rather than one species seen at different stages in its evolution. There is, in fact, no direct evidence that any other galaxy is significantly older than our own Galaxy (which is thought to be about 12 thousand million years old), and none younger apart from the miniscule *extragalactic HII regions. All galaxies appear to contain a mixture of stellar *populations.

See also galaxies, formation and evolution; recession of the galaxies.

galaxies, classification. Galaxies may be classified according to a variety of criteria, including morphology (shape, concentration, and structure), spectroscopy (integrated spectrum, etc.) colorimetry (integrated colour, etc.), photometry (luminosity, etc.), and unusual activity at various wavelengths (radio, infrared, x-ray, etc.). The *Hubble classification is based on morphology while *Morgan's classification is basically a quantitative spectroscopic classification.

galaxies, formation and evolution. Galaxies must have condensed out of the gases expanding from the *big bang, beginning at a time when the average density of the universe was roughly the same as the current mean density of a galaxy. Details of the formation of galaxies are still highly uncertain, as is their subsequent evolution. Astronomers are not agreed, for example, on the extent to which the different types of *galaxy have been

determined by conditions at formation or by later evolution.

The primordial gas must have contained slight fluctuations in density, but *radiation pressure prevented these from growing until after the time of *recombination, when the universe was 300 000 years old. The initial fluctuations then grew by *gravitational instabilities to form *protogalaxies. The fluctuations could be of two kinds, giving rise to two rival theories of galaxy formation.

The isothermal theory, developed largely by R. H. Dicke and P. J. E. Peebles, considers the growth of density fluctuations that have the same temperature as the surrounding gas. Each denser region collapses to a relatively small fragment with a mass of about a million solar masses. Some remain isolated to become dwarf galaxies, but most amalgamate with neighbours to build up galaxies of normal mass (10^9 to 10^{11} solar masses). These are attracted together to form *clusters of galaxies; the clusters then form *superclusters.

On the pancake theory (or adiabatic theory), the average energy is everywhere the same so the denser fluctuations are cooler than elsewhere. The smaller fluctuations are averaged out, according to Ya. B. Zel'dovich and his Moscow school of cosmologists, so that the gas collapses into much larger clouds, each with a mass of about 10^{14} solar masses. The collapse is nonsymmetrical and the gas ends up in very flattened structures, or pancakes. Each pancake becomes a supercluster of galaxies. Internal shock waves produced in the collapse break the structure of the pancake down into regions the size of clusters of galaxies and, within these, into condensations that become individual galaxies. *Dark matter in the form of massive neutrinos, for example, promotes the formation of pancakes, and computer simulations have shown the evolution of large-scale structures very similar to the filaments and voids that are observed in the universe (*see* filamentary structure).

Elliptical galaxies probably formed in the densest regions of the original fluctuations. Rapid star formation converted almost all the available gas to stars in less than a thousand million years. The most distant ellipticals should contain a proportion of younger bluer stars, and observations with the UK Infrared Telescope have recently revealed this effect. Spiral galaxies formed by the slower accumulation of fragments (isothermal theory) or collapse of larger clouds (pancake theory) in less dense regions, and where turbulence caused the protogalaxy to rotate. Fairly rapid star birth during formation produced the old stars of the halo and central regions; the remaining gas settled into a disc, where stars continued to form much more slowly and interstellar gas remains to the present day.

The most dramatic examples of galaxy evolution are caused by external factors. *Cannibalism can produce supergiant elliptical galaxies, and it is also possible that some ordinary ellipticals formed from similar collisions between spiral galaxies. The pool of hot intergalactic gas in the centre of a cluster of galaxies can strip gas from a spiral galaxy, to leave it as a gas-free lenticular or S0 *galaxy.

Galaxy (Milky Way System). The giant star system to which the sun belongs. The Galaxy has a spiral structure and, like other spiral galaxies, is highly flattened. It is estimated to contain of the order of 100 billion (10^{11}) stars, the bulk of which are organized into a relatively thin *disc* with an ellipsoidal bulge, or *nucleus*, at its centre. This system is embedded in an approximately spherical *halo* of stars and *globular clusters (see illustration). The radius of the disc is approximately 20 kiloparsecs and its maximum thickness (at the centre) is about 4 kpc. The halo is more sparsely populated than the disc and its full

extent is uncertain, although its radius is known to be greater than that of the disc. There may also be a very much larger *dark halo, or corona, of unseen matter stretching out to a radius of 100 kpc or more (see below). The characteristic *spiral arms*, which contain many of the brightest stars in the Galaxy, wind outwards from the nuclear region, in or close to the central plane of the disc.

The sun is situated only a few parsecs north of the central plane, near the inner edge of one of the spiral arms. Our distance from the centre is nominally taken as 10 kpc although the actual value is probably close to 8.5 kpc.

The entire Galaxy is rotating about an axis through the centre, the disc rotating fairly rapidly, the halo more slowly (*see* galactic rotation). At the sun's distance from the centre, the systematic rotation of the disc stars is between 220 and 250 km s^{-1}, whereas the halo system is rotating with a speed of only about 50 km s^{-1}.

Many astronomers now believe that our Galaxy (and others) is enveloped in a huge dark halo. This is undetected (so far) except by its gravitational effect on the more distant globular clusters in the halo and on nearby dwarf galaxies, but it would contain ten times as much mass as the stars of the Galaxy. It may contain black holes resulting from *population III stars, but most of the matter, like other dark matter, is likely to consist of some kind of elemenary particle.

The objects in the halo are old stars or clusters of old stars, i.e. globular clusters, that belong to *population II. They increase in density towards the centre of the Galaxy but show little concentration towards the galactic plane. Stars in this system are believed to have condensed early in the life of the Galaxy, maybe 12 billion years ago, when the gas cloud from which it formed was still almost spherical. The bulk of the stars in the disc, and probably in the nucleus al-

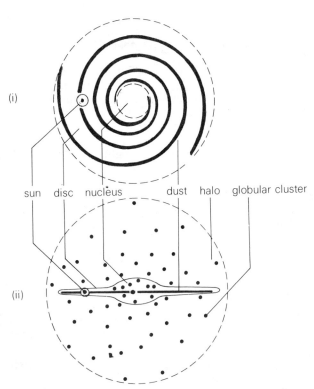

sun   disc   nucleus   dust   halo   globular cluster

Schematic views of Galaxy, (i) from above (ii) from the side

so, are stars of intermediate age (3–5 billion years) belonging to the disc *population. The young *population I stars are mainly confined to a layer about 500 pc thick along the centre of the disc.

Most of the interstellar gas and dust detected in the Galaxy lies in or close to the galactic plane, and about half of it is concentrated into very dense *molecular clouds distributed along the spiral arms. The youngest stars in the Galaxy, notably the *T Tauri stars and the very bright, short-lived *O and *B stars, are also largely confined to the spiral arms and it is almost certain that stars are still being formed from the molecular clouds there. Tracing of the spiral structure is complicated by our position in the disc. Three relatively nearby sections of arms have been traced optically – principally by mapping the distribu-

tion of O and B stars and associated *emission nebulae. These are the *Orion arm*, in which the sun is located; the *Perseus arm*, located about 2 kpc further out along the plane; and the *Sagittarius arm*, which lies about 2 kpc nearer the centre. Another section, the *Carina arm*, may be a continuation of the Sagittarius arm, the concatenation being called the *Sagittarius-Carina arm*. *Interstellar extinction makes optical tracing impossible beyond the first few kiloparsecs of the sun in most directions. More distant spiral features have been traced by radio-astronomy techniques, mainly by mapping the distribution of neutral hydrogen in *HI regions and, more recently, the *carbon monoxide in molecular clouds. The analysis is not simple, however, and it is even uncertain whether the Galaxy is a two-armed or four-armed spiral.

Interstellar extinction also obscures the central nucleus of the Galaxy at optical wavelengths. Information about this region has been derived mainly from radio, infrared, and x-ray observations. *See* galactic centre.

The total mass of the Galaxy (ignoring the dark halo) is estimated to be a little less than 2×10^{11} solar masses, of which about 10% exists in the form of interstellar matter. Where our Galaxy fits in the *Hubble classification is still open to argument. Some observations suggest a type Sb structure, others a type Sc; there is even tentative evidence of a central bar, although this evidence is not accepted by all astronomers.

Galilean satellites. The satellites *Io, *Europa, *Ganymede, and *Callisto of Jupiter, discovered in 1610 by Galileo and, independently, by Simon Marius. They are bright enough to be seen with the aid of binoculars and have been studied in detail by *Pioneer and *Voyager spacecraft. *See also* Jupiter's satellites; Table 2, backmatter.

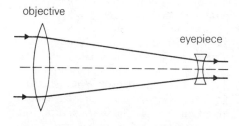

Ray path in Galilean refractor

Galilean telescope. The first type of astronomical telescope, developed in 1609 by Galileo from Hans Lippershey's 'magic tubes'. It is made up from a single long-focus object lens and a powerful diverging lens used as an eyepiece (see illustration). This optical system, which gives an upright image, survives in the modern opera glass. Galileo's best telescopes had a magnification of about 30 times; although very imperfect they led to the great achievements of 17th-century astronomy.

Galileo. A NASA mission to Jupiter, scheduled for launch in May 1986. The 2550 kg spacecraft consists of an orbiter and a small probe and will be carried into earth orbit by space shuttle then sent to Jupiter by a modified Centaur upper stage. The probe will separate from the orbiter about 150 days before arrival at Jupiter in Aug. 1988 and will follow an independent path to the planet. The probe, entering Jupiter's atmosphere at very considerable speed, will have to be greatly decelerated. It is expected to make measurements of the physical and chemical characteristics of the atmosphere over a period of about 60 minutes. The data will be relayed to the orbiter for transmission to earth. A few hours before probe entry the orbiter will fly within 1000 km of Io. In addition to making observations of the satellite, this will enable the orbiter to reduce its speed and go into orbit about Jupiter. It will observe the satellites, planet, and surrounding environment over a 20 month period and is designed to complete 11 orbits of the planet.

gamma (γ). The third letter of the Greek alphabet, used in *stellar nomenclature usually to designate the third-brightest star in a constellation or sometimes to indicate a star's position in a group.

Gammy I. A French-Soviet-Italian mission in which a medium-energy gamma-ray telescope built by the French and Soviets is to be launched on a Soviet satellite, probably in 1987. The telescope detectors will operate over the energy range 2 keV to 5 GeV and should be able to pinpoint sources within 10 arc minutes, a great improvement on earlier γ-ray satellites. It is similar to *SAS-2 and *COS-B in the sense that its central element is a multilayer *spark chamber system, but differs in as much as a *coded mask has been added at the top of the instrument in order to improve angular resolution.

gamma-ray astronomy. The study of radiation from space at the extreme short-wavelength end of the electromagnetic spectrum (less than 0.01 nanometres) and with the largest photon energies (usually exceeding 100 000 electronvolts (eV), i.e. >100 keV). The immense range of gamma-ray energies (*see* gamma rays) has led to a variety of detection instruments and techniques, including *scintillation counters, *spark chambers, *drift chambers, *coded-mask telescopes, *Compton telescopes, and solid-state detectors. Semiconductor materials such as lithium-doped silicon, lithium-doped germanium, and high-purity germanium are used for high-resolution γ-ray spectroscopy for energies up to 20 MeV.

Gamma rays from space cannot penetrate the earth's atmosphere so that γ-ray observations only became possible when instruments could be carried above the atmosphere in satellites. The first cosmic γ-rays were detected with a scintillation counter on the *OSO-3 satellite in 1968. More detailed observations confirmed their origin in a narrow band along the galactic plane, being particularly strong within 30° of the galactic centre; these observations were provided by the NASA *SAS-2 satellite launched in Nov. 1972 and suggested that the γ-rays come from interactions in interstellar gas. The spark-chamber detector on SAS-2, sensitive in the energy band 30–1000 MeV, also found discrete γ-ray sources coincident with two *pulsars, the Crab and Vela pulsars.

The European Space Agency's COS-B satellite, launched in Aug. 1975, also carried a large spark chamber and operated successfully for nearly seven years, providing improved exposure and angular resolution of medium-energy (70–5000 MeV) γ-rays. In addition to the Crab and Vela pulsars 22 other galactic sources, plus the bright *quasar 3C 273, were detected by COS-B. Identification of the remaining γ-ray emitters (e.g. *Geminga) with known celestial objects has not, however, been possible. Detailed study of the COS-B positional *error boxes has shown that none of these medium-energy γ-ray sources can have an x-ray luminosity more than a fraction of the γ-ray luminosity; this indicates that an understanding of their physical nature has to be found in the interpretation of γ-ray data and, furthermore, that they could represent a new class of cosmic object. At least two of the COS-B sources are coincident with *molecular cloud regions (rho Ophiuci and the *Orion molecular cloud). The observed positional correlation between the γ-ray *isophotes and the distribution of *extinction from these regions suggests γ-ray production by means of particle interactions with the matter of the cloud.

Apart from the localized sources in our Galaxy, a clearly defined band (±15° latitude) of enhanced γ-ray continuum emission lying along the galactic equator has been observed; it has been interpreted as being mainly the result of interaction of *cosmic-ray electrons (via *bremsstrahlung) and cosmic-ray nuclei (via neutral *pion, π^0, decay) with the interstellar gas. There is also evidence favouring local structure in the γ-ray emission coincident with the *Gould Belt as well as the Doliditz system. The Galaxy is also a rich source of low-energy (0.5–10 MeV) γ-ray emission, which encompasses a diffuse component as well as a number of discrete sources. The *galactic centre is a powerful emitter of electron-positron *annihilation radiation. Detailed studies with germanium detectors of high spectral resolution have revealed that the line is narrow and shows some evidence for three-photon *positronium continuum below 511 keV.

Although the quasar 3C 273 was detected as a medium-energy γ-ray source, other types of *active galaxies (i.e. *Seyferts and *radio galaxies) have been discovered to be strong emitters of low-energy (0.1–10 MeV) γ-rays. The γ-ray power output is found to dominate all emissions at other wavelengths. This high luminosity coupled with the timescale of the γ-ray variability (a few months) may be taken as evidence that the emissions are intimately related to the region containing the central power house.

A diffuse, and as far as can be measured isotropic, cosmic γ-ray flux has been observed to extend from 0.1 MeV to more than 100 MeV. At the present time it is unclear if this *cosmic background flux is derived from particle interactions throughout the universe, or whether it is derived from the contributions of a large number of active galaxies.

A growing number of galactic objects have been detected as very high energy ($> 10^{12}$ eV) γ-ray emitters. In two cases a *neutron star is almost certainly involved and the emission follows the period of pulsations at other wavelengths.

The measurement of γ-ray photons permits the study of the largest transfers of energy occurring in astrophysical processes. The extreme penetrating power of these high-energy photons offers a unique opportunity to probe deeply into the heart of violent galactic and extragalactic systems. The next generation of more sensitive γ-ray telescopes will reveal much new information on both the properties of astronomical objects and the high-energy processes that govern the dynamics of their evolution.

gamma-ray bursts. Intense flashes of hard x-rays or gamma rays, detected at energies up to one million elec-

tronvolts. They were discovered by US Air Force satellites in 1967 but not declassified until 1973. There are sharp temporal features in the burst time profile; this allows the measurement of differences in arrival times of wavefronts of the order of a few milliseconds over baselines separated by hundreds of light-seconds. For the strongest and most rapidly varying bursts, such measurements yield angular resolutions of the order of arc seconds. The positional *error box (0.1 arcmin²) on the most intense burst observed so far lies within the supernova remnant N49 in the Large *Magellanic Cloud. Generally, however, γ-ray bursts are thought to be Galactic in origin. γ-ray emission lines in their spectra may be related to *annihilation radiation redshifted by the strong gravitational field of a neutron star, and γ-ray absorption features to cyclotron absorption (see synchrotron emission) in intense magnetic fields. The rapid temporal structure, including the recently discovered periodic emission, is generally assumed to point to *neutron star origins for γ-ray bursts. The most probable energy source is thought to be either gravitational or nuclear in origin.

Gamma Ray Observatory (GRO). A satellite under construction for NASA and due to be launched by shuttle as a free-flying mission in 1988. The nominal circular orbit will be about 400 km with an inclination of 28.5°. A comprehensive study will be made of the gamma-ray portion of the electromagnetic spectrum, encompassing six orders of magnitude in photon energy (30 keV to 30 GeV). The four separate scientific instruments will survey γ-ray sources and diffuse emissions over the spectral range between 100 keV to 30 GeV, study nuclear γ-ray lines, and monitor a large fraction of the sky for a wide range of transient γ-ray events.

gamma rays (γ-rays). Very high energy *electromagnetic radiation, i.e. radiation with the shortest wavelengths and the highest frequencies: γ-ray wavelengths are less than 10^{-11} metres. There is no sharp cutoff between the γ-ray region and the adjacent x-ray region of the electromagnetic spectrum.

Gamma rays, like x-rays, are usually described in terms of photon energy, hv, where v is the frequency of the radiation and h is the Planck constant. The γ-ray region of the electromagnetic spectrum spans many decades of photon energy – from about 10^5 electronvolts (eV) to more than 10^{15} eV. The range is customarily subdivided into a number of energy bands that are related to changes in telescope technology:
low energy γ-rays, $10^5 - 10^7$ eV;
medium energy γ-rays, $10^7 - 10^9$ eV;
high energy γ-rays, $10^9 - 10^{11}$ eV;
very high energy γ-rays, $10^{11} - 10^{14}$ eV;
ultra high energy γ-rays, $>10^{14}$ eV.
Low energy γ-ray photons ($hv \sim 10^6$ eV) are the most penetrating photons available to the astronomer. Traditionally γ-ray source spectra have been measured in units of photons per unit area per second per unit energy interval, i.e. treating the photons as individual events. At present the general move is to unify the measurement system by use of the *jansky.

gamma-ray sources. See gamma-ray astronomy.

Ganymede. The brightest and largest of the four giant Galilean satellites of Jupiter. With a diameter of 5256 km it is the largest satellite in the solar system. It has an albedo of 0.4 and a density of 1.93 g cm^{-3}. The surface features form two distinct geological units: ancient darkish heavily cratered terrain and long parallel grooved areas. The grooves are typically 5 to 10 km in width. Ganymede is currently the only world other than the earth known to exhibit lateral faulting.

Some of the craters are ghostly (almost totally buried) while others are fresh with ray systems (bright radiating streaks) associated with them. See also Jupiter's satellites; Table 2, backmatter.

Gargantuan Basin. See Procellarum, Oceanus.

Garnet star (μ Cep). A red supergiant in the constellation Cepheus. It is a *semiregular variable with a magnitude that ranges from 3.6 to 5.1 although it is usually about 4.5. It is also a triple star. Spectral type: M2 Ia.

gas scintillation proportional counter (SPC). See proportional counter.

gauss. Symbol: G. The c.g.s. (electromagnetic) unit of *magnetic flux density. One gauss is equal to 10^{-4} tesla.

Gaussian gravitational constant. Symbol: k. A constant equal to
$$0.017\ 202\ 098\ 95$$
It defines the astronomical system of units of length (*astronomical unit), mass (*solar mass), and time (*day) by means of Newton's form of *Kepler's third law:
$$n^2a^3 = k^2(1 + m)$$
n is the mean motion of a planet in radians per day, a is the semimajor axis of its orbit in astronomical units, and m is its mass in terms of solar mass. The dimensions of k^2 are those of Newton's gravitational constant.

gegenschein (counterglow). A very faint patch of light that can be seen on a clear moonless night on the ecliptic at a point 180° from the sun's position at the time of observation. It is part of the *zodiacal light.

Geminga. An intense gamma-ray source in the constellation Gemini (hence the name), first discovered by the SAS-2 satellite. More than 99% of its power output is observed in the γ-ray spectral range. Despite a series of determined searches at other wavelengths, Geminga defies positive identification with known counterparts.

Gemini (Twins). A conspicuous zodiac constellation in the northern hemisphere near Orion, lying partly in the Milky Way. The brightest stars are the 1st-magnitude *Pollux (β) and the somewhat fainter multiple star *Castor (α), the 2nd-magnitude spectroscopic binary Alhena (γ), and several of 3rd magnitude. There are many interesting objects including variable stars, such as the *Cepheid Zeta Geminorum and the prototype of the U Geminorum *dwarf novae stars, binary stars, such as Eta Geminorum, the open cluster M35 visible to the naked eye, and the planetary nebula NGC 2392. Abbrev.: Gem; genitive form: Geminorum; approx. position: RA 7h, dec +25°; area: 514 sq deg.

Geminids. An important and active winter *meteor shower that maximizes on Dec. 13, meteors being visible between Dec. 7–15. The shower has a radiant of RA 113°, dec +32°, and a *zenithal hourly rate of 58; the meteoroids hit the atmosphere with a velocity of about 36 km s^{-1}. The *meteor-stream orbit has a low semimajor axis (1–5 AU) and matches almost exactly the orbit of the Apollo asteroid 1983 TB discovered by the infrared satellite IRAS. This could be the decayed parent comet. The shower has yielded a fairly constant rate during the last century.

Gemini project. A series of US space missions that extended the knowledge gained from the *Mercury project and preceded the *Apollo programme. The project demonstrated that man could function effectively over long periods of weightlessness, both inside and outside a spacecraft, and that two spacecraft could be made to rendezvous and dock while in orbit. Gemini 3 was the first manned flight,

launched in Mar. 1965, with two astronauts on board. America's first spacewalk was made by Ed White in Gemini 4, launched in June 1965. Geminis 6 and 7 rendezvoused in Dec. 1965, and in Mar. 1966 Gemini 8 made the first successful docking with an unmanned target. Gemini 7 remained in orbit for a record 330.6 hours. Gemini 12, launched Nov. 1966, completed the project.

Gemma. *See* Alphecca.

general precession. *See* precession.

general relativity. *See* relativity, general theory.

geocentric coordinate system. A *coordinate system, such as the *ecliptic coordinate system, in which the position of a celestial body is referred to the centre of the earth. *Compare* heliocentric coordinate system; topocentric coordinates.

geocentric parallax. *See* diurnal parallax.

geocentric system. A model of the solar system or of the universe, such as the *Ptolemaic system, that has the earth at its centre.

geocentric zenith. *See* zenith.

geocorona. A tenuous cloud of hydrogen and some helium extending more than 50 000 km from the earth's outermost *atmospheric layer to the *magnetosphere. It scatters Lyman-alpha radiation (*see* hydrogen spectrum) from the sun, producing a glow that is visible on long-distance far-ultraviolet photographs of the earth. It also interferes with far-ultraviolet astronomical observations from earth orbit.

geodesy. The study and measurement of the size, shape, and gravitational field of the earth, using both the measurements of the precise form of the *geoid and observations of artificial satellites, such as the *GEOS satellites.

geoid. The form of the earth obtained by taking the average sea level surface and extending it across the continents. It is an *equipotential surface defined by measurements of the variation of the earth's gravitational attraction with latitude and longitude and the acceleration produced by the earth's rotation. The geoid differs from a sphere in that the equatorial diameter (12 756.32 km) is greater than the diameter through the poles (12 713.51 km). The *flattening* of the earth corresponds to the difference between these diameters as a fraction of the equatorial diameter and has a value of 1/298.257.

Studies of the *perturbations affecting the orbits of artificial earth satellites have shown that the geoid departs from an oblate spheroid, or ellipsoid. The true geoid is elevated by 17 metres at the north pole, 81 metres to the north of Australia and by 60 metres south of South Africa and west of the British Isles; in addition there is a 27 metre depression at the south pole and 50 metre depressions south of New Zealand and both east and west of the USA. The greatest departure is a 113 metre depression to the south of India.

geology. The study of the history, structure, and composition of the *earth's crust. The principle divisions of geology include *physical geology, historical geology,* and *economic geology.*

Physical geology includes mineralogy, which is the study of minerals; petrology, the study of the formation, composition, and structure of different types of rock; structural geology, concerned with how these rocks combine to form the crust; and geomorphology (or physiography), the study of the relation between geographical features and sub-surface geological structures.

Historical geology includes stratigraphy, which is concerned with the chronology of rock strata; palaeogeography, the study of the distribution of geographical features (seas, deserts, mountains) during early periods of the earth's history; and palaeontology, dealing with the development of life as revealed by the fossil record on different geological strata.

Economic geology is concerned with the study of valuable mineral deposits, such as ores, coal, and oil.

geomagnetic storms. Sudden alterations in and subsequent recovery of the earth's magnetic field due to the effects of solar *flares. The variations are complex in the auroral zone and polar regions but at middle latitudes the horizontal component of the field shows four distinct phases.

The first is *Storm Sudden Commencement* (SSC), when a sharp rise in field strength (over 2.5 to 5 minutes) is caused by compression of the *magnetosphere by a flare-generated shock wave. The second is the *Initial Phase* (IP), lasting from 30 minutes to several hours, when the earth is surrounded by the high-speed post-shock plasma and field, and is effectively isolated from the interplanetary magnetic field. The surface field strength is higher than the pre-SSC value. In the *Main Phase* (MP) an increase in particle population, or particle acceleration due to reconnection of the geomagnetic and interplanetary fields, or in magnetospheric fluctuations produces a *ring current at three to five earth radii. This generates a magnetic field opposed to the earth's and causes a decrease of the surface field strength of 50–400 nanoteslas, which lasts from a few hours to more than a day. During the fourth phase, *Recovery Phase* (RP), which is typically longer than the MP, the ring current decays by diffusion of the trapped particles and plasma instabilities. The surface field strength may return to, or just below, the pre-SSC value.

Storms are usually accompanied by ionospheric and auroral activity, and some may recur after 27 days due to the persistence of a particular solar *active region or *coronal hole.

geomagnetism. The earth's magnetic field (or its study), which at the earth's surface approximates that of a bar magnet at the centre of the earth with its axis inclined by 11.4° to the earth's rotation axis and somewhat off-centred: the north magnetic and geographical poles are much closer together than the south poles. Both sets of poles wander in position. The strength of the magnetic field varies from 0.6 gauss near the magnetic poles to 0.3 gauss near the equator, i.e. from 60–30 microtesla, but can depart by up to 20% from the average without any correlation with major surface features. The dipole field changes only slowly with time but there are larger local variations in strength and direction. Violent short-term fluctuations occur during *geomagnetic storms. Studies of magnetized rocks show that the entire magnetic field has reversed in direction about every 15 million years in the past 100 million years. Complete reversals (i.e. north pole switching from pointing towards geographic north to pointing south, or vice versa) can occur within a few thousand years. The source of the field is believed to lie in a complex dynamo action in the earth's liquid iron-rich outer core. Convective motion in this rotating electrically conducting fluid maintains and regenerates the magnetic field. The field existed at least 2.5, probably 3.5 thousand million years ago. *See also* ring current; magnetosphere.

geometrical libration (optical libration). *See* libration.

geophysics. The physics of the earth. It includes the study of the history, motion, and constitution of the planet; movements within it, such as those associated with continental drift, mountain-building, earthquakes, glaciers, tides, and atmospheric circulation; geomagnetism; and the interaction between the earth and its interplanetary environment.

GEOS. The first all-scientific satellite in geostationary orbit, built by the European Space Agency for the purpose of studying the earth's *magnetosphere. GEOS–1, launched in April 1977, failed to reach its planned orbit; it was replaced by GEOS–2, which was successfully launched on July 14 1978. Having become redundant it was boosted into a higher orbit in 1984.

geostationary orbit. *See* geosynchronous orbit.

geosynchronous orbit. An earth orbit made by an artificial satellite (moving west to east) that has a period – 23 hours 56 minutes 4.1 seconds – that is equal to the earth's period of rotation on its axis. If the orbit is inclined to the equatorial plane the satellite will appear from earth to trace out a figure-of-eight, once per day, between latitudes corresponding to the angle of orbital inclination to the equator. If the orbit lies in the equatorial plane, and is circular, the satellite will appear from earth to be almost stationary; the orbit and orbiting body are then termed *geostationary*. A geostationary orbit has an altitude of 35 784 km.

A geosynchronous or geostationary orbit is very difficult to achieve, requiring a very high orbital velocity. Satellites in such orbits are used for communications and navigation and also for certain types of earth observations. Most communications satellites are now geostationary, with groups of three or more, spaced around the orbit, giving global coverage. The *International Ultraviolet Explorer (IUE) satellite has been to date the only astronomical satellite placed in a geosynchronous orbit, which has facilitated its operation as an observatory facility with real-time ground station interaction.

German mounting. *See* equatorial mounting.

German-Spanish Astronomical Centre. A recently built observatory on the mountain Calar Alto in Almeria, Spain at an altitude of 2160 metres. A joint project of Germany and Spain, it is an outpost of the Max Planck Institute for Astronomy (MIPA), Heidelberg. The instruments include a 3.5 metre reflector, operational 1984, a 2.2 metre reflector, operational 1979 and identical to MPIA's 2.2 metre telescope at the European Southern Observatory, a 1.5 metre reflector of Madrid University operational (independently) since 1979, a 1.2 metre reflector operational 1975, and a 0.8 metre Schmidt moved from Hamburg Observatory.

ghost crater. A *crater that has been almost totally buried by lava. Ghost craters survive in the shallow irregular *maria on the moon, where they can be made use of to calculate mare depths.

Giacobini-Zinner. A short-period (6.42 year) comet with perihelion near 1 AU and aphelion near 6 AU, discovered in 1900 by M. Giacobini and rediscovered in 1913 by E. Zinner. It has been seen 12 times since discovery, every other apparition leading to a good view from earth. Ground-based viewing in 1985 will be excellent. The *International-Sun-Earth-Explorer satellite (ISEE–3) has been reorbited so that it will intersect the comet on 11 Sept. 1985 when it is near perihelion. It will principally investigate the interaction between the comet and the *solar wind and will pass downstream of the cometary nucleus through the inner tail region. Earth's passage through the Giacobinid *meteor stream produced in Oct. 1946 one of the most spectacular meteor displays this century, about 70 meteors per minute being the maximum rate. The tracks were very short indicating that the meteoroids were extremely fluffy with a density of about 0.01 g cm^{-3}.

giant. A large highly luminous star that lies above the *main sequence on the *Hertzsprung–Russell diagram. Giants are grouped in luminosity classes II and III (*see* spectral types) and generally have absolute *magnitudes brighter than 0. Despite their great size, they are not necessarily more massive than typical main-sequence stars; they have dense central cores, but their atmospheres are very tenuous – a feature that shows in their spectra.

Giants represent a late phase in *stellar evolution, when the central hydrogen supplies have been exhausted and the star is 'burning' other nuclei in concentric shells near its core. As these nuclear processes change, the star's size, luminosity, and temperature gradually alter, and it moves about in the *giant region* and *horizontal branch region of the H–R diagram. Most stars cross the instability strip (*see* pulsating variables) at least once, and are then *Cepheid or *RR Lyrae variables. In its final stages, a giant becomes rather brighter and moves to the *asymptotic giant branch* just above the giants on the H–R diagram. Capella and Arcturus are typical examples of giant stars. *See also* red giant; supergiant.

giant molecular cloud (GMC). *See* molecular clouds.

giant planets. The planets Jupiter, Saturn, Uranus, and Neptune, which have diameters between 3.8 and 11.2

times that of the earth and masses of between 14 and 318 earth masses. They orbit the sun at mean distances ranging from 5.20 AU for Jupiter to 30.06 AU for Neptune in periods from 11.86 to 164.79 years. All have low densities – from 0.7 to 1.7 times that of water – and are probably composed largely of hydrogen in its molecular or metallic state. Their visible surfaces are thought to be clouds of ammonia or methane. Jupiter, Saturn, and Uranus have ring systems, while all four giant planets share more than 46 satellites between them (*see* Table 2, backmatter).

giant radio galaxy. *See* radio galaxy.

giant star. *See* giant.

gibbous moon. *See* phases, lunar.

giga-. Symbol: G. A prefix to a unit, indicating a multiple of 10^9 of that unit. For example, one gigahertz is 10^9 hertz.

Giotto. An ESA spacecraft scheduled for launch in July 1985 into earth orbit from where it will be fired into a trajectory that will intercept the path of Halley's comet in Mar. 1986. It is hoped that Giotto can be steered deep into the comet's coma to within 1000 km of the cometary *nucleus, with data gathered over a period of about 4 hours. Its 10 instruments will take pictures, measure the composition of the coma and the mass, distribution, and composition of the dust tail, and will analyse the dynamics of the cometary plasma and its interaction with the solar wind. *See also* Veha.

glitch. *See* pulsar.

globular cluster. A spherically symmetrical compact *cluster of stars, containing from several tens of thousands to maybe a million stars that are thought to share a common

origin. An example is the *Great Cluster in Hercules. A few globular clusters, such as Omega Centauri, appear to be slightly flattened. The concentration of stars increases greatly towards the centre of the cluster, where the density may be as much as 1000 stars per cubic parsec. About 125 globular clusters are known. They appear to move in giant and highly eccentric elliptical orbits about the centre of our *Galaxy. Unlike *open clusters, they are not concentrated towards the galactic plane; instead they show a roughly spherical distribution in the galactic halo.

Globular clusters are *population II systems: all the stars within them are relatively old (older than the sun) and have a very low metal content; the metal abundance of the clusters decreases with their distance from the galactic centre. The brightest occupants are *red giants. Although very few ordinary binary stars are observed in globular clusters, many contain strong x-ray sources typical of *x-ray binaries or *cataclysmic variables, i.e. systems containing a neutron star or a white dwarf, respectively. The distribution and other characteristics of globular clusters suggest that they were formed early in the life of the Galaxy, possibly some 12 to 13 million years ago, before the main body of the galactic disc had evolved. Since most of the member stars will have evolved away from the main sequence, the *Hertzsprung-Russell diagram for stars of a globular cluster differs greatly from the conventional H–R diagram (see illustration). The *turnoff point from the main sequence gives a measure of the age of a cluster. Distances to globular clusters are usually calculated from the apparent magnitudes of the *RR Lyrae stars within them. *See also* open cluster.

globule. *See* dark nebula; Bok globule.

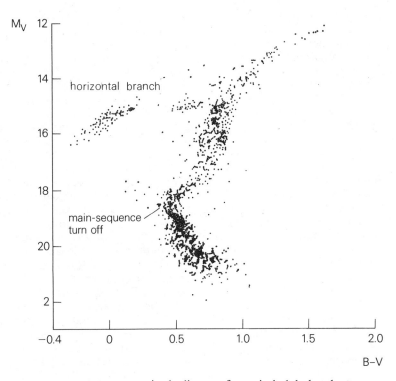

M_V

horizontal branch

main-sequence
turn off

B–V

Colour-magnitude diagram for typical globular cluster

gnomon. A device used in ancient times to measure the altitude of the sun and hence to determine the time of day and time of year. It consisted of a vertical shaft of known height that cast a shadow of measurable length and direction on a graduated horizontal base. The ratio of height to shadow length gave the tangent of the altitude angle. The term is also applied to the metal projection on a sundial, used for the same purpose.

Goldstone. Site of the *Jet Propulsion Laboratory, California.

Gould Belt (Local System). A local formation of stars and clouds of gas and dust that appears to be a spur attached to the lower edge of the Orion arm of the *Galaxy. It contains many of the apparently brightest stars in the sky, which follow the projec-

tion of the system across the sky in a 'belt' at 16° to the line of the Milky Way. The system is about 700 parsecs wide and 70 parsecs thick and includes the II Persei, Scorpio-Centaurus, and Orion OB-associations (*see* association). The sun lies approximately 12 parsecs north of the Belt's equatorial plane and about 100 parsecs from its centre.

grading function. *See* aperture synthesis.

gram. A unit of mass equal to one thousandth of a *kilogram.

grand unified theories (GUTs). *See* fundamental forces.

granulation. A network of convective cells in the solar *photosphere. It consists of bright irregularly shaped (often polygonal) *granules,* separated by dark narrow intergranular lanes. The individual granules are about 1000 km in diameter, have an upward velocity of about 0.5 km s^{-1}, and exhibit a horizontal flow of material – outward from the centre – with a velocity of about 0.25 km s^{-1}; their average lifetime is about 8 minutes. They represent the changing tops of currents from the *convective zone, bringing hot gases to the photosphere, where the gases cool and then return via the intergranular lanes.

Granulation is readily visible under favourable conditions, when it gives the photosphere a mottled appearance. It may be best seen near the centre of the sun's disc, where foreshortening is not significant. *See also* supergranulation.

graticule. A glass plate or cell bearing a grid, cross wires, or graduated scale that is set in the focal plane of the eyepiece of a telescope and is used for positioning or measuring.

grating. *Short for* diffraction grating.

grating response. *See* array.

grating ring. *See* aperture synthesis.

grating spectrograph. A *spectrograph in which a *diffraction grating is used to produce the spectra.

gravitation. The ability of all material bodies to attract each other. This mutual attraction, or *gravitational force*, is the most familiar force of nature and was first expressed in mathematical form by Isaac Newton. *Newton's law of universal gravitation* was published in 1687 in *Principia*. It states that the force of attraction, F, between two bodies is proportional to the product of their masses, m_1 and m_2, and inversely proportional to the square of their distance apart, d:
$$F = Gm_1m_2/d^2$$
The constant of proportionality, G, is the *gravitational constant. The force

experienced by the mass m_1 is equal to but in the opposite direction to that felt by the mass m_2: the two masses are attracted towards each other. Newton showed that any body behaves, gravitationally, as if its mass were concentrated at its centre. Thus gravitational force acts along a line joining the centre of two bodies. In a system in equilibrium, such as the solar system or a star like the sun, gravitational force is balanced by an equal force acting in the opposite direction.

The region of space surrounding a massive body and in which the gravitational force is appreciable is the *gravitational field* of that body. The magnitude of the field at a particular point is the gravitational acceleration (in the direction of the massive body) that would be experienced by any object at that point; this is equivalent to the force that would be experienced by an object of unit mass at that point. A gravitational field depends on the distribution of matter that causes it. Its effect is on another distribution of matter. The *gravitational potential* at a point in a gravitational field is the work done in bringing unit mass to this point from a point infinitely distant from the cause of the field; it is thus the potential energy of a particle of unit mass arising from the mass of a material body.

Gravitation is the weakest force known. It is many orders of magnitude smaller than the other *fundamental forces of nature – the strong, electromagnetic, and weak interactions. It is, however, the only means by which bodies can interact over immense distances.

Newton's theory of gravitation has proved adequate in most circumstances but was challenged by the more complex general theory of *relativity put forward by Albert Einstein in 1915. According to general relativity, gravitational fields change the geometry of *spacetime: both space and time are curved or warped around a massive body. Matter tells spacetime how to curve and spacetime tells matter how to move. *See also* Kepler's laws; quantum gravitation.

gravitational collapse. Contraction of a body arising from the mutual gravitational pull of all its constituents. Although there are several examples of such contraction processes in astronomy, 'gravitational collapse' usually refers to the sudden collapse of the core of a massive star at the end of nuclear burning, when its internal gas pressure can no longer support its weight. This may initially result in a *supernova explosion, removing much of the star's mass. The eventual degree of gravitational collapse is determined by the mass that remains. The three most likely end-products (in order of increasing mass) are *white dwarfs, *neutron stars, and *black holes.

gravitational constant. Symbol: *G*. A universal constant that appears in Newton's law of *gravitation and is the force of attraction between two bodies of unit mass separated by unit distance. It is equal to 6.672×10^{-11} N m^2 kg^{-2}. Predictions that *G* is decreasing very slightly with time (by less than one part in 10^{10}) are not supported by experimental evidence.

gravitational field. *See* gravitation.

gravitational force. *See* gravitation.

gravitational instability. The tendency for a nonuniform medium exceeding a critical size (the *Jeans length*) to become increasingly irregular with time because of gravitational attraction. This process is thought to be responsible for the formation of galaxies (*see* galaxies, formation and evolution).

gravitational lens. A gravitating body that bends light and other radiation from a more distant object, usually producing two or more images of the latter. The term is generally used to mean a galaxy or *cluster of galaxies

that focuses the light of more distant quasars lying behind it. The first example was discovered in 1979. It involves the quasar $0957+561$, often known as the *double quasar* or *twin quasar*. The quasar itself lies at a *redshift of 1.41; a foreground galaxy, redshift 0.36, focuses the quasar's light into two images of equal brightness that are about 6″ apart and have identical spectra. Five gravitational lenses are now (1984) known; in two, the 'lens' galaxy is visible as well as the multiple images of the quasar. There are generally two images but the quasar $1115+080$ appears to be quadruple. The appearance of an even number of images in all known cases is not understood since the theory of gravitational lenses (based on general *relativity) predicts an odd number.

gravitational mass. *See* mass.

gravitational potential. *See* gravitation.

gravitational radiation. *See* gravitational waves.

gravitational redshift (Einstein shift). The *redshift of spectral lines that occurs when radiation, including light, is emitted from a massive body. In order to 'climb out' of the body's gravitational field, the radiation must lose energy. The radiation frequency must therefore decrease and its wavelength λ, shift by Δλ towards a greater value. The redshift is given by
$$\Delta\lambda/\lambda = Gm/c^2r$$
G is the gravitational constant, *m* and *r* the mass and radius of the massive body, and *c* is the speed of light. Gravitational redshift was predicted by Einstein's general theory of *relativity and although extremely small has been detected in the spectra of the sun and several *white dwarfs. The redshift of the earth's gravitational field has been determined very accurately using beams of radiation travelling upwards through a tall

building. Predicted and measured values agree very closely.

gravitational waves (gravitational radiation). Extremely weak wavelike disturbances that were predicted by Einstein's general theory of *relativity (1915). They are produced when massive bodies are accelerated or otherwise disturbed. They are ripples in the curvature of *spacetime that travel at the speed of light, with a wide range of frequencies, and carry energy away from the source. They should affect all matter: gravitational waves hitting a suspended body, for example, should make it vibrate slightly. The interactions are very small, however. No conclusive direct evidence of the existence of gravitational waves has yet been forthcoming from the various highly sensitive experiments designed to detect them.

These waves should be emitted during supernova explosions or energetic events in the cores of active galaxies associated with black holes. They should also be emitted by two massive stars in close orbit. Recent observations of the *binary pulsar PSR 1913+16 show that its orbital period is decreasing by 10^{-4} seconds per year. The observed value corresponds almost exactly with the decrease predicted to result from the emission of gravitational waves and is at present the best indirect evidence for their existence.

gravity. *Another name for* gravitation.

grazing incidence. X-rays reflect from surfaces for the same physical reason that light reflects, mainly due to coherent scattering in the surface material. However, since x-ray photon energies are large, the *refractive index, n, is slightly less than unity and reflection occurs only up to a critical angle, θ, where $\cos \theta = n$. The angle θ is usually small and thus x-ray reflection is only high for rays that graze

the surface; hence 'grazing incidence'. For example θ is about 1° for a gold surface and x-rays of wavelengths of about 0.3 nm (4 keV). θ is proportional to x-ray wavelength and also increases with the *atomic number of the reflecting surface. These *x-ray mirrors* are finding increasing application in *x-ray astronomy.

The space station *Skylab was the first space mission to carry a large grazing incidence x-ray telescope, in fact two, both of which produced many high-resolution x-ray images of the solar corona in 1973–74. With two coaxial reflectors (usually a paraboloid/hyperboloid pair) high-quality x-ray images are produced over a field of view of order of the grazing angle. The great potential of such optical systems for x-ray astronomy was demonstrated with the flight of a 60 cm diameter grazing incidence telescope on the *Einstein Observatory in 1978–81. Future x-ray astronomy satellites that will employ this technique are *ROSAT and *AXAF.

grazing occultation. A lunar *occultation in which the astronomical body is only momentarily occulted by mountain peaks along the moon's limb.

Great Bear. Common name for *Ursa Major.

great circle. The circular intersection on the surface of a sphere of any plane passing through the centre of the sphere. A sphere and its great circles are thus concentric and of equal radius. *Compare* small circle.

Great Cluster in Hercules (M13, NGC 6205). The finest *globular cluster to be seen from the northern hemisphere, discovered in Hercules by Edmund Halley in 1714. It has a dense broad centre that is just visible to the naked eye; a 7.5 cm telescope will show some of the outer stars.

greatest elongation (GE). *See* elongation.

Great Red Spot (GRS). An immense oval feature centred 22° south of Jupiter's equator; it has a north-south width of 14 000 km and a variable east-west width of 24 000 to 40 000 km. Often visible in small telescopes, it has been observed for more than 100 years and may be identical with a feature recorded in the 17th century. Its prominence has varied, with colour fluctuations between pale pink and brick red. Early explanations involved a solid island adrift in Jupiter's atmosphere or an atmospheric disturbance above a Jovian mountain or basin; the latter was disputed by evidence that it drifts erratically in longitude.

Observations by *Pioneers 10 and 11 revealed the GRS as a turbulent vortex of cold anticlockwise-rotating clouds. The Voyager observations have shown that the GRS is not unique, since it possesses the same meteorological characteristics of other large-scale anticyclonically rotating vortices, such as the white ovals. There is no complete theory to explain its presence, although it would appear to be a manifestation of the weather systems. Its unique colour may be associated with the conversion of phosphene into red phosphorus in the GRS region. The lifetime of the GRS is unknown, although it is now half the size observed at the beginning of the century. It could eventually disappear.

Great Rift. A chainlike complex of large *dark nebulae that more or less obscures the light from a narrow but extensive band of the Milky Way between the constellations Cygnus and Sagittarius.

Great Square of Pegasus. *See* Pegasus.

Green Bank. Site of the *National Radio Astronomy Observatory, West Virginia.

greenhouse effect. A phenomenon whereby an environment is heated by the trapping of infrared radiation. Although the effect may be of secondary importance in the heating of greenhouses, it is responsible for the high surface temperature of Venus and operates also on earth, in the deep atmospheres of the giant planets, and on Saturn's large satellite Titan. Sunlight that filters to the Venusian surface through the dense clouds is absorbed and re-radiated at (longer) infrared wavelengths. The mainly carbon dioxide atmosphere is relatively opaque to infrared radiation, which it absorbs and re-emits. By the time the influx of sunlight to the surface is balanced by the escape of infrared radiation from the atmosphere, enough infrared radiation is being exchanged between the atmosphere and the surface to maintain both at a high temperature.

Greenwich mean astronomical time (GMAT). *Greenwich mean time beginning at noon. It was the time system in astronomical use before Jan. 1, 1925.

Greenwich mean sidereal time (GMST). The Greenwich *hour angle of the mean equinox of date (*see* mean equator and equinox). *Universal time is defined in terms of GMST by a mathematical formula. The *local sidereal time* at a particular location is equal to the GMST plus 4 minutes for every degree of longitude that the location is east of Greenwich.

Greenwich mean time (GMT). A timescale that is equivalent to UTC (coordinated *universal time), and thus the timescale available from broadcast time signals. The term was also used earlier in the sense of universal time, UT. Since 1925 GMT has been counted from midnight rather than from noon. GMT is no longer used in astronomy.

Greenwich Observatory. *See* Royal Greenwich Observatory.

Greenwich sidereal date (GSD). A concept analogous to that of the *Julian date. It is the number of *sidereal days that have elapsed at Greenwich since the beginning of the sidereal day that was in progress on Julian date 0.0. The integral part of the date is the *Greenwich sidereal day number.*

Gregorian calendar. The calendar that is now in use throughout most of the world and that was instituted in 1582 by Pope Gregory XIII as the revised version of the *Julian calendar. The simple Julian four-year rule for leap years was modified so that when considering century years only one out of four, i.e. only those divisible by 400, were to be leap years: 1700, 1800, and 1900 were not leap years. There are therefore 365.2425 days per year averaged over 400 years. This greatly reduced the discrepancy between the year of 365.25 days used in the Julian calendar and the 365.2422 days of the *tropical year, which had resulted in the accumulation of 14 days over the centuries.

The revision came into effect in Roman Catholic countries in 1582, the year being brought back into accord with the seasons by eliminating 10 days from October: Thursday Oct. 4 was followed by Friday Oct. 15. The vernal *equinox, which should have occurred on Mar. 11 and which had originally fallen on Mar. 25 in Julius Caesar's time was thus adjusted to Mar. 21. Gregory also stipulated that the New Year should begin on Jan. 1. Non-Catholic countries were slow to accept the advantages of the Gregorian reform. Britain and her colonies switched in 1752 when an additional day had accumulated be-

tween old and new calendars: Sept. 2, 1752 was followed by Sept. 14 and New Year's Day was changed from Mar. 25 to Jan. 1, beginning with the year 1752. The very slight discrepancy between the Gregorian year and the tropical year amounts to about three days in 10 000 years.

Gregorian telescope. The first compound *reflecting telescope to be devised, designed by the Scottish mathematician James Gregory. It has a small concave secondary mirror that is mounted beyond the focal plane and reflects the light back through a central hole in the primary mirror (see illustration). The design, published in 1663, required a secondary mirror of ellipsoid figure. The design is now little used, the similar but more compact *Cassegrain configuration being preferred.

Grigg-Skjellerup comet. A comet that, after Encke, has the second shortest cometary period, 4.91 years.

ground state. *See* energy level.

Grus (Crane). A fairly conspicuous but isolated constellation in the southern hemisphere, the two brightest stars, Alpha and Beta Gruis (a red giant), being of 2nd magnitude. Abbrev.: Gru; genitive form: Gruis; approx. position: RA 22.5h, dec −50°; area: 366 sq deg.

G stars. Stars of *spectral type G that are yellow stars with a surface temperature (T_{eff}) of about 5000 to 6000 kelvin. The lines of ionized calcium (Ca II, H and K) dominate the spectrum and there are a large number of strong neutral metal lines (strengthening) and ionized metal lines (weakening) from G0 to G9. Molecular bands of CH and CN appear. The sun and Capella are G stars.

guide telescope. A telescope carried upon the same *equatorial mounting

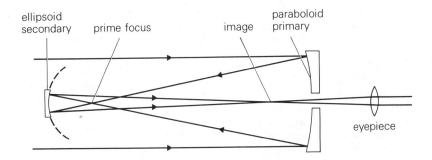

ellipsoid
secondary   prime focus   image paraboloid
                                      primary

eyepiece

Ray path in Gregorian telescope

as an astronomical camera and used to ensure that the camera is accurately guided for the whole of an exposure, which may be of many hours duration. For this duty a high magnification and an illuminated graticule are required. The telescope drives must also be provided with slow motions in both declination and right ascension so that corrections can be applied manually as the selected *guide star* begins to drift from the cross wires. The camera will often be a larger telecope with a photographic plate exposed at its prime focus.

In a modern well-equipped observatory the tedious duty of manual guidance is now taken over by automatic devices that monitor the position of the guide star image and adjust the drive to hold it steady.

Gum nebula. An immense *emission nebula that dominates the Milky Way in the southern constellations Puppis and Vela. It is an expanding shell of ionized gas some 35° (250 parsecs) in diameter and about 400 parsecs distant. It has been suggested that it is a fossil *Strömgren sphere produced by a supernova that exploded maybe up to a million years ago and gave off sufficient ultraviolet radiation to ionize interstellar gas out to an immense distance. Several very hot stars within the shell are now ionizing the gas thus preventing it from returning to a neutral state. The *Vela pulsar and its supernova remnant lie in the same direction as the Gum nebula.

GUT. *Abbrev. for* grand unified theory. *See* fundamental forces.

gyrofrequency. *See* synchrotron emission.

H

HA. *Abbrev. for* hour angle.

Hadar. *See* Beta Centauri.

Hadley Rille. A sinuous *rille in Palus Putredinis, adjacent to the Apennine (third) ring of the *Imbrium Basin. It was sampled by the crew of *Apollo 15.

hadron. *See* elementary particles.

Hakucho. The first Japanese x-ray astronomy satellite, launched in Feb. 1979 and expected to continue operating until re-entry in late 1984. Its payload includes several small nonimaging detectors well matched to the study of variability in the brighter galactic x-ray sources. Extended observations of x-ray pulsars and of x-ray burst sources were the major scientific contributions. *See also* Tenma; ASTRO–C.

Hale Observatories. The group of observatories comprising the *Mount Wilson Observatory, *Palomar Observatory, and *Big Bear Solar Observatory, all in California, and *Las Campanas Observatory, Chile. The group is named after George Ellery Hale, who played a great part in the development of the first two. The observatories are sponsored by the Californian Institute of Technology and the Carnegie Institution of Washington.

Hale telescope. *See* Palomar Observatory.

half-life. The time taken for the number of atoms of a radioactive isotope to be reduced, by radioactive decay, to one half. The *mean life* is the average time before decay of a large number of similar elementary particles or atoms of a radioisotope. Mean life is equal to 1.44 times the half-life.

half-wave dipole. *See* dipole.

half-width. Half the width of a spectral line, measured at half the height. In some branches of spectroscopy it is customary to use the term to mean the full width at half height. *See also* line profile.

Halley's comet. A comet that was first positively recorded in 240 BC and has a period of 76 years on average, one return coinciding with the battle of Hastings (1066) and leading to the comet's representation on the Bayeux Tapestry. The orbit is retrograde, the comet moving in the opposite direction to the planets. It is named after Edmund Halley who showed that comets move around the sun in accordance with Newton's theory of gravitation. He noticed that the orbits of the bright comets of 1531, 1607, and 1682 were very similar and concluded that they were actually one and the same comet. He calculated its orbit and predicted that this comet should return in 1758. Unfortunately Halley died 16 years before the comet returned.

Halley's comet can be seen with the naked eye in October and November 1985 and in early spring 1986: *perihelion is on Feb. 9 1986 and *perigee on Nov. 27 1985 and Apr. 11 1986. Detailed observations will be made using instruments on earth and in three scheduled spacecraft: ESA's *Giotto, the Soviet-led mission *Veha, and the Japanese *Planet A. It is hoped that pictures can be obtained by Giotto of the 6 km nucleus and of structure and activity near the nucleus. The comet's development as it rounds the sun will be recorded and studies made of cometary composition, of processes producing and occurring in the coma and tails, and the interaction between comet and *solar wind.

halo. 1. *See* Galaxy.
2. *See* dark halo.
3. *See* Donati's comet.

Hamal (α Ari). An orange giant that is the brightest star in the constellation Aries. m_v: 2.00; M_v: 0.3; spectral type: K2 III; distance: 22 pc.

h and Chi Persei (Sword-Handle; Double Cluster in Perseus). A fine double cluster in the constellation Perseus. Both the constituent *open clusters, NGC 869 (h Persei) and NGC 884 (Chi Persei) are visible to the naked eye. They are relatively young clusters, about 10 million years old, and appear to be associated with the surrounding Zeta Persei OB-association. *See* association; OB-cluster.

H and K lines. Lines in the spectrum of singly ionized calcium, occurring at the visible/ultraviolet wavelengths of 393.4 nm (H) and 396.8 nm (K). They are very prominent *Fraunhofer lines.

Haro galaxies. Galaxies, found in the southern sky by G. Haro, that are all blue objects with sharp emission lines. They are believed to be spiral or irregular galaxies that have recently had a burst of star formation.

Harvard classification (Henry-Draper system). *See* spectral types.

harvest moon. The full moon that occurs closest to the autumnal *equinox. At this time the *retardation is small (so that the moon appears to rise at the same time on successive evenings) because of the low inclination of the plane of the moon's orbit to the horizon.

Hawking radiation. *See* black hole.

Hayashi track. A near-vertical downward-directed line on the *Hertzsprung–Russell diagram that is believed to be followed by *protostars as they contract towards the *main sequence. C. Hayashi (1965) calculated these tracks as the evolutionary paths of stars that are convective throughout (*see* energy transport), as protostars are expected to be.

HD. A prefix used to designate a star as listed in the *Henry Draper Catalogue.

H-D system. *Short for* Henry-Draper system. *See* spectral types.

head. The coma and nucleus of a *comet when seen together. In this context 'nucleus' means the diffuse starlike luminous condensation sometimes observed in the coma. The head generally contracts as *perihelion is approached and expands again afterwards. Remarkable changes in size, luminosity distribution, and the number of observed 'nuclei' can take place inside the head in a few hours.

HEAO. *Abbrev. for* High Energy Astrophysical Observatory. Any of three large (about 3000 kg) earth satellites built by NASA for second-generation studies of *x-ray sources, *gamma-ray sources, and *cosmic rays.

HEAO–A, or HEAO–1 after launch, was put into a 400 km, 22° inclination orbit in Aug. 1977 and carried a payload of large x-ray detectors with which to conduct a more sensitive sky survey than that of *Uhuru and *Ariel V. HEAO–1 completed two and a half surveys of the sky and several extended pointings at individual sources before re-entry early in 1979. The eventual outcome was a catalogue of approximately 1000 x-ray sources, brighter than ∼1 microjansky, together with the best current broad-band spectra of many active galaxies, clusters, and galactic sources.

HEAO–2, the *Einstein Observatory, operated from launch in Nov. 1978 until loss of control gas in Apr. 1981. The unique capabilities of its 60 cm *grazing incidence telescope revolutionized x-ray astronomy, in the same way Uhuru had done a decade earlier.

HEAO–3 was launched in Sept. 1979. It carried a number of large cosmic-ray instruments to study the mass and charge distribution of primary *cosmic rays, particularly the nuclei of the heavier elements, over a wide range of energies. The gamma-ray spectroscopy experiment was designed specifically to search for cosmic sources of narrow-line emission in the energy range 50 keV to 10 MeV. It consisted of a cluster of germanium solid-state detectors surrounded by an active caesium iodide scintillator shield. Apart from the study of γ-ray line emission from solar *flares and 511 keV *annihilation radiation from the galactic centre region (*see* gamma-ray astronomy), the spectrometer detected two strong γ-ray line emissions from the extraordinary galactic object *SS 433. The various experiments on HEAO–3 ceased operations in the period 1980–82.

heavy elements. All elements heavier than hydrogen and helium. The *cosmic abundance of the heavy elements is very much lower than that of hydrogen and helium.

heavy-metal stars. *See* S stars.

heliacal rising (or setting). The rising or setting of a star or planet at or just before sunrise so that it can be seen in the morning sky.

heliocentric coordinate system. A *coordinate system referring usually to the plane of the *ecliptic but specifying position with respect to the centre of the sun rather than that of earth (*see* geocentric coordinate system). The coordinates – *heliocentric latitude* and *longitude* – are used to determine the relative positions of objects in the solar system.

heliocentric parallax. *See* annual parallax.

heliocentric system. A model of the solar system or of the universe that has the sun at its centre. Aristarchus of Samos proposed a heliocentric system in the third century BC in which the planets moved at uniform rates around the sun in circular orbits. The theory was abandoned due to large discrepancies between prediction and observation, and the geocentric *Ptolemaic system was subsequently favoured until the establishment of the heliocentric *Copernican system.

heliographic latitude. The angular distance from the sun's equator, measured north or south along the meridian. The position of a point on the sun's disc may be defined in terms of this and its *heliographic longitude.

heliographic longitude. The angular distance from a standard meridian on the sun (taken to be the meridian that passed through the ascending *node of the sun's equator on 1854 Jan 1 at 12h 0m UT and calculated for the present day assuming a uniform *sidereal period of rotation of 25.38 days; *see* sun), measured from east to west (0° to 360°) along the equator. The position of a point on the sun's disc may be defined in terms of this and its *heliographic latitude.

heliometer. An obsolete refracting telescope with a split objective lens that was used to measure the angular distance between two stars and to measure position angles. The two sections of the objective were moved until the images of the stars produced in the two sections were made to coincide. The amount of movement gave a precise measure of the angular distance.

heliosphere, heliopause. *See* solar wind.

Helios probes. Two 370 kg interplanetary spacecraft designed to observe the sun and the *solar wind. Constructed in West Germany, they were launched by the US in Dec. 1974 and Jan. 1976 into orbits taking them within 45 million km of the sun: closer than any previous probes.

heliostat. A *coelostat mirror mounted on a synchronously controlled drive mechanism so that it may reflect sunlight in a fixed direction as the sun moves across the sky. Together with additional optical elements it is a basic component of *solar telescopes.

helium. Chemical symbol: He. A chemical *element with an *atomic number of two. In its most abundant form it consists of a nucleus of two protons and two neutrons, around which orbit two electrons. This nucleus, i.e. the positively charged helium ion, is exceptionally stable; it is called an alpha particle. A second far less abundant isotope, helium-3, has two protons and one neutron as its nucleus. Two radioactive (i.e. unstable) isotopes also exist.

Helium is the second most abundant element in the universe (after *hydrogen): about 25% by mass and 6% or more by numbers of atoms. All but about 1% of this approximate 25% cosmic abundance is now considered to have been synthesized in the first few minutes of the universe (see big bang theory). Helium is also synthesized, from hydrogen, by *nuclear fusion reactions in the centres of main-sequence stars. It is by these reactions – the *proton-proton chain reaction and the *carbon cycle – that the energy of the stars is generated. When the hydrogen supplies in the stellar cores are exhausted, the helium is itself consumed by further fusion processes to form carbon (see triple alpha process; helium flash; giant).

Helium is not easy to detect. It was first discovered in 1868 by Norman Lockyer: a yellow emission line of a then unknown element was observed in the spectrum of the sun's chromosphere, recorded during a solar eclipse. High temperatures are required for helium to emit or absorb radiation. Its spectral lines do not appear, for example, in the sun's absorption spectra, which originates in the photosphere. Even greater temperatures are required to ionize helium, i.e. to remove one or both of its electrons. Singly ionized helium does however occur in hot *O stars and in *emission nebulae. Helium, although a minor component in the inner planets, is a major constituent in the atmospheres of Jupiter and Saturn and possibly of Uranus and Neptune.

helium burning. The *nuclear fusion of three helium nuclei to form carbon. See triple alpha process.

helium flash. The explosive onset of helium burning (by the *triple alpha process) in the degenerate core of an evolved star (a *giant) when the core temperature reaches about 10^8 kelvin.

The sudden increase in energy production causes a rapid rise in temperature, in turn increasing the reaction rate, but this runaway process eventually removes the degeneracy and the reaction once again becomes sensitive to the gas pressure.

helium star. 1. One of a rare class of stars (excluding white dwarfs) whose outer layers contain more helium than hydrogen. These are evolved stars that have lost their hydrogen-rich envelope, possibly by the effect of a companion in a close *binary system.
2. Obsolete name for B star.

Helix nebula (NGC 7293). A *planetary nebula lying in the southern hemisphere about 140 parsecs away in the direction of Aquarius. It has the largest apparent diameter (about one degree) of any planetary but is too faint to be seen with the naked eye.

Hellas Planitia. A 1600 km diameter and 3 km deep basin in Mars' southern hemisphere. It can be seen as a bright area from the earth and is probably an impact feature. See Mars, surface features.

Helmholtz – Kelvin contraction. See Kelvin timescale.

Henry Draper Catalogue. Nine volumes of star catalogues compiled by Annie J. Cannon and E. C. Pickering of the Harvard College Observatory and published 1918–24 as part of the Harvard Annals. The catalogue (a memorial to the astronomer Henry Draper) lists the stellar spectra of 225 300 stars, each allocated a *spectral type according to the Harvard classification.

Henry-Draper system (H-D system). Another name for Harvard classification. See spectral types.

Herbig emission-line stars. See Be stars.

Herbig–Haro objects (HH objects). Small peculiar bright nebulae, containing concentrations of gas and dust, that have been lit up by the flux of radiation from, for example, *T Tauri stars or shocked into excitation in gas outflows from T Tauri stars or other protostellar objects. The latter may give rise to a string of HH objects, aligned along a *bipolar outflow.

Hercules. An extensive but rather faint constellation in the northern hemisphere near Ursa Major. The brightest stars, Beta and Zeta Herculis, are of 3rd magnitude with many others of 3rd and 4th magnitude. There are several binary stars, including Zeta and *Ras Algethi (α) and the x-ray binary *Hercules X-1; globular clusters include the *Great Cluster in Hercules (M13), visible to the naked eye, and the smaller slightly fainter M92. Hercules A is a distant and very powerful radio galaxy. Abbrev.: Her; genitive form: Herculis; approx. position: RA 17h, dec +30°; area: 1225 sq deg.

Hercules X-1. An *x-ray binary star, the second to be established as such (after *Centaurus X-3), based on *Uhuru satellite observations. It exhibits a complex behaviour, with regular 1.24 second x-ray pulsations, eclipses (by the binary companion star) every 1.75 days, and longer-term modulation of the x-ray intensity over a 35 day cycle. Her X-1 was the first x-ray binary to be optically identified; the optical counterpart, HZ Her, varies through the binary period from *spectral type A/F to B due to x-ray heating. Its mass (about 1.5 times that of the sun) and the 1.24 second pulsations strongly suggest the x-ray component to be a rotating *neutron star. It is thus often referred to as an x-ray pulsar. The x-ray pulsations are believed to arise from channelling of the accreting gas onto the magnetic poles of the neutron star. The discovery in 1976 of a hard x-ray emission line at 55 keV, attributed to cyclotron radiation from near the star's surface, provides support for this view and gives a direct measure of the intense polar magnetic field (about 10^8 tesla).

Hermes. A *minor planet discovered in 1937 when it approached within 0.006 AU (780 000 km) of the earth. A very small member of the *Apollo group, it was lost because of uncertain knowledge of its orbit. *See* Table 3, backmatter.

Herschel. *See* Mimas.

Herstmonceux. Site of the *Royal Greenwich Observatory.

hertz. Symbol: Hz. The SI unit of frequency, defined as the frequency of a periodic phenomenon that has a period of one second. The frequency range of *electromagnetic radiation is about 3000 Hz (very low frequency radio waves) to about 10^{22} Hz (high-frequency gamma rays).

Hertzsprung gap. A region on the *Hertzsprung–Russell diagram, to the right of the main sequence, in which few stars are found. This is because of a star's rapid evolution through this region, away from the main sequence. This occurs during the period when hydrogen is being burnt in a shell around the core of helium, before the onset of helium burning (*see* stellar evolution). The gap can be easily seen on H–R diagrams of open clusters.

Hertzsprung–Russell diagram (H–R diagram). A two-dimensional graph that demonstrates the correlation between *spectral type (and hence temperature) and *luminosity of stars discovered independently by the Danish astronomer E. Hertzsprung in 1911 and the American astronomer H. N. Russell in 1913. Instead of a uniform distribution the stars are

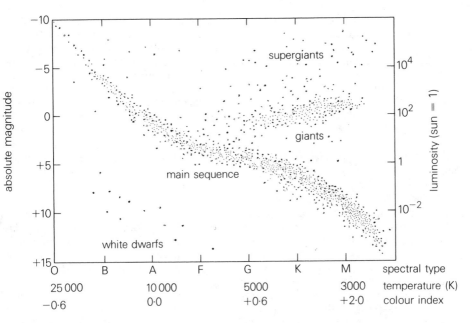

Hertzsprung–Russell diagram for bright stars

found to form well-defined groups or bands on the graph of spectral type plotted against absolute visual *magnitude: about 90% lie along a diagonal band known as the *main sequence; the somewhat brighter *giant stars, belonging to the later spectral types, form another sequence as do the very brightest and relatively rare *supergiants; *white dwarfs and other groupings can also be distinguished.

There are several forms of H-R diagram. In most observational applications the absolute visual magnitude, M_v, is the vertical axis and the *colour index, $B-V$, the horizontal axis, colour index being related to but more easily measured than spectral type. These are *colour-magnitude H-R diagrams*. When studying a *cluster, whose stars are all at the same distance, apparent rather than absolute magnitude is used. Other studies use bolometric *magnitude against *effective temperature – *theoretical H-R diagrams* – or luminosity against colour index – *colour-luminosity H-R diagrams*.

The H–R diagram is of great importance in studies of *stellar evolution. Diagrams obtained on the basis of theoretical calculations can be tested against observationally determined diagrams. They can be drawn for the brightest stars (see illustration), for stars in a particular locality such as the solar neighbourhood (mainly small cool main-sequence stars), for *pulsating variables, for *globular clusters, etc. The two broad stellar populations – *populations I and II – can be demonstrated by the H–R diagrams of a young *open cluster (no giants), a somewhat older open cluster (a few giants), and a much older globular cluster (many giants and supergiants).

The H–R diagram can also be used for distance determination by both *main-sequence fitting for stellar clusters and by spectroscopic *parallax for individual main-sequence stars, the star's spectral type fixing its position on the diagram and thus indicating its absolute magnitude and hence its *distance modulus.

Hevelius formation. The *ejecta blanket of the *Orientale Basin. *See also* highland light plains; Fra Mauro formation.

hexahedrite. *See* iron meteorite.

HH object. *Short for* Herbig-Haro object.

Hidalgo. *Minor planet number 944 that was discovered in 1920 and has an eccentric 14.04 year orbit that carries it between 2.00 AU and 9.64 AU from the sun. Until the discovery of *Chiron in 1977, it had the furthest aphelion and longest period of any known minor planet. From the similarity of its orbit to those of the periodic *comets, it has been suggested that Hidalgo represents the remains of a large comet nucleus. *See* Table 3, backmatter.

hierarchical universe. A *cosmological model, originally due to Charlier in 1908, in which inhomogeneities exist on increasingly larger scales: galaxies occur in *clusters (as observed), which in turn are clustered into *superclusters (as observed), which may in turn aggregate as clusters of superclusters, and so on. As the level of clustering increases then the mean density of matter must decrease in the progressively larger volumes considered.

highland light plains (Cayley formations). The widespread relatively level 'ponds' of light-coloured materials that have accumulated in depressions in the lunar *highlands. Originally thought to be highland volcanics they are now believed to be rubble derived from impacts, possibly from the *Orientale Basin.

highlands, lunar. The topographically elevated and more rugged regions of the *moon. They have a higher *albedo – and thus appear brighter – than the *maria and consist largely of anorthosite, a calcium-aluminium silicate rock. Lunar highland areas include most of the farside and southern part of the *nearside of the moon. *Craters are larger and more abundant here than in the maria; this reflects the greater age of the highland crust, parts of which date back 4600 million years to the formation of the moon. *See also* mountains, lunar; moonrocks.

high-velocity stars. Very old (*population II) stars in the galactic halo that are relatively near the sun but do not share in the common circular motion around the galactic centre of the sun and other stars in the sun's neighbourhood (*see* local standard of rest). They are actually moving slower than the sun in elliptical orbits that often carry them far out of the galactic plane. Their apparently high velocities (>65 km s^{-1}) relative to the sun result from this difference in relative motions. High velocity provides an easy means of identifying individual old stars near the sun, i.e. stars with a low abundance of heavy elements.

HII region. A region of predominantly ionized hydrogen in interstellar space, existing mainly in discrete clouds. The ionization is usually caused by *photoionization by ultraviolet photons in regions of recent star formation, but cosmic rays, x-rays, or shock waves in the medium may sometimes be responsible. In comparison with the 21 cm radio emission of neutral hydrogen in *HI regions, the ionized hydrogen of HII regions emits radio waves by *thermal emission and *recombination line emission; the ionized hydrogen also emits recombination lines in the infrared, ultraviolet, and optical, the latter making an HII region appear as an *emission nebula. HII regions are roughly spherical with a sharply delineated boundary (*see* Strömgren sphere). Their size is usually less than 200 parsecs, the largest being relatively constant in diameter from galaxy to galaxy. By studying the apparent diameters of the HII regions in a distant galaxy, the distance to the galaxy can be estimated. *See also* interstellar medium; extragalactic HII region.

Himalia. A satellite of Jupiter, discovered in 1904. *See* Jupiter's satellites; Table 2, backmatter.

Hind's nebula (NGC 1554–5). A *reflection nebula in the constellation Taurus, illuminated by the star T Tauri. The nebula has varied considerably in brightness in the past century.

Hipparcos. An astrometric satellite of the ESA to be launched in 1988. Its 30 cm telescope will measure the positions, *proper motions, and trigonometric parallaxes (*see* annual parallax) of about 100 000 selected stars with an accuracy (0.002 arc sec) that is not possible on earth. The mission will last about 2.5 years. It will provide a whole-sky stellar catalogue suitable for astrometric and astrophysical studies.

Hirayama families. Groupings of *minor planets whose members orbit at approximately the same distance from the sun and have very similar orbital characteristics (period, *eccentricity, *inclination). More than 40 families are recognized, some containing more than 70 members. They are named after the Japanese astronomer K. Hirayama who discovered the first nine families in 1928. It is believed that each family represents debris resulting from a collision between larger minor planets in the past. In contrast, the *Amor group and the *Apollo group each comprises minor planets with a range of orbital characteristics. A large number of minor planets are members of Hirayama families.

HI region. A diffuse region of neutral, predominantly atomic, hydrogen in interstellar space. A typical HI region is about 5 parsecs across, contains 50 solar masses of gas, and is at a temperature of 70 K. Between these clouds is more tenuous neutral hydrogen gas at a temperature of a few thousand kelvin. Although the temperature is too low for optical emission, neutral hydrogen emits radio radiation at the spot frequency of 1420.4 megahertz, which corresponds to a wavelength of about 21 cm. The emission, termed the *hydrogen line* or *21 cm line emission*, is associated with a *forbidden transition between two closely spaced *energy levels of the ground state, related to the relative electron and proton spin orientation in the hydrogen atom. Predicted in 1944 it was first detected in 1951.

This radio emission has allowed the distribution and relative velocity of neutral hydrogen to be studied both in our own and in nearby spiral *galaxies, using *line receivers. Motions within the galaxy cause the observed frequency (1420.4 MHz) to be displaced by the *doppler effect. The velocity is determined from the frequency shift. *See also* interstellar medium; HII region.

H line. *See* H and K lines.

Hoba meteorite. The world's largest single meteorite mass, found in 1920 near Grootfontein, Namibia. It is an *iron meteorite weighing 60 tonnes and measures $2–7 \times 2–7 \times 1$ metres. It still lies at its original resting place and produced no crater, being only partly buried in the ground.

Hohmann transfer (or orbit). *See* transfer orbit.

homogeneity. Uniformity in space. A medium or process that is not homogeneous is called *inhomogeneous*. *See* cosmological principle; isotropy.

Hooker telescope. *See* Mount Wilson Observatory.

horizon. 1. The horizontal plane perpendicular to a line through an observer's position and his *zenith. The great circle in which an observer's horizon meets the celestial sphere is the *astronomical horizon*. *See also* cardinal points.
2. *See* event horizon.
3. *See* horizon distance.

horizon distance (particle horizon). The total distance that a light signal could have travelled since the beginning of the universe, and hence the maximum distance that can be seen in a cosmological model.

horizontal branch. A horizontal strip on the Hertzsprung–Russell diagram of a *globular cluster, to the left of the red giant branch. It consists of low-mass stars that have lost mass during the red-giant phase; they have absolute magnitudes of about zero. Where the instability strip associated with *pulsating variables crosses the horizontal branch, the stars are *RR Lyrae stars; in a conventional H-R diagram of nonvariable stars, this appears as a gap in the horizontal branch.

horizontal (or horizon) coordinate system. A *coordinate system in which the fundamental reference circle is the astronomical *horizon and the zero point is the north point (*see* cardinal points). The coordinates are *altitude* and *azimuth* (see illustration).
The altitude (h) of a celestial body is its angular distance north (counted positive) or south (counted negative) of the horizon; it is measured along the vertical circle through the body and ranges from 0°, when the object rises or sets, to 90°, when the object is directly overhead at the zenith. The *zenith distance (ζ) is the complement of the altitude ($90° - h$) and is frequently used instead of altitude in the

horizontal system. The azimuth (A) of a body is its angular distance measured eastwards along the horizon from the north point (or sometimes the south point) to the intersection of the object's vertical circle.
The horizontal system is simple but is strictly a local system. At any given moment a celestial body will have a unique altitude and azimuth for a particular observation point, the coordinates changing with observer's position. The coordinates of a star, etc., also change with time as the earth rotates and the observer's zenith moves eastwards among the stars.

horizontal parallax. *See* diurnal parallax.

horn antenna. *See* waveguide.

Horologium (Clock). An inconspicuous constellation in the southern hemisphere near the star Achernar. The brightest stars are of 5th magnitude apart from one of 4th magnitude. Abbrev.: Hor; genitive form: Horologii; approx. position: RA 3.5h, dec −50°; area: 249 sq deg.

Horsehead nebula (NGC 2024). A *dark nebula in Orion located near the star Zeta Orionis. It is seen as a dark region against the bright background of an emission (HII) nebula.

horseshoe mounting. *See* equatorial mounting.

hot big bang. *See* big bang theory.

hot spot. *See* radio source structure.

hour angle (HA). Symbol: t. The angle measured westwards along the *celestial equator from an observer's meridian to the *hour circle of a celestial body or point. It is usually expressed in hours, minutes, and seconds from 0h to 24h. It is thus measured in the same units but in the opposite direction to *right ascension.

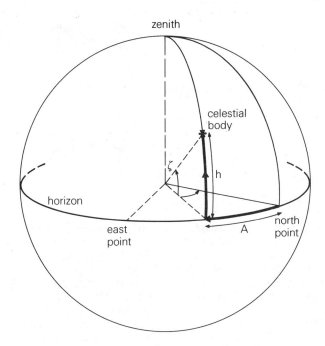

Horizontal coordinate system

The angle measured eastwards along the equator from the meridian is sometimes called the *meridian angle.* Due to the daily apparent rotation of the celestial sphere, a celestial body's hour angle increases daily from 0h at the meridian; after six hours the hour angle is 6h and 24 hours later the celestial body again crosses the meridian. *See also* sidereal hour angle.

hour circle. Any of a series of great circles on the *celestial sphere along which *declination is measured. The hour circle of a particular celestial body or point passes through the body or point and the north and south celestial poles. *See also* hour angle.

H–R diagram. *Short for* Hertzsprung–Russell diagram.

Hubble classification. A classification scheme for galaxies introduced by Edwin Hubble in 1925. Although more complex and refined schemes have since been devised the simple but slightly revised Hubble scheme is still widely used. The scheme recognizes three main types of *galaxy: elliptical, spiral, and barred spiral. Each is divided into subtypes according to gross observable characteristics, the principle one being shape. These subtypes are arranged into a morphological sequence (see illustration).

Elliptical galaxies are denoted by the letter E followed by a number from 0 to 7 that indicates the apparent degree of flattening; this number is the nearest integer to

$$10(1 - b/a)$$

where a and b are respectively the semimajor and semiminor axes of the observed profile. Thus E0 galaxies are almost circular in outline and E7, the most elliptical, have an $a{:}b$ ratio of about 3:1.

The spirals (type S) and barred spirals (type SB) are separated into subtypes a, b, and c along a sequence

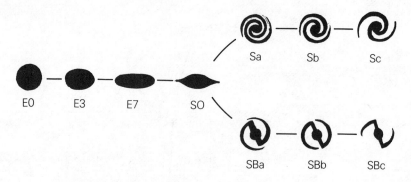

Hubble's 'tuning fork' diagram

that follows a progressive decrease in importance of the nucleus and an increase in the openness and prominence of the spiral arms. An additional subtype, d, has been introduced. Galaxies with intermediate characteristics can be classified as Sab, SBab, Sbc, SBbc, etc. The S0 galaxies, placed at the junction of the elliptical and spiral sequences, resemble the spirals in general shape but lack spiral arms. This type was predicted by Hubble in 1925 but only identified later.

The less commonly observed irregular galaxies were designated Irr I if their contents were predominantly blue *population I objects and Irr II if they contained mostly population II objects. Irregular galaxies were not included in Hubble's original diagram; they are now generally placed to the right of the S or SB sequences. Current terminology refers to Hubble's Irr I galaxies as irregulars; the Irr IIs are examples of perturbed galaxies and are now reclassified as *interacting galaxies, *starburst galaxies, etc.

Hubble constant. Symbol: H_0. The rate at which the expansion velocity of the universe changes with distance; it is commonly measured in $km\,s^{-1}$ megaparsec^{-1}. Current estimates place H_0 between 55 and 100 $km\,s^{-1}\,Mpc^{-1}$.

The value of H_0 is derived from the ratio of recession velocity to distance for galaxies beyond the *Local Group. The velocity can be measured accurately from the *redshift in the galaxy's spectrum. It must be corrected for the sun's motion (see galactic rotation); more recently the discovery that the Virgo cluster's gravity affects the motion of the Galaxy has introduced another correction. The main source of dispute over the value of H_0 comes from the uncertainty of the distances to far-flung galaxies. Refinements in distance measurements have reduced the value since Hubble first determined it (as about 500 $km\,s^{-1}$ Mpc^{-1}). Uncertainties in large extragalactic distances by a factor of two still leave H_0 in doubt by the same factor (see distance determination). A value of 75 $km\,s^{-1}$ Mpc^{-1} is often used when galaxy and quasar distances are determined from their redshift.

The inverse of H_0 has the dimensions of time and is a measure of the *age of the universe – the *Hubble time*. In an evolving universe the Hubble 'constant' actually changes with time at a rate dependent upon the *deceleration parameter, q_0. It is only independent of time in a *steady-state universe where q_0 equals -1. In terms of the *cosmic scale factor, R,

$$H_0 = (dR/dt)/R$$

Hubble diagram. A plot of the *redshift of galaxies against their distance (see illustration) or of redshift against

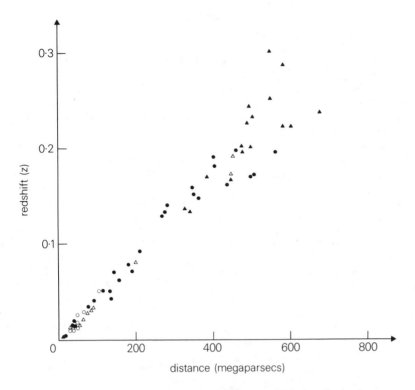

Hubble diagram compiled for several kinds of galaxies

apparent magnitude. Apparent magnitude is a crude measure of galactic distance, the value for the brightest or the third or tenth brightest member of a *cluster of galaxies commonly being used. *See* Hubble's law; Hubble constant.

Hubble flow. *See* Hubble's law.

Hubble nebula (NGC 2261). A fan-shaped *reflection nebula in the constellation Monoceros that is illuminated by the star R Monocerotis. R Mon is an A–F peculiar variable star and causes the brightness and outline of the nebula to vary.

Hubble-Sandage variables. The most massive and luminous stars ($M_v < -9$ and all spectral types) occurring in the nearby galaxies M31 and M33. Like other massive and *supermassive stars, they are *variable stars. The variations in brightness, discovered by E. Hubble and A. Sandage in 1953, were in most cases slow, small, and irregular although some displayed much greater variation. They are thought to be young unstable stars that cannot evolve normally owing to their huge mass, which could exceed 50 solar masses. *Eta Carina and *P Cygni are possible examples in our galaxy.

Hubble's law. The law, first proposed by the American astronomer Edwin Hubble in 1929, stating that the recession velocity, v, of a distant extragalactic object (one outside the *Local Group) is directly proportional to its distance, D. The constant of proportionality is known as the *Hubble constant, H_0, thus

$$v = H_0 D$$

The law is a direct consequence of a uniformly expanding isotropic universe. The separation of points on a balloon, as measured along the surface of the balloon, behaves in just this manner. Variations from a linear relationship – or *Hubble flow* – are detectable. These are due to the *velocity dispersion in the Local Group and in nearby *clusters of galaxies, and to the gravitational effect of the *Virgo cluster on the Local Group.

Hubble Space Telescope (HST). The 2.4 metre (94.5 ft) telescope to be launched into earth orbit from a space shuttle in 1987. Previously known as the *Space Telescope*, it has been named after Edwin P. Hubble. It is primarily a NASA enterprise but with collaboration from the ESA, who will share 15% of the observing time. The untraviolet, optical, and near-infrared observations will range over wavelengths from 115 to 1010 nanometres. There are two spectrographs, a high-speed photometer, and two cameras – the wide field/planetary camera and ESA's faint object camera. Fine-guidance sensors will track the stars and other objects; only two sensors are needed at one time so that a third can make astrometric measurements down to magnitude 17. High-gain radio antennas will permit communications to and from earth via relay satellites.

It is expected that the HST will revolutionize observational astronomy. Free from the absorption and distortion effects of the earth's atmosphere, its resolution will be 0.1 arcsecs, considerably better than ground-based telescopes. Its pointing control system is so accurate that it will be able to lock on to and study a target for up to 20 hours. It should be able to detect objects some 50 times fainter than those detectable at present from the surface of earth; its limiting visual magnitude is about 28.

During its planned 15-year life the HST will be under *remote operation from earth. In-orbit checks and maintenance will be performed by shuttle crews.

Hubble time. *See* Hubble constant.

Huygens eyepiece. *See* eyepiece.

Hyades. An *open cluster of about 100 stars in the constellation Taurus. It is about 46 parsecs away. Its brightest members form a V-shaped group that is visible to the naked eye. The Hyades is a *moving cluster. Its age is about 500 million years.

hybrid ring. A four-port device by which two input signals A and B are combined to give two output combinations $A + B$ and $A - B$, with a high degree of isolation between one input port and the other and between one output port and the other, and such that identical cables are matched (*see* matching) at all four ports.

Hydra (Sea Serpent or Monster). The largest constellation, straggling between Leo in the northern hemisphere and Centaurus in the south. Its stars are rather faint apart from the solitary 2nd-magnitude *Alphard. It contains several types of variable stars, the just-visible open cluster M48, the globular cluster M68, the bluish planetary nebula NGC 3242, and the spiral galaxy M83. Abbrev.: Hya; genitive form: Hydrae; approx. position: RA 8h to 15h, dec $+7°$ to $-35°$; area: 1303 sq deg.

hydrogen. Chemical symbol: H. The simplest chemical *element, with an *atomic number of one and the lowest density of all elements. In its most abundant form it consists of a *proton orbited by a single electron. Two other forms, i.e. isotopes, exist: *deuterium* has a proton plus a neutron as its nucleus; *tritium*, which is radioactive (i.e. unstable), has a proton plus two neutrons.

Hydrogen is the most abundant element in the universe: 94% by numbers of atoms, 73% by mass. It is generally agreed that all this hydrogen is primeval in origin, created in the earliest phase of the universe (see big bang theory). The amount of deuterium formed relative to the amount of hydrogen would have been very sensitive to the density of matter at the time of formation: deuterium readily combines with an additional neutron to form tritium, which rapidly decays into an isotope of helium. Thus if the universe was very dense in its first few minutes, most of the deuterium would have been converted to helium. Deuterium is not easy to detect but recent measurements of the ratio of deuterium to hydrogen give an upper limit of about 2×10^{-5} for interstellar gas.

Hydrogen can exist in atomic, molecular, and ionized forms. Ionized hydrogen, H^+, more usually denoted HII, results when neutral atoms are stripped of their electrons. It occurs at high densities and temperatures, as in the centres of stars. *Nuclear fusion of hydrogen ions, i.e. protons, in the stellar core generates the energy of main-sequence stars by either the *proton-proton chain reaction or the *carbon cycle. Ionized hydrogen can also be produced by *photoionization, as in *emission nebulae, and may then be detected by its *recombination line emission (see also HII region).

In addition to the positive ion, the negative ion, H^-, can occur when an electron attaches itself loosely to a neutral atom. A continuous spectrum of radiation is emitted in the process. The negative ion is not very stable and breaks up easily with the second electron escaping with any amount of energy: this results in a continuous absorption spectrum. These two processes, producing continuous emission and absorption of both light and infrared radiation, take place concurrently, as happens in the sun's *photosphere. The H^{--} ion has recently been discovered.

Neutral hydrogen, usually denoted HI, occurs throughout interstellar space as filaments and clouds (*HI regions) of varying density. It is detected by means of the 21 cm hydrogen line emission. At low temperatures and when the hydrogen density is sufficiently high, pairs of hydrogen atoms can combine to form molecular hydrogen, which exists in discrete *molecular clouds. Molecular hydrogen is also a major constituent of the atmosphere of the *giant planets, with liquid hydrogen forming the bulk of the interiors of the giants. See also hydrogen spectrum.

hydrogen alpha line (Hα line). See hydrogen spectrum.

hydrogen line (21 cm line). See HI region.

hydrogen region. See HI region; HII region; interstellar medium.

hydrogen spectrum. Emission and absorption spectra of hydrogen are relatively simple compared with spectra of heavier elements. Atomic hydrogen (HI) has a *line spectrum in which several series of lines can be distinguished. The Swiss physicist Johann Balmer showed, in 1885, that lines in the visible region of the hydrogen spectrum formed a series represented by the equation
$$1/\lambda = R(1/4 - 1/m^2)$$
λ is the wavelength of the line and m an integer greater than two. The constant R has a value of $1.096\,78 \times 10^7$ m^{-1}, and is called the *Rydberg constant*.

This series of spectral lines is called the *Balmer series*. Its existence was first explained by Niels Bohr in his theory of the atom (1913). Bohr postulated that only a discrete number of orbits is allowed to an electron in an atom and that when the electron jumps from one orbit to another, a photon of radiation is emitted or absorbed (see energy level). The Balmer series is produced by transitions of the electron between the second permitted orbit, in order of distance from the nucleus, and higher orbits. When $m = 3$, the spectral line is produced by a transition from the second to the third orbit (or from the third to the second orbit). This line is referred to as Hα (the *hydrogen alpha line*) and has a wavelength of 656.3 nm, i.e. it falls in the red region of the visible spectrum. The next line, Hβ ($m = 4$), is at 486.1 nm in the blue. Hγ and Hδ occur at 434.2 nm and 410.2 nm, respectively. The lines in such a series get closer together at shorter wavelengths and the Balmer series converges to a limit at 364.6 nm in the ultraviolet region of the spectrum. Emission lines are produced by transitions from higher levels to the second orbit; absorption lines result from transitions from the second orbit to higher orbits.

In general, lines in the hydrogen spectrum can be represented by the equation

$$1/\lambda = R(1/n^2 - 1/m^2)$$

where n and m are integers; n is called the *principal quantum number* and can have values from one to infinity. The Balmer series is formed for $n = 2$, i.e. the second orbit of the atom. Transitions from or to the first orbit, i.e. the ground state ($n = 1$), produce the *Lyman series* of spectral lines. This is in the far ultraviolet region. Lyman alpha, denoted Lyα ($n = 1$, $m = 2$), occurs at 121.6 nm; the series Lyα, Lyβ, Lyγ..., converges to a value of 91.2 nm. The *Paschen series* is produced by transitions to or from the third permitted orbit ($n = 3$) and occurs in the near infrared region of the spectrum. The *Bracket series* occurs at longer infrared wavelengths involving the fourth orbit ($n = 4$).

hydrostatic equilibrium. A state of equilibrium in which the inwardly directed gravitational force in a star just balances the outwardly directed gas and radiation pressure. The star is thus held together but supported against collapse. The equation of hydrostatic equilibrium is given at *stellar structure.

hydroxyl radical. The chemical radical OH, consisting of an oxygen and hydrogen atom bound together. It was the first interstellar molecule to be detected (1963) in the radio region of the spectrum and is also associated with certain red and infrared stars (*see* OH/IR star; maser source). The spectral lines – OH lines – occur at 18, 6.3, 5.0, 3.7, and 2.2 cm.

Hydrus (Water Snake). A small constellation in the southern hemisphere lying between the Large and Small Magellanic Clouds. Its brightest star, Beta Hydri, is of 3rd magnitude and is the nearest conspicuous star to the south pole. Abbrev.: Hyi; genitive form: Hydri; approx. position: RA 2h, dec $-70°$; area: 243 sq deg.

hyperbola. A type of *conic section that has an eccentricity greater than one. *See also* orbit.

hyperboloid reflector. A reflector whose surface is hyperboloid in shape, i.e. its curvature is that obtained by rotating a hyperbola about its axis. The secondary mirror in a *Cassegrain telescope is hyperboloid.

Hyperion. A heavily cratered elongated satellite of Saturn, discovered in 1848. *See* Table 2, backmatter.

hypersensitization. The subjection of an unexposed *photographic emulsion to one or more methods of treatment in order to improve its response to light or other radiation. This allows the emulsion to be used at the extremely low intensities of light, etc., encountered in astronomy without an excessively long exposure time. Alternatively in a set period of telescope time the quantity or quality of information obtained can be increased by hypersensitization. Various techniques are used. One particularly suitable for light-sensitive emulsions involves heating the emulsion in either pure nitrogen gas or in nitrogen with a small admixture of hydrogen. The gas removes oxygen and water vapour from the emulsion; these impurities act as desensitizers. Heating speeds up this leaching process and can also increase the chemical sensitization, i.e. the possibility that grains in the emulsion if hit by light will be activated.

Hz. *Symbol for* hertz.

I

Iapetus. A satellite of Saturn, discovered in 1671. The diameter of Iapetus is 1440 km so that in size it is the twin of the satellite *Rhea but its density, 1.1 g cm^{-3}, is much lower. It is a strange satellite with a dark (albedo 0.04–0.05) leading hemisphere compared with a bright trailing edge of albedo 0.5. The demarcation between the regions is not abrupt and there is a transition region 200–300 km in width with a meandering boundary. Craters occur in both types of terrain. The difference in the brightness may not be due to external material. It is possible that the dark material could have extruded from the interior of the body itself. *See also* Table 2, backmatter.

IAT. *Abbrev. for* International Atomic Time.

IAU. *Abbrev. for* International Astronomical Union.

IC. *Abbrev for* Index Catalogue. *See* NGC.

Icarus. *Minor planet number 1566 that was discovered in 1949 and passed only 0.04 AU from the earth in 1968. It belongs to the *Apollo group and has one of the smallest *perihelion distances (0.19 AU), well within the orbit of Mercury. *See* Table 3, backmatter.

IGY. *Abbrev. for* International Geophysical Year.

Ikeya-Seki comet (1965 VIII). A *comet that was discovered by two amateur Japanese comet searchers. At one time it was actually thought that it would hit the sun but it turned out to be a sungrazer with a perihelion distance of 0.0078 AU, missing the sun's surface by 0.68 solar radii. The orbit is retrograde with an eccentricity of 0.99915.

image converter. *See* image tube.

image intensifier. *See* image tube.

image photon counting system. *See* IPCS.

image processing. *See* imaging.

image tube (image intensifier). An evacuated electronic device that is used to intensify a faint optical image. The beam of light (or near-ultraviolet radiation) falls on a *photocathode so that electrons are liberated by the photoelectric effect. The electrons are accelerated by an electric field and may be detected and recorded by a variety of methods. They can be focused on a positively charged phosphor screen so that an optical image is produced, many times brighter than that of the original beam. This image can be made to fall on the photocathode of a second image tube and the intensification process repeated. Several image tubes can be linked in this way to form a *multistage image tube* but the coupling between phosphor screen and subsequent photocathode must be very efficient. The image on the final phosphor screen may then be photographed. The *electronographic camera and the *IPCS (image photon counting system) use slightly different techniques.

In *image converters* a beam of infrared or ultraviolet radiation or x-rays is incident on the photosensitive surface. The liberated electrons can then produce a pictorial representation of a 'nonvisible' image.

imaging. The representation, by means of TV pictures, photographs, graphs, etc., of an object or area by the sensing and recording of patterns of light or other radiation emitted by, reflected from, or transmitted through the object or area. Two broad classifications are *chemical imaging*, i.e. *photography, and *electronic imaging*. Both are important in astronomy, a variety of photographic emulsions and electronic devices being available for different frequency bands of the electromagnetic spectrum.

The majority of information gathered by ground-based and orbiting telescopes is in digital form so that it can be manipulated by computer. This information can be derived directly from electronic devices, such as *CCD, *IPCS, or *photovoltaic detectors, associated with the telescope. These devices respond to radiation by converting it to an electrical signal. They are more sensitive than photographic emulsions, responding to lower levels of intensity and/or producing an image in a shorter time. Photographic plates do, however, provide an image of a much greater area of the sky than existing electronic devices. Machines such as *COSMOS have therefore been built to measure data on photographic plates rapidly and automatically and produce results in digital form.

An electronic detector can be moved across the focused image of an astronomical object or area of the sky, or the image can be moved across the detector. The electrical signal from the detector is sampled in such a way that an array of values corresponding to an array of portions of the complete image is obtained. Alternatively the image falls on a large number of closely packed detectors, all producing a signal. In each case the result is a set of numbers corresponding to some property of the individual image portions, e.g. the intensity at a particular wavelength. The individual portions into which the image is divided are called *pixels* (short for *picture elements*). The greater the number of pixels per image, the higher the resolution, i.e. the greater the detail seen.

This numerical version of the image will normally reside in a computer system, and can be manipulated in different ways in order to highlight different aspects of the original image; the manipulative techniques are known as *image processing*. The final form of the display can be a TV monitor, plotting device, or photographic film, and information derived from the image can also appear in graphs and tables, and be subjected to statistical and numerical analysis.

imaging proportional counter. *See* proportional counter.

Imbrium Basin. The second youngest lunar *basin, of which much of the second and third rings survive as mountain arcs (*see table at* mountains, lunar). The *ejecta blanket of the basin covers much of the *nearside and provides the major reference horizon for lunar stratigraphy. The basin was formed 3900 million years ago and was flooded with basalt lavas during at least the following 600 million years. This produced *Mare Imbrium* (*Sea of Rains*), which was visited by Lunokhod 1, and also *Palus Putredinis*, which was sampled by Apollo 15. *See also* Fra Mauro formation.

immersion. Entry of a celestial body into a state of invisibility during an *eclipse or *occultation.

inclination. 1. Symbol: *i*. The angle between the orbital plane of a planet or comet and the plane of the *ecliptic. It is one of the *orbital elements and varies between 0 and 180°, being less than 90° for a body with *direct motion.
2. The angle between any two planes, such as that between the orbital plane of a satellite and the primary's equatorial plane.
See also Tables 1–3, backmatter.

Index Catalogue (IC). See NGC.

Indus (Indian). A constellation in the southern hemisphere near Grus, the brightest stars being of 3rd magnitude. Abbrev.: Ind; genitive form: Indi; approx. position: RA 21h, dec −60°; area: 294 sq deg.

inequality. An irregularity in the orbital motion of a celestial body. The inequalities of the moon's motion are periodic terms whose sum gives the variation of either the spherical coordinates or the osculating elements of the lunar orbit. The principal inequalities of the moon's motion are *evection and *variation.

inertia. The property of a body by which it resists change in its velocity. It is inertia that causes a body to continue in a state of rest or of uniform motion in a straight line (see Newton's laws of motion). The force required to give a specific acceleration to a body depends directly on its inertia. It is through the property of inertia that the concept of the *mass of a body (its *inertial mass*) arises. See also Mach's principle.

inertial frame. A *frame of reference in which an isolated particle placed at rest will remain at rest, and one in uniform motion will maintain a constant velocity. It is thus a frame of reference that is not subject to acceleration, as from a gravitational field, and in which Newton's first law of motion applies. According to Einstein's special theory of *relativity, which is concerned only with inertial frames, all laws of physics are the same in every inertial frame.

inertial mass. See inertia; mass.

inferior conjunction. See conjunction.

inferior planets. The planets Mercury and Venus, which lie closer to the sun than the earth.

inflationary universe. A possible phase in the very early universe when its size increased by an extraordinary factor, perhaps by up to 10^{50}, in an extremely brief period. The theory was proposed in 1980 by Alan Guth and has since been modified by Guth and others. At an age of 10^{-35} seconds, the state of the universe had to change as the electromagnetic and strong nuclear forces 'froze out' to different values (see fundamental forces). The energy released by this phase change is calculated to have caused the universe to expand, or inflate, catastrophically. The inflationary phase ended at some time before 10^{-30} s. The phenomena occurring during the course of the inflation are still very speculative. After 10^{-30} s, however, the inflationary model coincides with the standard big bang description of the universe.
The inflationary phase means that the observed universe is only a very small fraction of the entire universe. In addition distant parts of the universe would have been much closer in the period before inflation than has been previously calculated. The theory can thus explain the isotropy of the *microwave background radiation, which requires distant parts of the universe to have been in causal contact in the past.

infrared astronomy. The study of radiation from space with wavelengths beyond the red end of the visible spectrum, i.e. from about 0.8 micrometre (μm) to about 1000 μm. Observations are possible from the ground through several *atmospheric windows up to about 20 μm but longer-wavelength observations require balloon-, rocket-, or satellite-platforms. The equipment used includes reflecting telescopes (*see* infrared telescope) and solid-state infrared detectors, such as *photovoltaic and *photoconductive devices and *bolometers. The detectors need to be cooled to low temperatures, and in some cases the telescope optics are also cooled, to minimize the instrumental thermal emission and *noise. The first satellite-borne infrared telescope, *IRAS, was launched on Jan. 20 1983. IRAS has made an all-sky survey in the 10–100 μm waveband as well as investigating in more detail a wide range of astronomical sources. The European Space Agency is currently planning to launch the *Infrared Space Observatory (ISO) in the mid-1990s. ISO will use a cooled telescope to make observations in the 2–200 μm waveband.

Cosmic *infrared sources emit infrared radiation in a variety of ways. They may emit thermally as approximate *black-body radiators; such sources include the stars themselves, *cosmic dust grains in the circumstellar shells around stars, and *molecular clouds, and have temperatures in the range 20–1000 K. In *HII regions electrons moving in the thermally heated and ionized gas emit infrared radiation by a process known as thermal bremsstrahlung (*see* thermal emission). Infrared emission can also be produced by a nonthermal process in which high-energy electrons moving in a magnetic field emit *synchrotron radiation.

infrared background radiation. Unresolved radiation from space in the infrared waveband 10–200 μm. The diffuse radiation is the sum of several components: *zodiacal light, the infrared flux from stars, emission from *cosmic dust in the Galaxy, *infrared cirrus, and the infrared flux from external galaxies. Observations from the satellite *IRAS show that zodiacal emission dominates the infrared background at 12–60 μm except near the galactic plane. The zodiacal emission is also present at 100 μm but diffuse galactic emission is prominent at this wavelength. *See also* cosmic background radiation.

infrared cirrus. Extended sources of 60–100 μm infrared emission superimposed on the *infrared background radiation. Such sources were first detected by the satellite *IRAS. At high galactic latitudes some of these sources are associated with clouds of atomic hydrogen containing infrared-emitting dust at a temperature of about 30 K. Others are only poorly correlated with known interstellar gas clouds and may be due to cold material in the solar system.

infrared excess. *See* infrared sources.

infrared radiation. *Electromagnetic radiation lying between the radio and the visible bands of the electromagnetic spectrum. The wavelengths range from about 0.8 micrometre (μm) to about 1000 μm. The shorter wavelengths, i.e. those close to the red end of the visible region, are often called *near infrared*, with *far infrared* being applied to the longer wavelengths.

infrared sources. A variety of objects emitting strongly in the infrared (*see* infrared radiation). Many stars exhibit an *infrared excess* in that they radiate more strongly in the 1–20 μm waveband than would be expected from their *spectral type. In most cases the excess arises from circumstellar shells containing *cosmic dust grains that are heated by the visible and UV radiation from the central star. *Protostars and their associated dust clouds are also strong infrared emitters. Many *T Tauri stars emit infrared radiation, in some cases by *thermal bremsstrahlung and/or emission from heated dust. At longer wavelengths (60–200 μm) cool *molecular clouds containing dust are found to be infrared sources.

Infrared emission is also found in extragalactic sources (see IRAS) and is strongly correlated with galaxy type: elliptical and lenticular galaxies emit little or no infrared while most spiral galaxies are infrared emitters. Active galaxies, such as *Seyfert type II, *BL Lac objects, and *quasars, exhibit infrared emission from their nuclei. It is unclear at the present time whether this is thermal emission by dust in the nuclei or is due to nonthermal *synchrotron emission.

Infrared Space Observatory (ISO). An ESA satellite to be launched into an elliptical 12 hour orbit in the mid-1990s. It will carry a 60 cm telescope cooled by liquid helium and three focal-plane detectors – an imaging device, photometer, and a pair of spectrometers – for observations in the infrared wavelength range 2–200 μm. Unlike the less-sensitive survey satellite *IRAS, ISO will be able to make detailed observations over several hours of selected sources. Data will be transmitted continuously to two tracking stations in Spain and Australia. ISO has a planned mission life of at least 1½ years.

infrared telescope (flux collector). A reflecting telescope used to detect and study infrared radiation from space; refractors cannot be employed because glass is opaque at these wavelengths. The mirror surface does not, however, need to be made to the same accuracy as one used for optical work due to the longer wavelengths of infrared radiation. Since neither the

eye nor the photographic plate is sensitive to long-wavelength infrared radiation, special detectors are required, such as *photovoltaic and *photoconductive detectors and *bolometers. The detectors, and in some cases the telescope optics, must be cooled to liquid-helium or liquid-nitrogen temperatures to reduce thermal emission from the instruments and hence *noise. Infrared telescopes can detect sources partially or totally obscured at optical wavelengths by interstellar dust, and can identify those that are true infrared sources rather than highly reddened ordinary stars. Because water vapour in the earth's atmosphere absorbs infrared radiation (see atmospheric windows), the most advanced work is carried out with specially designed telescopes, such as the *UK Infrared Telescope, at high-altitude observatories. Infrared telescopes can operate day and night.

Infrared Telescope Facility (IRFT). See Mauna Kea.

infrared windows. See atmospheric windows.

initial mass function (IMF). See stellar mass.

injection (insertion). The process of boosting a spacecraft into a particular orbit or trajectory. It is also the time of such action or the entry itself.

inner Lagrangian point. The Lagrangian point, L_1, through which *mass transfer occurs. See also equipotential surfaces.

inner planets. The terrestrial planets Mercury, Venus, Earth, and Mars, which lie closer to the sun than the main belt of minor planets.

insolation. The exposure of any surface or body to solar radiation. It is usually quoted as the radiant flux received from the sun per unit area per unit time; it can therefore be considered as the heating effectiveness of solar radiation. Outside the earth's atmosphere it is equal to the *solar constant, varying by 1–2% with the sunspot cycle and by ±3.5% around the earth's orbit.

The area of the earth's surface over which a given amount of solar radiation is spread depends primarily on the sun's altitude: the greater the altitude the smaller the area; it also depends on the topography of the area and on atmospheric conditions. The sun's altitude varies both with time and latitude. The insolation will therefore be greatest in the middle of the day, will be greater during summer than in winter, and will be greater in equatorial regions than in higher latitudes.

Long-term changes in the amount of solar radiation received by the earth, or particular latitudes of the earth, arise from slow periodic changes in the *eccentricity of its orbit and in the tilt and direction of its axis (*see* obliquity of the ecliptic; precession). Changes would also result from any variation in the sun's energy output and from any variation in the energy penetrating the earth's atmosphere.

instability strip. *Short for* Cepheid instability strip. *See* pulsating variables.

INT. *Abbrev. for* Isaac Newton Telescope.

integrated magnitude. *See* total magnitude.

integration time. The period for which a noisy signal is averaged in order to improve the signal to noise ratio in an electronic system. *See* sensitivity.

intensity. 1. (radiant intensity). The *radiant flux emitted per unit solid angle by a point source in a given direction. It is measured in watts per steradian.

2. *Obsolete name for* field strength of a magnetic or electric field.

intensity interferometer. *See* interferometer.

interacting galaxies. Two, three, or (more rarely) four or more galaxies that show signs of mutual disturbance, such as tidal distortions or extruded filaments of luminous material (which can sometimes link to form bridges between the galaxies). Approximately 20% of *peculiar galaxies appear to be members of interacting groups. The infrared satellite IRAS found that many interacting galaxies are powerful infrared sources: the interaction apparently stimulates star formation. There is also evidence that some quasars lie in interacting galaxies.

interference. 1. Unwanted signals picked up by a *radio telescope or other electronic equipment; the signals usually arise from terrestrial sources but occasionally come from the sun or Jupiter.

2. The interaction of two sets of waves to produce patterns of high and low intensity (*interference fringes*). For this to happen the two beams of light or other radiation must be coherent, i.e. they must be in *phase and have the same *wavelength. They must intersect at a fairly small angle and not differ too much in intensity. In addition, they must not be polarized in mutually perpendicular planes.

The interference pattern of fringes formed at a particular position is the sum of the intensities of the two interacting waves at that position. The fringes occur because of differences in pathlength between interacting waves, i.e. because of unequal distances from source to interaction point. If the difference is a whole number of wavelengths, then wave peaks (or troughs) of the two interacting waves coincide and the waves reinforce one another, producing a bright fringe when light waves are involved; this is termed *constructive interference*. If the path difference is an integral number of half wavelengths a peak coincides with a wave trough and a dark fringe results; this is *destructive interference*.

Both constructive and destructive interference of light can be produced by means of thin films of uniform thickness, as used in *interference filters. Waves of selected wavelengths are reflected from the front and back surfaces of the film and by a suitable choice of film composition and thickness the waves will either be reinforced or will cancel each other. Using pairs of *radio telescopes the waves from a radio source can be made to interfere. *See also* interferometer.

interference filter. An optical, ultraviolet, or infrared *filter in which unwanted wavelengths are removed by destructive *interference rather than by absorption. It produces a very narrow band of wavelengths, typically between 1 and 10 nanometres. One form, the *multilayer filter*, consists of a stack of transparently thin films with different refractive indices. The films allow only a waveband of 1 nm or less to be transmitted right through. The remaining wavelengths are reflected and caused to interfere destructively.

interference fringes. *See* interference; radio telescope.

interferometer. An instrument or system in which a beam of light or other radiation is split into two or more parts and subsequently reunited after traversing a different pathlength so that *interference fringes are produced. The *radio interferometer* is one of the two basic forms of *radio telescope.

In optical astronomy *stellar interferometers* are used to measure small angles, such as the angular separation of a double star or more often the apparent diameter of a single star. In A. A. Michelson's original stellar interferometer the telescope objective was covered by a screen. The screen was pierced with two parallel slits whose distance apart was almost equal to the objective diameter and was also adjustable. The two components of a double star will each produce a set of interference fringes as a result of the double slit. By varying the distance, d, between the slits, the bright fringes of one pattern can be made to coincide with the dark fringes of the second so that a continuous line of light is seen. This occurs when the slit separation is equal to $\lambda/2\alpha$, where λ is the wavelength of the light and α is the angular separation of the stars. For a nearby giant or supergiant star, such as Betelgeuse or Antares, the two halves of the stellar disc produce the two interference patterns, which coincide when the apparent diameter equals $1.22\lambda/d$. With a few modifications Michelson used this technique in the late 1920s to measure the diameters of several stars down to about 0.01 arc seconds.

The now-dismantled *intensity interferometer* at Narrabri, Australia, was a much more sensitive instrument, capable of measuring the extremely small angles subtended by main-sequence stars with a precision of about $\pm 5\%$. The faintest star that could be measured was one of magnitude $+2.5$. The minimum diameter was about 5×10^{-4} seconds of arc. Two very large (but low-quality) optical reflectors, 6.7 metres in diameter, were mounted on a circular track, 188 metres across, and guided by computer. The starlight was focused by each reflector on to a separate *photocell, having passed through a narrow-band filter, and was converted into an electric current. The two currents were brought together and their fluctuations studied. A small component of the fluctuations corresponded to fluctuations in the intensity of the starlight. When the two reflectors were close together there was a high correlation between the intensity fluctuations – they were almost identical. As the distance between the reflectors increased, the correlation decreased. The apparent diameter of the star could be determined from measures of correlation as a function of reflector spacing.

See also speckle interferometry.

intergalactic medium. The matter contained in the space between the *galaxies, about which very little is known. It exists in large amounts only in *clusters of galaxies where it is completely ionized and constitutes a significant proportion of the total observed mass of the cluster. Absorption lines in quasar spectra often indicate many clouds of intergalactic hydrogen along the line of sight but their total mass is very small.

International Astronomical Union (IAU). An international association of astronomers that is the controlling body of world astronomy. It was founded in Brussels in 1919. A general assembly is held every three years, the first one occurring in 1922. In addition the IAU organizes periodic symposia for the discussion of current astronomical problems, produces several publications, and runs a telegram service for the rapid dispersal of information relating to transient phenomena such as nova and comets.

International Atomic Time (TAI). The most precisely determined timescale now available, set up by the Bureau Internationale de l'Heure in Paris, following analysis of *atomic time standards in many countries. It was adopted for all timing on Jan. 1 1972. The fundamental unit is the SI *second. It is a more precise scale than can be determined from astronomical observations. Coordinated *universal time is based on TAI and is used for all civil timekeeping. See also dynamical time.

International Geophysical Year (IGY). The period from July 1 1957 to Dec. 31 1958 that was timed to coincide with an epoch of maximum *solar activity. It saw a concerted international effort to study the *geophysics of the earth and its environment, including the effects of enhanced solar activity, particularly regarding the earth's upper atmosphere. The first artificial satellites were launched during this period.

International Solar Polar Mission (ISPM). A proposed joint ESA/NASA space mission to study for the first time the properties of the interplanetary medium and solar wind away from the plane of the ecliptic and over the polar regions of the sun. An ESA spacecraft, to be launched in 1986 by NASA, will travel towards Jupiter, accelerate into a high-velocity semi-orbit of the planet, and then, using the slingshot action of Jupiter's gravitational field, enter a polar orbit of the sun. About 3.5 years after launch it will pass over the sun's pole at a distance of about 300 million km, and will subsequently recross the ecliptic plane and pass over the opposite pole of the sun. The nominal mission will then be completed, having lasted about 5 years. On-board instruments will study the particles, gas, dust, and magnetic fields in the higher latitudes of the sun's environment.

International Sun-Earth Explorer (ISEE). A joint project of NASA and ESA for studying the structure and interactions within the earth's *magnetosphere and the solar phenomena producing such effects. The two satellites ISEE–1 and ISEE–2 were launched in tandem, in Oct. 1977, into a highly eccentric earth orbit at a controllable distance apart. ISEE–3 was launched in Aug. 1978 into an orbit lying outside the magnetosphere and measured the *solar wind, solar *flares, and *sunspots, unperturbed by the earth's influence. It was thus a reference point for simultaneous observations made by ISEE-1 and ISEE-2. In 1982/83 the trajectory of ISEE-3 was altered so that it will intercept the orbit of the comet *Giacobini-Zinner in Sept. 1985. It should pass within about 3000 km of the head.

International Sunspot Number. *See* Zürich relative sunspot number.

International Ultraviolet Explorer (IUE). A satellite that is a joint project of NASA, ESA, and SERC (the UK's Scientific and Engineering Research Council) and was launched by NASA into an elliptical *geosynchronous orbit on Jan. 26 1978. Its 45 cm Ritchey-Chrétien telescope focuses ultraviolet radiation on either of two *echelle grating spectrographs providing about 0.01 nm spectral resolution; they operate in the wavelength range 115–190 nm and 180–320 nm. In addition lower resolution (about 0.6 nm) spectrographs are available for observations of faint sources. The satellite is controlled from ground stations in Maryland and Madrid by real-time data link, which permits operation in a manner similar to that of a ground-based observatory. In the first 6 years of operation over 40 000 spectra have been obtained on a wide variety of astronomical sources including planets, comets, interstellar dust and gas, stars of most spectral types, galaxies and galactic halos, Seyfert galaxies, and quasars (*see* ultraviolet astronomy). It is anticipated that IUE will have an operational lifetime of at least 10 years.

International Years of the Quiet Sun (IQSY). The period 1964–65, when *solar activity was at a minimum, during which an intensive international programme was mounted by obser-

vatories all over the world to study solar and geophysical phenomena.

interplanetary medium. The matter contained in the solar system in the space between the planets. This space is filled with a tenuous ionized gas – the *solar wind – that streams outwards from the sun at supersonic speeds and in which is embedded the sun's magnetic field wound into a spiral by the sun's rotation. (This is often called the *interplanetary magnetic field*.) At the earth the medium has a density of about eight atoms per cm^3 moving at a velocity of 300 to 400 $km\ s^{-1}$, and a magnetic flux density of about 5×10^{-5} gauss (5×10^{-9} tesla). The earth moves at supersonic speed relative to the interplanetary medium, creating a shock front called the *bow shock*, rather like that set up by an ocean liner on the surface of the sea. Irregularities in the density of the solar wind cause interplanetary scintillations (*see* scattering) on small *radio sources viewed through the medium.

interplanetary scintillations (IPS). *See* scattering; interplanetary medium.

interpulse. The weaker pulse of radio emission received after the main pulse and before the next main pulse in the signals from some *pulsars.

interstellar absorption. *See* interstellar extinction.

interstellar dust. *Cosmic dust occurring in the *interstellar medium.

interstellar extinction (interstellar absorption). The reduction in brightness, i.e. the *extinction, of light (and other radiation) from stars as a result of absorption and scattering of the radiation by interstellar dust. Although dust makes up only a small proportion of the *interstellar medium, its affect on starlight is considerable. The amount of extinction, usually measured in *magnitudes, depends on the direction of observation. It is a maximum (about one magnitude per kiloparsec distance for visible wavelengths) towards the centre of the Galaxy, where the density and extent of the dust is greatest. The extinction varies inversely with wavelength. Red light is thus less affected than blue light so that starlight appears reddened when observed through a dust cloud. Similarly, radio and infrared wavelengths can pass through the interstellar medium with ease compared with optical wavelengths. The presence of obscuring matter between the stars was first conclusively demonstrated by Robert Trumpler in 1930.

interstellar medium (ISM). The matter contained in the region between the stars of the *Galaxy, constituting about 10% of the galactic mass. It is largely confined to a thin layer in the galactic plane and tends to be concentrated in the spiral arms. Several different constituents have been observed in the ISM, including clouds of ionized hydrogen – *HII regions – and smaller relatively cool (100 kelvin) clouds of neutral hydrogen – *HI regions – surrounded by more tenuous regions of hot gas (1000–10 000 K). The presence of some very much hotter regions has been inferred from a diffuse background of soft x-ray emission and by ultraviolet absorption lines. There are also very cool (10 K) dense clouds of molecular hydrogen – *molecular clouds – in which a number of other molecules and radicals have been observed.

In addition to the gas, *cosmic dust is to be found throughout the medium. The dust comprises small solid grains that are between about 0.01 to 0.1 μm in size and are thought to be composed of carbon, silicates, or iron material, with icy mantles when they occur in dense molecular clouds. The dust causes dimming and reddening of starlight (*see* interstellar extinction) and also partial *polarization of starlight. The interstellar grains can undergo a complex life cycle, passing into and out of molecular clouds: the ISM is constantly churned up by shock waves from expanding *supernova remnants, by *stellar winds, and by other forces. While in a molecular cloud the grains may be incorporated in a newly forming star, and may later be ejected back into the ISM in a stellar wind or supernova explosion.

The medium is permeated by a flux of *cosmic rays that spiral along the field lines of the *galactic magnetic field of a few microgauss and cause the *synchrotron emission that is the *galactic radio background radiation*. *Radio maps show that this radiation is largely confined to the galactic plane but has several *spurs*, in particular the *north galactic spur*, radiating away from it. The north galactic spur is the most prominent segment of a huge fragmentary ring of gas detected at radio and x-ray wavelengths. This giant gas shell is probably a nearby old *supernova remnant.

interstellar molecules. See molecular-line radio astronomy.

interstellar polarization. The partial *polarization of starlight caused by dust in the interstellar medium. Since only nonspherical particles with a nonrandom orientation can polarize light, the dust grains must be elongated and are thought to be partially aligned by a weak *galactic magnetic field.

interstellar reddening. See interstellar absorption; extinction.

interstellar scintillations. See scattering.

interstellar wind. See solar wind.

intrinsic colour index. See colour index.

intrinsic variable. *See* variable star.

invariable plane. The plane through the *centre of mass of the solar system and perpendicular to the direction of the *angular momentum vector of the solar system. At present the *ecliptic is inclined by 1.65° to the invariable plane.

inverse Compton emission. *See* Compton scattering.

Io. The innermost of the four giant *Galilean satellites of Jupiter. It keeps one face permanently towards the planet (*see* synchronous rotation) as it orbits at a distance of 422 000 km, within Jupiter's *magnetosphere. Io is yellow-brown in colour, has a high *albedo (0.63), and has a diameter of 3630 km. The orbital period of Io is a half that of *Europa; the two satellites are locked into a gravitational resonance, the interaction of the satellites having caused their orbital distances to adjust until their periods were common multiples. Early photographs from the *Voyager 1 probe showed the surface of Io to lack impact craters. An explanation for this was found when intense volcanic action was observed on the satellite: at least nine active volcanoes were seen to eject material over a very wide area; seven were still active when Voyager 2 arrived four months later. The surface is thus being continuously renewed.

Io is the most active volcanic body known in the solar system. The activity is thought to result from heating by tidal forces exerted by Jupiter. There are several types of activity: eruptive plumes, a few of which reach altitudes of hundreds of kilometres; huge calderas (volcanic collapse craters) with associated lava flows and/ or surface markings; some possible lava lakes. The first volcano observed by Voyager 1 was an eruptive plume ascending to 300 km and visible above the satellite limb; material was deposited in concentric rings up to ~1400 km across. This volcano has been named *Pele*. Ionized matter – sulphur, oxygen, and hydrogen – escapes from the tenuous volcanic atmosphere of Io and forms a torus centred on Io and encompassing the whole of Io's orbit. This matter interacts with Jupiter's magnetosphere: it controls the rate at which Jupiter radiates energy and probably affects aurorae on Jupiter, radio bursts from the planet, and other phenomena. *See also* Jupiter's satellites; Table 2, backmatter.

ion. An atom or molecule that has lost or gained one or more electrons, thus having a positive or negative charge.

ionization. Any of several processes by which normally neutral atoms or molecules are converted into *ions or by which one ion of an atom or molecule is changed to another ionic form, as by losing a second electron. Positive ions are commonly formed by impact of high-energy photons or particles with neutral atoms or molecules. The minimum energy required to remove an electron from an atom or molecule is the *ionization potential*, I, of that atom or molecule, and for ionization to occur the energy of the impacting photon or particle must exceed the ionization potential. It requires a higher energy (the *second ionization potential*) to remove a second more tightly bound electron, and an even greater energy to remove a third electron. For equilibrium to exist in a system, such as an *emission nebula, ionization processes must be balanced by recombination of positive ions and electrons.

Because of the high temperatures in stars, much of the matter present is in an ionized state; often more than one electron has been lost so that the atom or molecule is doubly or triply ionized, etc. In astronomy the neutral state of an element is conventionally denoted by I (HI, for example, denotes neutral hydrogen) whereas II denotes singly charged ions (HII is ionized hydrogen) and III indicates a doubly charged ion, and so on. *See also* Saha ionization equation; plasma.

ionization potential. *See* ionization.

ionization zone. *See* pulsating variables.

ionosphere A region in the earth's atmosphere that extends from about 60 km to over 10 000 km altitude and contains free electrons and ions produced by the ionizing action of solar radiation on atmospheric constituents. It disturbs the propagation of *radio waves through it by reflecting of refracting them and attenuating them. The degree of ionization varies with time of day, season, latitude, and state of *solar activity. Several distinct layers may be identified in the ionosphere where the density of free electrons is higher than average. The three most important are the *D-layer* at a height of between 60 and 90 km, the *E-layer* above about 120 km, and the *F-layer* above about 210 km. During the day the F-layer splits into two further layers labelled F_1 and F_2 but these combine again at night. Radio waves of sufficiently low frequency are reflected by the E- and F-layers. The D-layer, being lower in the atmosphere where collisions between molecules and ions are more frequent, is more an absorber than a reflector.

The ionosphere is important in radio communications, allowing long-distance propagation at frequencies up to about 30 megahertz by successive reflections between it and the ground. In *radio astronomy, however, it makes ground-based observations almost impossible below 10 megahertz, causes *phase and *amplitude scintillations (*see* scattering) on *radio sources viewed through it, and alters the observed *polarization of radio waves by *Faraday rotation.

Other bodies, including Mars and Venus and the Jovian satellite Io, are now known to possess ionospheres.

ionospheric scintillations. *See* scattering.

ion tail. *See* comet tails.

IPCS. *Abbrev. for* image photon counting system. A highly sensitive electronic system for detecting and accurately recording light or ultraviolet radiation; it was developed in 1973 by Alec Boksenberg. It enables very low intensities to be detected and recorded, or allows images to be recorded in a very short time. It is particularly useful for studying spectra. The light is first intensified by a high-gain multistage *image tube and then recorded by a special TV camera. The TV signal is fed to and stored in a computer, where extraneous noise can be removed. A sharp and accurately recorded two-dimensional image is thus built up in the computer memory. This can be displayed on a screen, analysed, or manipulated electronically.

IPS. *Abbrev. for* interplanetary scintillations. *See* scattering.

IR. *Abbrev. for* infrared.

IRAS. *Abbrev. for* Infrared Astronomical Satellite. A satellite that has made an all-sky survey in the infrared wavelength range 10–100 μm using liquid helium cooled optics and detectors. The project was developed and operated by the Netherlands Agency for Aerospace Programmes (NIVR), NASA, and the UK Science and Engineering Research Council (SERC). The satellite was launched into a 900 km orbit on Jan. 26 1983 and operated until Nov. 22 1983 when the helium supply was exhausted. By the end of the mission 95% of the sky had been surveyed with confirming scans,

3% surveyed only once, and the remaining 2% not surveyed at all.

IRAS consists of a spacecraft that pointed a liquid helium cryostat containing a 0.57 metre *Ritchey-Chrétien telescope cooled to less than 10 K. An array of cooled detectors was located in the focal plane at the Cassegrain focus of the telescope. The 62 *photoconductive detectors responded to one of four wavebands of 100, 60, 25, or 12 μm. They were arranged so that every source crossing the field of view could be seen by at least two detectors in each of the wavebands. The low resolution *spectrometer (LRS) covered the range 8–23 μm with a spectral resolution, λ/Δλ, of about 20. The chopped photometric channel (CPC) mapped infrared sources on a 9 × 9 arcmin raster at 50 and 100 μm with a spatial resolution of about 1 arcmin.

The operation centre for the satellite was at the SERC Rutherford-Appleton Laboratory, near Oxford, England where a preliminary analysis of the data was made. The final reduction of data was done in America and Holland. A catalogue of the IRAS observations, containing about 300 000 sources, was published in Nov. 1984.

The results obtained so far relate to almost every aspect of astronomy. Some of the highlights are: the discovery of a dust shell around *Vega, which may be an early planetary system; the discovery of *infrared cirrus in our Galaxy; the first detection from space using infrared of comets (IRAS discovered 5 new comets); the detailed mapping of the infrared emission from the galactic plane; observations on galaxies showing that most of the galaxies detected in the 60 and 100 μm wavebands are spirals; observations on low-luminosity *protostars in several *molecular cloud complexes.

IRAS–Iraki–Alcock. A comet detected in 1983 by the infrared satellite *IRAS and independently by G. Iraki and E. Alcock. It passed only 4.6 × 10⁶ km (0.03 AU) from earth and just before closest approach it brightened to 2nd magnitude and exceeded 3° across. The huge *coma was highly asymmetrical, with a fan of material spreading out sunwards from the off-centred star-like *nucleus (which reached magnitude 12). No normal tail could be seen. Infrared and radar observations provided information on the composition and rotation of the nucleus.

iron. *Short for* iron meteorite.

iron meteorite (siderite). One of the main classifications of *meteorites, containing nickel-iron with small amounts of accessory minerals. The composition averages 90% iron, 10% nickel. The three main subdivisions are *octahedrites*, which have more than 6% nickel, the less common *hexahedrites*, with less than 6% nickel, and the nickel-rich *ataxites*. *See also* Widmanstätten patterns.

iron-peak elements. *See* nucleosynthesis.

irregular galaxy. *See* galaxies; Hubble classification.

irregular variables. Stars, such as the majority of red giants, that are *pulsating variables showing slow luminosity changes either with no periodicity or with a very slight periodicity. The latter can be grouped into slow irregular variables of *spectral types F, G, K, M, C, and S that are usually giants, and slow irregular supergiants of spectral types M, C, and S.

Isaac Newton Telescope (INT). Originally, the 2.5 metre (98 inch) reflecting telescope that began operations in 1967 at the *Royal Greenwich Observatory in Sussex. It was dismantled in 1979 and, completely refurbished, resumed observations in 1984 at the *Roque de los Muchachos Observatory in the Canaries. It has a new 100 inch primary mirror that has an extremely accurately figured paraboloidal surface and that is made of the glass-ceramic *Zerodur so that its shape remains almost constant during temperature changes from day to night. The focal ratio is 2.9. There is an f/15 Cassegrain focus and an f/50 coudé focus. The change in latitude (51°N to 28°N) required the *equatorial mounting to be modified for the new site, tipping it up by another 23°. The latest developments in telescope instrumentation have been incorporated, including a *CCD detector at the prime focus and an *IPCS associated with a new spectrograph at the Cassegrain focus. Like all major modern telescopes the instrumentation is automatic and computer-controlled. It will eventually be under *remote operation from the RGO at Herstmonceux. The Dutch have a 20% share in the INT; 20% and 5% of the telescope's time are allocated to Spain and international teams respectively.

ISEE. *Abbrev. for* International Sun-Earth Explorer.

Ishtar Terra. *See* Venus.

ISM. *Abbrev. for* interstellar medium.

ISO. *Abbrev. for* Infrared Space Observatory.

isophote. A line on a diagram joining points of equal flux density or intensity.

isothermal theory. *See* galaxies, formation and evolution.

isotopes. Forms of an element in which the nuclei contain the same number of protons but different numbers of neutrons. For example, there are three isotopes of hydrogen: 'nor-

mal' hydrogen has a single proton, deuterium has one proton and one neutron, and tritium, which is a radioactive isotope, has one proton and two neutrons.

isotropy. The property by which all directions appear indistinguishable to an observer expanding with the universe. Isotropy about every point in space implies *homogeneity but the reverse is not necessarily true. *See* cosmological principle.

ISPM. *Abbrev. for* International Solar Polar Mission.

ISS. *Abbrev. for* interstellar scintillations. *See* scattering.

IUE. *Abbrev. for* International Ultraviolet Explorer.

J

jansky. Symbol: Jy. The unit of *flux density adopted by the IAU in 1973 and used throughout the spectral range, especially for radio and far-infrared measurements. One jansky is equal to 10^{-26} watts per metre squared per hertz. It is possible to calculate the equivalent flux density, $F(\nu)$, in jansky from a value of *magnitude at the effective wavelength, λ_{eff}, appropriate to a particular optical or infrared filter (see table). For example, an R magnitude of 3.2 corresponds to a flux density at 0.7 μm of

$$2770 \times 10^{-0.4 \times 3.2} \text{ Jy}$$

i.e. 145 Jy. (The table gives the flux level for 0.0 magnitude.)

Jansky noise. *Another name for* cosmic noise. *See* antenna temperature.

Janus. A postulated satellite of Saturn. In 1966 the presence of an inner satellite at a distance of 169 000 km was indicated from the observations

Filter	λ_{eff} (μm)	$F(\nu)$ (jansky)	$F(\lambda)$ (W/cm^2 μm)
U	0.365	1910	4.3×10^{-12}
B	0.44	4260	6.6×10^{-12}
V	0.55	3830	3.8×10^{-12}
R	0.70	2770	1.7×10^{-12}
I	0.90	2240	8.3×10^{-13}
J	1.25	1720	3.3×10^{-13}
K	2.20	645	4.0×10^{-14}
L	3.40	310	8.1×10^{-15}
M	5.0	183	2.2×10^{-15}
N	10.2	41.5	1.2×10^{-16}
Q	21	9.5	6.5×10^{-18}

of A. Dollfus at the Pic du Midi Observatory in France. Many observers were sceptical about the existence of the new satellite, which had been named Janus. No other confirmation was made of this difficult observation, which can only occur when the rings of Saturn are wide open to search for fainter satellites. The Voyager observations have since failed to locate this object.

JD. *Abbrev. for* Julian date.

Jeans length. *See* gravitational instability.

jet. A long thin linear feature of bright emission extending from a compact object, such as a *galaxy. Jets have been seen in radio, optical, and x-ray emission. They are sometimes broken up into a number of bright *knots. An example is that found in the giant elliptical galaxy M87 (*see* Virgo A).

Jet Propulsion Laboratory (JPL). An institution for space research and engineering at Goldstone, California, run by the California Institute of Technology under a contract from NASA. It is the site of NASA's Goldstone Complex, which is concerned with unmanned planetary spacecraft and control of planetary missions. Its chief instrument is a 64 metre (210 foot) radio dish.

Jewel Box (Kappa Crucis; NGC 4755). A brilliant *open cluster in the southern constellation Crux that contains over 100 stars of various colours. It appears to be a young cluster of an estimated age of 10 million years.

Jodrell Bank. The site at Macclesfield in Cheshire of the Nuffield Radio Astronomy Laboratories of the University of Manchester. The principal instrument is the 76.2 metre (250 foot) fully steerable radio dish, which has been in operation since 1957. It has an *altazimuth mounting. It was originally known as the *Mark I* but was renamed the *Mark IA* following modifications made in 1970/71. These included changes to the supporting structure and the fitting of a new reflecting membrane having an improved shape and longer focal length (22.9 m or 75 ft). The Mark IA can be used effectively at wavelengths down to 10 cm. There is also a smaller elliptical-shaped reflector, the *Mark II*, 38 × 25.4 metres. This was completed in 1964 and can be used effectively at wavelengths down to 3 cm. Both instruments, together with a 13 metre dish formerly at Woomera, Australia, are part of the radio network *MERLIN, whose component telescopes are linked to a computer at Jodrell Bank.

Johnson Space Center (JSC). A NASA establishment in Houston, Texas, concerned with the control of manned space missions and space-flight, such as the space shuttle.

joule. Symbol: J. The SI unit of energy, equal to the work done when the point of application of a force of one newton is moved one metre in the direction of the force.

Jovian planets. A term derived from the Latin name for Jupiter and applied collectively to the giant planets Jupiter, Saturn, Uranus, and Neptune.

Jovian satellites. *See* Jupiter's satellites.

JPL. *Abbrev. for* Jet Propulsion Laboratory.

Julian calendar. The calendar that was established in 46 BC in the Roman Empire by Julius Caesar, with Sosigenes of Alexandria as his chief advisor. It reached its final form in about 8 AD under Augustus and was in general use in the West up to 1582, when the *Gregorian calendar was instituted. Each year contained 12 months and there was an average of 365.25 days per year: three years of 365 days were followed by a leap year of 366 days. (Leap years were not correctly inserted until 8 AD). Since the average length of the year was about 11 minutes 15 seconds longer than the 365.2422 days of the *tropical year, a discrepancy arose between the calendar year and the seasons, with an extra day 'appearing' about every 128 years.

Julian century. *See* Julian year.

Julian date (JD). The number of days that have elapsed since noon GMT on Jan 1 4713 BC – the *Julian day number* – plus the decimal fraction of the day that has elapsed since the preceding noon. The consecutive num-

Julian day at noon, UT, on March 1st

year	Julian day	year	Julian day
1800	2378556	1940	2429690
1820	2385861	1950	2433342
1840	2393166	1960	2436995
1860	2400471	1970	2440647
1880	2407776	1980	2444300
1900	2415080	1990	2447952
1920	2422385	2000	2451605

bering of days makes the system independent of the length of month or year and the JD is thus used to calculate the frequency of occurrence or the periodicity of phenomena over long periods. The system was devised in 1582 by the French scholar Joseph Scaliger. A specific Julian date can be determined from astronomical tables of Julian day numbers: thus since March 1 1980, noon GMT, is tabulated as day number 244 4300 then March 17 1980, 6 p.m. GMT, is JD 244 4316.25 (see table). The *modified Julian date (MJD)* is found by subtracting 240 0000.5 from the Julian date.

Julian year. A period of 365.25 days (each containing 864 000 seconds, i.e. 24 hours). A *Julian century* is 100 Julian years. The Julian year has been used since 1984 for defining standard *epochs, replacing the *Besselian year. It is denoted by the prefix J, as in J2000.0.

June Lyrids. A minor *meteor shower, radiant: RA 278°, dec +35°, maximizing on June 16.

Juno. The third *minor planet to be discovered, found in 1804. With a diameter of 247 km it is the smallest of the first four minor planets to be found and numbered. It has a spectrum similar to that of the *stony-iron (siderolite) meteorites. See Table 3, backmatter.

Jupiter. The largest planet, orbiting the sun at a mean distance of 5.2 AU in 11.9 years. It has an equatorial diameter of 142 800 km (11 times that of the earth) and a polar diameter of 135 500 km. This *oblateness results from a rotation period of less than 10 hours, shorter than that of any other planet. Jupiter has a mass more than twice that of all the other planets combined but its average density, only 1.3 times that of water, suggests that it contains a high proportion of the lightest elements, hydrogen and helium. Orbital and physical characteristics are given in Table 1, backmatter.

*Oppositions recur at 13 month intervals; the planet then has, on average, an apparent diameter of 47 arc seconds and shines at about magnitude −2.4, brighter than every other night-time object other than the moon, Venus, and (rarely) Mars. Jupiter has 16 known satellites, including the four *Galilean satellites visible with a minimum of optical aid, and also a ring system (*see* Jupiter's satellites). Telescopes show the disc to be crossed by bands of light and dark clouds, called *zones* and *belts* respectively, running parallel to the equator. Irregular spots and streaks are seen, their motion across the disc indicating planetary rotation periods varying between about 9 hours 51 minutes in equatorial regions and 9 hours 56 minutes at high latitudes. Except for the *Great Red Spot and the white ovals, most of the markings are temporary, lasting for days or months.

Models for Jupiter's atmospheric and internal structure have been refined following the fly-bys of the spaceprobes *Pioneer 10 and 11 in 1973 and 1974 and the Voyager probes in 1979; earlier spectroscopic work had, however, detected the presence of hydrogen, ammonia, and methane, and also water vapour, ethane, acetylene, phosphine, germanium tetrahydride, and carbon monoxide. It appears that the white or yellowish zones are areas of higher clouds supported by upward convection of warm gases; the reddish-brown belts have descending gas flows and lower clouds. With a rapidly rotating planet the weather systems are primarily zonal, and it is this that produces the banded colourful cloud systems superimposed with spots of a variety of different shades of colours; the spots are anticyclonically rotating systems. Huge convective storms are found in the equatorial region of Jupiter, and cyclonic storms, called *barges*, are seen in the northern latitudes. Eddies may give rise to the spots. At latitudes greater than about 45°, the belt/zone system gives way to a mottled surface appearance corresponding to ascending and descending convection cells (gas columns). The higher cloud zones appear to comprise ammonia crystals while the lower cloud belts may contain sulphur compounds, hydrogen and ammonia polysulphides, or complex organic compounds, possibly formed in photochemical reactions energized by ultraviolet radiation and by lightning discharges in the atmosphere. The highest cloud features, such as the Great Red Spot, may be coloured red by traces of phosphorus brought up by convection from the lower atmosphere.

Jupiter radiates, as heat, about twice as much energy as it receives from the sun, indicating that there is an internal reservoir of thermal energy left over from its creation 4.6 thousand million years ago. This internal energy source aids in driving Jupiter's weather system. In addition to the rapid rotation of Jupiter, the outflow of energy from the planet and a *greenhouse effect ensure that there is little variation in temperature between the equator and poles, or between the day and night hemispheres. Atmospheric temperatures increase from −130°C at the cloud tops to about 30°C at the base of a lower cloud layer of water droplets and ice-crystals believed to lie about 70 km below the ammonia clouds. Over this same range the pressure rises from 0.5 to 4.5 (earth) atmospheres. All Jupiter's weather occurs in this upper skin of its 1000 km deep gaseous. atmosphere. At first sight the Jovian weather system appears quite different from that of the earth. The analyses of the Voyager data have shown, however, that Jupiter (and also Saturn) and the earth drive their meteorological systems in the same way, with energy being transported from the small-scale features into the main flow. The depth of the motions are unknown. However, they appear more stable than the specific cloud elements whose lifetimes vary from months, years, to decades.

The atmosphere of Jupiter consists primarily of hydrogen and helium in near solar proportions in the ratio 82 to 17. The hydrogen becomes liquid at a depth of 1000 km; this marks the interface between atmosphere and planetary interior. At a depth of about 25 000 to 30 000 km, under an estimated three million atmospheres pressure, the hydrogen becomes metallic: the molecules dissociate into protons and a sea of unattached electrons. Temperatures may increase from 2000°C at 1000 km, to 11 000°C at 25 000 km, and to 30 000°C at the planet's centre where there may be a rocky core 10 to 20 times more massive than the earth. The bulk of Jupiter's interior is thus composed of liquid hydrogen, most of which is in metallic form.

A magnetic field, probably arising from a dynamo action within Jupiter's conductive metallic hydrogen, has a total strength about 17 000 times the earth's magnetic field. Jupiter's field is tilted by 11° relative to its rotation axis and is reversed in polarity relative to that of the earth. It supports a *magnetosphere, about 15 million km across, in which the eight inner satellites orbit. The magnetosphere contains radiation belts possibly 10 000 times more intense than the earth's *Van Allen radiation belts.

Jupiter emits radio waves by three mechanisms: high-frequency thermal radio noise comes from the atmosphere along with infrared radiation; high-frequency nonthermal *synchrotron emission is generated by electrons in Jupiter's magnetic field; intense bursts of decametric radio waves are believed to arise from electrical discharges along Jovian magnetic field lines when *Io, with its conductive ionosphere, moves across them.

Jupiter's comet family. A *comet family whose distribution of aphelion distances (the comets' furthest points from the sun) correlates with the mean distance of the planet Jupiter. It contains about 70 members with aphelion distances of 4–8 AU. Jupiter is capable of substantially changing the orbits of comets that happen to pass close by. *Brooks 2* passed within two Jovian radii of Jupiter's surface in July 1886. Its period and aphelion distance were changed from 31 years and 14 AU to 7 years and 5.4 AU. The formation of comet families is caused by such gravitational perturbations. Jupiter is also capable of throwing comets out of its family into long-period orbits (using a process similar to the one used to deflect spacecraft out of the solar system). Over 90% of the Jovian family move in direct orbits. Comets with periods less than 10 years have a mean inclination to the ecliptic of 12°. After a comet has been captured by Jupiter it will take another 200 to 400 orbits (depending on perihelion distance) before it decays completely. No comet can remain in the Jupiter family for more than about 4000 years so the family is being replenished continually.

Jupiter's satellites (Jovian satellites). A system of at least 16 satellites, three of which were discovered in

1979/80 during the *Voyager mission. Their physical and orbital properties are given in Table 2, backmatter. They can be divided conveniently into three groups. There is an outer group of four small bodies (Ananke, Carme, Pasiphae, and Sinope) that move in loosely bound retrograde orbits (see direct motion) that are highly eccentric and are inclined to Jupiter's equatorial plane by an angle of ~150°–160°; they orbit the planet in periods of about 700 days and at mean distances of roughly 21 to 24 million km (and are thus strongly perturbed by the sun). An intermediate group of four small satellites (Leda, Himalia, Lysithea, and Elara) move in approximately 250 day orbits at mean orbital distances between 11 and 12 million km; the orbits are less eccentric than those of the outer group and are inclined at an angle of ~28° to Jupiter's equatorial plane. The inner group is comprised of the four large Galilean satellites – *Io, *Europea, *Ganymede, and *Callisto – together with *Amalthea and the three newly discovered satellites Thebe, Adrastea, and Metis; these all move in near-circular orbits that lie close to the plane of Jupiter's equator. The members of the outer and intermediate groups may be captured *minor planets, although the capturing process that would place objects into one group rather than the other has not been defined. Possibly some of them were former members of the *Trojan group of minor planets. All eight inner satellites lie within Jupiter's *magnetosphere and are effective in sweeping up the charged particles found there, becoming intensely radioactive in the process.

Few details of the irregularly shaped Amalthea were known until the Voyager probes visited Jupiter in 1979. The Galilean satellites however have been studied throughout the telescopic era and more recently by the *Pioneer 10 and 11 spaceprobes, which returned hazy photographs in 1973 and 1974. Much more detailed information came from the Voyager probes (see individual entries). The Galilean satellites are comparable in size with the small planets: Ganymede, which is the largest satellite in the solar system, is slightly larger than Mercury; except for Europa all are larger than the moon. Each is in *synchronous rotation, keeping one face turned towards the planet. Their maximum surface temperatures vary between the 120 K of Io to 167 K for Callisto, probably as a result of differences in *albedo; the albedo of Io is 0.63, of Europa 0.70, of Ganymede 0.43, and of Callisto as low as 0.17. Io has the greatest density, 3.53 times that of water, with Europa 3.03 times, Ganymede 1.93 times, and Callisto 1.79 times that of water. These density differences may reflect a loss of water by the two innermost satellites as a result of heating during the formation of Jupiter itself. There is considerable volcanic activity on Io. Ganymede and Callisto, unlike Io and Europa, are both heavily cratered bodies.

As Voyager 1 moved inside the orbit of Amalthea, a ring system of rocky boulders was discovered around Jupiter, lying within the Roche limit of the planet at about 55 000 km above the cloud tops.

K

Kapteyn's star. A red *subdwarf in the constellation Pictor that is a *high-velocity star with a radial velocity of 242 km/s. It is one of the nearest stars to the sun and is the nearest subdwarf. m_v: 8.8; M_v: 10.8; spectral type: sdM0; distance: 3.91 pc.

Kapteyn Telescope. See Roque de los Muchachos Observatory.

Karl Schwarzschild Observatory. An observatory at Tautenburg in East Germany at an altitude of 330 metres. It is part of the Central Institute for Astrophysics of the Academy of Sciences of the GDR. The chief instrument is a reflector with a 2.0 metre (79 inch) spherical mirror that in addition to Cassegrain and coudé facilities can be used with a 1.34 m correcting plate as a *Schmidt telescope. A second correcting plate is ground so as to produce spectra of celestial bodies, i.e. it acts as an *objective prism.

K-corona. The inner part of the solar *corona, responsible for the greater part·of its intensity out to a distance of about two solar radii. It consists of rapidly moving free electrons, and exhibits a linearly polarized *continuous spectrum due to Thomson scattering by the electrons of light from the *photosphere. The *Fraunhofer lines are not present in the spectrum: they have been sufficiently broadened by the large *doppler shifts resulting from the rapid random motion of the electrons so as to overlap and be indiscernible. The K-corona attains a temperature of around 2 000 000 kelvin at a height of about 75 000 km. It may exhibit considerable structure, with *streamers* (regions of higher than average density) overlying *active regions and quiescent *prominences, and *coronal holes (regions of exceptionally low density) overlying areas characterized by relatively weak divergent (and therefore primarily unipolar) magnetic fields.

Kellner eyepiece. *See* eyepiece.

kelvin. Symbol: K. The SI unit of temperature, equal to 1/273.16 of the *thermodynamic temperature of the triple point of water. A temperature in kelvin may be converted to one in degrees Celsius by the subtraction of 273.15.

Kelvin contraction. *See* Kelvin timescale.

Kelvin timescale. The time taken for a mass of gas to collapse under its own gravitation. It is given by
$$\tau_{KH} = GM^2/RL$$
where G is the gravitational constant and M, R, and L are the mass, radius, and luminosity of the object. Such a collapse – known as the *Helmholtz-Kelvin contraction* – was proposed in the late 19th century as the source of the sun's energy output. This theory was abandoned when the contraction timescale of only 10^7 years was shown to be too short. The energy source was subsequently shown to originate in nuclear-fusion reactions.

Kennedy Space Center. A NASA establishment at Cape Canaveral, Florida, concerned with satellite launching and space shuttle payload integration. The launching site on Merritt Island comprises launchpads, runways, and space vehicle assembly areas.

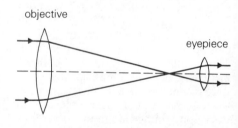

Ray path in Keplerian refractor

Keplerian telescope. The first major improvement of the *Galilean telescope, developed by the German astronomer and mathematician Johannes Kepler, in which a positive (convex) lens was used as the eyepiece in place of the negative lens that Galileo used (see illustration). This gave a larger though inverted field of view and higher magnifications.

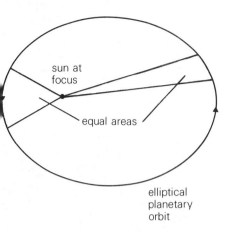

elliptical
planetary
orbit

Kepler's second law

Kepler's laws. The three fundamental
laws of planetary motion that were
formulated by Johannes Kepler and
were based on the detailed observations of the planets made by Tycho
Brahe, with whom Kepler had
worked. The laws state that

1. The orbit of each planet is an
ellipse with the sun at one focus of
the ellipse.

2. Each planet revolves around the
sun so that the line connecting planet
and sun (the radius vector) sweeps
out equal areas in equal times (see illustration). Thus a planet's velocity
decreases as it moves further from the
sun.

3. The squares of the sidereal periods of any two planets are proportional to the cubes of their mean
distances from the sun. If the period,
P, is measured in years and the mean
distance, a, in *astronomical units,
then $P^2 = a^3$ for any planet.

The first two laws were published
in 1609 in *Astronomia Nova* and the
third law in 1619 in *Harmonices
Mundi*. The third law, sometimes
called the *harmonic law*, allowed the
relative distances of the planets from
the sun to be calculated from measurements of the planetary orbital peri-

ods. Kepler's laws gave a correct
description of planetary motion. The
physical nature of the motion was not
explained until Newton proposed his
laws of motion and gravitation. From
these laws can be obtained Newton's
form of Kepler's third law:
$$P^2 = 4\pi^2 a^3 / G(m_1 + m_2)$$
where G is the gravitational constant,
m_1 and m_2 are the masses of the sun
and a planet, a is the semimajor axis
of the planet, and P is its sidereal period.

Kepler's star. A *supernova that was
observed in Oct. 1604 in the constellation Ophiuchus and could be seen
with the naked eye for over a year. It
was studied by astronomers in Europe, China, and Korea and its position was so accurately determined by
Kepler and Fabricius that a small
patch of nebulosity above the galactic
plane could be identified as the remnant of the original 1604 supernova
(Walter Baade, 1943). The light curve
plotted from the original observations
shows that the magnitude increased to
a maximum of over −2.5, dropping to
+4 in about 300 days, and that it
was a type I supernova.

**Kerr black hole, Kerr-Newman black
hole.** *See* black hole.

Keyhole nebula. *See* Eta Carinae nebula.

kilo-. Symbol: k. A prefix to a unit,
indicating a multiple of 1000 or 10^3
of that unit. For example, one kilogram equals 1000 grams.

kilogram. Symbol: kg. The SI unit of
mass. It is defined as the mass of a
prototype platinum-iridium cylinder
kept at the International Bureau of
Weights and Measures at Sèvres,
France. The spelling *kilogramme* is also used.

kinetic energy. Energy possessed by a
body by virtue of its motion, equal to

the work that the body could do in
coming to rest. In classical mechanics
a body with mass *m* and velocity *v*
has kinetic energy $\frac{1}{2}mv^2$. A body with
moment of inertia *I* rotating with angular velocity ω has kinetic energy
$\frac{1}{2}I\omega^2$.

Kirkwood gaps. Gaps in the distribution of orbits within the main belt of
*minor planets, corresponding to the
absence of orbits with periods that
are simple fractions (1/4, 1/3, 2/5, 1/2,
etc.) of the orbital period of Jupiter
(see illustration). Minor planets with
such orbital periods would be perturbed into other orbits by the regularly recurring gravitational pull of
Jupiter. Daniel Kirkwood first explained them in 1857.

**Kitt Peak National Observatory
(KPNO).** An observatory near Tucson, Arizona, at an altitude of 2060
metres, that has a very large assembly
of telescopes. Like the *Cerro Tololo
Interamerican Observatory, it is run
by the Association of Universities for
Research in Astronomy and began
operations in 1960. The chief instruments are a 4 metre (158 inch) reflecting telescope – the *Mayall
reflector* – which has a very low *focal ratio of f/2.6 and went into use in
1973, a 2.1 metre (84 inch) reflector
operational in 1964, and a huge 1.5
metre (60 inch) *solar telescope, the
McMath solar telescope, completed in
1962. An 11 metre (36 ft) radio telescope is operated by the *National
Radio Astronomy Observatory. First
used in 1967 this was the pioneering
instrument for *millimetre astronomy.
Since 1983 it has had a somewhat
larger (12 m) and more accurate reflecting surface. A 2.3 metre (90 inch)
reflector is operated by the *Steward
Observatory.

Kleinmann-Low nebula (KL nebula). A
diffuse source of infrared radiation
within the *Orion molecular cloud
(OMC-1). At far-infrared wavelengths

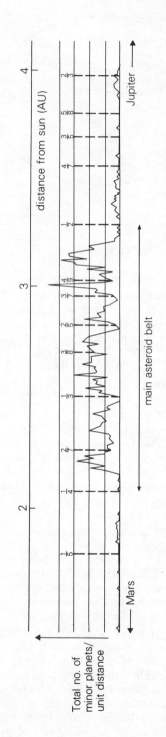

Kirkwood gaps, indicated by fractions

(100 μm) it is the brightest member of the infrared cluster in the OMC-1 complex, with a luminosity about 10^5 times that of the sun. The temperature of the dust within the nebula is found to be about 70 K.

K line. *See* H and K lines.

knot. A bright compact feature particularly in radio and x-ray *jets.

Kochab. *See* Ursa Minor.

Kohoutek (1973 XII). A comet discovered in March 1973 by Lubos Kohoutek when he was looking for Biela's comet. At discovery comet Kohoutek was 4.6 AU away from the sun and early predictions indicated that near perihelion, at its brightest, Kohoutek would reach a magnitude of about −12, comparable with the full moon. This would make it visible in daylight. Comets however are among the most unpredictable astronomical objects and Kohoutek only just attained naked-eye visibility. It was extensively investigated from Skylab, astronaut Gibson being impressed by its overall orange colour and its sunward spike.

Kraus-type radio telescope. A *radio telescope consisting of a long thin flat reflector that can be tilted about its long horizontal axis and that reflects radiation from the sky onto a large fixed curved reflector that forms part of a paraboloid. The radiation is collected at the focus of the paraboloid by the feed (*see* dish). The telescope thus achieves a large collecting area that can be steered in one direction.

KREEP. *See* moonrocks.

K stars. Stars of *spectral type K, orange-red in colour and with a surface temperature (T_{eff}) of about 3500 to 5000 kelvin. The spectral lines of calcium (Ca II and Ca I) are strong and neutral metal lines are very promi-

nent; molecular bands strengthen. Arcturus and Aldebaran are K stars.

Kuiper Airborne Observatory (KAO). NASA's instrumented Lockheed C 141 transport aircraft that carries a 0.9 metre telescope for high-altitude astronomical observations, mainly at infrared wavelengths. The telescope is in the fuselage and is subject to very low temperatures and pressures; the Cassegrain focus is in the aircraft cabin. The observatory became operational in 1975 and is named after Gerard P. Kuiper.

L

labyrinthus. Intersecting narrow depressions. The word is used in the approved name of such a surface feature on a planet or satellite.

Lacerta (Lizard). A small inconspicuous constellation in the northern hemisphere near Cygnus, lying partly in the Milky Way, the brightest stars being of 4th magnitude. It contains the prototype of the quasar-like *BL Lac objects. Abbrev.: Lac; genitive form: Lacertae; approx. position: RA 22.5h, dec +45°; area: 201 sq deg.

lacus. A small relatively smooth dark area. The word is used in the approved name of such a surface feature on a planet or satellite.

Lagoon nebula (M8; NGC 6523). An *emission nebula in the form of an HII region that lies about one kiloparsec away in the direction of Sagittarius. Its total apparent magnitude (5.8) makes it easily visible with binoculars.

Lagrangian points. Five locations in space where a small body can maintain a stable orbit despite the gravitational influence of two much more massive bodies, orbiting about a common centre of mass. They are named after the French mathematician J. L. Lagrange who first suggested their existence in 1772. A Lagrangian point 60° ahead of Jupiter in its orbit around the sun, and another 60° behind Jupiter, are the average locations of members of the *Trojan group of minor planets. The three other Lagrangian points in the sun-Jupiter gravitational field do not permit stable minor planet orbits owing to the perturbing influence of the other planets. In any system these three points lie on the line joining the centres of mass of the two massive bodies and are in unstable equilibrium (*see* equipotential surfaces). Lagrangian points 60° ahead of and 60° behind the moon in its orbit of the earth are proposed as the site of existing gas and dust clouds and of future earth-orbiting space colonies and solar power stations.

La Palma Observatory. *See* Roque de los Muchachos Observatory.

Laplace's nebular hypothesis. *See* nebular hypothesis.

Large Magellanic Cloud (LMC; Nubecular Major). *See* Magellanic Clouds.

Las Campanas Observatory. An observatory near La Serena, Chile, sited on Cerro Las Campanas, at an altitude of 2300 metres; the *seeing at this site is exceptional. The Observatory is owned by the Carnegie Institution of Washington and is operated as part of the *Hale Observatories. The principal instruments are the 2.5 metre (100 inch) *du Pont Telescope*, which went into operation in 1977, and the 1 metre (40 inch) *Swope Telescope*, which became operational in 1971.

last quarter (third quarter). *See* phases, lunar.

late-type stars. Relatively cool stars of *spectral types K, M, C (carbon), and S. They were originally thought, wrongly, to be at a much later stage of stellar evolution than *early-type stars.

latitude. 1. *Short for* celestial latitude. *See* ecliptic coordinate system.
2. *Short for* galactic latitude. *See* galactic coordinate system.

launch vehicle. Any system by which the necessary energy is given to a satellite, spaceprobe, space station, etc., in order to insert it into the desired orbit or trajectory. Expendable multistage rockets were used originally, and are still being used and developed: ESA's *Ariane and NASA's Thor Delta rockets are examples. The reusable *space shuttle was developed by NASA so that recovery of the vehicle is possible.

Rocket propulsion is a form of jet propulsion: all the propellant is carried in the vehicle at take-off and the hot combustion gases, resulting from the mixture of fuel with reactant, are ejected at high speed through a nozzle to produce the necessary force – termed *thrust* – to lift the vehicle off the ground. The thrust exerted by the rocket engine is equal (in vacuum conditions) to the rate of change of *momentum of the expelled propellant gas. An engine is rated in terms of the *specific impulse,* i.e. the impulse (thrust times time) delivered per unit weight of propellant expended; it is measured in seconds. The specific impulse is (ideally) equal to the jet velocity divided by the sea-level acceleration due to gravity, g. Although the thrust is held constant, the rocket velocity will increase as its fuel is expended and it becomes lighter.

Single rockets, carrying only a very small payload, would only just achieve earth orbit. *Multistage rockets* have therefore been developed: in a three-stage rocket, one rocket plus propellant plus spacecraft is used as the payload of a larger rocket, which in turn is the payload of a still larger rocket. The empty tanks, engine, and structure of first the largest and then the second largest rocket are discarded at relatively low altitudes and velocities, after each completes its firing. The final stage then carries the spacecraft into the desired orbit, the satellite separating from the rocket stage when orbital velocity is reached. During the flight the rocket path is tilted from the vertical so that when it reaches the orbit altitude it is parallel to the earth's surface. Modern launchers generally consist of two, three, or four stages; there are design variations in the type of fuel used, the means of carrying the fuel, and the process by which the tanks, etc., are discarded.

launch window. An interval of time during which a planetary probe, etc., must be launched in order to attain a desired position at a desired time.

leading spot. *Another name for* preceding (*p*-) spot. *See* sunspots.

leap month. *See* lunar year.

leap second. *See* universal time.

leap year. A year that contains one more day than the usual *calendar year so that the average length of the year is brought closer to the *tropical year of 365.2422 days or to the *lunar year. In the Julian calendar a leap year of 366 days occurred once every four years when the year was divisible by four. In the *Gregorian calendar this rule was modified so that century years are leap years only when they are divisible by 400. The additional day is now added at the end of February (Feb. 29).

Leda. A small satellite of Jupiter, discovered in 1974. *See* Jupiter's satellites; Table 2, backmatter.

Lemaître model. *See* cosmological models.

lens. A specially shaped piece of transparent material, such as glass, quartz, or plastic, bounded by two surfaces of a regular (usually spherical) but not necessarily identical curvature. A ray of light passing through a lens is bent as a result of *refraction. Spherical lenses may be either converging or diverging in their action on a light beam, depending on their shape (see illustration). *See also* focal length; aberration; objective; eyepiece; coating of lenses.

lenticular galaxy. *See* galaxies.

Leo (Lion). A large conspicuous zodiac constellation in the northern hemisphere near Ursa Major. The brightest stars are the 1st-magnitude *Regulus (α), three of 2nd magnitude including *Denebola (β) and the binary *Algeiba (γ), and several of 3rd magnitude. Regulus lies at the base of the *Sickle of Leo*, the other stars in the Sickle being Eta, Gamma, Zeta, Mu, and Epsilon Leonis. The area also contains the Mira star R Leonis, the dwarf nova X Leonis, and the spiral galaxies M65, M66, M95, and M96. Abbrev.: Leo; genitive form: Leonis; approx. position: RA 10.5h, dec +15°; area: 947 sq deg.

Leo Minor (Little Lion). A small inconspicuous constellation in the northern hemisphere near Leo, the brightest stars being of 4th magnitude. Abbrev.: LMi; genitive form: Leonis Minoris; approx. position: RA 10h, dec +35°; area: 232 sq deg.

Leonids (November swarm). A periodic *meteor shower that tends to be insignificant most years (ZHR about 10) but at 33 year intervals produces a spectacular meteor storm. The shower, radiant: RA 152°, dec +22°, maximizes on the two days around Nov. 17 each year. The associated

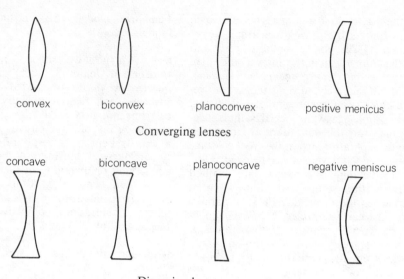

convex   biconvex   planoconvex   positive menicus

Converging lenses

concave   biconcave   planoconcave   negative meniscus

Diverging lenses

*meteor stream is the dust debris from comet *Tempel-Tuttle (1866 I) and storms occur when the earth intersects the stream near the parent comet. This comet has an orbital period of 33 years. The last storm occurred on Nov. 17, 1966 when observers in the central USA saw about 5000 meteors in twenty minutes. Similar storms occurred in 1799, 1833, and 1866; 1899 and 1933 were years of only mediocre displays.

Leo systems. Two members of the *Local Group of galaxies.

leptons. A class of *elementary particles that do not participate in strong interactions (*see* fundamental forces); it includes the *electron, *muon, the massive tau lepton, the *neutrinos, and their associated *antiparticles. Leptons show no evidence of any internal structure. The *lepton number* is the total number of leptons minus antileptons in a system.

Lepus (Hare). A constellation in the southern hemisphere near Orion. The brightest star is the very luminous remote Arneb (α), which is of 2nd

magnitude. Abbrev.: Lep; genitive form: Leporis; approx. position: RA 5.5h, dec $-20°$; area: 290 sq deg.

Leviathan. The 183 cm (72 inch) reflecting telescope built 1842–45 by William Parsons, 3rd earl of Rosse. It was dismantled in 1908. *See* reflecting telescope.

Lexell's comet. A comet that was discovered by Messier in June 1770 but was named after the St. Petersburg mathematician who calculated its orbit. Prior to 1767 Lexell's comet had a period of 11.4 years; a close approach to Jupiter in 1770 changed this to 5.6 years and the next close approach to Jupiter in 1779 changed this period to a calculated 174 years. This last approach perturbed the comet so that it has a perihelion distance of 5.4 AU and it has never been seen again, always being too far away from earth.

Libra (Balance). A fairly inconspicuous zodiac constellation in the southern hemisphere near Scorpius, the brightest star, Beta Librae, being of 2nd magnitude and apparently green in colour. Delta Librae is an *Algol variable (magnitude 4.8–6.1) with many other variables occurring in the globular cluster M5. Abbrev.: Lib; genitive form: Librae; approx. position: RA 15h, dec −15°; area: 538 sq deg.

Libra, first point of. *See* first point of Aries.

libration. The means by which 59% of the moon's surface is made visible, over a 30 year period, to the terrestrial observer despite the lunar day being of the same length as the *sidereal month (*see* synchronous rotation). *Physical libration is a real irregularity in the moon's rotation, arising from minor distortions in shape. *Geometrical librations*, including *diurnal librations, are apparent oscillations due to the moon being observed from slightly different directions at different times. A geometrical *libration in longitude* of 7°45′ results from the inconstancy of the moon's orbital velocity, which follows from the *eccentricity of the moon's orbit and *Kepler's second law. This libration reveals part of the eastern or western farside limbs at different times during one month. In addition the 5°9′ inclination of the moon's orbital plane to the ecliptic produces a monthly geometrical *libration in latitude* and makes the moon's polar regions more readily observable.

Lick Observatory. An observatory on Mount Hamilton, California, at an altitude of 1280 metres. Endowed by James Lick, it was given to the University of California on the completion in 1888 of its (then) powerful 91 cm (36 inch) refracting telescope. The chief instrument today is the 3 metre (120 inch) reflecting telescope, the *Shane telescope*, with a comparatively large focal ratio of f/5, which became fully operational in 1959. Its mirror was one of the spare bases for the 200 inch Hale telescope.

life in the universe. *See* exobiology; SETI.

light. *Electromagnetic radiation to which the human eye is sensitive; light is thus also called *visible* or *optical radiation*. The wavelength range varies from person to person but usually lies within the limits of 380 to 750 nanometres. Light therefore forms a very narrow band of the electromagnetic spectrum between the infrared and ultraviolet bands. As the wavelength decreases, the colour of the light changes through the spectral hues of red, orange, yellow, green, blue, to violet. These colours, which may be seen in a rainbow or produced by a prism, form the *visible spectrum*. The eye is most sensitive to greenish-yellow light of wavelength about 500 to 550 nm. The earth's atmosphere is transparent to light so that astronomy at optical wavelengths has not had the problems with detection and observation as studies at other wavelengths. *See* atmospheric windows.

light cone. The representation of a light flash in *spacetime. In three-dimensional space the wavefront of a light flash would be a sphere centred on the emitting point and growing in radius at the speed of light. In spacetime (with one spatial coordinate suppressed) the vertex of a cone represents the time and place at which a flash is emitted and the cone itself describes the history of the propagating flash. More generally a light cone defines the regions of the universe accessible from a given point in space and moment in time, i.e. from a given position in spacetime. This position is called an *event* and is the vertex of the light cone.

light curve. A graph of the brightness of a *variable star or other variable object plotted against time, the brightness usually being expressed in terms of apparent or absolute *magnitude. Different types of variable star can be distinguished by the shape of the curve; the period of the variable is one complete oscillation in brightness. The light curve of an eclipsing binary has two minima: a shallower one (the *secondary minimum*) and a deeper one (the *primary minimum*), which occurs when the brighter star is eclipsed. The minimum will have a flat base if the eclipse is total or annular.

light-gathering power (light grasp). A measure of the ability of an optical telescope to collect light and thus discern fainter objects. It is proportional to the area of the telescope *aperture, i.e. to the square of the diameter of the primary mirror or objective.

light mantle. An avalanche of lunar *highland material, possibly triggered by the formation of *secondary craters from *Tycho, that covers part of the *Taurus-Littrow region and was sampled by the crew of Apollo 17.

light pollution. One of the factors setting a limit on the faintness of stars that may be seen, photographed, or observed or recorded by telescope. Manmade light pollution includes street and domestic lighting. On a moonless night there are still three natural sources, of roughly equal contribution; these are *airglow and other atmospheric effects, the *zodiacal light arising from the scattering of sunlight by solar-system dust, and the background light of our Galaxy. *See also* seeing.

light-time. The time taken for light, radio waves, or other electromagnetic radiation to travel at the *speed of light $(3 \times 10^5$ km/s) between two points. The light-time along the mean distance between sun and earth is 499 seconds. More precisely, the light-time for *unit distance is a constant, τ_A, equal to 499.00478 seconds. Since the light-time between celestial bodies and

earth varies as the earth revolves in its orbit, a correction – the *equation of light* – must be made to calculations involving measurements of time. Light-time produces a delay between transmission and reception of command signals and data between earth and a spacecraft.

light-year (l.y.). A unit of distance equal to the distance travelled through space in one tropical year by light, radio waves, or any other form of electromagnetic radiation. Since all electromagnetic radiation travels in a vacuum at the speed of light, 299 792 km/s, one light-year equals 9.4605×10^{12} km. Distances expressed in light-years give the time that radiation would take to cross that distance. One light-year equals 0.3066 parsecs, 63 240 astronomical units, or a parallax of 3.259 arc seconds.

Analogous but smaller units of distance, such as the *light-month, light-week, light-day*, and *light-second* are also used, often to indicate the size of an object (such as the core of an *active galaxy) whose output is varying: the timescale of the variations imposes an upper limit on the size, the conditions being unlikely to change more quickly than the time it takes for light to travel across the region.

limb. The apparent edge of the sun, moon, or a planet, or any other celestial body with a detectable disc.

limb brightening. The increase in intensity at x-ray and radio wavelengths from the centre to the limb of the sun's disc. It arises at radio wavelengths for the same reason as the *limb darkening at visible wavelengths but has the opposite effect, since the radiation is emitted from the high-temperature upper *chromosphere/inner *corona rather than from the *photosphere. At x-ray wavelengths, where the material is less opaque, intensity is determined by path length rather than radiation temperature.

limb darkening. The decrease in intensity at visible wavelengths from the centre to the limb of the sun's disc. Light from the central portion of the disc is radiated radially towards us whereas light from closer to the limb has to pass more obliquely through a proportionally greater thickness of the solar atmosphere. For a given limit of penetration we can therefore see deeper into the *photosphere at the centre of the disc than at the limb, where our line of sight passes through cooler gases. The limb consequently appears less bright.

Limb darkening has been detected in other stars, principally by the use of three-colour *photometry of the *light curves of *eclipsing binaries (the effect being more pronounced in blue light than in red) and by *speckle interferometry.

A similar effect is observed in planets with deep atmospheres, notably Jupiter and Saturn, and arises from the scattering of sunlight by the atmospheric particles.

limiting magnitude. The faintest apparent *magnitude that may be observed through a telescope and/or recorded on a photographic plate or other device. It depends on the telescope aperture, on the sensitivity of the recording device, on atmospheric conditions, etc. Star catalogues usually list stars to a specific limiting magnitude.

linea. A line, straight or curved, on the surface of a planet or satellite. The word is used in the approved name of such a feature.

linear momentum. *See* momentum.

line broadening. The production of broadened spectral lines by various effects. The natural width of a spectral line is determined by quantum mechanical uncertainty. Other factors can, however, produce extra line width, including *doppler broadening, *pressure broadening, and the *Zeeman effect. Considerations of the *line profile can give information about the physical conditions of celestial objects.

line of apsides. *See* apsides.

line of cusps. The plane passing through the *cusps of a crescent moon or planet. *See also* equator of illumination.

line of inversion. *See* sunspots.

line of nodes. *See* nodes.

line-of-sight velocity. *See* radial velocity.

line profile. The plot of radiation intensity against wavelength or frequency for a spectral line. *See also* line broadening.

line receiver. A radio-astronomy *receiver that is either tuned to receive signals only in one narrow band of frequencies, the centre frequency being adjustable, or that has a broad *bandwidth divided up into many narrow channels of fixed frequency, the outputs of the channels being kept separate from one another. Line receivers are used in *molecular-line radio astronomy where the radio emissions are of narrow bandwidth. They are also used to study the distribution and relative velocity of neutral hydrogen (HI) in our own and in nearby *spiral galaxies (*see* HI region).

line spectrum. A *spectrum consisting of discrete lines (*spectral lines*) resulting from radiation emitted or absorbed at definite wavelengths. Line spectra are produced by atoms or ionized atoms when transitions occur between their *energy levels as a result of emission or absorption of photons. The *Fraunhofer lines of the sun are an example of an absorption line spectrum.

lithosphere. *See* earth.

LMC. *Abbrev. for* Large Magellanic Cloud. *See* Magellanic Clouds.

lobate ridge. A ramp-shaped feature occurring in the lunar *maria. Lobate ridges are the surviving remnants of the margins of thick lava flows. *Compare* wrinkle ridges.

lobate scarp. *See* Mercury.

lobe. *See* antenna.

Local Group. The *cluster of galaxies of which our *Galaxy is a member. It is a comparitively small cluster with 30–40 known members (1984) and some other possible members, most of which are dwarf ellipticals or irregular galaxies (see table). The most massive members are the Galaxy and the *Andromeda galaxy (M31). Maffei 1 (*see* Maffei galaxies), once considered a possible massive member, is now thought to lie outside the Local Group. The other major members are the *Triangulum spiral (M33) and the Large *Magellanic Cloud. The group is irregular in shape with two notable subclusters: one around our Galaxy and the Magellanic Clouds, the other around M31 and M33.

local mean time. *See* local time.

local oscillator. *See* receiver.

local sidereal time (LST). The *sidereal time at a particular location. The *local mean sideral time (LMST)* is equal to the *Greenwich mean sideral time plus four minutes for each degree of longitude that the location is east of Greenwich, four minutes being subtracted for each degree west. The *local apparent sideral time (LAST)* is obtained by adding the equation of the equinoxes (*see* sideral time) to the LMST. The *local hour angle (LHA)* is the LAST minus the apparent right ascension. The local sidereal time is

Main members of Local Group of galaxies

galaxy	position (1950) RA (h, m) dec (o)		type	approx. distance (kpc)	absolute magnitude	approx. mass (M☉)
Galaxy	17 42	−29.0	Sb or Sc		−20 (?)	2×10^{11}
LMC	05 24	−69.8	Irr I	50	−18.7	10^{10}
SMC	00 51	−73.2	Irr I	60	−16.7	2×10^{9}
M31 (And. gal)	00 40	+41.0	Sb	660	−21.1	3×10^{11}
M33 (Tri. gal)	01 31	+30.4	Sc	720	−18.8	10^{10}
M32 (NGC 221)	00 40	+40.6	E2	660	−16.3	3×10^{9}
NGC 205	00 38	+41.4	E5	660	−16.3	8×10^{9}
NGC 147	00 30	+48.2	Ep	660	−14.8	10^{9}
NGC 185	00 36	+48.1	Ep	660	−15.2	10^{9}
IC 1613	01 02	+01.8	Irr I	700	−14.8	2×10^{8}
NGC 6822	19 42	−14.9	Irr I	500	−15.6	3×10^{8}
Sculptor system	00 57	−34.0	E	85	−12	3×10^{6}
Fornax system	02 37	−34.7	E	190	−13	2×10^{7}
Leo I system	10 06	+12.6	E4	220	−11	4×10^{6}
Leo II system	11 11	+22.4	E1	220	−9.5	10^{6}
Draco system	17 19	+58.0	E	67	−8.5	10^{5}
UMi system	15 08	+67.3	E	67	−9	10^{5}

thus the LHA of a *catalogue equinox: a celestial object crosses the local meridian when the LST is equal to the object's *right ascension.

local standard of rest (LSR). The frame of reference, centred on the sun, in which the space velocities (*see* radial velocity) of all the stars average to zero. The *peculiar motions of these stars with respect to the LSR would also average to zero if the sun were at rest. This LSR is called the *kinematic LSR*. The sun however is moving, relative to the LSR, at about 20 km s^{-1} towards the solar *apex. The stars reflect this solar motion in addition to their peculiar motions in their *proper motions. A different LSR is sometimes useful. It is the reference frame, centred on the sun, that moves in a circular orbit around the galactic centre so that all neighbouring stars (i.e. those in circular orbits) are essentially at rest. This is called the *dynamic LSR*. The rotation velocity of the dynamic LSR is about 250 km s^{-1} (*but see* galactic rotation). These two definitions of the LSR do not exactly agree; the kinematic LSR lags behind the dynamic LSR.

Local Supercluster. A huge flattened cloud of *galaxies and *clusters of galaxies that is centred on or near the *Virgo cluster and contains the *Local Group (including our Galaxy) as an outlying member. Its radius is believed to be about 30 megaparsecs. Although its reality was for long in doubt, the discovery of other *superclusters has convinced most astronomers that the Local Supercluster is a true physical system.

Local System. *See* Gould Belt.

Long Duration Exposure Facility. A NASA satellite that was deployed from an orbiting space shuttle in Apr. 1984 and will remain in orbit until Feb. 1985. It carries 57 experiments on the effects of prolonged exposure to conditions in space, or that require long-term exposure. The largest of the investigations is the Ultra Heavy Cosmic Ray experiment designed jointly by the ESA and the Dublin Institute for Advanced Studies for the study of the rare very heavy components of *cosmic rays and the origin and propagation of cosmic rays in the Galaxy and the origin of different elements in stars.

long-focus photographic astrometry. A branch of photographic *astrometry in which (usually) refracting telescopes of long focal lengths of eight metres or more and focal ratios of f/15 to f/20 are used to produce sharp photographs of both the star under consideration plus three or more distant reference stars. Accuracies of about 0.01 arc seconds can be reached, distances being determined from *annual parallax. The technique is now also employed in the determination of orbital motion and masses of binary stars and of *secular parallax.

longitude. 1. *Short for* celestial longitude. *See* ecliptic coordinate system.
   2. *Short for* galactic longitude. *See* galactic coordinate system.

longitude of the ascending node. *See* orbital elements.

long-period variables. A group of variable stars that are principally *Mira stars but sometimes the red *semiregular stars, type SRa, are included in the group. The longest periods (up to 5 years) are found in the *OH/IR stars, normally observable only at infrared wavelengths.

loop prominence. *See* prominences.

Lorentz – FitzGerald contraction. The tiny contraction of a moving body in the direction of motion, put forward by H. A. Lorentz (1895) and independently by G.F. FitzGerald (1893) as an explanation for the result of the Michelson-Morley experiment (*see* ether). The contraction was later shown to be an effect of the special theory of *relativity (1905).

Lorentz transformation. A set of equations used in the special theory of *relativity to transform the coordinates of an event (x, y, z, t) measured in one inertial frame of reference to the coordinates of the same event (x', y', z', t') measured in another frame moving relative to the first at constant velocity v:

$$x = (x' + vt')/\beta$$
$$y = y'$$
$$z = z'$$
$$t = (t' + x'v/c^2)/\beta$$

β is the factor $\sqrt{(1 - v^2/c^2)}$ and c is the speed of light. When v is very much less than c, these equations reduce to those used in classical mechanics.

Lowell Observatory. An observatory at Flagstaff, Arizona, at an altitude of 2210 metres, set up by Percival Lowell in 1895.

lower culmination. *See* culmination.

LSR. *Abbrev. for* local standard of rest.

LST. *Abbrev. for* local sidereal time.

luminosity. Symbol: L. The intrinsic or absolute brightness of a star or other celestial body, equal to the total energy radiated per second from the body, i.e. the total outflow of power. Luminosity is related to the body's surface area and *effective temperature, T_{eff}, by a form of Stefan's law:

$$L = 4\pi R^2 \sigma T_{eff}{}^4$$

where σ is Stefan's constant and R is the radius. Thus two stars with similar T_{eff} (i.e. of the same *spectral type) but greatly different luminosities must differ in size: they belong to different luminosity classes within that spectral type, as determined from

their spectra. In the luminosity class of main-sequence stars, luminosity decreases as temperature decreases; the luminosity of *giant stars however increases with decreasing temperature: *red giants are much brighter than yellow giants; the luminosity of *supergiants drops and then rises with decreasing temperature. The luminosity of stars is in theory dependent on mass if their chemical composition is similar. It has been found that with the notable exception of highly evolved stars the *mass-luminosity relation is obeyed approximately.

The luminosity of a star or other body can be expressed as a multiple or fraction of the sun's luminosity, $L_\odot$, which is equal to 3.9×10^{26} watts. The ratio $L/L_\odot$ is given by

$$2.5 \log_{10}(L/L_\odot) = M_\odot - M$$

where $M_\odot$ and M are the absolute bolometric *magnitudes of sun and star; $M_\odot$ is equal to 4.75. There is a very great range in stellar luminosity from about a million times to less than one ten thousandth that of the sun's luminosity.

luminosity classes. *See* spectral types.

luminosity function. The relative numbers of objects of different luminosities in a standard volume of space (usually a cubic parsec or cubic megaparsec). The luminosity may be in the optical band, the radio band, or at any other defined waveband. Observationally the luminosity function is determined as a histogram; theoreticians fit this with an algebraic expression.

In optical astronomy the luminosity function is given the symbol $\Phi(M)$ and is the number of stars per cubic parsec or the number of galaxies per cubic megaparsec with absolute magnitudes $M \pm \frac{1}{2}$. The luminosity function for stars within 10 parsecs of the sun shows a peak at an absolute magnitude of about $+15$: there is thus a predominance of intrinsically faint stars in the solar neighbourhood. The luminosity function for any given magnitude interval is not the same everywhere in the Galaxy: for instance, the luminosity function of a *globular cluster is different from that of an *open cluster. Stars of all *spectral types contribute to the value $\Phi(M)$. The specific luminosity function Φ (M,S) considers one spectral type at a time, each showing a peak distribution at a different mean magnitude.

The luminosity function of *radio sources is the number of sources per cubic megaparsec with a given radio emission. This radio power is measured at a standard frequency in watts per unit solid angle. Similar specific definitions apply to other wavebands.

luminosity-volume test (V/V$_m$ test). A method of testing for evolutionary effects in a sample of extragalactic sources of known *redshift. All sources should be known over a particular region of the sky down to a certain flux density level. The volume, V, within the redshift of each object, together with the volume V_m within the maximum redshift out to which that object could have been and yet still remain in the sample, are computed from a cosmological model. If no evolution is present and the correct cosmological model is employed, then V ought to be uniformly distributed between 0 and V_m. The mean of V/V_m is then $\frac{1}{2}$. If the mean differs significantly from $\frac{1}{2}$ for all reasonable cosmological models, then evolution must be occurring; this is the case for *quasars.

Luna probes (Lunik probes). A series of Soviet lunar probes (see table) that included the first craft to reach the vicinity of the moon (1), the first to photograph the farside (3), the first crash lander (2), soft lander (9), orbiter (10), unmanned sample-return mission (16), and unmanned roving vehicle (17). *See also* Lunokhod;

Zond probes; Lunar Orbiter probes; Surveyor; Ranger.

lunar calendar. *See* lunar year; calendar.

lunar cycle. *See* Metonic cycle.

lunar day. The rotation period (27.322 earth days) of the moon, equal in length to the *sidereal month. The moon is thus in *synchronous rotation with the earth. *See also* libration.

lunar eclipse. *See* eclipse.

lunar grid. A postulated but questionable arrangement of preferentially aligned lineations that, if real, would imply global tectonism on the moon. The apparent pattern, however, may be the result of lighting artifacts and of *crater chains and *rilles associated with the major impact *basins, particularly *Imbrium.

lunar module (LM). *See* Apollo.

lunar month. *See* synodic month.

Lunar Orbiter probes. A series of five US space probes used for photographic reconnaissance of the Apollo landing sites (Lunar Orbiters 1, 2, and 3) and global mapping (4 and 5) of the moon from equatorial and polar orbits respectively. The probes were launched between Aug. 1966 and Aug. 1967. Each spacecraft was equipped with medium- and high-resolution cameras, the films from which were scanned in strips for transmission back to earth. Satellite tracking resulted in the discovery of *mascons. The spacecraft were crashed onto the moon on mission completion. *See also* Ranger; Surveyor; Luna probes; Zond probes.

lunar parallax. *See* diurnal parallax.

lunar rocks. *See* moonrocks.

Luna probes

Spacecraft	Launch date	Comments
Lunik 1	1959, Jan. 2	Missed moon by 6000 km
Lunik 2	1959, Sept. 12	Struck moon in Palus Putredinis
Lunik 3	1959, Oct. 4	First farside photographs
Luna 4	1963, Apr. 2	Missed moon by 8500 km
Luna 5	1965, May 9	Struck moon
Luna 6	1965, June 8	Missed moon by 160 000 km
Luna 7	1965, Oct. 4	Struck moon
Luna 8	1965, Dec. 3	Struck moon
Luna 9	1966, Jan. 31	Soft landed in Oc. Procellarum
Luna 10	1966, Mar. 31	Performed experiments in lunar orbit
Luna 11	1966, Aug. 24	Performed experiments in lunar orbit
Luna 12	1966, Oct. 22	Returned photographs from lunar orbit
Luna 13	1966, Dec. 21	Soft landed in Oc. Procellarum
Luna 14	1968, Apr. 7	Performed experiments in lunar orbit
Luna 15	1969, July 13	Struck moon in Mare Crisium
Luna 16	1970, Sept. 12	Returned 101 g soil from Mare Fecunditatis
Luna 17	1970, Nov. 17	Landed Lunokhod 1 in Mare Imbrium
Luna 18	1971, Sept. 2	Struck moon
Luna 19	1971, Sept. 28	Performed experiments in lunar orbit
Luna 20	1972, Feb. 14	Returned 30 g soil, Crisium highlands
Luna 21	1973, Jan. 8	Landed Lunokhod 2 in Mare Serenitatis
Luna 22	1974, May 29	Performed experiments in lunar orbit
Luna 23	1974, Oct. 28	Soft-landed but drill failure
Luna 24	1976, Aug. 15	Returned 170 g soil from Mare Crisium

Lunar Roving Vehicle (Lunar Rover; Rover; LRV). The battery-driven car used to extend the sampling capabilities of *Apollos 15, 16, and 17. Transported to the moon in a compact form in the Lunar Module descent stage, the 213 kg Rover was equipped with a remote-controlled television camera and high-gain antenna for direct radio communication with earth. It was left behind on the moon after use. See also Lunokhod; Modular Equipment Transporter.

lunar year. A year of 12 *synodic months, each of 29.5306 days, i.e. a year of 354.3672 days. A lunar calendar is based solely on the moon's motion; it has a year of 354 days with a *leap year of 355 days and is composed of 12 months of 29 or 30 days. A luni-solar calendar is a lunar calen-dar that is brought into step with the solar or seasonal calendar by the intercalation (addition) of a 13th leap month.

lunation. See synodic month.

Lunik probes. See Luna probes.

luni-solar calendar. See lunar year.

luni-solar precession. See precession.

Lunokhod. An eight-wheeled Soviet unmanned lunar roving vehicle, soft-landed on the moon by *Lunas 17 and 21. Lunokhod 1 was equipped with a laser reflector for lunar-ranging experiments; it was active in Mare Imbrium for 10 months during which time it completed a 10 km traverse, performing photographic tasks, magnetic field measurements, and chemical (x-ray) and cosmic-ray analyses. Lunokhod 2 travelled 37 km in four months in the vicinity of Le Monnier crater on the borders of Mare Serenitatis. *See also* Lunar Roving Vehicle.

Lupus (Wolf). A constellation in the southern hemisphere near Centaurus, lying partly in the Milky Way, with several stars of 3rd magnitude. There are many naked-eye double stars. Abbrev.: Lup; genitive form: Lupi; approx. position: RA 15.3h, dec −45°; area: 334 sq deg.

l.y. *Abbrev. for* light-year.

Lyman series. *See* hydrogen spectrum.

Lynx. A constellation in the northern hemisphere near Ursa Major, the brightest stars being one of 3rd magnitude and several of 4th magnitude. The area contains many faint double stars. Abbrev.: Lyn; genitive form: Lyncis; approx. position: RA 8h, dec +47°; area: 545 sq deg.

Lyot filter (Lyot-Öhman filter). *See* birefringent filter.

Lyra (Lyre). A constellation in the northern hemisphere near Cygnus, lying partly in the Milky Way, the brightest star being the blue zero-magnitude *Vega. It contains the prototype of the *RR Lyrae stars, the *eclipsing binary Beta Lyrae (*see* W Serpentis star), and the naked-eye pair Epsilon Lyrae, both of 4th magnitude and both double. It also contains the planetary *Ring nebula and a small globular cluster M56. Abbrev.: Lyr; genitive form: Lyrae; approx. position: RA 18.5h, dec +40°; area: 286 sq deg.

Lyrids (April Lyrids). A minor *meteor shower, radiant: RA 272°, dec +32°, that maximizes on April 21. In the past it was much more active, the last great Lyrid shower occurring in

1803. Observations of Lyrids have been traced back 2500 years, Chinese observers describing a remarkable display in 15 BC. The associated *meteor stream has the same orbit as comet 1861 I. *See also* June Lyrids.

Lysithea. A small satellite of Jupiter, discovered in 1938. *See* Jupiter's satellites; Table 2, backmatter.

M

Mach's principle. The idea put forward by Ernst Mach in the 1870s that the inertial properties of a particular piece of matter were in some way attributable to the influence of all the other matter in the universe. The *inertia of a body is its reluctance to change its state of rest or uniform motion in a straight line. According to Mach, inertia can be considered to arise from some (unspecified) interaction between the body and the background presence of distant stars and galaxies; a body in total isolation would not therefore possess inertia.

macula. A dark spot on the surface of a planet or satellite. The word is used in the approved name of such a feature.

Maffei galaxies. Two galaxies that lie in Cassiopeia close to the galactic equator and are heavily obscured by interstellar dust. They were discovered as infrared objects by P. Maffei in 1968. *Maffei 1*, the brighter of the two, is a large elliptical or S0 galaxy with a total mass of about 10^{12} solar masses. *Maffei 2* appears to be a medium-sized spiral galaxy. Although Maffei 1 was originally thought to be a member of the *Local Group, both galaxies are now believed to lie 2 to 3 Mpc away, in the Ursa Major-Camelopardalis group of galaxies.

Magellanic Clouds. Two comparatively small galaxies that are close neighbours of our own Galaxy. Both are naked-eye objects but, being close to the south celestial pole, they are only visible from the southern hemisphere. They were first recorded in 1519 by Ferdinand Magellan. The *Large Magellanic Cloud (LMC)* has a total mass of 10^{10} solar masses and a diameter of about 12 000 parsecs; it lies in the constellation Dorado at a distance of about 50 000 parsecs. The *Small Magellanic Cloud (SMC)* has a total mass of approximately 2×10^9 solar masses and a diameter of about 6000 parsecs; it lies in the constellation Toucan and is about 60 000 parsecs away. Observations made in 1983 have suggested that the SMC may consist of two galaxies superimposed along the line of sight: the *Mini Magellanic Cloud* would lie some 10 000 parsecs beyond the main galaxy.

Although the LMC and SMC are generally classified as irregular (Irr I) systems, there is some evidence of barred spiral structure (*see* Hubble classification), particularly in the LMC. Both are rich in *population I objects and contain a much greater proportion of gas than our own Galaxy. They are enveloped in a common cloud of cool neutral hydrogen, which extends into a narrow streamer. This *Magellanic Stream* stretches over 180° of the sky, extending towards the Galaxy. The Stream contains almost 10^9 solar masses of gas, probably ripped out of the Magellanic Clouds when they passed near the Galaxy about a thousand million years ago and now strewn along the orbit of the Clouds. If the Galaxy has a massive *dark halo, the Magellanic Clouds are gravitationally bound as satellites of the Galaxy and they have probably made several close approaches; otherwise, they have been involved in just one encounter, and the Magellanic Stream marks a hyperbolic *orbit.

Magellanic Stream. *See* Magellanic Clouds.

magnetic field. The region surrounding a magnet, a conductor carrying an electric current, a stream of charged particles, etc., in which such a body or system exerts a detectable force. This force will be experienced by another magnetic substance or by a moving charged particle, such as an electron. The *magnetic flux density is a measure of the strength of the field, usually quoted in teslas or sometimes in gauss. *See also* geomagnetism; sunspot cycle; galactic magnetic field.

magnetic flux density (magnetic induction). Symbol: *B*. A vector quantity that is a measure of the strength of a magnetic field in a particular direction at a particular point. It may be given in terms of the force, *F*, on a charge, *q*, moving in the magnetic field:

$$B = F/qv \sin\theta$$

where *v* is the velocity of the charge, which is moving at an angle θ to the direction of the field. Magnetic flux density is usually measured in teslas. The *magnetic flux*, symbol: Φ, is the surface integral of *B*, i.e.

$$\Phi = \int B.dA = \int B \cos\alpha \, dA$$

where α is the angle between *B* and a line perpendicular to the surface. The magnetic flux density is thus the magnetic flux passing perpendicularly through unit area. Magnetic flux is measured in webers.

magnetic induction. *See* magnetic flux density.

magnetic stars. Stars that have detectable and often very large magnetic fields, up to 10^{-3} tesla. Magnetic stars are found in spectral classes B to F, but most are peculiar A stars (*Ap stars). Many of these stars show either periodic or irregular variations of the field together with a reversal of the polarity; others show irregular field variations but constant polarity. Magnetic variability is usually accompanied by very small changes in brightness. Magnetic stars are studied by measurements of the Zeeman splitting (see Zeeman effect) of spectral lines caused by a magnetic field; these give the average magnetic field over the entire disc. Periodic variations in the magnetic field could result from the magnetic axis of the star being tilted with respect to the rotational axis so that areas of different magnetic field strength are presented to the observer as the star rotates: this is the *oblique rotator theory*. Evidence for chromospheric activity and starspots on late-type stars and RS CVn binaries indicates that they have weaker magnetic fields, similar to the sun's. A few *white dwarfs have fields up to 1 tesla while some *neutron stars detected as pulsars have fields of typically 10^4 tesla. *See also* spectrum variables.

magnetic storms. *See* geomagnetic storms.

magnetobremsstrahlung. *See* synchrotron emission.

magnetograph. An instrument used to map the distribution, strength, and polarity (direction) of magnetic fields over the sun's disc. It is usually an automatic scanning system employing instruments to detect and measure the Zeeman splitting of a selected spectral line (see Zeeman effect). The resulting diagram is called a *magnetogram*.

magnetohydrodynamics. The study of the behaviour of electrically conducting fluids, i.e. an ionized gas, a plasma, or some other collection of charged particles, in a magnetic field. The collective motion of the particles gives rise to an electric field that interacts with the magnetic field and causes the fluid motion to alter. This coupling between hydrodynamic forces and magnetic forces tends to make the fluid and the magnetic field lines move together.

magnetometer. Any of a variety of instruments used to measure the strength and direction of a magnetic field.

magnetopause. *See* magnetosphere.

magnetosheath. *See* magnetosphere.

magnetosphere. A region surrounding a planet, and in which ionized particles are controlled by the magnetic field of the planet rather than by the sun's magnetic field, which is carried by the solar wind (*see* interplanetary medium). The region is bounded by the *magnetopause* and includes any radiation belts of the planet, such as the *Van Allen belts of the earth. The earth's magnetosphere extends to 60 000 km on the sunward side of the planet but is drawn out to many times this distance on the side away from the sun by the solar wind (see illustration). Above the magnetopause is a layer of turbulent magnetic field, the *magnetosheath,* enclosed by a shock wave where the smooth flow of solar-wind particles past the planet is first interrupted. Mercury, Jupiter, Saturn, and Uranus are also known to possess magnetospheres.

magnifying power (magnification). The angle subtended by the image of an object seen through a telescope, divided by the angle subtended by the same object seen without telescopic aid. Where the object is too small or too distant to be resolved, the magnifying power may be calculated from the ratio of the focal lengths of the *objective (or *primary mirror) and *eyepiece, or from the *aperture divided by the diameter of the *exit pupil.

magnitude. A measure of the brightness of stars and other celestial objects. The brighter the object the

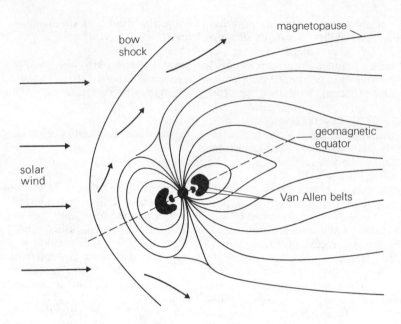

Earth's magnetosphere, shaped by the solar wind

lower its assigned magnitude. Expressions of magnitude are used primarily in the visible, near-infrared, and near-ultraviolet regions of the spectrum.

The *apparent magnitude*, symbol: m, is a measure of the brightness of a star as observed from earth. Its value depends on the star's *luminosity (i.e. its intrinsic brightness), its distance, and the amount of light absorption by interstellar matter between the star and earth. In ancient times the visible stars were ranked in six classes of apparent magnitude: the brightest stars were of first magnitude and those just visible to the naked eye were of sixth magnitude. This system became inadequate as fainter stars were discovered with the telescope and instruments became available for measuring apparent brightness. In the 1850s it was proposed that the physiological response of the eye to a physical stimulus was proportional to the logarithm of that stimulus (Weber-Fechner law). A difference in apparent magnitude of two stars is thus proportional to the difference in the logarithms of their brightness, i.e. to the logarithm of the ratio of their brightness.

In order to make the magnitude scale precise the English astronomer N.R. Pogson proposed, in 1856, that a difference of five magnitudes should correspond exactly to a brightness ratio of 100 to 1. (W. Herschel had shown this to be approximately true.) Hence two stars that differ by one magnitude have a brightness ratio of $\sqrt[5]{(100)}:1$, i.e. a ratio – known as the *Pogson ratio* – of 2.512. A star two

Pogson scale of magnitudes

m_1-m_2	I_2/I_1	m_1-m_2	I_2/I_1
0.2	1.202	3.5	25.12
0.4	1.445	4.0	39.81
0.6	1.738	4.5	63.1
0.8	2.089	5.0	100
1.0	2.512	6.0	251
1.5	3.981	7.0	631
2.0	6.310	8.0	1585
2.5	10.0	9.0	3981
3.0	15.85	10.0	10 000

magnitudes less than another is $(2.512)^2$, i.e. 6.3 times brighter, and so on. In general the apparent magnitudes m_1 and m_2 of two stars with apparent brightness I_1 and I_2 are related by:

$$m_1 - m_2 = 2.5 \log_{10}(I_2/I_1)$$

The Pogson scale, based on the Pogson ratio, is now the universally adopted scale of magnitude (see table). The zero of the scale was established by assigning magnitudes to a group of standard stars near the north celestial pole, known as the *North Polar Sequence*, or more recently by photoelectric measurements. The class of magnitude-one stars was found to contain too great a range of brightness and zero and negative magnitudes were consequently introduced: the higher the negative number, the greater the brightness. The scale also had to be extended in the positive direction as fainter objects were discovered: the greatest apparent magnitude that can currently be measured is about $+25$.

Originally apparent magnitude was measured by eye – *visual magnitude*, m_{vis} – usually in conjunction with an instrument by which brightnesses could be compared. It is now measured much more accurately by photometric techniques but previous to that photographic methods were used. *Photographic magnitudes*, m_{pg}, are determined from the optical density of images on ordinary film, i.e. film that has a maximum response to blue light. *Photovisual magnitudes*, m_{pv}, are measured using film that has been sensitized to light – yellowish green in colour – to which the human eye is most sensitive. In the early *International Colour System* these two magnitudes were measured with films having a maximum response to a wavelength of 425 nanometres (m_{pg}) and 570 nm (m_{pv}), the magnitudes being equal for A0 stars.

A suitable combination of *photocell and filter can select light or other radiation of a desired wavelength band and measure its intensity. These *photoelectric magnitudes* can be either *monochromatic* if a very narrow band is selected or *heterochromatic* for a broader band. The *UBV system* of stellar magnitudes is based on photoelectric photometry and has been widely adopted as the successor of the International System. The photoelectric magnitudes, denoted U, B, and V are measured at three broad bands: U (ultraviolet radiation) centred on a wavelength of 365 nm, B (blue) centred on 440 nm, and V (visual, i.e. yellowish green) centred on 550 nm. These magnitudes can also be written m_U, m_B, and m_V. The zero point of the UBV system is defined in terms of standard stars having a carefully studied and agreed magnitude. *See also* uvby system.

The UBV system has been extended by the use of magnitudes at red and infrared wavelengths. The photometric designations are R (700 nm, i.e. 0.7 μm), I (0.9 μm), J (1.25 μm), H (1.6 μm), K (2.20 μm), L (3.40 μm), M (5.0 μm), N (10.0 μm), Q (21 μm). The designations $J-Q$ relate to the infrared *atmospheric windows.

Apparent magnitude is a measure of the radiation in a particular wavelength band, say of blue light, received from the celestial body. Apparent *bolometric magnitude*, m_{bol}, is a measure of the total radiation received from the body. The *bolometric correction (BC)* is the difference between the bolometric and visual (V) magnitudes, apparent or absolute, and is always negative.

Apparent magnitude gives no indication of a body's *luminosity: a very distant very luminous star may have a similar apparent magnitude as a closer but fainter star. Luminosity is defined in terms of *absolute magnitude*, M, which is the apparent magnitude of a body if it were at a standard distance of 10 parsecs. It can be shown that the two magnitudes of a body are related to distance d (in parsecs) or its *annual parallax, π (in arc seconds):

$$M = m + 5 - 5 \log d - A$$
$$M = m + 5 + 5 \log \pi - A$$

A is the *interstellar extinction. As with apparent magnitude there are values of absolute photoelectric magnitudes: M_U, M_B, M_V, etc.; of bolometric magnitude: M_{bol}; and of photographic magnitudes: M_{pg} and M_{pv}. Knowledge of a body's absolute bolometric magnitude enables its *luminosity to be found. The flux density in *jansky of a body can be determined from a value of absolute magnitude at one of the photometric designations U–Q.

magnitude at opposition. A measure of the brightness of a planet or minor planet when at *opposition; it corresponds, normally, to the maximum brightness attained by the object during the *apparition since it is then viewed at its fullest phase and is near its closest point to the earth.

main lobe (main beam). See antenna.

main sequence. The principal sequence of stars on the *Hertzsprung-Russell diagram, running diagonally from upper left (high temperature, high luminosity) to lower right (low temperature, low luminosity) and containing about 90% of all known stars. A star spends most of its life on the main sequence. A newly formed star appears on the main sequence when it first achieves a stable state whereby its core temperature is sufficient for nuclear reactions to begin. It is then at age zero. The positions of the age-zero stars on the H-R diagram are specified by reference to the *zero-age main sequence*. As hydrogen is converted to helium, changes in chemical composition and stellar structure cause the star's position to shift slightly to the right from its zero-age position.

A star's position on the main sequence depends primarily on its mass,

the more massive and thus more luminous stars occurring higher up the sequence. This gives a well-defined *mass-luminosity relationship for main-sequence stars. The star's lifetime there also depends on its mass, the more massive stars having much shorter lifetimes: the lifetime is approximately

$$10^{10}(M/M_\odot)^{-2} \text{ years}$$

where M and $M_\odot$ are the stellar and solar mass. After the star has consumed most of the helium in its core (see stellar evolution) it evolves away from the main sequence: its radius and luminosity increase and it eventually becomes a *giant.

main-sequence fitting. A technique for measuring the distance modulus and hence distance for a star *cluster, especially an *open cluster: the main-sequence portion of a graph of colour index versus apparent magnitude for the cluster is superimposed on the main sequence of a *Hertzsprung-Russell diagram, the colour indices of the two curves being made to overlap. The difference between cluster magnitudes and the superimposed absolute magnitudes of the H-R diagram is approximately constant and equal to the distance modulus.

main-sequence star. A star lying on the *main sequence.

major axis. See semimajor axis.

major planets. The planets Mercury, Venus, Earth, Mars, Jupiter, Saturn, Uranus, Neptune, and Pluto, which have diameters significantly larger than the 1003 km diameter of Ceres, the largest *minor planet. See Table 1, backmatter.

Maksutov telescope (meniscus telescope). A *catadioptric telescope named after its Soviet inventor, who published its design in 1944. It differs from the *Schmidt telescope in that the *correcting plate is a deeply

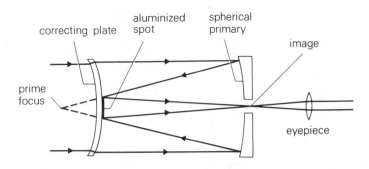

Ray path in Maksutov telescope

curved meniscus lens. Since the primary mirror is also spherical all three optical surfaces are simple to make. Its commonest form (see illustration) is a Cassegrain adaption in which the image is reflected by an aluminized spot on the meniscus lens back through a hole in the primary mirror. It performs well as a prime-focus camera. It can also be made in Newtonian and Gregorian configurations. Although having exceptional performance and compactness, the thick correcting plate limits it to relatively small sizes both because of its weight and the very thick blanks needed to make larger diameters.

manganese stars. *See* Ap stars.

mantle *See* earth

mare (plural: **maria**). A large relatively smooth dark area on the surface of a planet or satellite. The word is used in the approved name of such a surface feature.

maria, lunar (ring plains). Expanses of iron-rich basaltic lavas of low *albedo (i.e. comparatively dark) that were erupted onto the lunar surface between 3900 and 3000 million years ago. The maria are largely confined to the *nearside of the moon where the crust is thinnest. The youngest *basins were flooded to produce the *circular maria* with their associated

*mascons. The *irregular maria* are shallower and fill the older less well-defined basins. Chemical variations between the maria are revealed in *reflectance spectra and by *colour-difference photography. Age differences are apparent from their crater densities, which are invariably lower than in the older *highlands. Maria are mainly named after mental or meteorological states; the smaller ones are described as lakes, bays, and marshes (see table). *See also* moonrocks.

maria, Martian. *See* Mars.

Mariner 2. A 202 kg US probe that made the first successful planetary fly-by, 34 830 km from Venus, on Dec. 14 1962. Its flight confirmed the existence of the *solar wind, measured Venus' temperature, but did not detect a Venusian magnetic field.

Mariner 4. A 261 kg US spacecraft that was the first successful probe to Mars. It passed the planet at a distance of 9844 km on July 14 1965, returning 22 photographs revealing an arid cratered surface.

Mariner 5. A 245 kg US spacecraft that passed Venus at a distance of 3990 km on Oct. 19 1967. It measured Venus' temperature and atmospheric profile, accurately determined its mass and diameter, but failed to detect any magnetic field.

Lunar maria in approximate order of size

Oceanus Procellarum	Ocean of Storms	Mare Moscoviense	Sea of Moscow
Mare Imbrium	Sea of Rains	Mare Ingenii	Sea of Ingenuity
Mare Serenitatis	Sea of Serenity	Mare Anguis	Serpent Sea
Mare Fecunditatis	Sea of Fertility	Mare Vaporum	Sea of Vapours
Mare Tranquillitatis	Sea of Tranquillity	Sinus Medii	Central Bay
Mare Crisium	Sea of Crises	Sinus Aestuum	Bay of Heats
Mare Humorum	Sea of Moisture	Sinus Roris	Bay of Dews
Mare Nectaris	Sea of Nectar	Sinus Amoris	Bay of Love
Mare Frigoris	Sea of Cold	Sinus Iridum	Bay of Rainbows
Mare Orientale	Eastern Sea	Palus Putredinis	Marsh of Decay
Mare Australe	Southern Sea	Palus Somnii	Marsh of Sleep
Mare Cognitum	Sea of Knowledge	Palus Epidemiarum	Marsh of Epidemics
Mare Nubium	Sea of Clouds	Palus Nebularum	Marsh of Mists
Mare Marginis	Border Sea		
Mare Smythii	Smyth's Sea	Lacus Mortis	Lake of Death
Mare Spumans	Foaming Sea	Lacus Somniorum	Lake of Dreams
Mare Undarum	Sea of Waves	Lacus Veris	Lake of Spring
Mare Humboldtianum	Humboldt's Sea	Lacus Autumni	Lake of Autumn

Mariner 6, 7. Two identical 413 kg US probes that passed Mars at distances of 3412 km and 3424 km on July 31 and Aug. 5, 1969, respectively. Mariner 6 photographed part of Mars' equatorial region while Mariner 7 examined the southern hemisphere and southern polar cap. A total of 201 photographs were returned, while other experiments measured surface temperatures, atmospheric pressure and composition, and Mars' diameter.

Mariner 9. A 1031 kg US probe that became the first artificial satellite of another planet when it entered a 12-hour orbit of Mars on Nov. 13 1971. This orbit, ranging in altitude from 1650 km to 17 100 km approximately, allowed Mariner's two television cameras to return 7329 photographs before contact was terminated on Oct. 27 1972. The entire Martian surface was mapped and close-up views were obtained of Mars' two small natural satellites, *Phobos and *Deimos. Other equipment included infrared and ultraviolet spectrometers, used for atmospheric studies, and an infrared radiometer to measure surface temperatures.

The photographic reconnaissance was jeopardized by a violent planet-wide dust storm that developed in late September 1971. With the exception of four hazy spots, later found to be the summits of the principal Martian volcanoes of the *Tharsis Ridge, all surface features were obscured by dust when Mariner 9 arrived. As the dust settled during the following weeks, photographs revealed for the first time the extent of Martian volcanism and the existence of canyons, such as the *Valles Marineris, and channels, which suggest that water once flowed on Mars. The changing appearance of the *polar caps with the Martian seasons was also followed. See Mars; Mars, surface features.

Mariner 10. A US spacecraft that executed the first dual-planet mission by making close approaches to Venus and, three times, to Mercury. Weighing 503 kg, it carried two television cameras that returned the first close-up views of Venus and Mercury. Other instruments included an infrared radiometer to measure Mercury's surface temperature; an ultraviolet spectrometer to detect atmospheric airglow and another to measure atmospheric absorption of solar ultraviolet radiation; and two magnetometers to monitor magnetic field variations throughout the flight.

During the Venus fly-by at a distance of 5760 km on Feb. 5 1974, Mariner 10 returned 3500 photographs – those taken in ultraviolet radiation showing the planet's cloud cover and atmospheric circulation in unprecedented detail. Using the gravitational pull of Venus, the probe was placed on course for a Mercury fly-by on Mar. 29 1974. After course corrections two more visits to Mercury took place on Sept. 21 1974 and Mar. 19 1975, each after two solar orbits by Mercury and one by Mariner 10. The three Mercury encounters, at minimum distances of 740 km, 48 000 km, and 330 km respectively, provided more than 10 000 photographs of Mercury and led to the discovery of the planet's magnetic field. *See also* Mercury; Venus.

Mariner probes. A series of US planetary probes. Mariners 2 and 5 made close-up observations of Venus while *Mariner 10 became the first two-planet probe, visiting Venus and Mercury. *Mariners 4, 6, 7, and 9 were successful Mars probes, the latter becoming the first man-made orbiter of another planet. Mariners 11 and 12 were renamed *Voyagers 1 and 2. Mariners 1 (to Venus) and 8 (to Mars) suffered launch failure while Mariner 3 failed to achieve its intended trajectory to Mars.

Markab (α Peg). A white giant that is the brightest star in the constellation Pegasus and, with Algenib, lies on the S side of the Great Square of Pegasus. m_v: 2.49; M_v: 0.0; spectral type: B9.5 III; distance: 32 pc.

Markarian galaxies. A number of galaxies that were discovered by the Soviet astronomer B.E. Markarian in the 1970s to have a strong ultraviolet continuum. The ultraviolet sources are either concentrated in a bright nucleus or are distributed throughout a more diffuse object. The former group consists of *Seyfert galaxies, *N galaxies, and *quasars; the latter group are galaxies where star formation has recently taken place and are similar to *Haro galaxies.

Mars. The fourth *planet from the sun and the nearest superior planet to the earth. It orbits the sun every 687 days at a distance that varies between 1.38 and 1.67 AU. Its reddish 6794 km diameter disc is most favourably placed for observation during oppositions that occur with Mars near *perihelion (*see* Mars, oppositions). Mars rotates in 24h 37m 23s about an axis tilted by 24° to its orbital plane; these values help to make the Red Planet the most earthlike of all the other planets, even though it has only 28% of the earth's surface area. Mars has two small natural satellites: *Phobos and *Deimos. Orbital and physical characteristics are given in Table 1, backmatter.

Telescopes reveal an orange-red surface with indistinct darker markings, once thought to be seas and named *maria*. White polar caps expand and contract with the Martian seasons. When it became apparent that the maria also varied in intensity and shape with the seasons, it was suggested that they were areas of lichen-like vegetation that flourished when water became available from the poles during each Martian spring. *Canals, charted by some observers, were postulated as an artificial planet-wide water distribution network. Observations by *Mariner spacecraft showed that the canals were a myth and reveal the maria as areas of darker bedrock on which the Martian winds deposit varying amounts of lighter-coloured dust, so changing their appearance. For details of Martian topography *see* Mars, surface features; Mars, volcanoes; Mars, polar caps.

The Martian atmosphere has a surface pressure of about 7 millibars, or

0.007 times the average pressure at the earth's surface, and extends to include an ionosphere at altitudes between 100 and 300 km. Daytime surface temperatures rarely climb above 0°C except during summer in the southern hemisphere. Most areas experience minimum temperatures near −100°C before sunrise each morning.

Tests by the *Viking spacecraft, which landed in 1976, show an atmosphere comprising mainly carbon dioxide (95%), with nitrogen (2.7%), argon (1.6%), oxygen (0.15%), a variable trace of water vapour, and traces of carbon monoxide, krypton, and xenon. The water vapour sometimes freezes to form clouds of ice crystals, especially above high topographic features such as the volcanoes, or condenses as fog in low-lying areas, such as *Hellas Planitia and the *Valles Marineris. The rich variety of clouds systems – leewaves, cirrus, orographically produced clouds – are a major part of the planet's meteorology. More extensive surface obscurations occur during dust storms, which can engulf the entire planet. Such events happened in 1971 at the beginning of the Mariner 9 mission and in 1977 during the Viking mission. These global phenomena, which can occur both before perihelion and after this period in the Martian year when summer occurs in the southern hemisphere, will distribute material to altitudes of more than 40 km and last for several months. However, global storms do not seem to appear every Martian year.

Mars has no radiation belts and only a weak magnetic field, at most 0.2% as strong as the earth's; this suggests that it lacks a molten nickel-iron core and may have no iron core at all. The surface rocks, however, appear to be rich in iron, resembling the minerals haematite and magnetite and lending the planet its red coloration. *See also* Viking probes.

Mars, oppositions. *Oppositions recur at an average interval of 2 years 50 days, although this varies by 30 days either way because of the *eccentricity of Mars' orbit. Favourable oppositions for observation of surface features recur every 15–17 years when Mars is near *perihelion and may approach within 56 million km of the earth, attaining an apparent diameter of 25 arc seconds. The next favourable opposition will occur on Sept. 28 1988. At *aphelion oppositions Mars may be as much as 101 million km distant with a diameter of 14 arc seconds.

Mars, polar caps. Two variable white areas at the Martian poles, visible to earth-based observers but studied in detail by the *Mariner and *Viking Orbiter spacecraft. Each cap grows under a haze of cloud during winter for its hemisphere, reaching a latitude of 50° at its greatest extent. During summer it retreats to leave a small irregular patch about 500 km across. This residual cap probably consists of water ice, although the variable cap appears to be a frost of frozen carbon dioxide. At latitudes greater than 80°, the polar terrain exhibits a crater-free layered structure up to six km thick that possibly represents successive deposition of wind-borne dust.

Mars probes. A series of ill-fated Soviet probes to Mars. Contact was lost with Mars 1, launched in Nov. 1962, after it had travelled 106 million km. Mars 2 and 3 entered orbit around Mars in Nov. and Dec. 1971 after detaching lander capsules: that from Mars 2 crashed and the other stopped transmitting within two minutes of touchdown. The orbiters made surface and atmospheric studies. Mars 4, an intended orbiter, passed the planet in Feb. 1974 after retrorocket failure. Mars 5 entered orbit two days later but ceased functioning after a few days of photographic reconnaissance. Mars 6 flew past Mars in Mar.

1974, ejecting a lander capsule that crashed on the surface. Mars 7 passed Mars three days before Mars 6 but its intended lander missed the planet completely.

Mars, surface features. Impact *craters dominate Mars' southern hemisphere – and may be as old as 3500 million years – but are thinly scattered over the younger and mainly volcanic surface of most of the northern hemisphere. In addition there are white polar caps, volcanic features (*see* Mars, polar caps; Mars, volcanoes), and areas, particularly in equatorial regions, that show evidence of running water in the past.

Many of the Martian craters show the effects of erosion by dust, which is generally of a higher *albedo than uneroded surface rocks and forms bright deposits in low-lying terrain such as crater floors and the southern hemisphere impact basins of *Argyre Planitia and *Hellas Planitia. In places the dust is swept into ridges or dunes by the wind. Bright or dark streaks occur in the lee of some craters where the prevailing wind has piled up the dust or has scoured the surface to reveal the underlying darker bedrock. Variations in dust deposition are responsible for albedo and outline changes of the darker markings, or maria, once attributed to life processes.

Southeast of the volcanoes of the *Tharsis Ridge lies the complex equatorial canyon system of *Valles Marineris, which measures 4500 km from east to west and 150 to 700 km from north to south. Individual canyons are up to 200 km wide and 7 km deep, dwarfing the maximum 28 km width and 2 km depth of earth's Grand Canyon. They appear to result from faulting and collapse of the Martian surface – a process that may be continuing today. Other fault systems are found elsewhere, all of them apparently associated with areas of volcanic activity.

Smaller valleys, or channels, meander for up to 1000 km or more across equatorial areas of Mars. They appear to have been formed by running water derived, at least in part, from rain falling on the planet's surface. Some channels begin in areas of collapsed terrain as though the water they carried was originally frozen as a subsurface layer of permafrost that was subsequently thawed and released by volcanic heating. The enigma is that neither rain nor surface water can survive on Mars given present atmospheric constraints. Possibly the channels indicate that Mars has experienced spells of milder climatic conditions when the atmosphere was thicker than it is now. These spells, during which polar and subsurface ice may have been released, may be related to real fluctuations in the sun's output, to variations in the planet's axial tilt, or to periods of enhanced volcanic activity. Further evidence for the existence of a permafrost layer comes from the shape of the ejecta blankets surrounding some large craters.

Mars, volcanoes. Much of the northern hemisphere of Mars shows signs of volcanism with extensive lava plains sparsely pock-marked with impact craters: an indication of an age considerably less than the 3500 million years assigned to the heavily cratered southern hemisphere. This disparity may arise from the fact that the mean Martian surface lies at a lower level in the northern hemisphere.

The two main areas of volcanic activity, both situated on bulges in the Martian crust, are the *Tharsis Ridge – with the four principal volcanoes including the largest, *Olympus Mons – and the *Elysium Planitia. The volcanoes are of the shield type, having long gently sloping flanks. Counts of the numbers of impact craters on the slopes suggest ages between 200 million and 800 million years for the

Tharsis Ridge volcanoes, and about 1500 million years for those of Elysium Planitia. The extraordinary size of the volcanoes may be related to the absence of plate tectonics on Mars, so that the crust remains locked in relation to volcano-inducing 'hot spots' in the mantle below. *See also* Mars, surface features.

Martian probes. *See* Mariner 4; Mariner 6, 7; Mariner 9; Mariner probes; Mars probes; Viking probes; Zond probes.

mascons. *Short for* mass concentrations. Positive gravitational anomalies associated with the young circular lunar *maria, where relatively dense disc-shaped masses are now known to be located close to the surface: because the gravitational attraction is somewhat higher over these regions, mascons perturb the orbits of lunar satellites. They were produced either by the mare basalt itself as it flooded the basins or by the uplifting of higher-density mantle material during *basin excavation. Mascons imply the existence of a rigid lunar crust for at least 3000 million years. The removal of crust during crater formation may also produce negative gravitational anomalies; these are observed in the younger large *craters and around the *Orientale Basin, which is only partially lava-filled. Since the circular maria are topographic lows, mascons may be partially isostatically compensated, i.e. they may have sunk under their excess density until some sort of equilibrium depth was reached with relation to adjacent material.

maser amplifier. A type of negative-resistance amplifier in which amplification is achieved by stimulating excited molecules to release photons in phase with the passing radiation. Amplifiers of this type have extremely low *noise figures, limited in practice by losses in the input and output circuits, and they therefore find applica-

tion in *radio telescopes above one gigahertz.

maser source. A small celestial source of electromagnetic radiation, displaying narrow spectral lines and having an anomalously high *brightness temperature, in which the emission is thought to be by maser action: radiation at the maser frequency is amplified by stimulating excited molecules in the source to emit *photons in phase with the passing radiation. Several different types have been identified including *OH (hydroxyl) masers* radiating at 1.665 gigahertz, H_2O *(water) masers* at 22.235 GHz, and *SiO (silicon oxide) masers* at 86.243 GHz.

Maser sources have been detected, for example, in the atmospheres of old variable stars and in molecular clouds associated with newly formed or forming stars. The first maser source – an OH maser – was discovered in the Orion nebula in 1965. Hundreds of sources containing hydroxyl masers have since been found, some of which also have water masers. Masers are usually very *variable radio sources and produce highly polarized radio emission from regions that are typically about one parsec in diameter. *See also* OH/IR star; Mira stars.

mass. A measure of the quantity of matter in a body. The SI unit of mass is the kilogram. The astronomical unit of mass is the *solar mass, i.e. 2 × 10^{30} kg. Mass is a property of matter that determines both the *inertia of an object, i.e. its resistance to any change in its motion or state of rest, and the gravitational field that it can produce. The former is called *inertial mass*, defined by *Newton's laws of motion, and the latter is *gravitational mass*. Inertial and gravitational mass have been found to be equivalent. This led to Einstein's principle of *equivalence between inertial and gravitational forces. *See also* weight; stellar mass; relativity, special theory.

mass discrepancy. *See* missing mass.

mass loss. The loss of mass by a star during its evolution by one or more processes. Mass is lost at a high rate in *bipolar outflows during the formation of a star, and as a *T Tauri wind from some protostars. Mass is known to flow slowly into space from many stars, especially giants, in the form of a *stellar wind. Mass can also be ejected as the shell of a *planetary nebula or in a violent *supernova explosion. *Mass transfer can occur in close *binary stars, and mass can be lost from the binary in a *nova explosion.

mass-luminosity relation (M-L relation). An approximate relation between the mass and *luminosity of main-sequence stars, predicted by Eddington in 1924. Although having some basis in theory it is obtained empirically from a graph of absolute bolometric *magnitude against the logarithm of mass (in *solar units), i.e. $M/M_\odot$, for a large number of binary stars. Most points lie on an approximately straight line. Since a star's absolute bolometric magnitude is a function of the logarithm of its luminosity (in solar units), i.e. $L/L_\odot$, this line is represented by the M-L relation:

$$\log(L/L_\odot) = n \log(M/M_\odot)$$
$$L/L_\odot = (M/M_\odot)^n$$

n averages about 3 for bright massive stars, about 4 for sun-type stars, and about 2.5 for dim red dwarfs of low mass. The relation holds for main-sequence stars, which all have a similar chemical composition and have similar internal structures and nuclear power sources. It is not obeyed by *white dwarfs (degenerate matter) or *red giants (extended atmospheres).

mass number. Symbol: A. The number of protons and neutrons (i.e. nucleons) in the nucleus of a particular atom. It is the number closest to the mass of a nuclide in atomic mass units, i.e. in units equal to 1/12 of the mass of a carbon-12 atom. *See also* atomic number.

mass transfer. A process occurring in close *binary stars when one of the stars has expanded to such an extent that mass can be transferred to its companion. This mass may strike the more compact companion directly or, more usually, may orbit the star and eventually form a ring of matter – an *accretion disc*. The mass transfer takes place through the inner Langrangian point, lying between the stars, and occurs when the larger star fills its Roche lobe (*see* equipotential surfaces).

matching (impedance matching). Arranging electrical impedances so that maximum power is transferred from one device to another. This occurs when the input impedance of one device equals the output impedance of the other to which it is connected. *See also* feeder.

matter era. *See* big bang theory.

Mauna Kea. A dormant volcano on the Big Island of Hawaii whose high-altitude summit (4200 metres, 13 800 ft) lies above almost half of the earth's atmosphere. Conditions at the summit for astronomical observations, especially at infrared and submillimetre wavelengths, are extremely good. It is therefore the site of several major telescopes: the 3.8 metre *UK Infrared Telescope (UKIRT), the 3.6 metre *Canada-France-Hawaii Telescope (CFHT)*, used for optical and infrared studies, the 3.0 metre *Infrared Telescope Facility (IRTF)*, operated by NASA, all inaugurated in 1979, and the 2.2 metre telescope of the University of Hawaii, taken into service in 1970 for planetary studies.

Other telescopes are under construction or design. The 15 metre radio telescope, the UK-Dutch *Millimetre Wave Telescope*, should be ready in 1986 for observations at wavelengths of 0.3–1 mm (*see* submillimetre astronomy). The University of California is planning a huge 10 metre reflector with a segmented mirror. It will be called the *Maximillian E. and Marion Hoffman Observatory* after the major sponsors.

The overall site is administered, as *Mauna Kea Observatory*, by the University of Hawaii.

Maunder minimum. *See* sunspot cycle.

Max Planck Institute for Radio Astronomy (MPI). An observatory at Effelsberg, near Bonn. It has the world's largest fully steerable radio dish, a lightweight 100 metre (328 foot) paraboloid, which has been in operation since 1971. It has a computer-controlled *altazimuth mounting.

Maxwell Montes. *See* Venus.

Mayall reflector. *See* Kitt Peak National Observatory.

McDonald Observatory. An observatory of the University of Texas on Mount Locke, Texas, at an altitude of 2081 metres. It has a 2.7 metre (107 inch) and a 2.1 metre (82 inch) reflecting telescope, acquired in 1969 and 1939 respectively. When commissioned, the 2.1 metre *Otto Struve Telescope* was the second largest in the world; it has recently been modernized and equipped for infrared work.

McMath telescope. *See* Kitt Peak National Observatory; solar telescope.

MCP detector. *See* microchannel plate detector.

mean anomaly. *See* anomaly.

mean daily motion. The angle through which a celestial body would move in one day if its orbital motion were uniform.

mean density of matter. Symbol: ρ_0. The factor that determines the dynamical behaviour of the *universe, i.e. whether it is open or closed. The *density parameter*, Ω_0, is a dimensionless quantity given by
$$\Omega_0 = 8\pi G\rho_0/(3H_0^2)$$
It is related to the *deceleration parameter, q_0, by
$$\Omega_0 = 2q_0$$
If the universe is a continuously expanding open system, Ω_0 must be ≤ 1; if it exceeds unity the universe is closed and must eventually collapse. The *critical density* for which Ω_0 is unity is 5×10^{-27} kg m^{-3} for a *Hubble constant H_0 equal to 55 km s^{-1} megaparsec^{-1}. Various estimates involving the mass to luminosity ratios of galaxies lead to values for Ω_0 of about 0.01 due to galaxies. The missing-mass problem for *clusters of galaxies indicates the presence of *dark matter and suggests that the true value is about 0.3. A value of unity is not precluded by these observations; many theoretical cosmologists believe that $\omega_0 = 1$ precisely, since any deviation from unity in the early universe would have been immensely amplified in the proposed inflationary phase (*see* inflationary universe).

mean equator and equinox. The reference system determined by ignoring small variations of short period in the motions of the celestial equator; it is affected only by *precession. Positions in star catalogues are now usually referred to the mean catalogue equator and equinox (*see* catalogue equinox) of a standard *epoch. The coordinates of a celestial body on the (sun-centred) celestial sphere, referred to the mean equator and equinox of a standard epoch, are its *mean place*. See also true equator and equinox.

mean life. *See* half-life.

mean motion. The constant angular speed required for a celestial body to complete one revolution of an (undisturbed) elliptical orbit of a specified semimajor axis.

mean parallax. The *parallax obtained by statistical methods for a group of stars with different apparent magnitudes and proper motions.

mean place. *See* mean equator and equinox.

mean sidereal time. *See* sidereal time.

mean solar time. Time measured with reference to the uniform motion of the *mean sun. The *mean solar day* is the interval of 24 hours between two successive passages of the mean sun across the meridian. The mean solar second is 1/86 400 of the mean solar day. The difference between mean solar time and *apparent solar time on any particular day is the *equation of time. Since the mean sun is an abstract and hence unobservable point, mean solar time is defined in terms of *sidereal time. One mean solar day is equal to 24h 3m 56.555s of mean sidereal time.

Mean solar time was originally devised in order to provide a uniform measure of time based on the earth's rate of rotation, which was assumed, incorrectly, to be constant. The sun's mean daily motion does however conform closely to *universal time.

mean sun. An abstract reference point that was introduced to define *mean solar time. It has a constant rate of motion and is used in timekeeping in preference to the real sun whose observed motion is nonuniform. This nonuniformity results primarily from the earth's elliptical orbit around the sun, which causes the earth's orbital speed, and hence the sun's apparent speed, to vary (*see* Kepler's laws).

Secondly the direction of the sun's apparent motion is along the ecliptic and is thus inclined to the direction parallel to the celestial equator along which solar time is measured. An additional seasonal variation is therefore introduced when the sun's motion is measured relative to the equator.

The mean sun follows a circular orbit around the celestial equator so that it moves eastwards at a constant speed and completes one circuit in the same time – one *tropical year – as the apparent sun takes to orbit the ecliptic, if the slight secular acceleration of the sun is ignored. The mean sun is defined by an expression for its right ascension, which fixes its position among the stars at every instant.

mega-. Symbol: M. A prefix to a unit, indicating a multiple of 10^6 (i.e. one million) of that unit. For example, one megaparsec is one million parsecs.

meniscus lens. A lens in which the two surfaces have different or sometimes equal curvatures but, unlike a concave or convex lens, curve in the same general direction. A *positive meniscus* is thicker at the centre than at the edges; a *negative meniscus* is thinner at the centre. *See illustration at* lens.

meniscus telescope. *See* Maksutov telescope.

Menkalinan. *See* Auriga.

mensa. A flat-topped elevation with steep sides. The word is used in the approved name of such a surface feature on a planet or satellite.

Mensa (Table Mountain). A small inconspicuous constellation in the southern hemisphere, close to the south pole, the brightest stars being of 5th magnitude. It contains part of the Large Magellanic Cloud. Abbrev.: Men; genitive form: Mensae; approx.

position: RA 5.5h, dec $-78°$; area: 153 sq deg.

Merak (β **UMa**). One of the seven brightest stars in the *Plough and one of the two *Pointers.

Mercury. The innermost and, with a diameter of 4880 km, the second smallest planet, orbiting the sun in 88 days at an average distance of 0.39 AU. It has no natural satellite. The distance from the sun varies between 0.31 and 0.47 AU because of Mercury's high orbital *eccentricity of 0.21. Maximum *elongation from the sun varies between 28° and 18°, so that Mercury can be observed only when low in the twilight sky. Telescopes reveal a disc, between 5 and 15 arc seconds across, showing vague markings and *phases varying from full at superior conjunction to new at inferior conjunction. *Transits across the sun's disc recur at intervals of 7 or 13 years or in combinations of these figures. Orbital and physical characteristics are given in Table 1, backmatter.

Mercury's axial rotation period of 58.6 days, determined by radar in 1965, corresponds exactly to $^2/_3$ of its orbital period, probably because of tidal coupling between the extra gravitational force of the sun at *perihelion and irregularities in Mercury's almost perfectly spherical form. This period is close to half the planet's *synodic period; this explains why early observers, who saw the same face of the planet at favourable oppositions, concluded erroneously that it had a *synchronous rotation period of 88 days.

*Mariner 10, which made three close approaches to Mercury in 1974 and 1975, measured a maximum daytime temperature of 190°C; this may reach 450°C near perihelion. At night temperatures plunge to $-180°$C beneath the most tenuous of atmospheres, consisting mainly of minute traces of helium and argon. Mariner

found that Mercury has a *magnetosphere supported by a magnetic field about 1% as strong as the earth's; it is not clear whether this is generated within the planet's core or arises from permanent magnetization within the crust. Mercury's density, 5.4 times that of water, implies that it has a large nickle-iron core, perhaps 3600 km across.

Photography of half of Mercury's surface by Mariner 10 shows it to resemble the moon in being heavily cratered with intervening areas of lava-flooded plains, called *maria*. The largest impact feature detected is the 1300 km diameter *Caloris Basin,* similar to the moon's Mare *Imbrium. Antipodal to Caloris is an area of chaotic terrain created, it is believed, by shock-waves from the same impact. Some recent craters possess radiating systems of light-coloured surface streaks, similar to the lunar *rays. The *ejecta blankets surrounding many craters are half the size of those around lunar craters of similar diameter, probably because Mercury has twice the moon's surface gravity. Of particular interest, and not seen on the moon or Mars, are lines of cliffs, called *lobate scarps,* up to 3 km high and 500 km long. These may have resulted from wrinkling of the silicate-rich crust as Mercury's core cooled and contracted billions of years ago.

Mercury, perihelion. *See* advance of the perihelion.

Mercury project. The pioneering US space project to orbit a manned craft, to investigate man's reactions to and ability to adapt to the weightlessness of space, and to recover both men and spacecraft. Mercury MR3, launched May 5 1961 by a Redstone rocket, carried the first American, Alan Shepard, into the upper atmosphere in a suborbital hop. The first true orbital flight was made by John Glenn in Mercury MA6, launched

Feb. 20 1962 by an Atlas rocket. The longest flight, 34.3 hours, was made by Gordon Cooper in the last capsule in the series, Mercury MA9, launched May 15 1963. *See also* Gemini project; Apollo.

Mercury, transits. *See* transit.

meridian. 1. A great circle passing through a point on the surface of a body such as a planet or satellite, at right angles to the equator and passing through the north and south poles.
2. *Short for* celestial meridian. The projection of the observer's terrestrial meridian on to the *celestial sphere. It is thus the great circle passing through the north and south *celestial poles and the observer's *zenith and intersecting the observer's horizon at the north and south points (*see* cardinal points).

meridian angle. *See* hour angle.

meridian passage. The passage of a star across the meridian (*see* culmination) or of an inferior planet across the sun (*see* transit).

MERLIN. *Abbrev. for* Multi-Element Radio-Linked Interferometer Network (formerly known as MTRLI). A network of radio telescopes in England connected by microwave communication links to a computer system at *Jodrell Bank. In operation since 1980, it is used at wavelengths ranging from 1 metre to 1 centimetre approximately to build up high-resolution maps of *radio sources.

The telescopes in the network are Jodrell Bank's Mark IA, Mark II, or its 13 metre dish – the choice depends on wavelength, the 13 m being used at shorter wavelengths than the Mark II; the elliptical Mark III (identical in size to Mark II but of wire mesh) at Wardle, Cheshire; the 25 metre dish at Defford, Worcestershire;

three new identical 25 metre dishes at Pickmere and Darnhall in Cheshire and Knockin in Shropshire. An additional telescope, which would be built at Cambridge, is under negotiation. The existing telescopes are equipped to operate 15 *baselines ranging from 6 to 134 km in length.

Merope. *See* Pleiades.

mesons. *See* elementary particles.

mesosphere, mesopause. *See* atmospheric layers.

Messier catalogue. A catalogue of the brightest 'nebulae' prepared by the French astronomer Charles Messier and printed in final form in 1784. Messier compiled the list in order to avoid confusion of these cloud-like objects with comets, for which he was a keen searcher. 103 celestial objects, which appeared fuzzy and extended in telescopes of the time, were listed and given a number preceded by the letter M. This is the object's *Messier number,* an example being M42: the Orion nebula. Six more objects were listed in 1786. It was later found that most of the objects were not true *nebulae but galaxies and star clusters. *See* Table 8, backmatter; NGC.

Messier number. *See* Messier catalogue.

Me stars. Cool red stars whose spectra show emission lines of hydrogen in addition to the molecular bands of titanium oxide characteristic of *M stars. This occurs in both giant and dwarf (main-sequence) M stars, apparently for different reasons. The giant Me stars are mainly long-period variables, such as *Mira stars, whose distended atmospheres are losing matter to space. The emission lines from dwarf Me stars seem to be related to the presence of strong magnetic fields; they are often *flare stars.

meteor. The streak of light seen in the clear night sky when a small particle of interplanetary dust (a *meteoroid) burns itself out in the earth's upper atmosphere. On a clear moonless night at a time away from *meteor showers the dark-adapted eye can see about 10 per hour, the eye detecting meteors down to about fifth visual *magnitude. The most probable magnitude seen is +2.5, 75% of the meteors observed having magnitudes in the range 3.75 to 0.75. The visual rate maximizes at about 04.00 local time when the observer is on the leading side of the earth, 'ploughing' into the cosmic dust cloud. The meteoroid enters the atmosphere at velocities between 11 and 74 km s^{-1} depending on whether the earth just catches up the particle or they have a head-on collision. The ablating meteoroid not only leaves behind it a *train of excited atoms that de-excite to produce the brief blaze of light − a *visual meteor* − but it also produces a column of ionized atoms and molecules (both atmospheric and meteoric) that can act as reflectors of radar pulses transmitted from ground-based telescopes − a *radio meteor*. The meteoroid dissipates its energy, distributes its disaggregated atoms and molecules, and produces the visual and radio meteor in the region between 70 km and 115 km above the earth's surface. The mean altitude of maximum luminosity and electron density is 97 km.

A large percentage of the mass influx to the earth's atmosphere (some 16 000 tonnes per year) is made up of particles in the size range that produce visual and radio meteors, i.e. one millionth to one million grams. The meteoroid has a kinetic energy considerably in excess of the energy required to vaporize itself. Air molecules striking the meteoroid have a kinetic energy of about 400 electronvolts. As the binding energy of the atoms in the meteoroid is a few electronvolts, a hundred or so

meteoroid atoms boil off for each adsorbed air molecule. The meteoroid starts to ablate at about 115 km. These ablated atoms then collide with the surrounding air molecules and by losing energy in these collisions produce *excitation and *ionization. After ten or so collisions the energy of the ablated atom has dropped below the excitation potential of the air molecules and ionization and excitation no longer occur. The ablated atom has moved about 50 cm from the meteoroid path during this process. The small meteoroid is completely evaporated after intercepting an air mass about 1% of its own mass. Owing to these processes the meteoroid *train has a length of between 7 and 20 km and a width of about 100 cm. A 4th magnitude meteor contains around 10^{18} electrons and 10^{18} positive ions. The atoms quickly de-excite and electrons and positive ions recombine and become attached to other atmospheric molecules. The time duration is related to the magnitude but is usually less than a second.

The meteor spectrum is a low-excitation one, ionization potentials lying in the range 1.9 to 13.9 volts (equivalent to temperatures roughly 1600 to 4800 kelvin). The Ca II lines, H and K, are the main features of fast meteors whereas Na I, Mg I, and Fe I are dominant in slow ones.

meteorite types		finds %	falls %
chondrites ⎫ stones		43.7	84.3
achondrites ⎭		1.0	8.7
stony-irons		5.2	1.3
irons		50.1	5.7

meteorite. Interplanetary debris that falls to the earth's surface. The first documented one was the stone that fell near Ensisheim, Alsace in 1492 but it wasn't until 1803 that meteorites were accepted by the scientific community as being extraterrestrial. One that is seen to hit the ground is known as a *fall*; one that is discovered accidentally some time after having fallen is termed a *find*. Over 2300 meteorites have been recovered and roughly 6 falls and 10 finds are added to the list annually. About 3300 hit the earth each year; the majority go unrecorded, falling in oceans, deserts, and other uninhabited regions.

Meteorites can be crudely divided into three types: *stony meteorites, which subdivide into *chondrites and *achondrites, *iron meteorites, which subdivide into hexahedrites, octahedrites, and ataxites, and *stony-iron meteorites. Iron meteorites have densities of about 7.8 $g\,cm^{-3}$ and contain on average 91% iron, 8% nickel, and 0.6% cobalt. Stony meteorites have densities about 3.4 $g\,cm^{-3}$ and on average contain 42% oxygen, 20.6% silicon, 15.8% magnesium, and 15.6% iron, no other element exceeding 2%. Stony-irons are intermediate in composition. The relative percentages of each type are shown in the table. The fall percentages give a reasonable approximation to the actual meteorite population of the solar system. However the fragile *carbonaceous chondrites mostly disintegrate on entry.

The largest iron meteorite that has been found is the 60 tonne *Hoba meteorite, the largest stony meteorite being 1 tonne, part of the Norton County, Kansas achondrite. The found meteorite might only represent a small percentage of the in-space mass of the body. Meteorite entry is accompanied by a brilliant *bolide, the meteorite usually being retarded to *free-fall velocity at a height of about 20 km. It hits the ground at about 300 km per hour. The surface of the meteorite can easily be heated up to several thousand kelvin during entry. This usually produces a black glassy fusion crust, which covers the meteorite. Most meteorites fragment on entry, scattering pieces over an el-

liptical area: the carbonaceous chondrite meteorite *Allende*, which fell in 1969 in Mexico, scattered 5 tons of material over an area 48 km long by 7 km wide.

Meteorites are named after the nearest topographical feature to the fall point. Recognizing meteorite finds relies mainly on analysing chemically for the presence of nickel, which is present in all iron meteorites. Etched and polished irons also show *Widmanstätten patterns. Stony meteorites contain nickel-iron particles that can be extracted magnetically from a crushed sample.

Meteorites bring to our notice a type of rock existing in the solar system but different from anything that occurs in the outer shell of the planet earth. Most meteorites are believed to be fragments of *minor planets. Two problems arise: firstly, meteorites only seem to have had a short exposure to cosmic rays and secondly, the dynamics by which a meteorite gets from the asteroid belt to the earth's orbit is nontrivial. The carbonaceous chondrites are thought to originate as parts of cometary *nuclei. The ages of meteorites have been obtained by radiometric-dating methods and have been estimated at up to 4.7×10^9 years. This indicates that meteorites were formed at about the same time as the solar system.

meteorite craters. Circular impact structures produced by meteorite bombardment, usually known as *astroblemes if they occur in the earth's crust. *See also* craters.

meteoroid. The collective term applied to meteoritic material in the solar system, usually replaced by the terms *micrometeorite for particles with mass less than 10^{-6} gram and *meteorite for bodies with mass greater than about 10^5 grams. The majority of the mass of the meteoroid cloud around the sun is made up of particles with individual masses between

10^{-7} and 10^{-3} gram. Meteoroids are usually produced by the collisional fragmentation of *meteor stream particles. In the main they move around the sun in low-inclination direct orbits. The space density of meteoroids varies as about $1/r^{1.5}$, where r is the distance from the sun. Individual meteoroids in the mass range 10^{-6} to 10^4 grams are fragile crumbly rocky dust particles with a composition similar to carbonaceous chondrite meteorites. *See also* meteor.

meteor shower. The increase in observed rate of appearance of *meteors – *shower meteors* – when the earth passes through a *meteor stream; a particular shower occurs at the same time each year. The increase in rate differs from stream to stream: it is a function of mass of the originating comet, age of stream, position of intersection point with respect to stream axis, and orbital elements (which can vary slowly with time). With many showers the rate varies from year to year as the meteoroids are not evenly distributed around the stream. The ratio between shower rate and sporadic-meteor rate is a function of the mass of the meteoroids being observed. This ratio can be as high as 10 for visual meteors, is about 1–2 for radio meteors, and is negligible for satellite-observed meteors. Shower meteors appear to emanate from a *radiant, the shower being named after the constellation that contains the radiant. *See* Table 4, backmatter.

meteor storm (meteor swarm). A *meteor shower with an enormous rate of appearance that occurs when the earth intersects a new *meteor stream close to the originating comet. The most famous was the Leonid storm of Nov. 12–13, 1833 when the rate exceeded 10 000 visual meteors an hour, single observers often seeing 20 a second.

meteor stream. The annulus of *meteoroids around the orbit of a decaying *comet. Dust particles are pushed away from the cometary *nucleus by gas pressure at the time when the nucleus is near *perihelion. The particles then have slightly different orbital parameters to the comet. Stream-formation time depends on perihelion distance and meteoroid size, being anything from tens to hundreds of years. New streams only have meteoroids close to the comet and only produce *meteor showers periodically. Streams are about 10 times thicker at aphelion than at perihelion. They contain anywhere between 10^{12} grams and 10^{16} grams of dust and have volumes in the range 10^{21} to 10^{25} km^3.

Metonic cycle (lunar cycle). A period of 19 years (tropical) after which the phases of the moon will recur on the same days of the year: the period contains 6939.60 days, which is very nearly equal to 235 *synodic months, i.e. 6939.69 days. Since it is also almost equal to 20 *eclipse years, i.e. 6932.4 days, it is possible for a series of four or five eclipses to occur on the same dates at intervals of 19 years. The cycle was discovered by the Greek astronomer Meton in the fifth century BC and was used in determining how intercalary months could be inserted into a lunar calendar so that the calendar year and tropical (seasonal) year were kept in step.

metre. Symbol: m. The usual scientific unit of length and distance. It is defined (from Oct. 1983) as the length of the path travelled by light in vacuum during a time interval of

1/299 792 458 of a second

It is thus now defined in terms of the second and the fixed value of the *speed of light. The metre was formerly the length equal to 1 650 763.73 wavelengths in vacuum of the radiation corresponding to the transition between the levels $2p_{10}$ and $5d_5$ of the krypton-86 atom. This definition came into use in 1960. Before that, the metre was the length of a prototype metre bar.

Michelson interferometer. *See* interferometer.

Michelson–Morley experiment. *See* ether.

micro-. Symbol: μ. A prefix to a unit, indicating a fraction of 10^{-6} (i.e. one millionth) of that unit. For example, one micrometre is 10^{-6} metre.

microchannel plate (MCP) detector. A high-resolution device for recording x-ray, ultraviolet, or electron images. The incident photon (or electron) strikes the front surface of a thin glass plate consisting of a very large number of fine-bore tubes (usually aligned perpendicular to the plate surfaces). Secondary electrons are produced and are accelerated down the tubes by an applied voltage, striking the walls and creating further electrons. Charge multiplication factors of about 10^3 (or 10^6 in a 'cascaded pair' of MCPs) yield a signal that is easily read on a collector at the rear of the plate(s). Proximity focusing ensures retention of the initial photon (or electron) image and resolutions of $20-50$ μm have been achieved. MCP detectors have been or will be used in a number of astronomy satellites, including the Einstein Observatory, EXOSAT, and ROSAT.

microdensitometer. An optical instrument that measures and records small changes in the transmission density on a photographic plate, etc. If the density is linearly proportional to the amount of light that originally fell on the sample, as it is in an electronograph, this gives a direct recording of light intensity.

microgravity. The very low gravity encountered on spacecraft in earth orbit.

micrometeorite. A *cosmic dust particle of mass less than about 10^{-6} gram and diameter less than 0.1 mm. On impact with the earth's atmosphere the heat absorbed by the particle from atmospheric friction is insufficient to raise it to boiling point. The ratio of heat radiated to heat absorbed is proportional to the inverse of the radius of a particle: those larger than 10^{-6} gram ablate and form *meteors; smaller ones don't. The micrometeorite will be decelerated to a normal free-fall velocity and then drift to the surface of the earth. On the moon, however, no deceleration occurs and they impact the surface with the normal geocentric velocity.

Microscopium (Microscope). A small inconspicuous constellation in the southern hemisphere near Sagittarius, the brightest star being of 4th magnitude. Abbrev.: Mic; genitive form: Microscopii; approx. position: RA 21h, dec $-35°$; area: 210 sq deg.

microwave background radiation. The remnant of the radiation content of the primordial fireball associated with the hot *big bang. It is the dominant radiation of deep intergalactic space. A microwave background was first considered by George Gamow in the 1940s, and its existence with a temperature close to 5 K was predicted in 1948 by Ralph Alpher and Robert Herman. The concept was generally neglected and the prediction forgotten until the radiation was discovered in 1965 by Arno Penzias and Robert Wilson.

Observed as black-body radiation characteristic of a temperature of 2.9 K, the microwave background is very nearly equal in intensity from all directions of the sky. The earth's atmosphere complicates observations over the black-body peak, which occurs at a wavelength of about 1 mm (see illustration), necessitating balloon- or satellite-borne instruments. Early results indicated a possible deviation

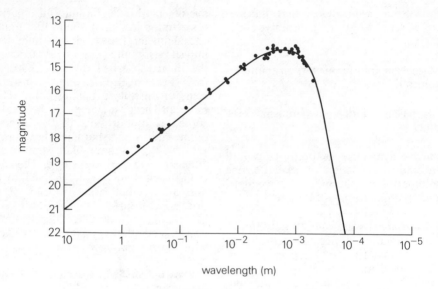

Microwave background radiation spectrum

from the 2.9 K black-body curve at these short wavelengths; if real this deviation is difficult to explain. More precise measurements have shown a slight variation in the microwave background around the sky, with the radiation 'hotter' in the direction of the *Virgo cluster. This anisotropy is attributed to the Virgo cluster's gravity attracting the Galaxy and the Local Group. There are slightly 'cooler' regions coinciding with distant clusters of galaxies, where the radiation has been *Compton scattered by hot intergalactic gas. A NASA satellite scheduled to be launched in 1988 will investigate the isotropy of the microwave background.

*Cosmological redshift implies that the photons comprising the background radiation were more energetic in the past, and at redshifts greater than about 1000 they were energetic enough to photoionize hydrogen (*see* big bang theory). The mean free path of the photons was then drastically reduced by scattering on the free electrons present at that time. This essentially prevents our learning anything about earlier times from the micro-

wave background, although it is hoped that small-scale variations in the microwave background may relate to fluctuations in the distribution of matter. Such fluctuations might later have become galaxies.

microwaves. Very high frequency radio waves. Their frequencies usually lie in the range 3×10^{11} hertz (just beyond the infrared band of the electromagnetic spectrum) to about 3×10^{8} hertz. The wavelength range is therefore about 1 mm to 1m.

Milky Way. A dense band of faint stars that extends right round the *celestial sphere, dividing it into roughly equal parts. Its central line marks the central plane of our *Galaxy, inclined at $62°$ to the celestial equator. Although the vast majority of the stars are too faint to be seen individually, they are collectively visible on a clear moonless night as a diffuse band of light. The Milky Way is seen because the sun lies close to the central plane of the galactic disc. Since this region of the Galaxy is highly flattened, a much greater depth of stars is visible in directions along the plane (i.e. towards the Milky Way) than in other directions. The Milky Way has a distinctly patchy appearance; it also varies considerably in width and brightness and is noticeably brighter towards Sagittarius (the direction of the *galactic centre). Many of the apparent gaps are due to *dark nebulae, such as the Coalsack, along our line of sight, which prevent stars behind them from being seen.

Milky Way System. *See* Galaxy.

milli-. Symbol: m. A prefix to a unit, indicating a fraction of 10^{-3} (i.e. 1/1000) of that unit. For example, one millisecond is 10^{-3} second.

millimetre astronomy. A branch of radio astronomy covering the wavelength range from 1 to 10 millimetres approximately. When studies began in the late 1960s it was the highest frequency range, 30–300 gigahertz, in which radio astronomy could be carried out. *Submillimetre astronomy has since become feasible. Both millimetre and submillimetre wavelengths are ideal for observations of giant *molecular clouds. A large variety of molecules form at the high densities in the clouds. The spectral lines emitted by the molecules are mainly in the millimetre and submillimetre range, and give information on chemical composition, relative abundances of isotopes, chemical reactions, and temperatures, densities, and velocities in the clouds (*see also* molecular-line radio astronomy).

Millimetre wave instrumentation – usually parabolic radio dishes plus line receivers – is similar to that used for submillimetre astronomy.

Millimetre Wave Telescope. *See* Mauna Kea.

millisecond pulsars. A newly discovered class of *pulsars that produce pulses with a period of only a few milliseconds, and are thus *neutron stars rotating hundreds of times per second. Unlike the fast 'normal' pulsars (e.g. the Crab and Vela pulsars), the millisecond pulsars' fast rotation is not a result of youth; they have almost certainly been 'spun up' by mass transfer in a close *binary star at an earlier evolutionary stage. Rapid rotation of a neutron star is normally slowed precipitously because the star's strong magnetic field radiates away the rotational energy, as with the Crab pulsar (*see* Crab nebula); the millisecond pulsars detected so far, however, have only a very gradual rate of slowing down, probably because their magnetic fields are comparatively weak (10^4–10^5 tesla).

PSR 1937+21 was the first millisecond pulsar to be found, and is currently the fastest known (period 1.56 ms, i.e. 642 rotations per second). The second to be found, PSR 1953+29 (6.1 ms), is a member of a *binary pulsar system; it orbits its unseen companion in 120 days. The old but rapidly spinning binary pulsar PSR 1913+16 also belongs in this class, although its period is rather longer (59 ms).

Mills cross antenna. An interferometer (*see* radio telescope) using two long narrow *antennas mounted in the form of a cross and producing a single pencil beam; it is named after B.Y. Mills. One arm of the cross may be shortened to give a *T-antenna*, which has an antenna pattern very similar to that of the full cross but whose *sensitivity is less.

Mimas. The innermost of the larger Saturn satellites, discovered in 1789. It has a diameter of only 390 km and a density of 1.20 g cm^{-3}. The most striking feature is the crater *Herschel*, 130 km in diameter and nearly centred on the leading hemisphere. The walls rise on average to a height of 5 km above the floor. In the centre of the crater is an enormous peak 20 ×

30 km at its base. The diameter of the crater is a third of Mimas itself, and is probably close to the maximum size of impact crater that a body can sustain without being broken up. All other craters on the surface are small, being less than 50 km in diameter. There are valleys too on the surface, which is scored by grooves. The surface of the satellite is icy and there is every reason to suppose that the ice makes up much of the entire body, which now bears the scars of past bombardments. *See also* Table 2, backmatter.

Mimosa. *See* Beta Crucis.

mini black hole. *See* black hole.

Mini Magellanic Cloud. *See* Magellanic Clouds.

minimum-energy orbit. *Another name for* Hohmann transfer. *See* transfer orbit.

minor axis. The least distance across an ellipse, crossing the *major axis perpendicularly at the centre of the ellipse. The length is usually given as 2*b*.

minor-planet families. *See* Hirayama families.

minor planets (asteroids; planetoids). Small rocky solar-system bodies that orbit the sun. Most (perhaps 95%) orbit in a main belt – the asteroid belt – between the orbits of Mars and Jupiter at distances of between 2.17 and 3.3 AU from the sun and with periods between 3 and 6 years. The minor planets also include members of the *Amor, *Apollo, and *Trojan groups, as well as the unique object *Chiron, which have orbits very different from those in the main belt. The distribution of orbits within the belt is modified by *Kirkwood gaps. Like the planets most minor planets follow nearly circular orbits close to

the plane of the *ecliptic, though the average minor planet orbit does have a larger *eccentricity (0.15) and a greater *inclination (9.7°) than those of most planets. A few objects, such as *Hidalgo, follow highly eccentric and inclined orbits more akin to those of some periodic *comets.

Minor planets shine by reflected sunlight but only one – Vesta – is bright enough to be seen with the naked eye. More than 100 000 minor planets may be bright enough for photographic observation and discovery but only about 3000 have received official minor planet numbers. A permanent number is given only when the orbit has been calculated accurately; the object's position can then be determined for any desired time. Many more, with less well-determined orbits, have been assigned temporary designations. Most numbered minor planets have received names also, usually the prerogative of the discoverer.

Until recently the diameters of even the largest minor planets could only be roughly estimated. Since 1970 studies of the polarization of reflected sunlight and of the brightness at visible and infrared wavelengths for a large number of minor planets has led to improved values for their *albedos; from these values the diameters have been recalculated. Minor planets vary in size from the largest and first to be discovered – the 1003 km diameter *Ceres – to objects smaller than 1 km: there probably exists a continuous distribution down to the size of *meteoroids. Most are thought to be irregular in shape, causing them to vary in brightness as they rotate every few hours. Smaller objects predominate: there are only some 200 objects larger than 100 km diameter and about 10 larger than 300 km.

The total mass of the minor planets is estimated at 2.4×10^{21} kg (approximately 0.0004 of the earth's mass). This indicates that the hypothesis that they are the remains of a single large planet which exploded is untenable.

Probably they represent the debris from collisions between a handful of small bodies that condensed between Mars and Jupiter during the formation of the solar system.

As well as providing improved values for their diameters, studies of minor-planet spectra and polarizations show that the surface rocks are diverse and are similar to those of *meteorites. It is now believed therefore that most meteorites are fragments of minor planets. The majority of minor planets resemble *carbonaceous chondrites while others appear to have a metallic or a silicaceous surface. The latter are more common near the outer edge of the asteroid belt. This is an important clue to the composition of the pre-solar nebula from which the proto-minor planets condensed. The minor planets are being considered as targets for *in situ* studies by future space missions. *See* Table 3, backmatter; Hirayama families.

Mintaka. *See* Orion.

minute of arc (arc minute; arc min). *See* arc second.

Mira or **Mira Ceti (the Wonderful; o Cet).** A red giant in the constellation Cetus that is a long-period *variable star with an average period of 331 days. It was known to early astronomers for its regular appearances and disappearances and was established in 1596 as a variable star. It is the prototype of the *Mira stars. Its radius varies by 20 per cent during its

cycle and at maximum size and brightness is over 330 times the sun's radius. The surface temperature varies from about 2600 K at maximum luminosity down to about 1900 K. The visible light emitted varies considerably, usually by about 6 magnitudes, over the cycle (see illustration). The average maximum apparent magnitude, m_v, lies between 3 and 4 although it may reach 2nd magnitude (2.1 in 1969) and reached 1.2 in 1779. At minimum it is about 8th to 10th magnitude. Mira is also an *infrared source, the emission arising from dust grains in the expanding envelope of gas.

Mira is both a visual binary with a faint peculiar companion that is also variable, and an optical double. M_v: −1.0; spectral type: M6e–M9e III; distance: 40 pc.

Mirach (β And). A red giant that is the second-brightest star in the constellation Andromeda. It forms an optical double with a 10th-magnitude star at 91″ distance. m_v: 2.07; M_v: 0.2; spectral type: M0 III; distance: 24 pc.

Miranda. The innermost and smallest of the five satellites of Uranus, discovered in 1948. *See* Table 2, backmatter.

Mira stars (Mira Ceti variables). Long-period *pulsating variables, either red giants or red supergiants, that have periods ranging from about 80 to over 600 days and a range in

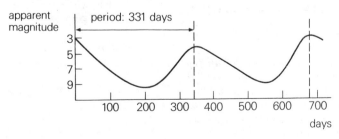

Light curve of Mira

brightness beginning at about 2.5 magnitudes and sometimes exceeding 10 magnitudes. They are numerous, *Mira Ceti being the prototype. The shape of the *light curve is not constant, the maximum brightness varying quite considerably between periods (see Mira). Because of the large amplitude they are easily recognizable, their high luminosity permitting detection at great distances.

Although their visual range is large, the range in bolometric and infrared *magnitudes is very much less, many of them being infrared sources. In terms of their spectra 90% can be classified as *Me stars, the rest being either *carbon stars (Ce) or zirconium (Se) stars, i.e. there are bright emission lines present in the spectra in addition to molecular bands. The pulsations of these huge stars are not very stable. There is evidence of shock waves developing within the tenuous atmosphere and travelling outwards, thus heating the gas and causing the production of emission lines. The expanding envelopes often contain condensed dust grains, which produce detectable infrared emission, and simple molecules: Mira stars often show *maser emission from hydroxyl, water, and silicon monoxide molecules in the outer atmosphere. See also OH/IR star.

Mirfak (α Per). A remote creamy-yellow supergiant that is the brightest star in the constellation Perseus. m_v: 1.80; M_v: −4.1; spectral type: F5 Ib; distance: 150 pc.

mirror. An optical element consisting of a glass or glass-like substrate on which is deposited a thin but uniform coating of highly reflective material. Aluminium is now usually used in telescope mirrors (see aluminizing). A very high quality reflecting surface can also be made from a multilayered stack of very thin films. The films have alternately high and low *refractive indices and are of such a thick-

ness and spacing that almost all the incident light is reflected.

Mirror shapes may be concave (converging) or convex (diverging) and are generally spherical or paraboloid in form. The mirror base must be accurately ground and the surface polished to within one quarter of the wavelength used, i.e. to about 0.1 micrometres, if good quality images are to be produced. Paraboloid mirrors do not suffer from *spherical aberration but for wide fields have severe *coma. Spherical mirrors are free of coma and astigmatism if a suitable stop is used; they do however suffer from spherical aberration. See also prime mirror.

Mirzam. See Canis Major.

missing mass. The *dark matter in a cluster of galaxies. It was first detected as a discrepancy between the mass of a cluster inferred by dynamical studies and the mass inferred by summing the contributions of luminous objects in the system.

Mizar (ζ UMa). A white star that is one of the brighter stars of the *Plough. It was the first *spectroscopic binary to be identified (E.C. Pickering, 1889) and has a period of 20.5 days; the B component is also a spectroscopic binary, period 361 days. It forms an optical double with the 4th-magnitude *Alcor, separation 14″. m_v: 2.12; M_v: 0.0; spectral type: A2 V; distance: 26 pc.

MK system (MKK system; Yerkes system). See spectral types.

M-L relation. See mass-luminosity relation.

MMT. Abbrev. for Multiple Mirror Telescope.

modified Julian date (MJD). See Julian date.

modulus of distance. *See* distance modulus.

moldavite. *See* tektites.

molecular clouds (giant molecular clouds, GMC). Huge clouds of cool dense interstellar matter that are the main sites for star formation in our Galaxy and others. Within these dense regions atoms tend to be combined into molecules. The clouds are composed principally of molecular hydrogen (H_2), with between 300 and 2000 molecules/cm³. There is also a small admixture of *cosmic dust comprising ~1% of the mass. The average linear dimension of a molecular cloud is about 40 parsecs, total mass about 5×10^5 solar masses, with gas temperatures between 10 and 20 K. Hydrogen molecules are not easy to detect, and molecular clouds were only discovered in the mid-1970s in surveys of radio emission from *carbon monoxide, CO, which is 10 000 times less common than hydrogen molecules in the clouds. The 2.6 and 1.33 mm CO lines are still the prime means of mapping and investigating the clouds. A wide variety of molecules, in addition to H_2 and CO, have been found in the clouds (*see* molecular-line radio astronomy); over 60 types have been detected in the largest clouds, such as *Sagittarius B2.

Several core regions, with about 10^5 hydrogen molecules/cm³ and mass about 10^2 to 10^3 solar masses, may be found within one molecular cloud. These contain *infrared sources, *HII regions, *maser sources, and the peak CO temperatures, which suggest that these regions are sites of massive star formation. There are also smaller clouds, containing about 500 solar masses of molecular hydrogen, throughout which low-mass stars are forming.

It is estimated that there are between 4000 and 5000 molecular clouds in the Galaxy, implying a total mass of roughly 2×10^9 solar masses. This represents a large proportion of the Galaxy's interstellar gas and is of importance in theories of galactic evolution and structure.

Some observed interstellar molecules

H_2	NH_3	CH_3NH_2
OH	H_2CO	CH_3CHO
CH	H_2CS	CH_3C_2H
CH^+	H_2C_2	CH_2CHCN
CN	HNCO	HC_5N
CO	HNCS	
CS	C_3N	$HCOOCH_3$
CS^+		CH_3C_2CN
C_2	CH_4	
NO	CH_2NH	CH_3OCH_3
NS	CH_2CO	CH_3CH_2OH
SO	NH_2CN	CH_3CH_2CN
SiO	HCOOH	HC_7N
SiS	HC_3N	
	HC_4	HC_9N
H_2O		
H_2S	CH_3OH	$HC_{11}N$
HCO	CH_3CN	
HCO^+	NH_2CHO	
HCN		
HNO		
HN_2^+		
HC_2		
SO_2		
OCS		

molecular-line radio astronomy. The study of celestial regions containing molecules by means of the radio emissions from the molecules themselves; unlike optical emissions, radio emissions are not obscured by material in the interstellar medium. A *radio telescope using a *line receiver is best suited to the purpose. Many different molecules have now been detected in the Galaxy in *molecular clouds. The first interstellar molecules – CH, CH^+, and CN – were detected optically in the late 1930s. This list has been greatly extended by radio astronomy measurements (see table). Most of the molecules are organic, i.e. based on carbon, but some inorganic species, such as silicon oxide (SiO), have also been detected. Many molecules occur in different isotopic forms, e.g. with deuterium rather than hydrogen.

The radio emissions usually come from transitions between different rotational states of the molecule and have wavelengths normally in the submillimetre and millimetre range, i.e. they lie in the gigahertz region of the radio spectrum. Some clouds of molecules appear to emit by maser action (*see* maser source); for example, the OH line at 1.667 gigahertz in some radio sources is observed to be very narrow and highly *polarized: characteristic features of a maser. Molecular-line radio astronomy is especially useful in determining cosmic isotope ratios.

Molonglo Radio Observatory. An observatory at Hoskinstown, New South Wales, run by the University of Sydney. The main instrument is a *Mills cross antenna, operational since 1960, although this has recently been modified for use as an *aperture synthesis instrument (the *Molonglo Observatory Synthesis Telescope, MOST*) by abandoning the north-south arm and phasing the east-west arm. The arms are about 1.6 kilometres long and consist of a series of 12 metre parabolic reflectors.

moment of inertia. Symbol: *I*. A property of any rotating body by which it resists any attempt to make it stop or change speed. It is given by the sum of the products of the mass, *m*, of each particle in the body and the square of the particle's perpendicular distance, *r*, from the axis of rotation, i.e.

$$I = \Sigma mr^2$$

The summation is effected by integration for a solid body. If the body is rotating with angular velocity ω, the kinetic energy of the body is $\frac{1}{2}I\omega^2$ and its angular momentum is $I\omega$.

moment of momentum. *See* angular momentum.

momentum. 1. (linear momentum) Symbol: *p*. A property of a body moving in a straight line. It is equal in magnitude to the product of the mass and velocity of the body. It is a vector quantity, *mv*, having the same direction as the velocity. *See also* conservation of momentum; Newton's laws of motion.
   2. *See* angular momentum.

Monoceros (Unicorn). An inconspicuous equatorial constellation near Orion, lying in the Milky Way, the brightest stars being of 4th magnitude. It contains the triple star Beta Monocerotis and the double Epsilon and several variables including R Monocerotis, which lights up the *Hubble nebula. There are many clusters, including the open cluster NGC 2244, and also the *Rosette nebula. Abbrev.: Mon; genitive form: Monocerotis; approx. position: RA 7h, dec 0°; area: 482 sq deg.

monochromatic. Of or producing a single wavelength or very narrow band of wavelengths.

monochromator. An instrument in which one narrow band of wavelengths is isolated from a beam of light or other radiation. This is usually achieved by means of a narrow-band interference filter, or by a diffraction grating or prism, together with an exit slit through which the desired waveband may pass. Changes in the intensity of the monochromatic beam can then be investigated.

mons (plural: **montes**). Mountain. The word is used in the approved name of such a surface feature on a planet or satellite.

month. The period of the moon's revolution around the earth with reference to some specified point in the

month	reference point	length (days)
tropical	equinox to equinox	27.321 58
sidereal	fixed star to fixed star	27.321 66
anomalistic	apse to apse	27.554 55
draconic	node to node	27.212 22
synodic	new moon to new moon	29.530 59

sky (see table). The differences in the monthly periods result from the complicated motion of the moon.

moon (or **Moon**). The only natural satellite of the earth, visible by virtue of reflected sunlight. It has a diameter of 3476 km, lies at a mean distance from the earth of 384 400 km, and completes one orbit around the earth in 27.322 days (*see* sidereal month; month). Its mass in relation to that of the earth is 1 : 81.3, i.e. 0.0123. It is in *synchronous rotation, i.e. it keeps the same face – the *nearside – towards the earth, although more than 50% of the moon's surface can be seen as a result of *libration. The same face is not however always turned towards the sun, the length of the solar day on the moon being equal to the moon's *synodic period, i.e. 29.530 59 days. During this period – the *synodic month – the moon exhibits a cycle of *phases, reaching an apparent magnitude of −12.7 at full moon. It can also undergo *eclipses and produce *occultations.

The moon's mean orbital velocity is 10 km s⁻¹. Its orbital motion is however subject to *secular acceleration and periodic *inequalities due to the

Moon's physical and orbital characteristics

mass	7.35×10^{25} g
diameter	3476 km
mean density	3.34 g cm⁻³
surface gravity	0.016 53 g
mean albedo	0.07
mean visual opposition magnitude	−12.7
interior structure: crust	60 km
mantle	1200 km
core radius	500 km
approximate age	4600 million years
mean horizontal parallax	3422″.46
moment of inertia	0.395 ± 0.005
mean distance	384 400 km
maximum	405 500 km
minimum	363 300 km
mean orbital velocity	10.1 km s⁻¹
escape velocity	2.38 km s⁻¹
mean eccentricity	0.0549
inclination of orbital plane to ecliptic	5° 8′ 43″
inclination of equator to ecliptic	1° 32′ 1″
sidereal month	27.321 66 days
anomalistic month	27.554 55 days
draconic month	27.212 22 days
synodic month	29.530 59 days

gravitational attraction of the sun, earth, and other planets (*see also* evection; variation; annual equation; parallactic inequality). In addition its *perigee and *nodes move eastward and westward with periods of 8.85 years and 18.61 years, respectively: this results in different lengths for the sidereal, *anomalistic, and *draconic months. Physical and orbital characteristics are given in the table.

Although earth-based observations have revealed much lunar information, the moon has now been carefully photographed, measured, and sampled by the *Apollo spacecraft and by *Ranger, *Lunar Orbiter, *Surveyor, *Luna, and *Zond spacecraft. It has an extremely tenuous atmosphere: a collisionless gas in which helium, neon, argon, and radon were detected by Apollo instruments. This near lack of atmosphere, together with the absence of an appreciable global magnetic field, exposes the *regolith to extremes of temperature ($-180°C$ to $+110°C$) and to the *solar wind, *cosmic rays, *meteorites, and *micrometeorites. Satellite magnetometer surveys and detailed analysis of lunar rocks have revealed that the moon's magnetic field was originally considerably stronger and might have been generated internally by a fluid core rather than produced by some external means.

The moon's interior structure is investigated from seismograms of *moonquakes and meteoritic and artificial impacts. The temperature of the central core may reach 1500°C but a metallic iron composition is precluded by the moon's moment of inertia. The moon's *centre of mass is displaced towards earth by 2.5 km because of a thick farside crust.

The moon was formed out of refractory materials about 4600 million years ago. Its outer layers were melted and differentiated to produce mafic cumulates (iron magnesium silicates) and a feldspathic crust (calcium aluminium silicates), below which ra-dioactive elements were concentrated. The first 700 million years of lunar history constituted a period of intense bombardment, culminating in a cataclysm during which the *basins and larger *highland craters were formed. The eruption of radioactive basalts (KREEP), or their excavation during formation of the Imbrium Basin, resulted in a concentration of natural radioactivity in the western hemisphere, as measured by gamma-ray spectrometers. A zone of melting then moved inwards towards the moon's centre to generate a sequence of basaltic lavas that produced the *maria. Since then the moon has been relatively quiescent.

The origin of the moon is still debatable. In the *fission hypothesis*, the moon formed from one or more fragments that broke off from the earth; this is now considered unlikely. *Capture theories* propose formation elsewhere and subsequent capture by the earth's gravitational field; these are improbable but could explain the absence of a metallic core and why the earth has such a massive companion. It seems more likely that the moon may have formed as part of a double planet or may have coalesced from a swarm of minor earth satellites.

See also moon, surface features; moonrocks.

moonquakes. Localized disturbances inside the moon, detected by the *Apollo seismic network. Rare but strong moonquakes occur in the crust, although they are much less intense than earthquakes by a factor of about 1000. Weaker but more numerous moonquakes occur at depths of 900–1000 km and are triggered at *perigee by lunar *tides. Seismic signals on the moon are extended by scattering in the dry and porous lunar crust. Deep moonquakes provide information, through their arrival times, of the structure of the lunar interior: a 60 km crust, a 1200 km mantle, and a central partially molten core 500 km in radius.

moonrocks (lunar rocks). Samples of the moon returned by *Apollo and *Luna spacecraft. Lunar soils, or fines, consist of glasses, aggregates, and meteorite debris in addition to recognizable fragments of larger rock types. Hand specimens include mare basalts (low titanium content from the western *maria, high titanium content from the eastern maria, and aluminous basalts from Mare Fecunditatis), non-mare basalt (also known as *KREEP* because of its high content of potassium, rare earth elements, and phosphorus), and anorthositic rocks from the *highlands. Rare lunar-rock types include granite, ultramafic green glass, pyroxenite, norite, troctolite, and dunite. Most highland rocks have been impact metamorphosed (to produce cataclastic, recrystallized, and impact melt textures), but cumulates are also observed and these date back to the formation of the first lunar crust.

Lunar rocks are depleted in volatile and siderophile elements. Strong correlations exist between elements that are geochemically similar to one another (e.g. iron and manganese) or that are both excluded by major minerals (e.g. potassium and lanthanum). The major lunar minerals are calcic plagioclase, clino- and ortho-pyroxene, olivine, and ilmenite. Minor phases include spinels, K-feldspar, troilite, metallic iron, quartz, tridymite, cristobalite, apatite, and whitlockite. Uniquely lunar minerals include armalcolite, tranquillityite, and pyroxferroite. No hydrated phases, highly oxidized minerals, diamonds, or mineralization have yet been found.

moon, surface features. The surface of the moon is characterized by light-coloured *highlands interspersed with much lower *albedo (i.e. darker) and less rugged *maria. The oldest and largest features on the moon are the vast impact *basins. These grade into large *craters, with which highland areas are saturated. There is no evidence for plate tectonics on the moon (*but see* lunar grid) but *faults, *rilles, *wrinkle and *lobate ridges, *domes, *ghost craters, and volcanic plateaus are the expressions of past tectonic and igneous activity. Impact features include *rays, *dark halo craters, *secondary craters (including *crater clusters and *crater chains), and textured *ejecta blankets. *See also* mountains, lunar.

Moreton wave. *See* flares.

Morgan's classification. A classification scheme for galaxies, devised by W. W. Morgan, in which galaxies are grouped according to the composite spectra of their component stars and also by their form, i.e. spiral, barred spiral, elliptical, irregular, etc., by their degree of concentration, and by their orientation, which varies from face-on to edge-on.

morning star. A term normally applied to *Venus when it shines brightly in the eastern sky before sunrise. Such periods last while Venus moves between inferior *conjunction and superior conjunction and is located west of the sun. The term occasionally refers to spells of morning visibility of Mercury or other planets.

MOST. *Abbrev. for* Molonglo Observatory Synthesis Telescope. *See* Molonglo Radio Observatory.

mountains, lunar. Raised regions on the moon, most of which consist of crustal blocks uplifted to define the multiple ring structures of impact *basins; these survive either as *ring mountains,* such as the Apennines, or as isolated peaks, such as La Hire (see table). In comparison the lunar *highlands are topographically higher than the *maria but are characterized

lunar mountains	location
Alps	N Imbrium, part of second ring
Altai	SW Nectaris, part of third ring
Apennines	SE Imbrium, part of third ring
Apollonius	S Crisium, part of third ring
Aristarchus Plateau	Procellarum, Volcanic province
Carpathians	S Imbrium, part of third ring
Caucasus	W Imbrium, part of third ring
Cordillera	Orientale basin, third ring
Haemus	S Serenitatis, part of second ring
Harbinger	SW Imbrium, part of third ring
La Hire	Imbrium, isolated peak, part of first ring
Marius Hills	Procellarum, Volcanic province
Piton	Imbrium, isolated peak
Pyrenees	W Nectaris, part of first ring
Riphaeus	W Cognitum
Rook	Orientale basin, second ring
Rümker Hills	Sinus Roris, volcanic province
Spitzbergen	N Imbrium, part of first ring
Straight	N Imbrium, part of first ring
Taurus	E Serenitatis, part of second ring
Tenerife	N Imbrium, part of first ring

more by the elevated rims of craters and the rolling hills of basin *ejecta blankets. Discrete mountains not associated with impact basins include the central peaks of *craters, such as Tsiolkovsky, and volcanic domes and plateaus. Despite their jagged appearance, few mountains have slopes that exceed 12°.

Mount Hopkins Observatory. *See* Fred Lawrence Whipple Observatory.

mounting. The structure that supports an astronomical telescope, designed so that the telescope may be pointed at almost every part of the heavens, preferably without forcing an observer into awkward positions. All telescope mountings must therefore be constructed so that the tube can turn about two axes at right angles. There are two main types: the *equatorial mounting, with one axis parallel to the earth's axis, and the *altazimuth mounting, with one axis vertical. Since high magnifications exaggerate any movement, the support must be

rigid. In addition, because the earth is turning through 0.25° per minute, means must be provided to move the telescope smoothly in the opposite direction; this counterbalances the diurnal motion of a star and keeps it in the field of view.

Mount Stromlo Observatory. An observatory on Mount Stromlo, near Canberra, at an altitude of 770 metres. It is run by the Australian National University. Its principal instrument is a 1.9 metre (74 inch) reflecting telescope acquired in 1953.

Mount Wilson Observatory. An observatory on Mount Wilson near Pasadena, California, at an altitude of 1740 metres. It was founded in 1904 by G. E. Hale and is one of the *Hale Observatories. The first major instrument was a 60 inch (1.5 metre) reflector, completed in 1908, which was followed by a 100 inch (2.5 metre) reflector. The 100 inch, named the *Hooker telescope* after its benefactor, began regular operation in 1919 as the world's largest telescope. Its great success led Hale to consider an even larger telescope, eventually sited at *Palomar Observatory. There is now considerable *light pollution at Mount Wilson, which has restricted the observing programme.

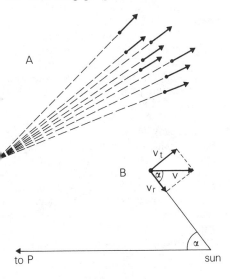

Moving cluster parallax

moving cluster. An *open cluster, such as the *Hyades or the *Ursa Major cluster, for which a distance can be derived from the individual radial velocities and proper motions of the member stars. The method is often termed *moving-cluster parallax.* The stars in a cluster share a common space motion and, if the system is sufficiently close to earth, their paths will appear to diverge from or converge towards a point in space (point P in the illustration). The direction of P is parallel to the space motion of the cluster. Considering each star in turn (illustration B):
$$v_r = v \cos \alpha$$
$$v_t = v \sin \alpha = 4.74\mu/p$$
where v, v_r, v_t are respectively the space, *radial, and tangential velocities of the star in km/s, μ is the proper motion in arc seconds per year, and p is the *parallax in arc seconds. Determination of α, μ, and v_r for each star gives a value for the parallax of the system.

moving-cluster parallax. *See* moving cluster.

MS stars. *See* S stars.

M stars. Stars of *spectral type M, cool red stars with surface temperatures (T_{eff}) less than about 3500 kelvin. Molecular bands are prominent in the spectra, with titanium oxide (TiO) bands dominant by M5. Lines of neutral metals are also present. Some show emission lines too, and are classed as *Me stars. Antares and Betelgeuse are M stars. *See also* carbon stars; S stars.

MTRLI. *Abbrev. for* Multi-Telescope Radio-Linked Interferometer, renamed *MERLIN.

Mullard Radio Astronomy Observatory. The radio observatory of the Cavendish Laboratory, Cambridge University, near Cambridge. The chief instruments are antenna arrays: the Five Kilometre Telescope (*see* aperture synthesis) with its eight paraboloid reflectors, which was in operation by 1972, and the One Mile Telescope of two fixed and one movable paraboloid, which came into use in 1964. There is also an aperture synthesis array of 60 *Yagis with a baseline of 5 km and an array of 4096 *dipoles covering an area of 36 000 m^2 (9 acres), both for use at low frequencies.

multilayers. *See* coating of lenses; interference filter; mirror.

multimission modular spacecraft (MMS). A spacecraft designed by NASA so that its basic structure can be used on a variety of missions in low earth orbit. The *Solar Maximum Mission was the first satellite to use the design. The MMS consists of a three-sided steel frame to which are attached three modules for attitude control, spacecraft commands and data handling (i.e. communications), and power supply; a propulsion unit can be added for orbital changes, if necessary. The payload of instruments to be taken into space is mated to the main platform of the MMS. The design is compatible with that of the *space shuttle, MMS plus payload fitting into the 14 metre wide cradle in the cargo bay.

Multiple Mirror Telescope (MMT). A reflecting telescope of totally new design, sited at the *Fred Lawrence Whipple Observatory, on Mount Hopkins, near Tucson, Arizona at an altitude of 2382 metres. It is jointly operated by the Smithsonian Institution and the University of Arizona's Steward Observatory. The telescope was fully operational in 1980. It comprises six identical mirrors, 1.8 metres (72 inches) in diameter, that are arranged in a hexagonal array on an *altazimuth mounting. The MMT has the light-gathering power of a 4.5 metre (176 inch) single mirror telescope and is thus the third largest telescope in the world; however it cost considerably less than a single 4.5 metre mirror to construct. It is designed for use at both visible and infrared wavelengths.

The mirrors are mounted symmetrically about a central axis. Their images are sent to a six-sided hyperbolic secondary mirror system on the central axis, which produces a single image at a common 'Cassegrain' focus; there is one secondary system for visible wavelengths and another, of identical figure but smaller size, for infrared wavelengths. An alternative focus can also be used. In perfect *seeing conditions star images of less than 0.5 arc seconds can be produced. The six mirrors remain focused at the same point by means of a feedback mechanism: the position of six laser beams, reflected by the six main mirrors are monitored by detectors linked to a computer. Any discrepancies arising from thermal or gravitational stresses are corrected by computer-controlled repositioning of the secondary mirror systems. The whole telescope can be smoothly guided to very great accuracies by computer-controlled drive mechanisms.

The multiple mirror concept, with its complicated alignment and guidance system, has proved reliable and is the design proposed for several future giant telescopes.

multiple star. A group of two or more stars, such as *Castor or *Alpha Centauri, that are held together by their mutual gravitational attraction. Orbital motions within multiple systems are complex. For instance, systems containing two or more close binaries in orbit about each other appear to be quite common. *See also* binary star.

multistage rocket. *See* launch vehicle.

muon. Symbol: μ. An *elementary particle (a lepton) with the same charge and *spin as an *electron but with a mass that is 207 times greater. It decays into electrons and neutrinos with a lifetime of two microseconds. *Cosmic-ray showers detected on earth are comprised mainly of muons produced by *pion decay. Muons do not interact strongly with matter and hence a large proportion of them reach the earth's surface. The *antiparticle of the muon is the antimuon.

mural quadrant. *See* quadrant.

Musca (Fly). A small constellation in the southern hemisphere near Crux, lying in the Milky Way, the brightest stars being of 3rd magnitude. Abbrev.: Mus; genitive form: Muscae; approx. position: RA 13h, dec −70°; area: 138 sq deg.

N

nadir. The point on the *celestial sphere that lies directly beneath an observer and is therefore unobservable. It is diametrically opposite the observer's *zenith and with the zenith lies on the observer's *meridian.

naked singularity. *See* singularity.

Nançay. Site near Vierzon, south of Paris, of the Nançay Radio Astronomy Station of the Paris Observatory. The chief instrument is a *Kraus-type radio telescope comprising a 305 metre (1000 foot) partially steerable (meridian transit) array, which has been in operation since 1964.

nano-: Symbol: n. A prefix to a unit, indicating a fraction of 10^{-9} of that unit. For example, one nanometre is 10^{-9} metres.

Naos. *See* Puppis.

NASA. *Abbrev. for* National Aeronautics and Space Administration. The US civilian agency that was formed in Oct. 1958 and is responsible for all nonmilitary aspects of the US space programme. Its major achievements so far have been the reusable *space shuttle, the *Apollo programme, which followed the *Mercury and *Gemini projects, the *Skylab space station, and a large number of artificial *satellites and *planetary probes. It collaborates on projects with the *European Space Agency (ESA) and/ or with individual countries. NASA offers a satellite launcher service to other countries.

Nasmyth focus. Either of two fixed focal points on the altitude axis of a telescope with an *altazimuth mounting. Light reflected from the secondary mirror of the telescope is diverted sideways by a third mirror mounted at 45° to the altitude axis. The light is brought to a focus at one of the two Nasmyth points on either side of the telescope, chosen by turning the third mirror. Bulky equipment can be mounted on observation platforms to use these foci. The arrangement was devised by James Nasmyth in the 1840s but has only recently come to prominence.

National Geographic Society-Palomar Observatory Sky Survey. *See* Palomar Sky Survey.

National New Technology Telescope (NNTT). The proposed US 15 metre reflector that will be the world's largest optical telescope. A four-mirror design has been selected in favour of a single segmented mirror. The design is similar to that of the *Multiple Mirror Telescope. The telescope should be operational by the early 1990s.

National Optical Astronomy Observatories (NOAO). A US organization consisting of *Kitt Peak National Observatory, Arizona, *Cerro Tololo Interamerican Observatory, Chile, and the *National Solar Observatory. The headquarters are in Tucson, Arizona.

National Radio Astronomy Observatory (NRAO). The US radio observatory at Green Bank, West Virginia. The principal instruments include a partially steerable (meridian transit) 91 metre (300 foot) paraboloid dish, completed in 1962, a fully steerable 43 metre (140 foot) dish, completed in 1965, and a four-element 1.6 km baseline interferometer. The Observatory also operates the Very Large Array in New Mexico (*see* aperture synthesis) and a millimetre wavelength radio telescope at *Kitt Peak, Arizona.

National Solar Observatory. A US organization consisting of the solar instruments of Kitt Peak, Arizona and those at Sacramento Peak, New Mexico.

nautical twilight. *See* twilight.

n-body problem. Any problem in celestial mechanics that involves the determination of the trajectories of *n* point masses whose only interaction is gravitational attraction. The bodies in the solar system are an example if it is assumed that the masses of the planets, etc., are concentrated at their *centres of mass. A general solution exists for the *two-body problem and in special cases a solution can be found for the *three-body problem. The complete solution for a larger number is normally considered insoluble.

neap tide. *See* tides.

nearest stars. The stars nearest to the sun. They are either comparable in luminosity to or fainter than the sun. Most are less massive than the sun and a large proportion belong to binary or triple star systems. *See* table 7, backmatter.

nearside. The hemisphere of the moon that is permanently turned towards the earth because of the tidal synchronization of the *lunar day and the *sidereal month. The lunar *librations allow the terrestrial observer also to see 9% of the *farside,* the hemisphere that is permanently turned away from the earth. The farside was first photographed by the Soviet *Luna 3 in 1959; it has no major *maria.

nebula. A cloud of interstellar gas and dust that can be observed either as a luminous patch of light – a bright nebula – or as a dark hole or band against a brighter background – a dark nebula. Interstellar clouds cannot usually be detected optically but

various processes can cause them to become visible.

With *emission nebulae ultraviolet radiation, generally coming from nearby hot stars, ionizes the interstellar gas atoms and light is emitted by the atoms as they interact with the free electrons in the nebula. Emission nebulae can be in the form of *HII (ionized hydrogen) regions, *planetary nebulae, or *supernova remnants. With *reflection nebulae light from a nearby star or stellar group is scattered (irregularly reflected) by the dust grains in the cloud. Reflection and emission nebulae are bright nebulae. In contrast *dark nebulae are detected by what they obscure: the light from stars and other objects lying behind the cloud along our line of sight is significantly decreased or totally obscured by *interstellar extinction. Dark nebulae contain approximately the same mixture of gas and dust as bright nebulae but there are no nearby stars to illuminate them.

The term 'nebula' was originally applied to any object that appeared fuzzy and extended in a telescope: over 100 were listed in the 18th-century *Messier catalogue. The majority of these objects were later identified as *galaxies and star *clusters.

nebular hypothesis. A theory for the origin of the solar system put forward by the French mathematician Pierre-Simon de Laplace in 1796 and similar to a suggestion of the philosopher Immanuel Kant in 1755. It was proposed that the solar system formed from a cloud, or nebula, of material collapsing under its own gravitational attraction. Rotation of the nebula caused it to form a disc that, because of the law of conservation of *angular momentum, rotated more rapidly as the contraction progressed. Laplace suggested that rings of material became detached from the spinning disc when the velocity at its edge exceeded a critical value, and that the material in these rings later coalesced to form the planets. The central product of the contraction is the sun, while the planetary satellites may have formed from further rings shed by the condensing planets.

The hypothesis, popular throughout the 19th century, went out of favour when it could not explain why the sun has almost 99.9% of the mass of the solar system but only 3% of the total angular momentum, and because calculations showed that the rings would not condense to form planets. However, in a modified form, it is the basis of most modern ideas for the formation of the sun and planets. *See* solar system, origin. *Compare* encounter theories.

Neptune. The eighth major planet, orbiting the sun every 165 years at an almost constant distance of 30.1 AU. It is the most distant *giant planet and resembles *Uranus in many respects; it does however have a slightly smaller diameter (48 600 km), a greater mass (17.2 earth-masses), and consequently a higher density (1.76 times that of water). Its two satellites, *Triton and *Nereid, move in unusual orbits. Orbital and physical characteristics are given in Table 1, backmatter.

Neptune was discovered in 1846 by J.G. Galle at the Berlin Observatory on the basis of predictions supplied by the Frenchman Urbain Leverrier, who had analysed observed perturbations in the motion of Uranus. Similar predictions had been made by the English mathematician John Couch Adams.

The planet returns to *opposition two days later each year, appearing as a magnitude 7.7 object visible in binoculars. No markings are seen on its greenish disc, only 2.3 arc seconds across. The presence of methane and hydrogen has been revealed by spectroscopic study of sunlight reflected from what is probably a visible surface of methane clouds. The rotation period, about an axis tilted by 29° to

the vertical, is now thought to be about 18 hours.

The atmospheric and internal structure may be almost identical to those of Uranus, with a rocky core 16 000 km in diameter encased in an 8000 km thick ice jacket and another 8000 km of, mainly, molecular hydrogen. To account for the greater overall mass, the core may be more massive than that of Uranus.

Neptune has an effective temperature of −220°C, a few degrees warmer than is to be expected at its solar distance and indicating that it, like Jupiter and Saturn, radiates more energy than it receives from the sun. Possibly this heat is generated by resistance to tides raised by its nearby massive satellite, Triton.

Nereid. A small satellite of Neptune with a highly eccentric orbit that takes it between 2 and 10 million km from the planet. It was discovered in 1949. *See* Table 2, backmatter.

neutrino. Symbol: ν. A stable *elementary particle with zero *charge and *spin ½. Three types of neutrinos are known: one is produced in association with an *electron (ν_e) and another in association with a *muon (ν_μ); a third neutrino is associated with the massive tau lepton. The *antiparticle of the neutrino is the *antineutrino ($\bar{\nu}$). Neutrinos were originally thought to have zero *rest mass. There is some experimental evidence that the electron neutrino may possess a finite rest mass of about 20–30 electronvolts.

Wolfgang Pauli first postulated the existence of the neutrino in the 1930s to explain the violation of the laws of conservation of energy and momentum in *beta decay. Neutrinos are leptons and thus only have a weak interaction with matter. Neutrinos produced in nuclear reactions in the centres of stars can thus escape without further collision. *See also* neutrino detection experiments.

neutrino detection experiments. Experiments to detect *neutrinos released during energy-producing nuclear reactions occurring in the centres of stars. Neutrino interaction with other particles is highly improbable so that the great majority can escape from the star and travel unhindered through space. They thus provide direct evidence of conditions in the stellar core.

An experiment to detect *solar neutrinos* was first set up by Raymond Davies, Jr., in the 1960s and is still in progress. It is being conducted 1500 metres underground in the Homestake Gold Mine near Lead, South Dakota. A large quantity of the dry-cleaning fluid perchloroethylene, C_2Cl_4, is exposed for several months to the presumed flux of solar neutrinos. A very small (but calculable) number of the high-energy particles should react with the chlorine atoms:
$$^{37}Cl + \nu_e \rightarrow ^{37}Ar + e^-$$
The amount of radioactive argon, ^{37}Ar, produced by the reaction gives a measure of the neutrino flux. The flux is measured in *solar neutrino units (SNU)*, where one SNU is the flux giving rise to one neutrino-induced event per second per 10^{36} atoms of chlorine. The average flux detected is 1.8 ± 0.2 SNU. This is much lower than the predicted value of 7.5 ± 0.5 SNU.

The expected neutrinos arise from one of the rarer pathways in the *proton-proton chain reaction whereby hydrogen is converted into helium. (The proton-proton chain is the most important energy-producing reaction in the sun.) The neutrinos are emitted with high energy during the decay of boron–8, which in turn results from a reaction between nuclei of helium–3 and helium–4. The rate of neutrino production has been calculated from the expected conditions in the solar core and is very temperature-dependent.

The discrepancy between the observed and predicted flux of solar neutrinos has not yet been explained.

It could imply that the conditions in the sun's interior and/or the nuclear reactions taking place there are not fully appreciated. Alternatively it could be that the characteristics of the neutrino itself are not fully appreciated and that its possible possession of rest mass could provide an explanation. Experiments are now planned and underway in the US, the USSR, and elsewhere to detect the less energetic but more numerous neutrinos that are released in the earliest step of the proton-proton chain. For example, these neutrinos can be absorbed by nuclei of gallium–71 to produce germanium–71; the amount of ^{71}Ge formed will then be a measure of the rate of a more fundamental step in the energy-producing reaction.

neutron. Symbol: n. An *elementary particle (a baryon) that is present in the *nucleus of all atoms except hydrogen. It has zero *charge, *spin $\frac{1}{2}$, and a *rest mass of 1.6749×10^{-24} grams – slightly greater than that of the *proton. The absence of charge enables the neutron to penetrate atoms easily since it has a negligible electromagnetic interaction with the constituents of the atom. Unlike free protons, free neutrons are unstable: they *beta decay into protons, electrons, plus antineutrinos with a mean life of approximately 930 seconds. The small difference between the neutron and proton mass provides the energy for this decay. The neutron is stable, however, when bound in the nucleus of a nonradioactive atom. The *antiparticle of the neutron is the *antineutron*. See also* isotopes; nucleosynthesis; s-process; r-process; neutron star.

neutron star. A star that has undergone *gravitational collapse to such a degree that most of its material has been compressed into *neutrons. Neutron stars were postulated in the 1930s by a number of astronomers including Landau and Zwicky. They are thought to form when the mass of the stellar core remaining after a *supernova explosion exceeds the *Chandrasekhar limit, i.e. about 1.4 solar masses. Such a core is not stable as a *white dwarf star since even the pressure of *degenerate electrons cannot withstand the strong gravity. Collapse continues until the mean density reaches 10^{17} kg m^{-3} and the protons and electrons coalesce into neutrons, supporting the star against further contraction. Accretion of gas on to an existing white dwarf, raising its mass above the Chandrasekhar limit, may also lead to its collapse to become a neutron star.

With diameters of only 20–30 km, intense magnetic fields (10^8 tesla), and extremely rapid spin, young neutron stars are believed to be responsible for the *pulsar phenomenon. Their strong gravity is also thought to give rise to rapid heating of material observed in some *x-ray binary systems. Older pulsars, 'spun up' by accretion are detected as *millisecond pulsars, the fastest rotating neutron stars. Models of the structure of neutron stars have been derived from sudden changes in pulsar spin rates – so called glitches. A typical neutron star may have an atmosphere only a few centimetres thick, under which is a crystalline crust about one kilometre in depth. Beneath the crust the material is thought to act like a superfluid of neutrons (having zero viscosity) all the way through to a solid crystalline core.

New General Catalogue. See NGC.

new moon. See phases, lunar.

New Style (NS). The current method of measuring dates using the *Gregorian calendar as distinct from the former method – *Old Style (OS)* – that was based on the *Julian calendar. Britain and the American colonies switched calendars on Sept. 14 1752,

the 11 days from Sept. 3 to Sept. 13 having been eliminated from that year.

New Technology Telescope (NTT). The 3.5 metre reflecting telescope under construction for the *European Southern Observatory. Its major feature is an active optics system whereby small corrections can be made continuously to the shape of the primary mirror, and to the orientation of the secondary mirror. Image motion arising from atmospheric *seeing effects should thus be corrected, leading to a diffraction-limited performance at all times (see Airy disc). It will have an *altazimuth mounting.

newton. Symbol: N. The SI unit of force, equal to the force that gives a mass of one kilogram an acceleration of one metre per second per second.

Newtonian telescope. The first *reflecting telescope to be built, developed in about 1670 by Isaac Newton from the ideas of Zucchi and Gregory: Newton thought that there was no way in which the *chromatic aberration of refracting telescopes could be corrected. The Newtonian has a paraboloid *primary mirror and a flat *diagonal mirror* that reflects the light into an *eyepiece mounted on the side of the tube (see illustration). A 45° prism can be used as an alternative to the diagonal mirror. The Newtonian is still the form used by many amateurs although somewhat inconvenient to use in larger form. Newton's original telescope had a mirror only 2.5 cm in diameter and made of *speculum metal; it had a focal length of about 15 cm and magnified some 30 times.

Newton's law of gravitation. See gravitation.

Newton's laws of motion. The three fundamental laws concerning the motion of bodies that were formulated

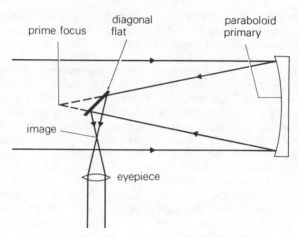

prime focus   diagonal flat   paraboloid primary

image

eyepiece

Ray path in Newtonian telescope

by Isaac Newton and published together with the law of *gravitation in *Principia*, 1687. The laws are

1. Every body continues in a state of rest or of uniform motion in a straight line until that state is changed by the action of a force on the body.

2. The rate of change of linear *momentum is proportional to the applied force, F, and occurs in the same direction as that of the force, i.e.

$$F = \mathrm{d}(mv)/\mathrm{d}t = m(\mathrm{d}v/\mathrm{d}t) = ma$$

where m is the mass, v the velocity, and a the resulting acceleration of the body.

3. Every action is opposed by a reaction of equal magnitude that acts in the opposite direction to the action.

The first law was conceived by Galileo who first realized the falsity of the Greek notion that a force is required to maintain a body in motion. Newton's laws of motion and of gravitation are fundamental to *celestial mechanics.

N galaxy (N-type galaxy). A distant galaxy with an exceedingly bright (and sometimes variable) nucleus; they were originally identified by W. W. Morgan in 1958 as one of the op-

tical counterparts of powerful radio sources. Many radio-quiet N galaxies have since been discovered. X-ray emission has also been detected from several N-types. The luminosities of N galaxy nuclei put them between *quasars and *Seyfert galaxies in the scale of *active galaxies.

NGC. Short for *New General Catalogue of Nebulae and Clusters of Stars,* prepared by the Danish astronomer J.L.E. Dreyer at Armagh Observatory, Ireland, and published in 1888. It was based on an earlier catalogue compiled by the Herschel family. The original work listed 7840 currently known nebulae, clusters, and galaxies (not yet identified as such). Each object is numbered by *right ascension and its position and description given. The number is preceded by the letters NGC; this is the object's *NGC number,* an example being NGC 1976: the Orion nebula. Some of the brightest objects also appear in the *Messier catalogue.

Dreyer listed further discoveries in two supplements to the NGC. These were the first and second *Index Catalogues (IC),* published in 1895 and 1908. The three catalogues, now published as one volume, cover the whole sky and contain over 13 000 objects.

nightglow. *See* airglow.

Nix Olympica. *Earlier name for* Olympus Mons, assigned by telescopic observers to the bright-albedo feature they saw in its position.

nm. *Short for* nanometre, i.e. 10^{-9} metre.

Nobeyama Radio Observatory. A Japanese observatory run by the University of Tokyo and situated in the Nobeyama highland. It is concerned mainly with *millimetre astronomy. Its 45 metre parabolic dish became operational in 1982 and is the first antenna of this size to be used efficiently at millimetre wavelengths. There is also an *aperture synthesis telescope of five 10 metre parabolic dishes that can be moved along two 500 metre baselines with different orientations. Test observations began in 1982.

nodes. The two points at which the orbital plane of a celestial body intersects a reference plane, usually the plane of the *ecliptic or *celestial equator. When the body, such as the moon or a planet, crosses the reference plane from south to north it passes through the *ascending node;* crossing from north to south the body passes through its *descending node.* The line joining these two nodes is the *line of nodes.*

A celestial body might not cross the reference plane at the same points every time: the line of nodes then moves slowly forwards or backwards around the orbital plane, i.e. in the same or in the opposite direction to the body. This is termed *progression* or *regression of the nodes.* The moon's nodes regress along its orbital plane, making one complete revolution in 18.61 years. This period is equal to the period of nutation.

nodical month. *See* draconic month.

noise. 1. Unwanted fluctuations that occur in the output of any electronic or electrical device and, in a measuring instrument, ultimately limit its sensitivity. Noise can be caused by many different factors, for example by insufficient regulation of the voltage of a power supply, but the most important source is usually that produced by random agitation of electrons (Brownian motion) in a resistance, i.e. *thermal noise. White noise* is completely random and uncorrelated noise that is uniform in energy over equal intervals of frequency.

In electrical circuits a resistance at temperature T produces an exchangeable *noise power* in *bandwidth B of
$$P = kTB$$
where k is the Boltzmann constant. If this noise power is generated in the first stages of the measuring instrument it is amplified along with the signal, which results in uncertainty in the measurement. The contribution to the noise power of any part of an electric circuit can be specified by its *noise figure* or *noise factor, F,* which is defined by
$$F = (P_1 + P_0)/P_0$$
where P_1 is the extra noise power generated by the circuit and P_0 is the noise power present at the input terminals of the circuit due to the source resistance, i.e. the thermal noise. Thus if the circuit adds no extra noise, it has a noise figure of unity. F is usually expressed in decibels. *See also* sensitivity.

2. Fluctuations of a random nature in which the amplitude of the deflection from a given reference level cannot be predicted fom one moment to the next but whose statistical properties may be well-defined. Signals from *radio sources are of this kind, and in a *radio telescope the problem is to detect these random signals in the presence of noise generated in the measuring instruments. *See also* radiosource spectrum.

noise figure (noise factor). *See* noise.

noise power. *See* noise.

noise temperature (system temperature). *See* sensitivity.

nonrelativistic. Describing any phenomenon, object, etc., for which the effects of general or special *relativity can be disregarded or do not apply.

nonthermal emission (nonthermal radiation). Electromagnetic radiation, such as *synchrotron emission, that is produced, like *thermal emission, by the acceleration of electrons or other charged particles but is nonthermal in origin, i.e. its spectrum is not that of a perfect *black-body radiator.

noon. The time of day at which the sun crosses the local meridian and is at its highest point above the horizon. This time differs from 12.00 *local time by up to about 15 minutes before or after midday, the amount – the *equation of time – depending on the time of year.

Norma (Rule; Level). A small inconspicuous constellation in the southern hemisphere near Centaurus, lying in the Milky Way, the brightest stars being of 4th magnitude. Abbrev.: Nor; genitive form: Normae; approx. position: RA 16h, dec −50°; area: 165 sq deg.

North America nebula (NGC 7000). An *emission nebula nearly 2° across lying in the constellation Cygnus. It is an *HII region apparently associated with the exceptionally luminous and remote A2 supergiant Deneb.

north celestial pole. *See* celestial poles.

Northern Cross. *See* Cygnus.

north galactic pole. *See* galactic poles.

north galactic spur. *See* interstellar medium.

north point. *See* cardinal points.

north polar distance (NPD). The angular distance of a celestial object from the north *celestial pole, measured along the object's hour circle. It is the complement of the object's *declination $(90° − \delta)$ and is sometimes used instead of this coordinate in the *equatorial coordinate system.

North Polar Sequence. *See* magnitude.

North Star. *See* Polaris.

nova. A close *binary star system in which there is a sudden and unpredictable increase in brightness by maybe 10 magnitudes or more (see table). Novae are a class of *cataclysmic variable. In a typical spiral galaxy like our own, there are maybe 25 nova eruptions per year. The brightness increases to a maximum within days or sometimes weeks and then declines to a value probably close to its faint pre-nova magnitude, indicating that the eruption did not disrupt the bulk of the star. *Fast novae* usually increase in brightness by a factor of 10^5 in a few days, remaining at peak brightness for less than a week; they then decline steadily, initially quite rapidly, over several months. *Slow novae* reach maximum brightness more slowly and erratically, the increase being less than in fast novae, and then decline much more slowly. The total energy released, however, is about the same in both cases. Hydrogen-rich gas is ejected from the star resulting in the tremendous outflow of heat and light. The ejected matter forms a rapidly expanding shell of gas that can become visible as the nova fades. At later stages of the eruption the spectra of most novae show the bright forbidden lines characteristic of very low density *emission nebulae.

nova	type	m_v	
		before	max
Nova Persei 1901 (GK Per 2)	fast	13.5	+0.2
Nova Aquilae 1918 (V603 Aql 3)	fast	10.6	−1.1
Nova Pictoris 1925 (RR Pic)	slow	12.7	+1.2
Nova Herculis 1934 (DQ Her)	slow	14.3	+1.4
Nova Cygni 1975 (V1500 Cyg)	fast	> 20	+1.8

Like other cataclysmic variables, a nova is a close binary system in which one component is a *white dwarf. The other is a main-sequence star that is expanding to fill its Roche lobe (see equipotential surfaces) and is hence losing mass to the white dwarf: some of its gases 'overflow' to form a disc surrounding the white dwarf. The rate of overflow is about 10^{-9} solar masses per year, about ten times higher than in a *dwarf nova system, and as a result the disc is always as bright as a dwarf nova at maximum. Before a nova explosion the hot turbulent disc is the brightest part of the system, and its light makes a pre-nova appear as a bluish irregularly fluctuating 'star'. The hydrogen in the disc spirals down on to the surface of the white dwarf, and after a period of some 10 000 to 100 000 years enough has accumulated to react in a thermonuclear explosion – the nova outburst. The explosion leaves the system fundamentally unchanged, however, and the flow of gas resumes to reestablish the accretion disc around the white dwarf.

Novae are now named initially by constellation and year of observation; before 1925 they were numbered in order of observation. They are later given a variable star designation, as for example with DQ Herculis, i.e. Herculis 1934. See also recurrent nova; x-ray transients.

November swarm. See Leonids.

NPD. Abbrev. for north polar distance.

N stars. See carbon stars.

N-type galaxy. See N galaxy.

Nubecular Major, Minor. Alternative names for Large and Small Magellanic Clouds, respectively. See Magellanic Clouds.

nuclear fission. A process in which an atomic nucleus splits into two smaller nuclei, usually with the emission of neutrons or gamma rays. Fission may be spontaneous or induced by neutron or photon bombardment. It occurs in heavy elements such as uranium, thorium, and plutonium. The process is accompanied by the release of large amounts of energy.

nuclear fusion. A process in which two light nuclei join to yield a heavier nucleus. An example is the fusion of two hydrogen nuclei to give a deuterium nucleus plus a positron plus a neutrino; this reaction occurs in the sun. Such processes take place at very high temperatures (millions of kelvin) and are consequently called thermonuclear reactions. With light elements, fusion releases immense amounts of energy: fusion is the energy-producing process in stars.

Iron and lighter elements evolve energy in fusion reactions whereas heavier elements require an input of energy to maintain the reaction. Thus although most elements up to iron can be formed by fusion reactions in stars (*see* nucleosynthesis), heavier elements must be synthesized by other nuclear reactions. *See also* proton-proton chain reaction; carbon cycle.

nuclear reactions. Reactions involving changes in atomic nuclei. Nuclear reactions include *nuclear fusion and fission and transformations of nuclei by bombardment with particles or photons. The energy released by nuclear reactions is equivalent to a decrease in mass according to Einstein's equation ($E = mc^2$) and appears in the form of kinetic energy of the particles or photons of electromagnetic radiation.

nuclear timescale. The total nuclear-energy resources of a star divided by the rate of energy loss. For typical stars, the nuclear timescale is of the order of 10^{10} years, although the rate of energy loss may change during the star's lifetime if the reaction processes alter.

nucleon. One of the constituents of the atomic nucleus, i.e. a proton or a neutron.

nucleosynthesis. The creation of the *elements by nuclear reactions. To unravel the complex situation, the measured *cosmic abundances of the elements are interpreted in terms of their nuclear properties and the set of environments (temperature, density, etc.) in which they can be synthesized. Nucleosynthesis began when the temperature of the primitive universe had dropped to about 10^9 kelvin (K). This occurred approximately 100 seconds after the big bang (*see* big bang theory). Protons fused with neu-

trons to form deuterium nuclei and deuterium nuclei could then fuse to form helium. Most of the helium in the universe was formed at this time, along with deuterium and lithium, but very little of the heavier elements.

It is now known that helium and the heavier elements are synthesized in stars; this idea was first developed in 1956/57 by Fowler, Hoyle, and the Burbidges (*see* B^{2}FH). Nucleosynthesis has occurred continuously in the Galaxy for many thousands of millions of years as a by-product of *stellar evolution. While a star remains on the *main sequence, hydrogen in its central core will be converted to helium by the *proton-proton chain reaction or the *carbon cycle; the core temperature is then about 10^7 K.

When the central hydrogen supplies are exhausted, the star will begin to evolve off the main sequence. Its core, now composed of helium, will contract until a temperature of 10^8 K is reached; carbon-12 can then be formed by the *triple alpha process, i.e. by helium burning. In stars more than twice the sun's mass a sequence of reactions, involving further nuclear fusion, produces elements with a mass number, A, that is a multiple of four: ^{16}O, ^{20}Ne, ^{24}Mg, and then, at temperatures increasing up to about 3.5 $\times$ 10^9 K, ^{28}Si to ^{56}Fe. Even higher temperatures will trigger reactions by which almost all elements up to a mass number of 56 can be synthesized. The *iron-peak elements*, i.e. ^{56}Fe, ^{56}Ni, ^{56}Co, etc., represent the end of the nucleosynthesis sequence by nuclear fusion: further fusion would require rather than liberate energy since nuclei with this mass number have the maximum binding energy per nucleon.

The formation of nuclei with $A \geq 56$ requires nuclear reactions involving neutron capture: neutrons can be captured at comparatively low energies owing to their lack of charge. If there is a supply of free neutrons in a star, produced as by-products of nuclear-fusion reactions, the *s-process can slowly synthesize nuclei up to ^{209}Bi. An intense source of neutrons allows the *r-process to generate nuclei up to ^{254}Cf, or higher, in a very short period. Such intense neutron fluxes arise in *supernovae. The newly synthesized elements are precipitated into the *interstellar medium by various *mass-loss processes; these include stellar winds from giant stars, planetary nebulae, and nova explosions for elements up to silicon, and supernovae for the iron-peak elements and heavier nuclei.

nucleus. 1. The kilometre-sized body that is the permanent portion of a *comet and is thought by most researchers to be the fount of all cometary activity. The density is about 0.5 g cm^{-3}. It contains about 75% by mass ice (mainly water ice, 85%, but with liberal amounts of CO_2, CO, CH_4, and NH_3) and 25% by mass dust thought to have a composition similar to *carbonaceous chondritic *meteorites. It is often described as a *dirty snowball*. A comet of mass 10^{18} grams would have a nucleus of diameter about 6 km. A positive identification of the nucleus of one comet, *IRAS-Iraki-Alcock, has recently been achieved using radar. As yet no one has seen a cometary nucleus. Even comet Halley, which has a calculated mass of 10^{17} grams and a nucleus diameter of 6 km, did not exhibit a nucleus when it *transited the sun in 1910. It is hoped a nucleus can be observed by the spacecraft scheduled to encounter *Halley in 1986. The nucleus of dirty ice is surrounded by a thin (a few cm) layer of insulating dust from which the ice has sublimated. This enables it to survive over 200 perihelion passages. At each passage the nucleus loses on average a layer of material one metre thick. This material forms the *coma and *comet tails and also any associated *meteor stream.

A less likely model for the nucleus, which was prevalent up to the 1950s, regarded it as a vast collection (a swarm) of tiny particles. In this model a 10^{18} gram comet was thought to contain about 10^{25} particles, each of about 0.1 mm diameter, separated by tens of metres and occupying a volume of space about 100 000 km across. *See also* head.

2. The small core of an *atom, consisting of *protons and *neutrons bound together by strong nuclear forces. The nucleus has a positive charge equal to Ze, where e is the magnitude of the electron charge and Z the number of protons present – the atomic number. The total number of protons plus neutrons is the mass number, A. A given element is characterized by its atomic number but may, within limits, have different numbers of neutrons in its nuclei, giving rise to different *isotopes of the element. The total mass of the protons and neutrons bound together in a nucleus is less than when the particles are unbound. This mass difference is equivalent to the energy required to bind the particles together. Nuclei are represented by their chemical symbols to which numbers are attached, as with $^4_2$He where 2 is the atomic number of helium (2 protons) and 4 the mass number of the particular isotope.

3. The central region of a galaxy.

Nuffield Radio Astronomy Laboratories. *See* Jodrell Bank.

nutation. A slight periodic but irregular movement superimposed on the precessional circle (*see* precession) traced out by the celestial poles. It results from the varying distances and relative directions of the moon and sun, which continuously alter the strength and direction of their gravitational attraction on the earth. The primary component is lunar nutation, which causes the celestial pole to wander by ±9 arc seconds from its mean position in a period of 18.6 years.

Changes in *right ascension and *declination arising from precession are perceptibly modified by nutation.

O

OAO. *Abbrev. for* Orbiting Astronomical Observatory.

OB-association. *See* association.

OB-cluster. An *open cluster that contains main-sequence O and B stars of high luminosity. These bright and very massive stars evolve more rapidly than less massive stars; their presence in a cluster suggests that it is comparatively young (10 million years or less) and that it may contain stars still in the process of formation. Examples are NGC 2264 and *h and Chi Persei. The young low-mass *T Tauri stars are often observed in these clusters. *See also* association.

Oberon. The outermost satellite of Uranus, discovered in 1787. *See* Table 2, backmatter.

objective (object lens). The lens or lens system in a *refracting telescope that faces the observed object. The focal plane in which the image forms is termed the *prime focus*. *See also* aperture.

objective prism. A narrow-angle prism placed in front of the primary mirror or objective lens of a telescope. The prism disperses (*see* dispersion) the incident light very slightly so that each star image is spread out into a small *spectrum. The spectra of a field of stars can thus be recorded in a single photographic or electronic image and the chief spectral features of the stars can be quickly assessed (*see* spectral types).

oblateness (ellipticity). The degree of flattening of a celestial object, such as a planet, from a true spherical form, largely as a result of rotation. It is numerically equal to the difference between the equatorial and polar diameters of the object divided by the equatorial diameter. Saturn, with an oblateness of 0.108, is the most oblate planet.

oblique rotator theory. *See* magnetic stars.

obliquity. The angle between the equatorial and orbital planes of a celestial body.

obliquity of the ecliptic. Symbol: ε. The angle at which the *ecliptic is inclined to the *celestial equator. This angle, equal to the tilt of the earth's axis, is about $23°26'$. *Precession and *nutation cause this angle to change between the extremes of about $22.1°$ and $24.5°$, altering (at present diminishing) by about $0''.47$ per year. Its value on 2000 Jan. 1 will be $23°26'21''$.

observatory. A structure built primarily for astronomical observation, equipped nowadays with optical, radio, and/or infrared telescopes and, in the larger observatories, the associated equipment with which spectrographic, photometric, and other such measurements are made. The sites of modern optical and infrared observatories are selected very carefully so that there is maximum transmission of signals by the atmosphere (i.e. the sky is free from clouds and dust), minimal atmospheric turbulence (i.e. the *seeing is optimal), and minimum *light pollution (i.e. the night sky is dark). Most are sited in mountainous areas or on volcanic islands where the atmosphere is very thin and the absorption effects of water vapour are reduced. Radio observatories, less hampered by seeing conditions, clouds, and light pollution, must still be isolated from terrestrial radio and electrical interference. Observatories are now being built in both northern and southern hemispheres. *See* Table 11, backmatter.

Instruments can now be carried into space, away from the disturbing and absorbing effects of the earth's atmosphere. The greatest hopes for future astronomical studies lie in the manned space observatories and in unmanned remote-controlled equipment on *infrared, *ultraviolet, *x-ray, and *gamma-ray astronomy satellites and the proposed *Hubble Space Telescope.

OB stars. Stars of spectral type O or B, i.e. *O stars or *B stars, all of which are *ultraviolet stars.

occultation. Complete or partial obscuration of an astronomical object by another of larger apparent diameter, especially the moon or a planet. A solar eclipse is strictly an occultation. The precise timings of occultations provide information about planetary atmospheres, the dimensions of extended visible, radio, and x-ray objects, and the positions of objects, such as distant radio sources. *See also* grazing occultation; eclipse.

oceanus. A very large relatively smooth dark area on the surface of a planet or satellite. The word is used in the approved name of such a feature.

octahedrite. *See* iron meteorite.

Octans (Octant). An inconspicuous constellation containing and surrounding the south celestial pole, the brightest stars being of 4th magnitude. The 5th-magnitude Sigma Octantis is the nearest naked-eye star to the south pole. Abbrev.: Oct; genitive form: Octantis; approx. position: RA 0 – 24h, dec −80°; area: 291 sq deg.

Of stars. Young massive O stars belonging to earlier subdivisions than O5. In addition to a well-developed absorption spectrum they show selectively enhanced emission lines of ionized helium (He II) and nitrogen (N III) that can vary in intensity in an irregular manner. The emission lines arise in an unstable atmosphere that is being lost from the star. Of stars are the hottest, most luminous, and probably the most massive stars in the Galaxy: the current record holder is HD 93129A, near *Eta Carinae, with a temperature of 50 000 K, a luminosity of 3 million suns, and a mass of probably 120 solar masses (*see also* supermassive stars).

OGO. *Abbrev. for* Orbiting Geophysical Observatory.

OH. *See* hydroxyl radical.

OH/IR star. A very large extremely cool giant or supergiant star that is losing mass very rapidly and is detected only by its infrared radiation and by its hydroxyl (OH) maser emission (*see* maser source). The first OH/IR stars were found in a survey of OH maser sources in 1973; many more were detected by *IRAS's infrared survey in 1983. These stars are usually very long period *variables. *Mira stars often show similar but weaker maser and infrared emission. The OH/IR stars, however, are losing mass a hundred times faster than the Mira stars. They are surrounded by a very dense shell of *cosmic dust that shows absorption typical of silicates. The shells have temperatures ranging from 1000 K to only 100 K. Since the stars lose a solar mass of gas and dust in only 10 000 to 100 000 years, they must be changing very rapidly to their next evolutionary state, probably a *planetary nebula.

Olbers' paradox. Why is the sky dark at night? Heinrich Olbers in 1826, and earlier J.P.L. Chesaux in 1744, pointed out that an infinite and uniform universe, both unchanging and static, would produce a night sky of the same surface brightness as the sun: every line of sight would eventually strike a star, a typical example of which is the sun. This theoretical argument is obviously in disagreement with observation. The observable universe is, however, neither uniform, unchanging, nor static and does not extend infinitely back into the past. The paradox is then resolved since the *redshift of extragalactic radiation (i.e. the diminution in its energy) and in particular the youth of our universe make the background radiation field at optical wavelengths very low indeed – less than about 10^{-21} times the surface brightness of the sun.

Old Style (OS). *See* New Style.

Olympus Mons. The largest volcano on Mars and possibly the largest in the solar system. With an estimated age of 200 million years, it may be the youngest of the volcanoes on the *Tharsis Ridge. It has a base 550 km across and rises to an 80 km diameter caldera at an elevation of 25 km above the surrounding plains. Around much of the base is an escarpment up to 6 km high. For comparison, the largest volcano on earth, Hawaii's Mauna Kea, rises only 9.3 km from its 120 km diameter base on the floor of the Pacific Ocean. *See also* Mars, volcanoes.

Omega Centauri. An impressive *globular cluster in the constellation Centaurus that has a surprising range of metal content and hence age in its members.

Omega nebula (Swan nebula; M17; NGC 6618). An *emission nebula with a conspicuous bar that lies in the direction of Sagittarius, very close to Scutum. Its apparent magnitude is 7. It is a double radio source.

Onsala Space Observatory. A radio observatory of Chalmers University of Technology, Onsala, Sweden. It has a 20 metre dish for observations at millimetre wavelengths.

Oort cloud. A cloud of *comets that move in orbits round the sun with perihelia in the range 5–30 AU and aphelia in a zone with heliocentric distances between 30 000 and 100 000 AU, i.e. in a zone far beyond Pluto's orbit. Since the nearest star to the sun – Proxima Centauri – is 270 000 AU away these comets can be perturbed by passing stars; this changes their orbits and send them in closer towards our sun (or out towards the perturbing star). These perturbed comets have periods of hundreds of thousands of years. The Oort cloud is so distant that it acts as a refrigerated cometary reservoir.

Its existence was proposed in 1950 by J.H. Oort who thought that comets originated from the disintegration of a planet some time in the early history of the solar system. After this the major planets, Jupiter in particular, drove the majority of the comets either into the inner solar system, where they decayed, or into the outer regions. Of this latter group 98% would have been lost from the solar system, only 2% remaining in the Oort cloud. The number of comets in the cloud is uncertain but it could be somewhere in the range 10^7 to 10^{11} with a total mass in the region 10^{25} to 10^{28} grams. The orbits are oriented at random. A cometary cloud of radius 100 000 AU has a half life of at least 1.1×10^9 years. The existence of the cloud shows up very clearly when the number distribution of comets is considered as a function of the reciprocal of the semimajor axes of the orbits.

Oort's constants. *See* Oort's formulae.

Oort's formulae. Two mathematical expressions derived by Jan H. Oort that describe the effects of differential *galactic rotation on the radial velocities (v_r) and tangential velocities (v_t) of stars at an average distance r from the sun. If each star is moving in a circular orbit about the galactic centre, and if r is small in comparison with the sun's distance from the centre, then

$$v_r = Ar \sin 2l$$
$$v_t = Br + Ar \cos 2l$$

where l is the galactic longitude (see galactic coordinate system) and A and B are *Oort's constants*. Since v_t is not directly measurable the second equation is usually replaced by the expression for *proper motion, μ:

$$\mu = 0.211(B + A \cos 2l)$$

The equations are valid for μ measured in arc seconds per year, velocities in km s^{-1}, and distances in parsecs. The most widely accepted values for Oort's constants are

A: 15 (km s^{-1})/kiloparsec
B: −10 (km s^{-1})/kpc

Most stars, including the sun, do not move in precisely circular orbits but once a correction has been made for the sun's velocity with respect to the average velocity of stars in the solar neighbourhood (assumed to be circular), the measured variation of v_r and μ with l is sufficiently close to that predicted by the equations to support the theory that the general rotation pattern is circular and galactocentric.

opacity. A measure of the ability of a gaseous, solid, or liquid body to absorb radiation. It is the ratio of the total radiant energy received by the body to the total energy transmitted through it. For gaseous material of a particular chemical composition, opacity depends on both temperature and density. Various processes contribute to the opacity: energy absorption by electrons bound to an atom or ion, allowing them to jump to a higher level or to escape from the atom as free electrons (bound-bound or *bound-free absorption); absorption by free electrons (*free-free absorp-

tion); scattering of photone of radiation by free electrons or atoms (*Compton scattering); absorption of γ-ray photons by ambient 'gas' photons (*photon-photon absorption). The negative *hydrogen ion, H$^-$, is a particularly important source of opacity in a star like the sun, and the solar opacity is seen to drop rapidly to zero at the sun's limb.

open cluster (galactic cluster). A loose *cluster of stars that contains at most a few hundred stars and sometimes less than twenty. Examples visible to the naked eye are the *Hyades and the *Pleiades. About 1000 open clusters are known. They are *population I systems and occur in or close to the plane of the Galaxy. The brightest stars in an open cluster can be either red or blue giants, depending upon its age. Stars in the older clusters, such as M67 in Cancer, are similar in appearance to those in *globular clusters, though with some subtle differences due to the higher metal content of the material from which they were formed. Open clusters are more loosely bound systems than globular clusters and they tend to be gradually dispersed by the combined effects of the differential rotation of the Galaxy and perturbations due to close encounters with interstellar clouds. Calculations suggest that many will not survive more than one or two circuits of the Galaxy. Hence most open clusters are comparatively young systems. Some, such as NGC 2264, are less than 10 million years old and in these clusters star-formation is probably still taking place (see OB-cluster). See also moving cluster; Hertzsprung-Russell diagram; turnoff point.

open universe. See universe; deceleration parameter.

Ophiuchids. A minor *meteor shower, radiant: RA 260°, dec −20°, maximizing on June 20.

Ophiuchus (Serpent Bearer). A large equatorial constellation near Scorpius, lying partly in the Milky Way. The brightest star is the 2nd-magnitude white giant Ras Alhague (α), which lies close to *Ras Algethi (Alpha Herculis). The area contains the supernova remnant of *Kepler's star, the dark *Ophiuchus nebula, and many globular clusters. Abbrev.: Oph; genitive form: Ophiuchi; approx. position: RA 17h, dec −7°; area: 948 sq deg.

Ophiucus nebula. A *dark nebula in the constellation Ophiucus, about 250 parsecs away.

opposition. The moment at which a body in the solar system has a longitude differing from that of the sun by 180°, so that it lies opposite the sun in the sky and crosses the meridian at about midnight (see elongation). The term also applies to the alignment of the two bodies at this moment. Although the *inferior planets cannot come to opposition, it is the most favourable time for observation of the other planets since they are then observable throughout the night and are near their closest point for that *apparition. See also synodic period.

optical axis. The imaginary line passing through the midpoint of a lens, mirror, or system of such elements and on which lies the focal point of parallel paraxial rays.

optical depth. Symbol: τ. A measure of the absorption of radiation of a particular wavelength as it passes through a gaseous medium. If the initial radiation flux Φ_0 is reduced to Φ_x after a distance x through the medium then
$$\Phi_x/\Phi_0 = \exp(-\tau)$$
where τ is the optical depth for the radiation wavelength. If τ equals zero the medium is transparent; if τ is much greater or much less than one the medium is *optically thick* or *opti-*

cally thin respectively. *See also* opacity.

optical double. 1. *See* double star.
2. *See* double galaxy.

optical libration. *Another name for* geometrical libration. *See* libration.

optical pathlength. The distance, d, travelled by a light beam multiplied by the *refractive index, n, of the medium through which the light has passed. If the light traverses more than one medium, then the optical pathlength is the sum
$$(n_1 d_1 + n_2 d_2 + \ldots)$$
for each medium. It gives the effective pathlength in terms of wavelength of light.

optical pulsars. *Pulsars that flash at visible wavelengths. Two are known: the Crab pulsar (see Crab nebula) appears as a 16th magnitude star but is pulsing with a period of 0.033 seconds; the optical flashes of the *Vela pulsar, about 26th magnitude, were discovered by looking for faint pulses with exactly the period (0.089 seconds) of its radio emission. Optical pulses have also been detected from *x-ray binary systems.

optical temperature. *See* effective temperature.

optical wavelengths. Wavelengths in the visible region of the spectrum, ranging from about 380 to 750 nanometres. *See* light.

optical window. *See* atmospheric windows.

orbit. The path followed by a celestial object or an artificial satellite or spaceprobe that is moving in a gravitational field. For a single object moving freely in the gravitational field of a massive body the orbit is a *conic section, in actuality either elliptical or hyperbolic. Closed (repeated) orbits are elliptical, most planetary orbits being almost circular. A hyperbolic orbit results in the object escaping from the vicinity of a massive body. *See also* orbital elements; Kepler's laws.

orbital elements. The parameters that specify the position and motion of a celestial body in its orbit and that can be established by observation (see illustration). *Osculating elements* specify the instantaneous position and velocity of a body in a perturbed orbit (*see* osculating orbit). *Mean elements* are those of some reference orbit that approximates the actual perturbed orbit. The shape and size of the orbit are specified by the *eccentricity, e, and for an elliptical orbit by the *semimajor axis, a. The orientation of the orbit in space is specified firstly by the *inclination, i, of the orbital plane to the reference plane, usually that of the ecliptic, and secondly by the *longitude of the ascending node*, Ω; the latter is the angular distance from the dynamical or vernal *equinox, Υ, to the ascending *node, N. The orientation of the orbit in the orbital plane is usually specified by the angular distance, ω, between the periapsis, P (*see* apsides), and the ascending node. For an orbit round the sun the periapsis is the *perihelion and the angular distance ω is the *argument of the perihelion*. It is measured along the direction of motion.

The eccentric *anomaly is used to determine the position of the body in its orbit. To calculate the position as a function of time an additional orbital element is used. This is the *time of periapsis* (or for a solar orbit *perihelion*) *passage*, t_0, by means of which Kepler's equation (*see* anomaly) can be solved. To determine the orbit of a binary star system of unknown mass, the *period must be established.

The earth's orbital elements vary with time due to gravitational effects of the moon and planets. The changes, approximately periodic, during the past several hundred thousand years have been calculated very accurately.

orbital velocity. The velocity of a satellite or other orbiting body at any given point in its orbit. It is also the velocity required by a satellite to enter an orbit around a body. The orbital velocity, v, is given by the expression

$$v = \sqrt{[gR^2(2/r - 1/a)]}$$

where R is the radius of the orbited body, r is the distance from the *centre of mass of the system (i.e. from the approximate centre of the primary), a is the *semimajor axis of the orbit, and g is the standard acceleration of gravity. For a circular orbit,

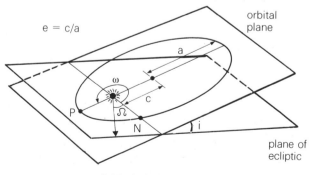

$e = c/a$

Orbital elements

$r = a$ and the circular velocity is given by

$$v = \sqrt{(gR^2/r)}$$

To escape from an orbit a must tend to infinity and the escape velocity is then given by

$$v_e = \sqrt{(2gR^2/r)}$$

The orbital period for an elliptical orbit is given by

$$P = 2\pi a^{3/2}/\sqrt{gR^2}$$

orbit determination. The derivation of the orbit of a celestial object, such as a comet, from several observations of its position by using the laws of celestial mechanics. At least three observations are needed to determine the *orbital elements and many methods have been used.

Orbiter probes. *See* Lunar Orbiter probes.

Orbiting Astronomical Observatory (OAO). Any of a series of four US satellites, two of which were successfully launched, that were primarily concerned with investigations in the ultraviolet (UV) region of the spectrum. OAO–2 was launched on Dec. 7 1968 and continued operating for over four years. In one group of experiments a UV survey of about one sixth of the sky was achieved by means of four telescopes plus UV-sensitive television camera tubes, the *Celescope experiment*, which produced pictures in the wavelength range 115–320 nanometres. A catalogue of bright UV sources was compiled from the data. In the other experimental package measurements were made of the UV luminosity and spectra of a large number of preselected targets. OAO–3 was launched on Aug. 21 1972 and was named *Copernicus after launch.

Orbiting Geophysical Observatory (OGO). Any of a series of US geophysical satellites, first launched in Sept. 1964, for studying the earth's atmosphere, ionosphere, magnetic field, radiation belts, etc., and how these are influenced by the *solar wind and other phenomena. Moving mainly in highly elliptical orbits, many studies could be made simultaneously for prolonged periods.

Orbiting Solar Observatory (OSO). Any of a series of US satellites designed to study the sun and solar phenomena, especially solar flares, from earth orbit. The first one, OSO–1, was launched in March 1962, OSO–8 being orbited in 1975. The quality and range of the observations improved as the series progressed, with high spectral and spatial resolution in ultraviolet and x-ray spectral regions in later craft. Instruments have included a *coronagraph, *spectroheliograph, ultraviolet and x-ray *spectrographs, *photometers, gamma-ray detectors, and particle-flux sensors.

Orientale Basin. The youngest lunar *basin, only partially filled by mare lava. The concentric ring structure of *mountains and the *ejecta blanket – the Hevelius formation – are exceptionally well preserved. Orientale is situated on the western limb of the moon and is only visible from earth at times of favourable *libration.

Orion. A very conspicuous constellation on the celestial equator, close to the Milky Way, that is visible as a whole from most parts of the world and can be used to indicate the positions of many bright neighbouring stars (see illustration). All the leading stars, apart from *Betelgeuse, are very hot, white, and luminous. The 1st-magnitude *Rigel (β) and Betelgeuse (α) and 2nd-magnitude *Bellatrix (γ) and Saiph (κ) form the quadrilateral outline inside which are the three remote 2nd-magnitude stars of *Orion's Belt,* Alnilam (ε), Alnitak (ζ), and Mintaka (δ). South of the Belt is *Orion's Sword,* in which lies the naked-eye gaseous *Orion nebula and the

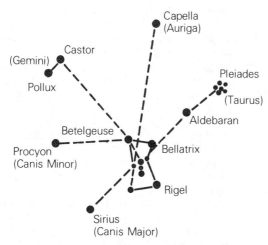

Orion and bright neighbouring stars

adjacent nebula M43. The area contains many double stars including Rigel, several variable stars, and the dark *Horsehead nebula. An immense shell of ionized hydrogen, *Barnard's loop*, surrounds most of Orion. Abbrev.: Ori; genitive form: Orionis; approx. position: RA 5.5h, dec 0°; area: 594 sq deg.

Orion arm. *See* Galaxy.

Orion association (I Orionis). A large OB-association (*see* association) that contains over 1000 stars and lies in the region of the Orion nebula. There are at least three clusters in the association: the Lambda Orionis cluster, the Belt group around Epsilon Orionis, and the *Trapezium cluster.

Orionids. An active *meteor shower, radiant: RA 96°, dec +15°, that maximizes on Oct. 21 with a *zenithal hourly rate of about 30. The activity has been observed to change by as much as a factor of four between one year and the next. The shower has been the centre of controversy firstly because it was thought that the *radiant was stationary during the three weeks of activity. This view has now been abandoned: all showers have

radiants that move with respect to the stars as the earth moves along its orbit. Secondly the shower is possibly associated with Halley's comet and the Eta *Aquarids because of similarities in orbital elements.

Orion molecular cloud (OMC-1). A dense cloud of neutral molecular hydrogen lying slightly behind, but in contact with, the *Orion nebula. It is part of a system of *molecular clouds in the Orion constellation. It has a total mass of about 500 solar masses and has a temperature of about 100 K. It is a very bright infrared source at a wavelength of 50 μm with a *luminosity 10^5 times greater than the sun's. Higher resolution measurements indicate that the centre of the cloud consists of a cluster of small infrared sources that may be evolving towards the stage the *Trapezium stars have already reached. It is likely that the Trapezium stars condensed from the same molecular cloud. Also associated with OMC-1 are the strong infrared sources, the *BN object and *Kleinmann-Low (KL) nebula, which together form the *BNKL complex*.

Orion nebula (M42; NGC 1976). One of the brightest *emission nebulae in the sky, about 500 parsecs distant. It is just visible to the naked eye as a diffuse luminous patch, 35 arc minutes across, in the centre of Orion's Sword. It is a complex region of ionized hydrogen (an *HII region) that is associated with the *Orion molecular cloud (OMC-1), part of a system of giant *molecular clouds in the Orion constellation. The Orion nebula is centred on four hot main-sequence stars, called the *Trapezium cluster, that excite and ionize the nebula so producing both radio and optical radiation. Their total *luminosity is about 3×10^5 times that of the sun. The nebula is also a source of x-rays. The region is one of active star formation, containing *T Tauri stars, *maser sources, *Herbig-Haro objects, the *BN object, and the *Kleinmann-Low nebula.

Orion's Belt. *See* Orion.

Orion's Sword. *See* Orion.

orrery. An instrument that demonstrates the movements of some or all of the planets around the sun, in their correct relative periods, by means of an elaborate, usually clock-driven system of gears. The movements of satellites, such as the moon, may be incorporated in the mechanism. The instrument was named after Charles Boyle, the 4th Earl of Orrery; they were previously called *planetaria*.

orthoscopic eyepiece. *See* eyepiece.

oscillating universe (pulsating universe). A closed *universe in which collapse is followed by a re-expansion.

osculating orbit. The truly elliptical orbit that a celestial object would follow if the perturbing forces of other bodies were to disappear so that it was subject only to the central gravitational field of a single massive body. In general elliptical motion is

an approximation to the real motion of members of the solar system. If all forces except the central gravitational force disappeared at time t, then the *osculating elements* would be the *orbital elements of the ellipse followed by a body at a given instant after time t. For a real body in perturbed motion the instantaneous osculating elements are not constant but are functions of time. The osculating elements are often used to describe the perturbed motion of a body.

OSO. *Abbrev. for* Orbiting Solar Observatory.

O stars. Stars of *spectral type O that are very hot massive blue stars, emitting copious quantities of ultraviolet radiation, and with surface temperatures (T_{eff}) from about 28 000 to over 40 000 kelvin. Strong lines of singly ionized helium (He II) dominate the spectra. Lines of doubly and triply ionized elements are also present with lines of neutral helium (He I) and hydrogen strengthening in later subdivisions. No O0, O1, or O2 stars have been discovered and only a few O3 and O4 stars are known. Most O stars are found in the spiral arms of galaxies, often in OB *associations, and are very fast rotators. They have short lifetimes of about three to six million years. *See also* Of stars; ultraviolet stars.

Otto Struve Telescope. *See* McDonald Observatory.

outburst, cometary. *See* Schwassmann-Wachmann 1.

outer planets. The planets Jupiter, Saturn, Uranus, Neptune, and Pluto, which lie further from the sun than the main belt of minor planets. With the exception of Pluto, all are *giant planets.

Owl nebula (M97; NGC 3587). A relatively large *planetary nebula in the constellation Ursa Major, about 600 parsecs away.

Ozma project. A pioneering project in the search for extraterrestrial intelligence. The 26 metre radio telescope at Green Bank, West Virginia, was used by Frank Drake in 1960 for about 150 hours in an unsuccessful attempt to detect signals – the 21 cm hydrogen line emission – from other civilizations. The targets were the two nearest and most sunlike stars Tau Ceti and Epsilon Eridani. *See also* SETI.

ozonosphere (ozone layer). *See* atmospheric layers.

P

pair production. The conversion of a high-energy (>1.022 MeV) gamma-ray *photon, usually in the field of an atomic nucleus, into an *electron and positron, which are formed simultaneously. *See also* annihilation; photon-photon absorption.

Pallas. The second *minor planet to be discovered, found in 1802. With a mean diameter of 540 km it is the second largest minor planet and has a spectrum similar to that of the *carbonaceous chondrite meteorites. *See* Table 3, backmatter.

Palomar Observatory. The observatory sited on Mount Palomar, California, at an altitude of 1710 metres. It is one of the *Hale Observatories. The site was chosen by G. E. Hale as suitable for a giant 200 inch (5 metre telescope, following the success of the 100 inch telescope at *Mount Wilson Observatory. The 200″ reflector, now known as the *Hale telescope*, saw first light in Dec. 1947. Regular observing began in 1949. It was the world's largest telescope until the Soviet 6

metre instrument was built. Now equipped with sophisticated electronic and computer systems, it remains one of the world's most powerful telescopes.

The 200″ mirror, ready for use in late 1947, almost 10 years after Hale's death, had been cast in 1934 after considerable design problems. It is of low-expansion Pyrex glass, with a reflecting surface of aluminium; it weighs 13.15 tonnes. The chosen *focal ratio of f/3.3 meant that a much shorter tube could be used than with previous telescopes, which traditionally had an f/5 ratio. The prime focus is 16.5 metres from the primary mirror; there is an observing cage at the focus where photographic work is done. A hyperboloid secondary mirror in the prime-focus cage can be used to provide a Cassegrain facility of focal ratio f/16 and two other similar mirrors with additional optics produce coudé systems of f/30.

In addition to the 200 inch there is a 48 inch *Schmidt telescope. This has a field of semiangle of three degrees, which produces a photographic plate about 35.5 cm square. It was used in the production of the *Palomar Sky Survey. It is one of the world's finest Schmidts.

Palomar Sky Survey. A photographic star atlas of the northern sky and part of the southern sky to a *declination of −33°. It was prepared at the Palomar Observatory, California, with the collaboration of the National Geographic Society, using the 48″ Schmidt telescope. It was released 1954–58. It consists of 935 pairs of photographic prints: one of each pair was made from an exposure on a blue-sensitive plate, the other was made from an exposure taken through a red filter. It includes stars to a limiting magnitude of 21 (20 in the southernmost plate). A new survey of the northern sky is now underway with the Palomar Schmidt, using much improved film. Objects four times fainter will be recorded. *See also* Southern Sky Survey.

palus (plural: **paludes**). A mixed smooth and rough dark area on the surface of a planet or satellite. The word is used in the approved name of such a feature.

pancake theory. *See* galaxies, formation and evolution.

parabola. A type of *conic section with an eccentricity equal to one. *See also* paraboloid.

parabolic velocity. The velocity of an object following a *parabolic trajectory* around a massive body. Its velocity at a given distance from the massive body is equal to the *escape velocity at that distance.

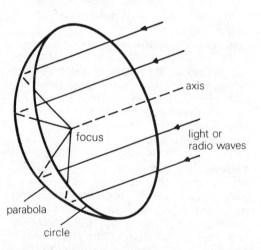

Paraboloid surface

paraboloid. A curved surface formed by the rotation of a parabola about its axis. Cross sections along the central axis are circular. A beam of radiation striking such a surface parallel to its axis is reflected to a single point on the axis (the focus), no matter how wide the aperture (see illustration). A paraboloid mirror is thus free of *spherical aberration; it does

however suffer from *coma. Paraboloid surfaces are used in reflecting telescopes and radio telescopes. Over a small area a paraboloid differs only slightly from a sphere. A paraboloid mirror can therefore be made by deepening the centre of a spherical mirror.

parallactic angle. *See* astronomical triangle.

parallactic ellipse. *See* annual parallax.

parallactic inequality. A minor periodic term in the moon's motion that results from the gravitational attraction of the sun and causes the moon to be ahead at first quarter and to lag behind at last quarter (*see* phases, lunar). The inequality can be used to calculate *solar parallax. *See also* evection; variation.

parallactic motion. *See* parallax.

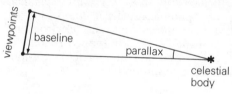

Parallax

parallax. The angular displacement in the apparent position of a celestial body when observed from two widely separated points. It is thus the angle that the baseline connecting two viewpoints would subtend at the object (see illustration). It is very small in value and is usually expressed in arc seconds. If the baseline is of a fixed length then as the distance to the celestial object decreases, its parallax will increase accordingly. If the parallax can be measured then so can the distance. 'Parallax' is thus often used synonymously with 'distance'.

The observer's position on earth changes with the daily rotation of the earth, the annual revolution of the earth around the sun, and the long-term motion of the sun and solar system relative to the background stars. Each motion produces a different measure of parallax: *diurnal parallax, *annual parallax, and *secular parallax, respectively. The continual change in the apparent position of a celestial object, produced by the observer's changing position, is termed *parallactic motion* and must be distinguished from the star's own *peculiar motion in space.

Methods used to determine the parallax and hence the distance of celestial bodies require an accurate knowledge of the baseline length. The baseline for diurnal parallax – the earth's radius – can only be used for distance measurements within the solar system. The baselines used in annual parallax – the semimajor axis of the earth's orbit – and secular parallax are sufficiently great for stellar distances to be measured. Other methods based on the concept of parallax and used in distance determination include *dynamical parallax, *moving-cluster parallax, and *rotational parallax.

Distances determined indirectly from stellar brightness are also sometimes called parallaxes: in *spectroscopic parallax* the absolute magnitude of a main-sequence star is deduced from the spectral type of the star using the *Hertzsprung-Russell diagram and this together with the apparent magnitude gives the *distance modulus and hence the distance. *See also* distance determination.

parametric amplifier. A type of negative-resistance amplifier that employs a nonlinear circuit element, such as a varactor diode, to act as a time-varying capacitance. These amplifiers have low *noise figures at high frequencies and are useful in *radio telescopes in the range 1–30 gigahertz.

paraxial. Lying close to the optical axis of a lens, mirror, or system of such elements.

Parkes. Site of the *Australian National Radio Astronomy Observatory. *See also* Australia Telescope.

parking orbit. An orbit in which a satellite or spacecraft can be placed before being injected into a final orbit or desired trajectory.

parsec. *Short for* parallax second. Symbol: pc. A unit of length normally used for distances beyond the solar system. It is the distance at which the semimajor axis of the earth's orbit subtends an angle of one arc second. It is thus the distance at which a star would have an *annual parallax of one arc second. A star with a parallax of p arc seconds is at a distance of d parsecs, given by $d = 1/p$ (accurate up to distances of about 30 pc). One parsec equals 30.857×10^{12} km, 206 265 astronomical units, and 3.2616 light-years.

partial eclipse. *See* eclipse.

particle. *See* elementary particle; alpha particle; beta particle.

pascal. Symbol: Pa. The SI unit of pressure, equal to a pressure of one newton per square metre.

Paschen series. *See* hydrogen spectrum.

Pasiphae. A small satellite of Jupiter, discovered in 1908. *See* Jupiter's satellites; Table 2, backmatter.

patera. An irregular crater with scalloped edges. The word is used in the approved name of such a surface feature on a planet or satellite.

pathlength. *See* optical pathlength.

path of totality (zone of totality). *See* eclipse.

Pauli exclusion principle. The principle that no two particles can exist in exactly the same quantum state. It is obeyed by *fermions but not by *bosons. The existence of *white dwarfs and *neutron stars is a consequence of this.

Pavo (Peacock). A constellation in the southern hemisphere near Grus, the brightest stars being the 2nd-magnitude spectroscopic binary Alpha Pavonis and two 3rd-magnitude stars. It contains several variables, including the bright Cepheid variable Kappa Pavonis (a W Virginis star), and the large bright globular cluster NGC 6752. Abbrev.: Pav; genitive form: Pavonis; approx. position: RA 19.5h, dec −65°; area: 378 sq deg.

Pavonis Mons. A 19 km high volcano on the *Tharsis Ridge of Mars. It has a base diameter of 400 km, the diameter of the summit caldera being 45 km. *See* Mars, volcanoes.

payload. 1. The total mass of a satellite, spacecraft, etc., that is carried into orbit by a launch vehicle. It is that part of the total launcher mass that is not necessary for the operation of the launcher. It is usually a small or very small fraction of the total launcher mass.
2. The mass of the experimental and operational equipment of a satellite, planetary probe, etc.

pc. *Symbol for* parsec.

P Cygni stars. A small group of hot variable stars whose prototype is the B1e star P Cygni. Their spectra show numerous strong emission lines, like those of *Be stars and *Wolf-Rayet stars, and sharp absorption lines that have a *blueshift. The blueshifted absorption components originate in an expanding shell of low-density matter continuously ejected from the hot star. Many, like P Cygni, suffer random outbursts. The characteristic shape of the spectral lines – blueshifted absorption on an emission feature – has led to the term *P Cygni profile* being used for similarly shaped spectral lines in other types of astronomical bodies. In all cases it is due to an outflow of absorbing gases. *See also* Hubble-Sandage variables.

peculiar galaxies. Galaxies that do not conveniently fit into the regular classification schemes of normal galaxies. These comparatively rare objects include *active galaxies, *starburst galaxies, Markarian galaxies, *Haro galaxies, *interacting galaxies, single irregular or disturbed galaxies, *dwarf galaxies, *ring galaxies, and spiral galaxies with peculiar spiral arms or even with one or three arms.

peculiar motion. 1. The motion of an individual star relative to a group of neighbouring stars.
2. That part of the *proper motion of a star that results from its actual movement in space; it is the star's proper motion after elimination of the effects of the sun's motion.

peculiar stars. Stars with spectral features that do not correspond exactly with the usual classification of *spectral types. In general they are designated by a 'p' after their spectral type, as with *Ap stars, but specific features are given other suffixes, e.g. 'e' for emission or 'm' for metallic lines.

Pegasus. An extensive conspicuous constellation in the northern hemisphere near Cygnus, with three 2nd-magnitude stars (Alpha, Beta, and Epsilon Pegasi) and four of 3rd magnitude. The stars *Markab (α), *Scheat (β), and *Algenib (γ) together with Alpha Andromedae (*Alpheratz) form the distinctive *Great Square of Pegasus*. The area also contains the globular cluster M15. Abbrev.: Peg; genitive form: Pegasi; approx. position: RA 23h, dec +20°; area: 1121 sq deg.

Pele. *See* Io.

pencil beam. *See* antenna.

penumbra. 1. *See* umbra.
   2. *See* sunspots.

penumbral eclipse. *See* eclipse.

periapsis. *See* apsides.

periastron. *See* apastron.

pericentre. The point in an orbit, e.g. of a planet or of a component of a binary star, that is nearest to the *centre of mass of the system. The furthest point is the *apocentre*.

pericynthion. The point in the lunar orbit of a satellite launched from earth that is nearest the moon. The equivalent point in the lunar orbit of a satellite launched from the moon is the *perilune*. *Compare* apocynthion.

perigalacticon. The point in a star's orbit around the *Galaxy that is nearest the galactic centre. The furthest point is the *apogalacticon*.

perigee. The point in the orbit of the moon or an artificial earth satellite that is nearest the earth and at which the body's velocity is maximal. Strictly, the distance to the perigee is taken from the earth's centre. The distance to the moon's perigee varies; on average it is about 363 300 km. *Compare* apogee.

perihelion. The point in the orbit of a planet, comet, or artificial satellite in solar orbit that is nearest the sun. The earth is at perihelion on Jan. 3. The time of year at which earth reaches perihelion varies over a period of about 20 000 years, getting progres-

sively later in the year. The perihelion distance q, and *aphelion distance Q, of major and minor planets are given in Tables 1 and 3, backmatter. *See also* advance of the perihelion.

perihelion distance. The distance from the sun to the *perihelion of a body in an elliptical orbit, or from the focus of a parabolic trajectory around the sun to the vertex.

perihelion passage, time of. *See* orbital elements.

perijove. *See* apojove.

perilune. *See* pericynthion.

perimartian. *See* apojove.

period. The time interval between two successive and similar phases of a regularly occurring event. The period of *rotation or of *revolution of a planet, etc., is the time to complete one rotation on its axis or one revolution around its primary. The period of a *binary star is the time observed for the companion to orbit the primary. The period of a regular intrinsic *variable star or an *eclipsing binary is the time between two successive maxima or minima on the *light curve.

period-density relation. *See* pulsating variables.

periodic comet. A comet that has been seen to orbit the sun more than once.

period-luminosity (P-L) relation. A relation showing graphically how the period of the light variation of a *Cepheid variable depends on its mean luminosity: the longer the period the greater the luminosity and hence mean absolute magnitude of the star. The relation was discovered by Henrietta Leavitt in 1912 and much subsequent work was done in calibrating the graph. It was found, in the 1950s, that there are two types of

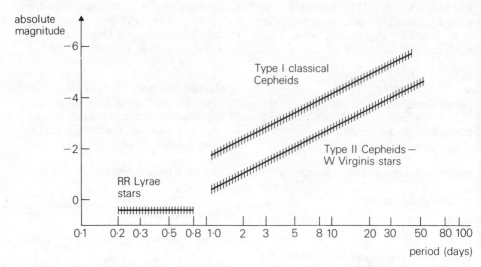

Period-luminosity relation

Cepheid – classical Cepheids and the less luminous W Virginis stars – whose P-L relations are approximately parallel, as shown on the graph. The relation results from the dependence of both luminosity and period of light variation, i.e. of pulsation, on stellar radius. For classical (Type I) Cepheids the relation is adequately represented by

$$\log(L/L_\odot) = 1.15 \log\Pi_d + 2.47$$

or

$$M = -2.8 \log\Pi_d - 1.43$$

where Π_d is the period in days, L and $L_\odot$ are the mean luminosities of the Cepheid and the sun, and M is the Cepheid's average absolute magnitude.

The relation is an invaluable means of determining the distances of Cepheids and hence of their surroundings: the absolute magnitude of a particular Cepheid is determined from the measured value of its period, and the distance can then be found from an independent measure of its apparent magnitude (*see* distance modulus). Cepheids occur in many star clusters within our own Galaxy and because of their great luminosity can be observed in nearby galaxies up to about seven million parsecs away. They can therefore be used for mea-

suring an immense range of distances. *See* distance determination.

perisaturnian. *See* apojove.

Perseids. A major *meteor shower, radiant: RA 46°, dec +58°, maximizing on Aug. 12 with a *zenithal hourly rate (ZHR) of about 70. Perseid meteors can be seen throughout the two weeks each side of maximum and are bright and flaring, with fine *trains. The geocentric velocity of the meteoroids is about 60 km s^{-1}. The *meteor stream is closely associated with comet 1862 III. Although the display is regarded as being constant in hourly rate there are exceptions, a ZHR of 250 being observed in 1921 and a ZHR of less than 10 in 1911 and 1912. The shower has been regularly observed for over 100 years.

Perseus. A fairly conspicuous constellation in the northern hemisphere near Cassiopeia, lying in the Milky Way. It has two 2nd-magnitude stars – *Mirfak (α) and the prototype eclipsing binary *Algol (β) – and several of 3rd magnitude. It also contains the naked-eye double cluster *h and Chi Persei, several open clusters including the just-visible M34, the *California nebula, and the galaxy NGC 1275, which is a powerful radio source and lies in the *Perseus cluster. Abbrev.: Per; genitive form: Persei; approx. position: RA 3.5h, dec +45°; area: 615 sq deg.

Perseus A. *See* Perseus cluster.

Perseus arm. *See* Galaxy.

Perseus cluster. A rich *cluster of galaxies lying about 70 megaparsecs away in the direction of the constellation Perseus. It contains several strong radio sources, three of which have been identified with the galaxies IC 310, NGC 1265, and NGC 1275. NGC 1275 is the most intense radio source (3C 84A or *Perseus A*) and occurs at the centre of the cluster. VLBI techniques show a central radio *jet, 3 pc long, growing at half the speed of light. It is a supergiant elliptical galaxy that is surrounded by a large x-ray emitting cloud of hot gas. The hot gas near to NGC 1275 is cooling and flowing into it in streamers that are visible optically. They increase the galaxy's mass by 300 solar masses per year. The *missing mass of the cluster is thought to be about 20 times the observed mass of the component galaxies.

persistent train. *See* train.

perturbation. A small disturbance that causes a system to deviate from a reference or equilibrium state. Periodic perturbations cancel out over the period involved and do not affect the system's stability. Secular perturbations have a progressive effect on the system that can cause it eventually to become unstable.

A single planet orbiting the sun would follow an elliptical orbit according to Kepler's laws; in reality a planet is perturbed from its elliptical orbit by the gravitational effects of the other planets. Likewise the revolution of the moon around the earth is perturbed mainly by the sun and to a much lesser extent by other bodies. The assumptions that the sun is the only perturbing body and that the earth orbits the sun in a fixed elliptical orbit lead to a simplified theory of lunar motion. The orbits of comets and minor planets are strongly perturbed when the body passes close to a major planet, such as Jupiter. The influence of a perturbing body can be calculated from the orbital elements of an *osculating orbit, to which corrections are made.

phase. 1. The appearance of the illuminated surface of the moon or a planet at a particular time during its orbit or of the sun during an eclipse. Specifically it is the fraction of the object's apparent disc (taken to be circular) that is illuminated. Unlike the superior planets, Mercury and Venus can exhibit those phases in which half or less of the illuminated hemisphere is visible from earth. *See also* phases, lunar.

2. The fraction of one complete cycle of a regularly recurring quantity that has elapsed with respect to a fixed datum point. The *phase difference*, ϕ, is the difference in phase between two electrical oscillations, wavetrains, etc., of the same frequency (i.e. coherent signals) and is usually expressed in terms of part of one complete cycle or wavelength. 'Phase difference' is often referred to simply as 'phase', as it is when considering the coherent signals in the two arms of a radio interferometer.

phase angle. The angle between the lines connecting the moon and sun and the moon and earth or connecting a planet and the sun and the planet and earth (see illustration). At full or new moon the phase angle is 0° or 180°, respectively. At first and last quarters it is 90°. For the superior planets phase angle is greatest at quadrature. For the inferior planets it

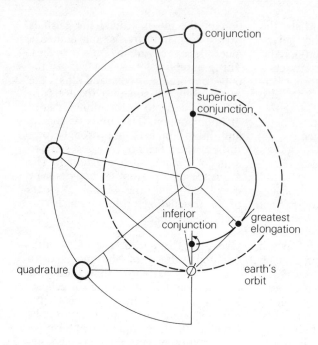

Phase angles for superior and inferior planet

is greatest (180°) at inferior conjunction and equals 90° at greatest *elongation.

phase defect. The angular amount by which the illuminated area of the moon differs from a circular disc at full moon due to the inclination of the moon's orbital plane to the *ecliptic.

phase difference. *See* phase.

phase rotator. A device that adds extra pathlength to one arm of an interferometer (*see* radio telescope), especially one used for *aperture synthesis, in order to keep the *phase difference close to zero as a *radio source moves across the sky.

phase-sensitive detector. An electronic device that detects a weak periodic signal in the presence of *noise. The signal plus noise is multiplied by a reference waveform whose shape and frequency are similar to those of the required signal. Components near zero frequency are produced. When the output is smoothed by a low-pass *filter a steady signal results, whose amplitude is related to the amplitude of the required signal and to the *phase difference between the required signal and the reference waveform. The device is used, for example, in a *phase-switching interferometer to detect the component at the switching frequency in the output of the first *detector.

phases, lunar. The *phases exhibited by the moon (see illustration). A *new moon* occurs when the moon is at *conjunction and the *nearside is totally unilluminated by the sun. As the moon moves eastwards in its orbit around the earth, the sunrise *terminator crosses the nearside from east to west to produce a *crescent moon*.

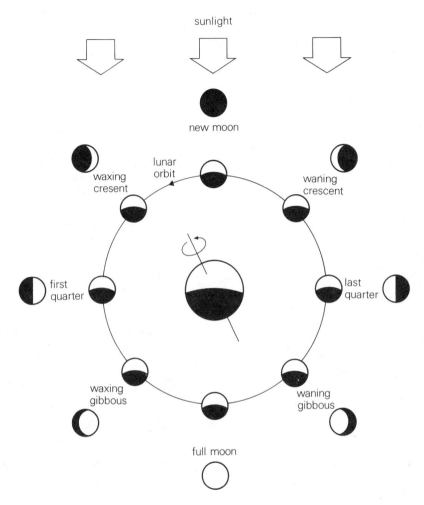

Moon's phases, outer circle showing phases as seen from earth

The crescent waxes to *first quarter* when the moon, at *quadrature, is half illuminated, through a *gibbous moon,* and finally to *full moon,* when the moon is at *opposition and the nearside is fully illuminated. The sunset *terminator then follows to produce a waning gibbous moon, *last quarter,* a waning crescent, and the next new moon. *See also* synodic month; eclipse.

phase-switching interferometer. A correlating interferometer (*see* correlation receiver) in which the *phase of the signal in one arm is periodically reversed before being added to the signal from the other arm. The phase reversal is achieved by means of a *phase switch,* which simply adds an extra half wavelength of path every alternate half cycle of the switching frequency, between 10 and 1000 hertz. The switching has no effect on noise signals, which are uncorrelated in the two arms. Correlated signals, however, add differently in phase and antiphase so that the first detector output in the *receiver contains a component at the switching frequency with an amplitude dependent on the power, and relative phase in the two arms, of the correlated signal.

This component is detected by means of a *phase-sensitive detector whose smoothed output contains no steady deflection unless a correlated signal is present. As a *radio source moves through the antenna beam, interference fringes appear in the output of the phase-sensitive detector; the fringes are similar to those in the output of an ordinary radio interferometer, except that in this case their average is zero rather than being superimposed on a steady background deflection. Phase-switching makes the interferometer much less sensitive to man-made *interference and reduces the effect on its output of changes in the gain and other parameters of the system.

Phecda or **Phekda** (γ UMa). One of the seven brightest stars in the *Plough.

Phobos. The innermost and larger of the two satellites of Mars, both discovered in 1877 by Asaph Hall. Observations from Mars-orbiting spacecraft show it to be an irregular potato-shaped body, measuring 19 × 22 × 27 km with a mass of 10^{16} kg and a density twice that of water. It orbits above the equator of Mars at a distance of 9380 km from the centre of the planet every 7.65 hours, keeping its long axis pointing towards Mars. The low density, together with a low *albedo of 0.06, suggest that it resembles the water-rich *carbonaceous chondrite meteorites, though it is difficult to see how such an object could have formed at the distance of Mars from the sun. Possibly Phobos and *Deimos, Mars' other small satellite, formed as more distant *minor planets and were later captured by Mars in some way. The orbit of Phobos is contracting because of tidal

drag and the satellite may decay from orbit within another 100 million years.

Phobos has a *regolith-covered cratered surface with the two largest craters, *Stickney* and *Hall,* having diameters of 10 and 6 km. A network of pitted grooves, typically 100–200 metres wide and 20–30 metres deep, radiate around the satellite from Stickney and are thought to have arisen from severe internal fracturing and heating caused by the impact event that formed the crater. Chains of small craters parallel to the orbit of Phobos possibly represent the secondary impacts of clumps of material swept up from orbit around Mars after being ejected from the satellite during earlier primary impacts. *See* Table 2, backmatter.

Phoebe. The outermost satellite of Saturn, discovered in 1898, and the only one to follow a retrograde orbit (*see* direct motion) around the planet. It has a diameter of 160 km and may possibly be a captured asteroid. The surface is dark, with an albedo of 0.05, but is less red than Iapetus. There is a suggestion that its rotation period may be about 9 hours, so Phoebe may not be in *synchronous rotation. *See also* Table 2, backmatter.

Phoenicids. A minor southern hemisphere *meteor shower, radiant: RA 15°, dec −55°, maximizing on Dec. 4.

Phoenix. A constellation in the southern hemisphere near Grus, the brightest star, Ankaa (α) being of 2nd magnitude. Beta Phoenicis is a binary, both components being of 4th magnitude. Abbrev.: Phe; genitive form: Phoenicis; approx. position: RA 1h, dec −50°; area: 469 sq deg.

photocathode. An electrode in an electronic device, such as a photocell, photomultiplier, or image tube, that emits electrons when a beam of electromagnetic radiation strikes the surface. By a suitable choice of photocathode material, a reasonable response may be obtained from near-infrared wavelengths to low-energy x-ray wavelengths. The electrons result from the *photoelectric effect. As many as 30% of the incident photons can be converted into electrons although the percentage is usually lower when taken over a wide spectral region. The current of resulting electrons increases linearly with radiation intensity over a wide range of intensities.

photocell. An electronic device that converts a light signal, or a signal in the infrared or ultraviolet, into an equivalent electrical signal. The earliest devices were *photoelectric cells*. These are evacuated and work by the *photoelectric effect. Electrons are emitted from a *photocathode when a beam of radiation falls on the surface. The electrons are attracted to a positively charged anode so that a current flows in an external circuit. The word photocell is now sometimes used to describe *photoconductive or *photovoltaic devices.

photoconductive cell. An electronic device that consists of a layer of semiconductor sandwiched between two electrical contacts. When light, infrared, or ultraviolet radiation falls on the sandwich its electrical conductivity increases markedly as a result of photoconductivity: the incident photons are absorbed in the material where they produce free electrical charge carriers that change the conductivity. A current therefore flows in the external circuit of the device. The current is proportional to the amount of radiation falling on the cell.

photoconductive detector. A *photoconductive cell used to detect photons in the infrared, usually made from either germanium or silicon doped with a variety of other metals, e.g. copper,

arsenic, antimony. These detectors are used at wavelengths from about 10 to 200 μm, with liquid-helium cooling. In contrast to *bolometers, there is no significant temperature change in the responsive element of photoconductive detectors.

photoelectric effect. An effect whereby electrons are emitted by material exposed to electromagnetic radiation above a certain frequency. This frequency usually lies in the ultraviolet region of the spectrum for most solids, but can occur in the visible region. The number of ejected electrons depends on the intensity of the incident radiation. The electron velocity is proportional to the radiation frequency.

photoelectric magnitude. *See* magnitude.

photographic emulsion. A light-sensitive layer of silver halide crystals suspended in gelatin and coated on a flexible or rigid base, from which a permanent record of a light or near-ultraviolet image, focused on the surface, may be produced. Both black-and-white and colour films or plates are used in astronomy. Emulsions highly sensitive to infrared and ultraviolet radiation and to x-rays are also now available.

The sensitivity of a particular type of emulsion to light or other radiation is known as its *speed*. This is usually expressed numerically as its *ASA number*; very high speed colour and black-and-white emulsions, of ASA 400 and 1000, are commercially available and are suitable for astronomical work. The sensitivity of an emulsion can be increased by *hypersensitization; this reduces the required exposure time or improves the image quality, sometimes both.

Photographic emulsions have high resolution, wide spectral response, good dimensional stability and permanence, and reasonably high signal/

noise ratio images. The light from a star, well below the visual threshold, can be recorded during a long exposure: exposures of over an hour may be used. Emulsions are, however, inefficient detectors compared to electronic recording devices – they can only record a very small number of the incident photons of radiation. In addition a linear relation between intensity of incident radiation and resulting image density exists for only a narrow range of exposures ('exposure' here means intensity of illumination times exposure time); at shorter or longer exposures the relation is nonlinear. *See also* photography; imaging.

photographic magnitude. *See* magnitude.

photographic zenith tube (PZT). A specially designed telescope that is used for the accurate determination of time. The telescope is fixed in a vertical position so that the observed objects are stars at or near the *zenith of the observatory. A photographic plate is driven across the field of view of the telescope at the exact speed of the star image. As the carriage moves, the plate is marked at several positions at known clock times. The positions of the star image, relative to these marks, gives the correction to the clock. The plate carriage is reversed by 180° halfway through the observation to remove any instrumental errors that might have occurred in the recording of the star image.

photography. There are three broad categories of astronomical photography: wide-angle photography of large areas of the sky, photography of comparatively near solar-system bodies, and the recording of spectra. The light or other radiation from a celestial body is of very low intensity, requiring long exposure times of the film or plate (from minutes to many hours). In order to maintain a steady image and a constant field of view, the camera is mounted in the focal plane of a telescope (or possibly in isolation) and the assembly mechanically driven at a constant usually sidereal rate (*see also* guide telescope). Photography of the brighter solar-system objects is fairly straightforward. Wide-angle studies are somewhat more complicated: camera lenses of short *focal ratio (less than f/5) should be used together with a high-speed photographic emulsion that is able to respond to the very low light levels and the long exposure times. Long exposures can be badly affected by stray light so that *light pollution must be minimal: the sites of both major observatories and small amateur instruments must be carefully selected to reduce pollution and to have the best possible *seeing.

Plates or films for long-exposure use must have fine-grained emulsions with low reciprocity failure: they must be able to respond uniformly to a wide range of light intensity. One such emulsion, highly suited for astronomical purposes, is Kodak IIIa-J emulsion (blue sensitive) and its panchromatic version, IIIa-F. The resulting photographic images are very uniform and have a very high resolution, especially when *hypersensitization techniques are used to increase the speed. Kodak IV-N emulsions can be used for near-infrared work and can become very much faster when hypersensitized. A large area of the sky can be photographed at one time, much greater than that 'seen' at present using electronic *imaging. Photographic plates are thus used in *Schmidt telescopes for sky surveys. Information, in digital form, can be extracted rapidly and automatically from the plates using machines such as *COSMOS.

photoionization. The *ionization of an atom or molecule by photons of electromagnetic radiation. A photon can only remove an electron if the photon energy exceeds the first *ionization potential of the atom or molecule. The excess energy is shared between the electron and the ion so that the electron can leave the atom with considerable velocity. If the radiation is of sufficiently high energy more strongly bound electrons will be removed, leaving the resulting ion in an excited state. *See also* recombination line emission.

photometer. An instrument by which the intensity of a source of light, or of infrared or ultraviolet radiation, may be measured and its variation with time studied. In a visual photometer the intensity of a light source is compared by eye to that of a standard light source. The eye can detect very slight differences in intensity (i.e. in brightness) so that fairly accurate measurements can be made. For greater accuracies or when the intensity is constantly changing, physical receptors such as a *photoconductive or *photovoltaic detector, *photomultiplier, or *bolometer must be used. In these devices the incident radiation is converted into an electrical signal whose magnitude can be determined very precisely.

photometric binary. A *binary star that can be detected by virtue of periodic changes in its light curve. *See* eclipsing binary.

photometry. The measurement (or study) of the brightness of a source of light or other radiation, and of how this brightness varies with time. At optical, infrared, and near-ultraviolet wavelengths, brightness is measured in terms of apparent *magnitude. At other wavelengths measurements of *flux density are made.

In some cases the eye can make comparisons, with a fair degree of accuracy (to 0.1 magnitude) between the known brightness of a reference star and the unknown brightness of the source under study. The greater accuracies now required in astronomy are achieved with electronic instruments or specially prepared photographic emulsions. Magnitudes are determined over a selection of wavelengths bands. Present systems, such as the internationally accepted UBV system (*see* magnitude), use a combination of a suitable *filter and a device such as a *photocell or *photomultiplier, to convert the radiation selected by the filter into an equivalent electrical signal, which is then amplified and measured. The U, B, and V magnitudes are determined over three relatively broad wavelength bands in the ultraviolet, blue, and yellow-green (visual) spectral regions. Magnitudes can also be measured for various infrared bands. Narrower wavelength bands are used in the *uvby system. Extremely narrow bands, now obtainable over the entire optical and infrared regions, can be studied by means of *interference filters: the distribution of, say, hydrogen in a source can be determined by isolating a particular spectral line of hydrogen.

Photoelectric photometry can be used to study an isolated star, region, etc. In photographic photometry, measurements can be made of objects on a photographic image of a wide angle of sky; the photographs are taken at wavelengths varying from near-infrared to near-ultraviolet (depending on emulsion). Highly resolved features can be registered, often of much greater quality than can be obtained electronically. The image, for a given photographic exposure time, varies in density (and size) by an amount that, within relatively wide limits, depends on the original brightness. Densities can be measured accurately by means of microdensitometers, etc.

photomultiplier. An evacuated electronic device used to convert a low-intensity light signal into an electrical signal and to amplify this signal very considerably. It contains a *photocathode from which electrons are liberated by the incident light photons.

The electrons are accelerated down a tube by a series of positively charged electrodes to which an increasingly high potential is applied. At each electrode a number of additional electrons are released by each impacting electron (by secondary emission). The electrons are all collected at the final electrode in the chain, which is held at a very high potential, and a large current pulse is produced.

photon. Symbol: γ. A quantum of *electromagnetic radiation. The photon can be considered as an *elementary particle with zero *rest mass, *charge, and *spin, travelling at the speed of light. A beam of electromagnetic radiation can thus be thought of as a beam of photons, with the intensity of the radiation proportional to the number of photons present. The energy of the photons, and hence of the radiation, is equal to the product, $h\nu$, of the *Planck constant and the frequency of the radiation. Photons with sufficient energy can ionize atoms and disintegrate nuclei. Although photons have no rest mass they do have momentum and thus exert a pressure, usually referred to as *radiation pressure.

photon-photon absorption. The absorption of gamma-ray photons on collision with ambient 'gas' photons to produce electron-positron pairs (*see* pair production). The *optical depth for this process can be written as:
$$\tau_{\gamma\gamma} \sim R \int \sigma(E_T) \, N_T(E_T) \, dE_T$$
if the density of the target photons, N_T, is assumed constant throughout the source region of size R. $\sigma(E_T)$ is the cross-section for pair production with a target photon of energy E_T.

photosphere. The 'visible' surface of the *sun and source of the *absorption spectrum that is characteristic of most stars. The photosphere of a star is considerably more dense than the atmospheric layers that lie above it, i.e. the *chromosphere and *corona.

The solar photosphere is a stratum several hundred kilometres thick, from which almost all the energy emitted by the sun is radiated into space. Within the photosphere the temperature falls from about 6000 kelvin just above the *convective zone to about 4000 kelvin at the *temperature minimum,* where the photosphere merges with the chromosphere.

The intensity of the solar photosphere, which decreases at visible wavelengths from the centre to the limb of the disc (*see* limb darkening), is due to the radiation emitted, principally by negative *hydrogen ions (H^-), at depths of up to a few hundred kilometres. At higher levels, where the density of H^- ions is too low for appreciable *opacity, the lower temperature gives rise to the absorption of radiation at discrete wavelengths. The *Fraunhofer lines of the resulting absorption spectrum have provided the key to determining the chemical composition of the photosphere, since a direct comparison can be made with the laboratory spectra of known elements under various conditions.

Regions of the solar photosphere (and lower chromosphere) several thousand kilometres in diameter rhythmically rise and fall with a period of about 5 minutes, attaining a maximum velocity of about 0.5 km s^{-1}. These vertical oscillations are thought to be produced by the outward propagation of low-frequency sound waves, generated by turbulence in the convective zone. A longer-period (2h 40m) pulsation that was announced is generally believed to be the result of the transit of individual supergranular cells (*see* supergranulation) across the field of view of the observing instrument rather than a global oscillation as such.

See also granulation; supergranulation; sunspots; faculae.

photovisual magnitude. *See* magnitude.

photovoltaic detector. An electronic device to detect *photons of radiation. It consists of a junction between two opposite-polarity semiconductors (a p-n junction). Photons absorbed at or near the junction produce charge carriers that are separated by the junction to produce an external voltage. The magnitude of the voltage is related to the number of incident photons. This type of detector is used in astronomy to detect ultraviolet and infrared radiation. To achieve optimum performance in the infrared, the detectors are cooled for example by liquid nitrogen to operate at 77 K or with liquid helium at about 4 K. In contrast to *bolometers, there is no significant temperature change in the responsive element of this type of detector.

physical double. 1. *See* double star.
  2. *See* double galaxy.

physical libration. Slight variations in the moon's rotation on its axis that are caused by minor irregularities in shape. They amount to less than two arc minutes but allow the moon's principal moments of inertia, and hence the distribution of mass within the moon, to be calculated. The mean orientation of the moon with respect to its physical librations defines the moon's *prime meridian. *See also* libration.

Pic du Midi Observatory. A French observatory in the Pyrénées at an altitude of 2860 metres. It was founded in 1930 for solar observations.

pico-. Symbol: p. A prefix to a unit, indicating a fraction of 10^{-12} of that unit. For example, one picosecond is 10^{-12} seconds.

Pictor (Painter). A small inconspicuous constellation in the southern hemisphere near the Large Magellanic Cloud, the brightest stars being the 3rd-magnitude Alpha Pictoris and two of 4th magnitude. Abbrev.: Pic; genitive form: Pictoris; approx. position: RA 5.5h, dec $-50°$; area: 247 sq deg.

Pioneer probes. A series of US solar-system probes. Pioneers 1–3 were intended lunar probes in 1958, although Pioneer 2 suffered launch failure and Pioneers 1 and 3 fell short of the moon but did measure the extent of the *Van Allen radiation belts. Pioneer 4 began to orbit the sun after passing the moon and was followed into solar orbit in 1960 by Pioneer 5, which returned data about the *solar wind and solar *flares.

Pioneer 6, launched in 1965, was joined in solar orbit by Pioneer 7 in 1966, Pioneer 8 in 1967, and Pioneer 9 in 1968. Together they formed a network of 'solar weather stations' between 0.75 and 1.12 astronomical units from the sun. Monitoring *solar activity and the solar wind, they provided early warnings of solar flares during the *Apollo programme.

Pioneer 10, launched on Mar. 3 1972, made the first fly-by of Jupiter at a distance of 130 300 km on Dec. 4 1973, after becoming the first probe to traverse the asteroid belt. It returned photographs of three of the *Galilean satellites of Jupiter (*Europa, *Ganymede, and *Callisto), as well as of the Jovian surface and the *Great Red Spot. Other experiments measured the strength and extent of Jupiter's radiation belts. Pioneer 11, launched on a similar mission on Apr. 5 1973, passed 42 940 km below Jupiter's south pole on Dec. 3 1974; it reached a top speed of 171 000 km per hour, faster than any previous man-made object. Nearly five years later it flew past Saturn, making its closest approach, 20 900 km above the clouds, on Sept. 1 1979. It twice crossed the plane of Saturn's rings. The probe, renamed *Pioneer Saturn*, returned 440 pictures and much new information on the planet, its satellites, and its rings. *See also* Pioneer Venus probes.

Pioneer Venus probes. Two US probes, an orbiter and a multiprobe, launched in May and Aug. 1978, respectively, towards Venus. The orbiter, Pioneer Venus 1, entered a highly inclined elliptical orbit around Venus on Dec. 4 1978, approaching to within 150 km of the surface. It was designed to make observations over 243 days (Venus' rotational period) and do radar mapping of the surface, cloud studies at ultraviolet and infrared frequencies, and magnetometer studies.

The multiprobe, Pioneer Venus 2, released one large probe on Nov. 15 of 316 kg and three small probes on Nov. 19, each of 93 kg. These probes penetrated the Venusian atmosphere on Dec. 9 at different locations and continuously gathered and relayed data as they approached and hit the surface. One small probe survived the hard landing and continued transmitting for a further hour. Pioneer Venus 2 – the interplanetary 'bus' – entered the atmosphere on Dec. 9, shortly after its payload, as an upper-atmosphere probe.

The probe measurements revealed a surprisingly large amount of primordial argon, krypton and neon in Venus' atmosphere, which relates to the early history of the planetary atmosphere. They also provided evidence of the layered structure of the clouds and confirmed the role of the *greenhouse effect in maintaining the high surface temperature. Radar maps of the surface showed evidence of massive plateaus, volcanic areas, and large rift valleys greater than those seen on earth. A major surprise was the evidence of lightning since it is not compatible with the meteorological observations. It is possible that volcanic activity at the time of the Pioneer mission may be related to this surprising observation.

pions. Symbol: π. A triplet of *elementary particles (mesons) that have *spin zero and exist in three states: neutral, positively charged, and negatively charged. The magnitude of the charge is equal to that of the electron. The charged pions have a mass of 2.4886×10^{-25} grams, the neutral pion mass being slightly less at 2.4006×10^{-25} grams. Charged pions decay into *muons and *electrons with a mean life of 2.6×10^{-8} seconds; the neutral pion converts, usually directly, into gamma-ray photons with an average lifetime of 8.4×10^{-15} seconds. *See also* cosmic rays.

Pisces (Fishes). A large but inconspicuous zodiac constellation mainly in the northern hemisphere near Pegasus. It contains the *first point of Aries, i.e. the vernal (or dynamical) *equinox. It has one 3rd-magnitude star and several of 4th magnitude. Abbrev.: Psc; genitive form: Piscium; approx. position: RA 1h, dec +15°; area: 889 sq deg.

Pisces Australids. A minor southern hemisphere *meteor shower, radiant: RA 340°, dec −30°, maximizing on July 30.

Piscis Austrinus (Southern Fish). A small constellation in the southern hemisphere near Grus, the brightest stars being of 4th magnitude apart from the 1st-magnitude *Fomalhaut. Abbrev.: PsA; genitive form: Piscis Austrini; approx. position: RA 22.5h, dec −30°; area: 245 sq deg.

Piscis Volans. *See* Volans.

pixel. *See* imaging.

plages (formerly sometimes termed **bright flocculi**). Bright patches in the solar *chromosphere, at a higher temperature than their surroundings, that occur in areas where there is an enhancement of the relatively weak vertical magnetic field. They are approximately coincident with photospheric *faculae (though unlike the faculae they are visible, in the monochromatic light of certain strong *Fraunhofer lines, anywhere on the sun's disc) and may be considered as defining the extent of *active regions.

Planck constant. Symbol: *h*. A fundamental constant with the value 6.626 196 × 10⁻³⁴ joule seconds. According to quantum theory, *electromagnetic radiation has a dual nature. Although many phenomena, including reflection and refraction, can be explained in terms of the wavelike nature of radiation, radiation may also be considered to be composed of discrete packets of energy called *photons, so that it acts as a stream of particles. The particle-like and wavelike properties are related by *Planck's law*, in which the Planck constant is the constant of proportionality:
$$E = h\nu = hc/\lambda$$
E is the energy of the photons and *v* and *λ* the frequency and wavelength of the wave; *c* is the speed of light in a vacuum. The Planck constant appears in most equations of quantum theory and quantum mechanics, including Planck's radiation law for *black bodies.

Planck's law. *See* Planck constant.

Planck's radiation law. *See* black body.

plane-polarized wave (linearly polarized wave). *See* polarization.

planet. A body that orbits the sun or another star and shines only by the light that it reflects. In the solar system there are nine known *major planets and numerous *minor planets, all of which probably share a common origin with the sun about 4600 million years ago (*see* solar system, origin). Planets of other stars are too faint to be seen from earth but very massive planets may be detected around the nearest stars by their perturbing effect on stellar motion, as is the case with *Barnard's star. The slight wobble that would be produced in a star's motion can be determined by precise measurements of a star's position over many decades, or from characteristic changes in a star's radial velocity determined from its doppler shift. Far-infrared observations can reveal dust shells around stars and it may be possible to detect a protoplanetary system in this way. *See* Vega; Fomalhaut.

planet–A. A Japanese spacecraft to be launched in Aug. 1985 to fly past *Halley's comet, closest approach occurring in Mar. 1986 at a distance of ∼150 000 km. It will take UV pictures, mainly of the comet's hydrogen cloud, and also measure the interaction between comet and solar wind. A similar probe, MS–T5, launched 6 months earlier, will study the undisturbed interplanetary medium for comparison purposes.

planetarium. An optical instrument by means of which an artificial night sky can be projected on the interior of a fixed dome so that the positions of the sun, moon, planets, stars, etc., may be shown and their motions, real or apparent, demonstrated. The instruments were first built by the Carl Zeiss firm from 1913. The term also refers to the building housing such an instrument.

planetary. *Short for* planetary nebula.

planetary aberration. The apparent angular displacement of the observed position of a planet, etc., produced by motion of the observer (*see* aberration) and by motion of the observed body (*see* light-time).

planetary alignment. The rough alignment along a solar radius of the outer planets, Jupiter to Neptune, that takes place approximately every 179 years. This occurred recently at the beginning of the 1980s.

This planetary configuration can be used to direct spaceprobes on a relatively short duration trajectory, swinging past two or more of the planets. A probe is given sufficient energy (i.e. velocity) to reach Jupiter. Jupiter's strong gravitational field swings the passing probe around the planet, away from its original direction of motion. With the correct initial velocity and flight path, the probe will be swung through an angle to take it towards Saturn or towards Uranus. Saturn, Uranus, and Neptune have a similar but smaller action on a passing probe. The energy and time required for such a mission is low, because of the planets' proximity, thus reducing the cost of the programme.

This 'slingshot action' has been successfully used in the current space exploration of the planets, and in particular the *Voyager mission. Voyager 1 flew by Jupiter in Mar. 1979 and the spacecraft trajectory was deflected so that the craft could encounter Titan and Saturn in Nov. 1980. The spacecraft has now completed its planetary encounters and is on a path out of the solar system. After the Voyager 2 encounter with Jupiter in July 1979, the spacecraft encountered Saturn in Aug. 1981 and is now en route for further planetary encounters – with Uranus in Jan. 1986 and Neptune in 1989. These multiple planetary encounters have been possible because of the unique planetary alignment. Unfortunately this 'Grand Tour' of the outer planets by Voyager 2 cannot visit Pluto. More modest slingshot paths have been used by *Pioneer 11 to Jupiter and Saturn and by *Mariner 10 to Venus and repeated encounters with Mercury.

planetary nebula. An expanding and usually symmetrical cloud of gas that has been ejected from a dying star. Most are believed to be the ejected envelopes of *red giant stars, shed as a result of instabilities late in their evolution. The gas cloud is ionized by the compact hot burnt-out stellar core that remains in the centre of the cloud; the cloud is detected by virtue of the resulting light emission. Planetary nebulae are therefore a class of *emission nebulae. They are usually ring-shaped or sometimes hourglass-shaped and are generally less than 50 000 years old. The name refers to their resemblance to planetary discs rather than point-like stars under low magnification. They have a large size range: the smallest objects have a starlike appearance on photographs – and are thus called *stellar planetaries* – but can be identified by the characteristic spectral emission lines. Planetary nebulae occur in isolation and usually lie close to the galactic plane, concentrated towards the galactic centre.

A planetary nebula is believed to form as part of the normal evolution of single stars with masses of up to 8 solar masses; the stage immediately preceding is probably a rapid mass loss *OH/IR star. Instabilities eject a succession of planetary nebula shells, reducing the mass of the star until the core (the nebula's central star) is only about 0.6 solar masses. This degenerate core becomes a *white dwarf. The recent discovery of planetary nebulae with close binary stars at the centre suggests that some planetaries form as a result of interactions in a double star system. One star has expanded sufficiently to cocoon both in a *common envelope, with the two star cores orbiting inside; frictional drag transfers energy from the orbiting stars to the surrounding gas and thus expels the envelope as a planetary nebula.

Although planetary nebulae are less massive and more symmetrical than *HII (ionized hydrogen) regions, their optical spectra are similar. There are bright emission lines of oxygen, hydrogen, nitrogen, and other components, the characteristic green of the inner region being due to doubly ionized oxygen and the red of the outer periphery resulting from singly ionized nitrogen and from hydrogen alpha emission. About 1000 planetary nebulae are known, the *Ring nebula in Lyra being a typical example. *See also* nebula.

planetary precession, *See* precession.

planetary probe. An unmanned spacecraft designed and equipped to study conditions on or in the vicinity of the planets, their satellites, and other members of the solar system, and in the interplanetary medium itself, and to transmit the information back to earth. At the time of writing, probes have made successful visits to all the major planets of the solar system outwards as far as Saturn. Voyager 2 is en route for Uranus. *See* Galileo; Mariner probes; Mars probes; Pioneer probes; Pioneer Venus probes; Venera probes; Viking probes; Voyager probes; Zond probes.

Most planetary and interplanetary probes have used solar panels to derive power for communications and internal operation. However, spacecraft that travel further from the sun than the orbit of Mars receive comparatively little solar energy and must rely on other power systems, such as the radioisotope thermoelectric generators carried by Pioneers 10 and 11 and Voyagers 1 and 2. Most probes carry a main rocket propulsion system for course corrections as well as smaller thrusters for attitude adjustments. Orientation is often maintained by reference to a bright star, *Canopus being a frequent choice.

planetary system. A system of *planets and other bodies, such as comets and meteoroids, that orbits a star. The sun and its planetary system together comprise the *solar system. Planetary systems may be common in the Galaxy, being formed along with stars from the gravitational contraction of interstellar clouds of gas and

dust. *See also* Barnard's star; Vega; Fomalhaut; protoplanet.

planetary weather. *See* atmosphere.

planetesimals. Bodies, ranging in size from less than a millimetre to many kilometres, that are thought to have formed the planets of the *solar system by a process of *accretion.

planetesimal theory. *See* encounter theories.

planetoids. *See* minor planets.

planetology. The study of the planets of the solar system – their history, internal and surface structures, and atmospheres – with particular regard as to how and why these differ from planet to planet.

planets, origin. *See* solar system, origin.

planet X. Unofficial designation for a hypothetical planet orbiting the sun beyond the orbit of Pluto. Various suggestions for its mass and distance have been made, most of them being based on analyses of apparent perturbations by it on the orbits of planets and comets. It now appears that the planets are not detectably perturbed by an unknown body; planet X, if it exists, must therefore be small and/or very remote. However, analysis of the orbits of very long period comets suggests that there may be two earth-mass planets at distances of about 53 and 83 AU (*see* comet family).

planisphere. A two-dimensional map projection centred on the northern or southern pole of the celestial sphere that shows the principal stars of the constellations, the Milky Way, etc., and is equipped with a movable overlay to indicate the stars visible at a particular time on any day of the year for a particular zone of terrestrial latitude.

planitia. A plain on the surface of a planet or satellite. The word is used in the approved name of such a feature.

planum. Plateau. The word is used in the approved name of such a surface feature on a planet or satellite.

Plaskett's star. *See* stellar mass.

plasma. A state of matter consisting of ions and electrons moving freely. A plasma is thus an ionized gas that can be formed at high temperatures (as in stars) or by *photoionization (as in interstellar gas). The properties of plasmas differ from those of neutral gases because of the effects of magnetic and electric fields on the charged species and because of interactions between the species. *See also* magnetohydrodynamics.

plate tectonics. *See* earth.

Platonic year. *See* precession.

Pleiades (Seven Sisters; M45; NGC 1432). A young *open cluster about 120 parsecs away in the constellation Taurus. It contains maybe 3000 stars of which six are visible to the naked eye: *Alcyone, Maia, Atlas, Electra, Merope, and Taygete; *Pleione could once have been brighter. The brightest stars are blue-white (B or Be) and highly luminous, the less bright ones being mainly A and F stars. Gas and dust surrounding the brighter stars reflect the starlight thus producing faint *reflection nebulae.

Pleione. One of the stars of the *Pleiades. It developed a surrounding envelope that was first observed in 1938, reached maximum intensity in 1945, and was just visible in 1954; another shell started to develop in 1972. It is thought to be unstable ow-

ing to its fast rotation rate. Spectral type: B8pe.

Plough (*US* Big Dipper). A group of stars in *Ursa Major that contains the seven brightest stars in that constellation, nearly all with similar magnitudes (m_v), and thus has a very distinctive shape (see illustration). The seven stars are *Alioth, *Dubhe, *Alcaid, *Mizar, Merak (m_v: 2.36), Phecda (m_v: 2.43), and Megrez (m_v: 3.31), in order of brightness. Alcaid and Dubhe have different values and directions of *proper motion than the other five so that the shape of the Plough is slowly but continuously changing.

P-L relation. *See* period-luminosity relation.

Pluto. The ninth and smallest major planet, discovered in 1930 by Clyde Tombaugh at the Lowell Observatory in Arizona. Its 248 year orbit has a higher *inclination to the ecliptic (17°) and a greater *eccentricity (0.25) than any other planet. Although it has the greatest mean planetary distance of about 39 AU from the sun, it approaches within 30 AU at *perihelion; between 1979 and 1999 Pluto will therefore be closer to the sun than Neptune. Orbital and physical characteristics are given in Table 1, backmatter.

*Oppositions occur one or two days later each year, but Pluto's 14th magnitude disc is too small to be measured using earth-based telescopes. Until 1978 the best estimate of its size was about 5900 km, a value based on negative observations of a predicted occultation of a star by Pluto during an *appulse in 1965. However, examination of photographs of Pluto in 1978 by James Christy of the US Naval Observatory revealed that it has a very close satellite, called *Charon*, orbiting at a distance of about 20 000 km every 6.4 days – a period previously identified in the

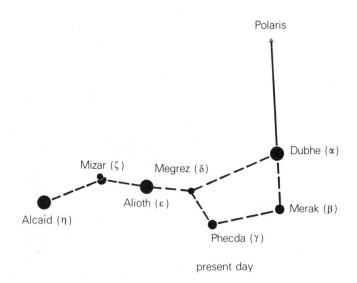

present day

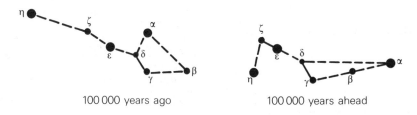

100 000 years ago   100 000 years ahead

The Plough

small light variation of the planet and ascribed to its axial rotation. The orbital parameters of Charon enable *Kepler's third law to be used to estimate Pluto's mass as about 0.25% that of the earth. Apparently Pluto is a low-density body about 3000 km across, while Charon may be 1000 km in diameter; it may thus be the largest satellite in proportion to the size of its planet in the solar system – a title previously credited to our own moon. Because of their proximity, both Pluto and Charon are probably in *synchronous rotation about their common centre of mass.

At a temperature of −230°C, Pluto is too cold and small to have an ap-

preciable atmosphere. Infrared spectroscopy has shown methane to be present but it is difficult to distinguish spectroscopically between gaseous and solid forms of methane at the temperature and pressure found on Pluto: part of the surface, at least, could have a covering of methane ice, possibly with a thin atmosphere of methane. It is possible that the whole globe may be a low-density ice-ball.

Pogson ratio. *See* magnitude.

Pointers. The two stars in the *Plough that point towards *Polaris and the north pole; Dubhe and Merak.

point source. A celestial object whose angular extent cannot be measured.

polar axis. 1. The axis in an *equatorial mounting that is parallel to the earth's axis and hence points towards the celestial poles.
2. *See* poles.

polar caps. White deposits of ices in the vicinity of the rotational poles of earth and Mars. Those of earth consist of water ice and snow, while the caps of Mars comprise both water ice and frozen carbon dioxide. The caps of both planets show seasonal variations. *See also* Mars, polar caps.

polar diagram. *Another name for* antenna pattern. *See* antenna.

polar distance. *Short for* north polar distance or south polar distance.

Polaris (North Star; α UMi). A remote creamy-yellow supergiant that is the brightest star in the constellation Ursa Minor. It lies very close to the north celestial pole and is the present *pole star. Its position is found by means of the *Pointers in the Plough. Polaris is a classical *Cepheid variable (period 3.97 days). It is also the primary (A) of a multiple star system: it is a spectroscopic binary (period 29.6 years) and also a visual binary with a 9th-magnitude F3 V companion (B). There are two other fainter components (C and D) of the system. m_v: 2.02; M_v: −4.5; spectral type: F8 Ib; distance: 200 pc.

polarization. A measure of the way in which light of other *electromagnetic radiation from a celestial body is affected by factors such as scattering due to cosmic dust or strong stellar or interstellar magnetic fields, or reflection from a surface. It is the degree to which the direction of the electric or magnetic vector in an electromagnetic wave changes in a regular fashion. The vector thus shows a less

random orientation in the plane perpendicular to the direction of wave motion than occurs in an unpolarized (normal) wave. Waves in which the electric vectors are entirely vertical or horizontal with respect to the direction of motion are described as vertically or horizontally *plane* (or *linearly*) *polarized*. In general, both polarizations are present and the wave is then *elliptically polarized* in the right-handed or left-handed sense accordingly as the resultant vector rotates clockwise or anticlockwise when viewed along the direction of motion of the wave.

Radio emissions from celestial sources are usually partially polarized, i.e. the waves can be considered to be composed of a completely unpolarized component plus a small polarized component. *Synchrotron emission, however, is strongly polarized. The general situation can be described by the four *Stokes' parameters* (I, Q, U, and V), which are defined in such a manner that specifying their four values uniquely describes the state of polarization. *See also* interstellar polarization; Faraday rotation.

polarization curve. A curve showing the relationship between the angle of reflection of light from a surface and the degree of *polarization of the reflected light. It depends on the surface characteristics (texture, composition). It is used to deduce the *albedo of the surfaces of minor planets so that their diameters may be calculated.

polar orbit. A satellite orbit that passes over the north and south poles of the earth or of some other solar-system body.

poles. The two points on a sphere lying 90° above or below all points on a particular great circle. The *polar axis* is the imaginary straight line between the poles. *See* celestial poles; galactic poles.

poles of the ecliptic. *See* ecliptic.

pole star. The star closest to the north or south celestial pole at any particular time. As a result of *precession the position of each pole traces out a circle in the sky in a period of about 25 800 years. There is thus a sequence of stars, including *Polaris, *Vega, and *Thuban, that have been and will become again the northern pole star.

The nearest naked-eye star to the south celestial pole is at present the 5th magnitude star Sigma Octantis. The altitude of the pole star is approximately equal to the local latitude.

Pollux (β Gem). An orange giant that is the brightest star in the constellation Gemini and is the nearest giant to the earth. It has several optical companions none of which are physically related to it. m_v: 1.15; M_v: 1.0; spectral type: K0 III; distance: 10.7 pc.

population I, population II. The two classes into which stars and other celestial objects within a galaxy can be divided; the distinction is basically that of age, and the most obvious distinguishing characteristics are the objects' space distribution and chemical content. The classification was first made by W. Baade in 1944. He proposed that population I stars are the young metal-rich highly luminous stars found in spiral arms of galaxies – i.e. strongly concentrated in the galactic plane – and associated with interstellar gas and dust. In contrast population II stars are old red stars found throughout elliptical and lenticular galaxies and in spiral galaxies located in the galactic centre, in the surrounding galactic halo, and in a disc coextending with the galactic plane but much thicker.

It is now known that in our Galaxy there is a continuum of populations existing between Baade's two classes (see table). The origin of population types is connected with the dynamic and chemical evolution of the Galaxy. Put very simply the earliest stars, formed from the original hydrogen-helium gas cloud, were *halo population II stars* followed, after further gas contraction produced a relatively thin extended disc, by *intermediate population II stars*; contraction of the remaining gas into an even thinner disc produced the *disc population*, which may be young population II stars or old population I stars. Successive generations of population I stars have subsequently formed from interstellar matter that has been slowly enriched with metals and other elements produced by *mass loss from stars and by *supernovae explosions. The youngest *extreme population II stars* contain 20 to 30 times the metal abundance of the oldest halo objects.

population III. A hypothetical population of supermassive stars that could have existed before the galaxies formed. They could have produced the large helium content of the universe and the microwave background radiation, both usually attributed to the early stages of the big bang. If population III stars did exist, cosmologists would need to severely modify current forms of the *big bang theory.

pores. *See* sunspots.

Porrima (γ Vir). A visual binary with cream components that lies in the constellation Virgo and has a highly eccentric orbit (period 171 years). The separation, about 5″ in the 1970s and 6″.6 at maximum, is decreasing and by 2010 the star will appear single except in the largest telescopes. Porrima is also an optical double. m_v: 3.6 (A and B), 2.75 (AB); spectral type: F0 V (A and B).

positional astronomy. *See* astrometry.

Characteristics of populations I and II in the Galaxy

population group	extreme population I	older population I	disc population	intermediate population II	halo population II
typical objects	interstellar gas, dust HII regions T Tauri stars OB associations O, B stars supergiants young open clusters classical Cepheids	sun strong-line stars A stars Me dwarfs giants older open clusters	weak-line stars planetary nebulae galactic bulge novae RR Lyrae stars (P < 0.4 days)	high-velocity stars long-period variables	globular clusters extreme metal-poor stars (subdwarfs) RR Lyrae stars (P > 0.4 days) Type II Cepheids
distribution	spiral arms: extremely patchy	patchy	smooth	smooth	smooth
*z, Z, age	none ⟵———— increasing ————⟶	little	considerable	strong	strong
concentration to Galactic centre	none	little	considerable	strong	strong
Galactic orbits	circular	approx. circular	slightly eccentric	eccentric	very eccentric
brightness, metals/H ratio	⟵———— increasing ————⟶				

Adapted from A. Blaauw

* z, Z are mean distance from Galactic plane, mean of velocity components perpendicular to Galactic plane

position angle. 1. The direction in the sky of one celestial body with respect to another, measured from 0° to 360° in an easterly direction from north. It is used, for example, to give the position of the fainter component of a *visual binary with respect to the brighter component.

2. The angle at which the axis of a planet or star, or some other line on a celestial body, is inclined to the *hour circle passing through the centre of the body; it is measured eastwards from 0° to 360° from north.

position circle. *See* circle of position.

position micrometer. *See* filar micrometer.

positron. The *antiparticle of an electron.

positronium. A positron-electron system that lasts for a measurable time before combining to produce *annihilation radiation. Positronium can be thought of as an atom analogous to that of hydrogen in which the electron and positron move in Bohr orbits about the centre of mass, which is halfway between them. During formation, approximately 25% is formed in a singlet spin state and the other 75% is formed in a triplet state. The annihilation radiation emitted by the combining of a positron-electron pair in a singlet state consists of two γ-ray photons emitted simultaneously with energies of 511 keV each; the annihilation radiation from the triplet state consists of three γ-ray photons emitted simultaneously each with energy less than 511 keV.

potential energy. The energy possessed by a body or system by virtue of its position or configuration. It is equal to the work done by the system changing from its given state to some standard state. In a gravitational field, a mass m placed at a height h above a standard level (say the surface of the earth) has potential energy mgh, where g is the acceleration of gravity.

power. The rate at which *energy is expended by a source or work is done. It is measured in watts, i.e. in joules per second.

power, optical. The reciprocal of the focal length in metres of a lens or spherical mirror. It gives a measure of the relative focusing or defocusing powers of optical elements. A converging lens or mirror has a positive power; one that is diverging in action has a negative power. The power of two thin lenses in contact is equal to the sum of the separate powers. For a thick lens, account must be taken of refractive index.

power pattern. *See* antenna.

power spectrum. A plot of the power contained in a signal versus the frequency, or of the power in a spatial distribution versus the wavenumber (reciprocal of wavelength). It may be obtained by taking a *Fourier transform of the signal or distribution and squaring the amplitude of each component in the result.

Poynting–Robertson effect. An effect whereby small interplanetary dust particles (radius 10^{-6} to 10^{-2} metres) slowly spiral towards the sun. The particles interact with solar radiation (by absorbing and re-radiating energy) so that they lose orbital momentum; this causes a decrease in orbital velocity so that they move into progressively smaller orbits around the sun. For a particle moving in a circular orbit that is d astronomical units from the sun, the time taken to fall into the sun is given by

$$t = 7 \times 10^6 \, \rho r d^2 \text{ years}$$

where r and ρ are particle radius (in cm) and density (in $g\,cm^{-3}$), respectively.

*Radiation pressure on the particles opposes the Poynting–Robertson effect and can overcome it for very small particles (radius less than 10^{-6} m). These particles are blown out of the solar system by the action of the radiation pressure aided by the *solar wind. Although these two processes remove dust from the solar system, the dust level is constantly replenished by disintegrating comets, etc.

p-p chain. *See* proton-proton chain reaction.

p-process. A process of *nucleosynthesis by which certain heavy proton-rich nuclei are thought to be produced.

Praesepe (Beehive; M44; NGC 2632). A large *open cluster in the constellation Cancer that contains over 100 stars. It can almost be resolved by the naked eye but is best viewed with a very low-power telescope or with binoculars.

preamplifier. An amplifier used in a *radio telescope to boost the signals from the *antenna before they are fed to the *receiver.

preceding spot (*p*-spot). *See* sunspots.

precession. The slow periodic change in the direction of the axis of rotation of a spinning body due to the application of an external force (a torque). The extremities of the axis trace out circles in one complete precessional period if the rotational speed and applied torque are both constant. The earth's axis precesses with a period of 25 800 years, the axial extremities, the celestial poles, tracing out circles about 23.5° in radius on the celestial sphere (see illustration). This radius is equal to the inclination of the earth's equatorial plane to the plane of the ecliptic. The precession results primarily from the gravitational attraction of the sun and moon on the *equatorial bulge of the nonspherical earth. The

sun and moon and also the planets pull on the bulge in an attempt to bring the equatorial plane into near coincidence with the plane of the ecliptic. The result of this gravitational action and the opposing rotation of the earth is a precessional force that acts perpendicularly to both these actions.

The combined effect of the sun and moon leads to *luni-solar precession*, which together with the much smaller *planetary precession* gives the *general precession* of the earth's axis and of the equinoxes. The *annual general precession* is given by

50.2564 + 0.0222*T* arcsecs

where *T* is the time in Julian centuries from 1900.0. Thus one precessional circle is completed about every 25 800 years, this period being the *Platonic year*. The precession of the earth's axis leads not only to the *precession of the equinoxes and the consequential change in stellar coordinates but also to the changing position of the celestial poles and hence to the slow change in the two *pole stars.

precession, constant of. The ratio of the luni-solar *precession to the cosine of the *obliquity of the ecliptic. It is about 54.94 arc seconds per year and varies very slightly with time.

precession of the equinoxes. The slow continuous westward motion of the *equinoxes around the *ecliptic that results from the *precession of the earth's axis. It was first described by Hipparchus in the second century BC. As the axis precesses, the celestial equator, which lies in the plane perpendicular to the axis, moves relative to the ecliptic. The two points at which the planes intersect – the equinoxes – thus move round the ecliptic in the precessional period of about 25 800 years, the annual rate being about 50.27 arc seconds. Coordinates such as *right ascension that refer to an equinox as their zero point thus

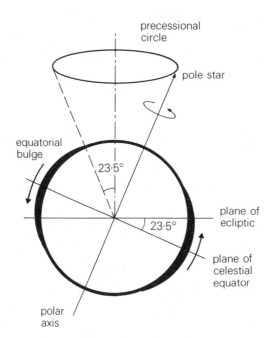

precessional
circle

pole star

equatorial
bulge

23·5°

plane of
ecliptic

23·5°

plane of
celestial
equator

polar
axis

Precession of the earth

change with time. Star catalogues therefore use a *catalogue equinox as their origin.

precision catalogue. *See* star catalogue.

pre-main-sequence star. *See* stellar birth.

pressure. The force per unit surface area at any point in a gas or liquid. The pressure of a gas is proportional to temperature and density: at constant temperature, as the density is increased the pressure increases accordingly. This law of classical physics does not apply to *degenerate matter.

pressure broadening. Broadening of spectral lines as a result of high density, and hence pressure, of the emitting or absorbing material: the greater the density and pressure, the greater the width of the spectral lines. The broadening is caused by collisions with other atoms while the atom is emitting or absorbing radiation. *See also* line broadening.

primary. The celestial body that is nearest to the centre of mass of a system of orbiting bodies. The other members, called *secondaries*, appear to orbit the primary, which is the most massive in the system. In fact all members move round the common centre of mass. The earth is the moon's primary. *See also* visual binary.

primary cosmic rays. *See* cosmic rays.

primary mirror. The optically worked mirror in a *reflecting telescope that faces the observed object. The focal plane in which its image forms is termed the *prime focus*. Modern mirror-substrate materials are glass/ceramic compounds, such as Zerodur or Cer-Vit, or glass-like materials, such as fused quartz or Pyrex, all of which have very low coefficients of thermal expansion. The substrate is usually *aluminized to obtain the reflecting surface. The surface must be accurately ground and polished to within one quarter of the wavelength used, i.e. to about 0.1 micrometres, if images of good quality are to be produced. The shape is usually *paraboloid in form to avoid *spherical aberration; this shape does however produce *coma in off-axis images. The mirror is supported over its whole rear face, either mechanically or by fluid suspension, and can maintain an accurate profile up to considerable diameters without sagging.

The new generation of ground-based reflectors will have very large apertures, up to 15 metres. Technological problems and cost limit the diameter of a single one-piece primary mirror to 7–8 metres. Designs for larger apertures include assemblies of several large lightweight mirrors, based on the *Multiple Mirror Telescope, or a single *segmented mirror* composed of many elements that are individually supported and manoeuvred. Another idea is the *array*, in which light from a group of telescopes is collected at a common focus. *See also* cell; mounting.

prime focus. *See* primary mirror; objective.

prime meridian. The arbitrary circle of zero longitude on the earth or some other planet or mapped celestial body. The earth's prime meridian is the *Greenwich meridian.*

primeval atom. The earliest phases of the universe. *See* big bang theory.

primeval nebula. The cloud of interstellar gas – mainly hydrogen and helium – and dust grains from which the solar system developed. *See* nebular hypothesis; solar system, origin.

prime vertical. The *vertical circle that passes through an observer's zenith at right angles to the *meridian and intersects the horizon at the east and west points (*see* cardinal points).

primordial fireball (ylem). The radiation-dominated phase of the universe. *See* big bang theory.

prismatic astrolabe. An astrometric instrument for determining the position of a celestial body by the accurate timing of its passage across a *vertical circle of the celestial sphere. It usually consists of an artificial horizon (such as a mercury surface) and a prism, which are placed in front of a telescope. The prism is fixed in position so that a fixed altitude, usually 45° or 60°, is used. Parallel rays of light falling on the artificial horizon and the prism are each reflected by the prism into the telescope (see illustration). As the celestial body moves towards the fixed altitude, the two images approach each other from top and bottom of the field of view. At the fixed altitude the two images lie close together on the same horizontal line through the centre of the field. The time is then recorded, usually electronically.

probe (spaceprobe). *See* planetary probe.

Procellarum, Oceanus (Ocean of Storms). A vast irregular *mare – possibly occupying an impact depression, the *Gargantuan Basin* – that covers a major part of the moon's western *nearside. It was the landing site for *Lunas 9 and 13, *Surveyors 1 and 3, and *Apollo 12. It is one of the youngest areas on the moon, parts of which may be less than 3000 million years old.

Procyon (α CMi). A conspicuous creamy-yellow star that is the brightest one in the constellation Canis Minor and is one of the nearest stars to the sun. It was found to be a visual binary in 1840 from observations of its variable proper motion.

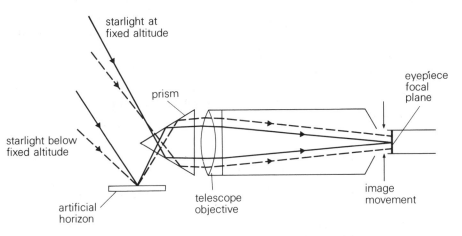

Optical path of prismatic astrolabe

The companion (B) was discovered in 1896 to be a *white dwarf. The period is 40.6 years, the separation 4″.6. There is a third 12th-magnitude optical component (C) at 81″. Procyon is also a spectroscopic binary, period 40.23 years. m_v: 0.4; M_v: 2.7; spectral type: F5 IV-V (A), wdF8 (B); mass: 1.74 (A), 0.63 (B) times solar mass; radius: 1.7 (A), 0.01 (B) times solar radius; distance: 3.5 pc.

Prognoz. A series of USSR unmanned observatories, first orbited in 1972. They have investigated the effect of solar activity on the interplanetary medium and the earth's magnetosphere by measuring solar radiation (both electromagnetic and corpuscular) and magnetic fields in near-earth space. They have also studied galactic ultraviolet radiation, x-rays, and gamma rays.

Progress. Any of a series of unmanned Soviet spacecraft specially designed to dock with the *Salyut 6 and 7 space stations and used to supply additional fuel, food and water, equipment, etc.; the first one was launched in Jan. 1978. Progress craft are 7.9 metres long with a mass of 7 tonnes and can ferry 1 tonne of fuel and 1.3 tonnes of cargo. Reloaded

with discarded articles, rubbish, etc., and undocked, they are destroyed on re-entry into the earth's atmosphere.

prominences. Clouds of gas in the sun's upper *chromosphere/inner *corona, with a higher density and at a lower temperature than their surroundings, that are visible (ordinarily only in the monochromatic light of certain strong *Fraunhofer lines) as bright projections beyond the limb. When viewed against the brighter disc they appear as dark absorption features and are termed *filaments*. They exhibit a great diversity of structure and are most conveniently classified according to their behaviour, as either *quiescent* or *active*.

Quiescent prominences are particularly long-lived and are among the most stable of all solar features. They may persist for several months before breaking up or, less frequently, blowing up and have been known to reform at the same location with an almost identical configuration. Typically they are a couple of hundred thousand kilometres long, several tens of thousands of kilometres high, and several thousand kilometres thick. They occur in high *heliographic latitudes, where they are supported by the horizontal magnetic field separating the polar field from the adjacent fields of opposite polarity that were formerly associated with the *f*-spots of *active regions (*see* sunspot cycle). They attain their greatest frequency a few years after the minimum of the sunspot cycle, when their average latitude is around ±50°. Thereafter they appear in increasingly high latitudes, reaching the polar regions shortly after sunspot maximum. Then, after a brief discontinuity, they reappear around latitude ±50° and remain there in small numbers until a few years after the next minimum, when they again progress poleward.

Active prominences are relatively short-lived and may alter their structure appreciably over a matter of minutes. There are many characteristic types, for example *surges and *sprays, in which chromospheric material is ejected into the inner corona, and *loop prominences* and *coronal rain*, in which the reverse occurs. Loop prominences are impulsive events that often accompany *flares, while coronal rain represents the return of flare-ejected material. In developing active regions *arch filaments* are usually present. These tend to connect regions of opposite polarity across the line of inversion and gradually ascend while material descends along both sides of the arch.

Intermediate between quiescent and active prominences are the so-called *active-region filaments*. These invariably lie along the line of inversion between the vertical magnetic fields of opposite polarity in well-developed active regions. Though long-lived, they may be distinguished from quiescent filaments by an almost continuous flow of material along their axis.

With the exception of surges and sprays, prominences may be regarded as an efficient (though limited) heat sink for the highly ionized corona, their material condensing out of it and then descending along the predominantly vertical magnetic field lines.

promontorium. Promontory. The word is used in the approved name of such a surface feature on a planet or satellite.

proper motion. Symbol: μ. The apparent angular motion per year of a star on the celestial sphere, i.e. in a direction perpendicular to the line of sight. It results both from the actual movement of the star in space – its *peculiar motion – and from the star's motion relative to the solar system. It was first detected in 1718 by Halley. The considerable distances of stars reduce their apparent motion to a very small amount, which for most stars is considered negligible. It is only after thousands of years that differences in the directions of proper motion cause groups of stars to change shape appreciably.

The proper motion of a star, usually a nearby star, can be determined if its cumulative effect over many decades produces a measurable change in the star's position. It is quoted in arc seconds per year, often in terms of two components: proper motion in right ascension (μ_α) and in declination (μ_δ). *Barnard's star has the greatest proper motion ($10''.3$ per year). If the distance, d (in parsecs) of the star is known then the velocity along the direction of proper motion – the *tangential velocity*, v_t – can be found:
$$v_t = \mu d \text{ AU/year} = 4.74 \mu d \text{ km s}^{-1}$$

proper time. Time measured by a clock that is sharing the observer's motion. Clocks moving with respect to the observer, or experiencing a different gravitational field, will measure time to flow at a different rate to that of proper time. *See also* relativity, special theory.

proportional counter. A gas-filled electronic device for the detection of x-rays and other ionizing radiation and the counting of charged particles. A potential difference is maintained across a pair of electrodes. Radiation entering the tube ionizes the gas along its path and the resulting electrons are attracted towards the positively charged electrode. The size of the output pulse of the counter is proportional to the number of electron-ion pairs produced by the ionizing event and thus to the energy of the radiation. It is possible to differentiate between various forms of ionizing radiation. In astronomy proportional counters are used mainly for x-ray studies.

Recently, *imaging proportional counters* have been developed, which record the position of each incident x-ray photon so that a two-dimensional image is built up. Linear resolutions of $200-500$ μm are obtained in the $0.2-5$ keV energy band. Such devices have been used successfully in a number of x-ray astronomy satellites, including the *Einstein Observatory and *EXOSAT. Another recent development is the *gas scintillation proportional counter (SPC)*, which has an energy resolution about twice that of conventional counters, and has been successfully used on the satellite *Tenma.

protogalaxy. An inhomogeneous gas cloud in the early stages of evolution into a galaxy. *See* galaxies, formation and evolution.

proton. Symbol: p. An *elementary particle (a hadron) that has a positive *charge, equal in magnitude to that of the *electron. It forms the nucleus of the hydrogen atom and is present in differing numbers in all nuclei. The rest mass of the proton is 1.6726×10^{-24} gram, approximately 1836 times that of the electron. It has spin $\frac{1}{2}$. It has a lifetime known to exceed 10^{31} years. The antiparticle of the proton is the *antiproton. See also* hydrogen; proton-proton chain reaction.

proton-proton chain reaction (p-p chain). A series of *nuclear fusion reactions by which energy can be generated in the dense cores of stars. The overall effect of the chain reaction is the conversion of hydrogen nuclei to helium nuclei with the release of an immense amount of energy. The reactions require a temperature of 10^7 kelvin and are therefore believed to be the major source of energy in the sun and in all main-sequence stars that are cooler than the sun. The *carbon cycle predominates in hot stars.

The sequence of steps of the p-p chain is shown in the illustration, with the principal sequence outlined heavily in black. The possible chains occur simultaneously but in what is thought to be 99.75% of the reactions, the sequence begins with two hydrogen nuclei (i.e. protons) combining to emit a positron and a neutrino and forming a nucleus of deuterium. The deuterium then combines with a proton to yield a nucleus of helium-3 and a photon. About 86% of the helium-3 nuclei will combine with another helium-3 to give helium-4, liberating two protons. *See also* neutrino detection experiments.

protoplanet. An evolving planet in the process of *accretion, together with any of its satellites forming by accretion at the same time in the vicinity of the planet. The dust cocoon around some *T Tauri stars has the attributes of a *protoplanetary system*. In the one typical case, HL Tauri, it forms a flattened disc, radius 160 AU, containing about one earth mass of dust; allowing for dust condensed into planetesimals and the system's gas content, it could condense into several planets. Infrared observations from the satellite *IRAS show dust around over 40 slightly older stars, including *Vega, indicating either a protoplanetary system or a swarm of comets.

protostar. A stage in the evolution of a young star after it has fragmented from a gas cloud but before it has

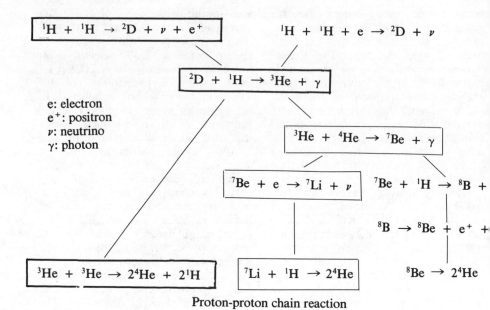

$$^1H + {}^1H \rightarrow {}^2D + \nu + e^+ \qquad\qquad {}^1H + {}^1H + e \rightarrow {}^2D + \nu$$

$$^2D + {}^1H \rightarrow {}^3He + \gamma$$

e: electron
e$^+$: positron
ν: neutrino
γ: photon

$$^3He + {}^4He \rightarrow {}^7Be + \gamma$$

$$^7Be + e \rightarrow {}^7Li + \nu \qquad {}^7Be + {}^1H \rightarrow {}^8B +$$

$$^8B \rightarrow {}^8Be + e^+ +$$

$$^3He + {}^3He \rightarrow 2{}^4He + 2{}^1H \qquad {}^7Li + {}^1H \rightarrow 2{}^4He \qquad {}^8Be \rightarrow 2{}^4He$$

Proton-proton chain reaction

collapsed sufficiently for nuclear reactions to begin. This phase may take from 10^5 to 10^7 years, depending on the mass of the star. Simple theoretical models of protostars are probably inaccurate, according to recent observations of *stellar birth regions. The satellite *IRAS has probably detected many protostars in its infrared survey, but their identification requires clear criteria for distinguishing true protostars from young stars cocooned in dust. *See also* Bok globule; Hayashi track; Herbig-Haro objects; T Tauri stars.

Proxima Centauri. The nearest known star to the sun. It is a *flare star that lies in the constellation Centaurus and is a component of the triple star *Alpha Centauri. It is an intense source of low-energy x-rays and high-energy ultraviolet – i.e. XUV wavelengths. m_v: 10.7; M_v: 15.1; spectral type: M5e V; mass: 0.1 times solar mass; radius: 0.093 times solar radius; distance: 1.31 pc. (4.26 ly).

p-spot. *Short for* preceding spot. *See* sunspots.

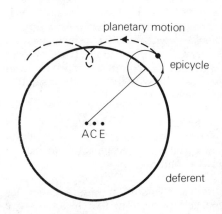

planetary motion

epicycle

ACE

deferent

Ptolemaic system

Ptolemaic system. A model that attempted to explain the observed motions of the sun, moon, and planets and predict their future positions. Originally proposed by Apollonius of Perga in the third century BC and developed by Hipparchus, it was completed by Claudius Ptolemaeus of Alexandria in the second century AD. Ptolemy assumed a *geocentric system, possibly for ease of calculation, in which the moon, Mercury, Venus, the sun, Mars, Jupiter, and Saturn moved around the earth in circular orbits of increasing radius. The position of the earth, E, was somewhat offset from the centre of the orbit, C (see illustration). The orbiting body moved at a uniform rate about a point A rather than about C. A, C, and E were colinear and AC = CE. This improved the accuracy of earlier theories in which Ptolemy had made the moon or planet move in a small circle (an *epicycle*) the centre of which moved around the circumference of a larger circle (the *deferent*), which was the body's orbit. By a suitable choice of the relative size of epicycle to deferent, the relative rates of motion in these circles, and the distance CE, Ptolemy was able to predict planetary positions to within about 1°. Comparison between prediction and observation were much simpler with a geocentric rather than a heliocentric system. It is not known however whether Ptolemy believed in an actual geocentric solar system, as did his later followers. The Ptolemaic system, with minor modifications, was in use until the *Copernican system was finally accepted.

Pulkovo Main Astronomical Observatory. An observatory of the Academy of Sciences of the USSR, built near Leningrad and opened in 1839. It was completely destroyed in World War II, after which it was rebuilt. It reopened in 1954 with the surviving instruments repaired and modernized and some new instruments.

pulsar. A very regularly 'pulsing' source of radiation, which almost certainly originates from a rotating *neutron star. Pulsars were originally discovered at radio wavelengths, and two of these have now been detected at optical and gamma-ray wavelengths; x-ray astronomy has turned up the related class of *x-ray pulsators or pulsars.

The first radio pulsar, PSR 1919+21, was detected by Antony Hewish and Jocelyn Bell (now Burnell) during a pioneering study of interplanetary scintillation (*see* scattering). Over 350 radio pulsars are now known, of which 155 were discovered in 1977 in a pulsar survey made at Molonglo Radio Observatory. The periods range from 1.56 milliseconds (PSR 1937+21) to 4 seconds and can be measured to accuracies of, typically, one part in 10^{10}. The width of the pulse is usually a few per cent of the pulsar's period but can reach 50%. Most pulsars are single, but a few *binary pulsars are now known.

The total number of pulsars in the Galaxy is estimated at over one million. Their rate of formation is now calculated as maybe 5 per century. Pulsars, together with a smaller number of black holes, are currently thought to originate in *supernova explosions. The estimated supernova production rate in the Galaxy is only 2–3 per century. The discrepancy is less than was thought a few years ago, but it may still be necessary to invoke the formation of some pulsars by another means (perhaps the accretion of gas on to a *white dwarf).

The received pulses occur when a beam of radio waves, emitted by a rotating neutron star, sweeps past the earth in an identical manner to the flashes produced by a lighthouse lamp. This beam of radiation is comprised of *synchrotron emission; it arises from electrons moving in the neutron star's strong magnetic field (about 10^8 tesla), whose direction differs from that of the pulsar's rotation axis. There is still dispute over the emission site of the beam: it may be near the star, at the magnetic poles, so that radio waves are beamed down the magnetic axis; alternatively it may be further out near the speed of light cylinder, where the magnetic field is rotating at almost the speed of light.

Despite their impressive regularity, the periods of all radio pulsars are very gradually increasing as the neutron star loses rotational energy. The central star of the *Crab nebula – the youngest known pulsar – is slowing at the rate of one part in a million per day. The slow-down rate of both this pulsar and the young *Vela pulsar are occasionally interrupted by *glitches* (or *spin-ups*): temporary changes in rotation rate. These are thought to be due to rearrangements of the crust or core of the neutron star. The Crab and Vela pulsars are both *optical pulsars. They are also among the brightest *gamma-ray sources in the sky. Both optical and gamma-ray luminosity seem to decrease rapidly with a pulsar's age. The Crab pulsar is the fastest of the young pulsars (having a period of 33 milliseconds) but the newly discovered *millisecond pulsars represent a much faster class of pulsar, probably 'spun-up' by mass transfer.

X-ray pulsars occur in close binary systems when gas from a companion is channelled on to the neutron star at its magnetic poles: the x-rays appear to pulse on and off as these hot gas patches are exposed by the star's rotation. The gas flow affects the neutron star's spin and as a result all x-ray pulsars are gradually speeding up. Although some have typical pulsar periods (up to a few seconds) there are many 'slow' x-ray pulsars (with periods of several minutes), whose rotation must somehow have been strongly 'braked'.

pulsating universe. *See* oscillating universe.

pulsating variables. *Variable stars that periodically brighten and fade as a result of large-scale and more or less rhythmical motions of their outer layers. The simplest motion is purely radial, a cycle of expansion and contraction in which the star remains spherical but changes volume. The idea that a periodic stellar expansion could lead to a variable light output

was proposed by Shapley in 1914, with Eddington presenting the theory in 1918. The period of light variation is equal to the period of pulsation and is normally approximately constant. The spectrum of the star also changes periodically due to changes in surface temperature. The pulsation cycle is demonstrated by variations in *radial velocity. All these variations are shown in the diagram at *Cepheid variables. The pulsation period, P, was shown by Eddington to be related to mean stellar density, ρ, by the *period-density relation* and hence to the star's radius, R, and mass, M:

$$P \propto 1/\sqrt{\rho} \propto \sqrt{(R^3/M)}$$

Since *luminosity is also proportional to radius, pulsation period should be related to luminosity (*see* period-luminosity relation). Types of pulsating variable include the *Cepheid variables, *RR Lyrae stars, *dwarf Cepheids, *Delta Scuti stars, long-period *Mira stars, *semiregular variables (including *RV Tauri stars), *irregular variables, and *Beta Cephei stars.

A star pulsates because there is a small imbalance between gravitational force and outward directed pressure so that it is not in hydrostatic equilibrium. When it pulsates it expands past its equilibrium size until the expansion is slowed and reversed by gravity. It then overshoots its equilibrium size again until the contraction is slowed and reversed by the increased gas pressure within the star. The fact that the pulsations do not die away as energy is dissipated means that some process is continuously converting heat energy into mechanical energy to drive the pulsations. The centre of the star is not involved in the pulsation. For most pulsating variables (probably excluding Beta Cephei stars) the driving force is a valve mechanism in the form of a region of changing *opacity near the stellar surface where the pulsation amplitude is greatest. This region is an *ionization zone* in which

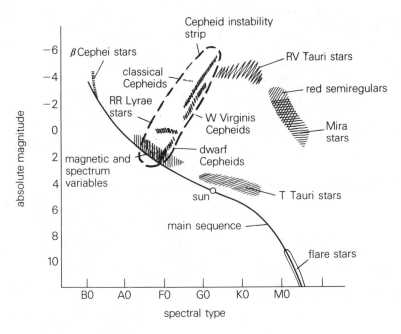

Hertzsprung-Russell diagram of pulsating and other variable stars

atoms, mainly helium or hydrogen, are partially ionized. Normally atoms become transparent, letting heat escape, as they are compressed. In an ionization zone the opacity increases with compression and the zone effectively acts as a heat engine: as long as it lies at the required depth below the surface it can drive the pulsations. If the zones are too close to the surface, as in a very hot star, or if too deep, then pulsations cannot occur. The period of pulsation is complicated by the fact that the gas is thought to be able to undergo oscillations (as in an open musical pipe) either in a fundamental mode or as a harmonic of the fundamental.

The distribution of pulsating variables on the *Hertzsprung-Russell diagram differs from that of normal stars. Most lie on a nearly vertical band – the *Cepheid instability strip* – that extends upwards from the main sequence (and possibly below it) and merges into a broader instability region at top right (see illustration).

Stars reach these instability regions by various evolutionary paths and may turn up there several times during their evolution. The existence of such instability regions indicates that with certain combinations of stellar luminosity and effective temperature a state of pulsation rather than of rest is favoured.

Puppis (Poop, Stern). A constellation in the southern hemisphere near Canis Major, lying partly in the Milky Way. It was once part of the constellation *Argo. The brightest star is the extremely remote and luminous 2nd-magnitude Naos (Zeta). It also contains several open clusters including the just-visible M47, the bright planetary nebula NGC 2438, and the supernova remnant *Puppis A,* which is a strong radio and x-ray source. Abbrev.: Pup; genitive form: Puppis; approx. position: RA 7.5h, dec −30°; area: 673 sq deg.

Puppis A. *See* Puppis.

Pyxis (Mariner's Compass). A small inconspicuous constellation in the southern hemisphere near Vela, lying partly in the Milky Way. It once formed part of the Mast of Argo, an obsolete division of *Argo. It has one 3rd-magnitude and two 4th-magnitude stars and contains the peculiar *recurrent nova T Pyxidis. Abbrev.: Pyx; genitive form: Pyxidis; approx. position: RA 9h, dec −30°; area: 221 sq deg.

PZT. *Abbrev. for* photographic zenith tube.

Q

QSO. *Abbrev. for* quasi-stellar object. *See* quasar.

quadrant. An instrument dating back to antiquity and used for measuring altitudes. It remained the most important astronomical instrument until the telescope was invented. It consisted of a 90° graduated arc (a quarter circle) with a swivelling arm to which a sighting mechanism was attached. In the *mural quadrant* the graduated arc, often very large, was attached to a wall and was orientated along the observer's meridian. The mural quadrant was therefore the forerunner of the *transit circle.

Quadrantids. A major *meteor shower, radiant: RA 232°, dec +50°, that maximizes on Jan. 3 when the *sun's longitude is 282.8°. The *zenithal hourly rate (ZHR) at maximum is about 110 and is constant from year to year. The *meteor stream is narrow, the ZHR being greater than half maximum for only 17 hours: the *meteor stream's cross-sectional diameter is about 1.7 million km compared with 20 million and 13 million km for those of the *Perseids and *Geminids. The stream has its ascending *node close to the orbit of Jupiter and there is a possibility that Quadrantid meteors could be seen from the surface of that planet too. Owing to gravitational and radiational perturbations the large 'visual' *meteoroids are in a slightly different orbit to the smaller 'radio' meteoroids. The shower is named after the obsolete constellation Quadrans Murali found in early 19th-century star atlases. The radiant is actually in Bootes.

quadrature. The position, or aspect, of a body in the solar system when its *elongation from the sun, as measured from the earth, is 90°, i.e. when the angle sun-earth-planet is a right angle.

quantum. The minimum amount by which certain properties of a system, such as its energy or angular momentum, can change. The value of the property cannot therefore vary continuously but must change in steps: these steps are equal to or are integral multiples of the relevant quantum. This idea is the basis of *quantum theory and quantum mechanics. The *photon, for example, is a quantum of *electromagnetic radiation.

quantum efficiency. A measure of the efficiency of conversion or utilization of light or other radiation. It is given by the proportion of incident *photons, of a particular frequency, that are converted by a device into a useful form, such as a permanent record. Photographic emulsions, at best, record only a few photons out of every thousand falling on them. Electronic detectors have a much higher efficiency.

quantum gravitation. Any of various theories of gravitation that are consistent with the theories of quantum mechanics and special relativity. It is postulated that the gravitational force between two particles is generated by the exchange of an intermediate particle. (This is similar to the explanation of the other *fundamental forces.) In earlier theories the force was transmitted by the exchange of particles called *gravitons* of zero charge, zero rest mass, and spin 2; other particles have also been suggested.

quantum mechanics. *see* quantum theory.

quantum theory. The theory put forward by the German physicist Max Planck in 1900. Classical physics regarded all changes in the physical properties of a system to be continuous. By departing from this viewpoint and allowing physical quantities such as energy, angular momentum, and action to change only by discrete amounts, quantum theory was born. It grew out of Planck's attempts to explain the form of the curves of intensity against wavelength for a *black-body radiator. By assuming that energy could only be emitted and absorbed in discrete amounts, called *quanta, he was successful in describing the shape of the curves. Other early applications of quantum theory were Einstein's explanation of the *photoelectric effect and Bohr's theory of the atom (*see* hydogen spectrum; energy level). The more precise mathematical theory that developed in the 1920s from quantum theory is called *quantum mechanics*. Relativisitc quantum mechanics resulted from the extension of quantum mechanics to include the special theory of *relativity. *See also* quantum gravitation.

quark. One of probably six different *elementary particles (evidence for a 6th quark was presented 1984) that are thought to be the structural units from which other particles (excluding leptons) are formed. In the usual scheme, quarks all have spin $\frac{1}{2}$, electric charges that are multiples of one third, and other innate qualities (colour). Corresponding to each quark is an *antiquark*, with opposite electric charge, etc. (*see* antiparticle). The theory of quark interactions predicts that isolated quarks can never exist in the free state; they only occur in particular pairs and triplets that form the class of particles called the hadrons. Despite intensive searches no quarks have been definitively observed.

At present there are tentative suggestions that states of condensed matter, such as the deep interiors of *neutron stars, may consist of a soup of free quarks.

quasar. A compact extragalactic object that looks like a point of light but emits more energy than a hundred supergiant galaxies. The name is a contraction of *quasi-stellar object (QSO)*, and early papers on the subject sometimes refer to *quasi-stellar galaxies (QSG)*. Although they are bright optical sources, quasars emit most of their energy as infrared radiation. They are also strong x-ray sources. About 1% of quasars are also radio sources.

Quasars, of which several thousand are known, were discovered in 1963 as the optical counterparts to some powerful radio sources. The spectra of these 'stars' were peculiar, with bright emission lines, apparently of an unknown element, superimposed on a continuum. Maarten Schmidt finally recognized the pattern of lines in one of these objects (3C 273) as the Balmer series of hydrogen redshifted to $z = 0.158$ (*see* redshift), showing that it was impossible for this to be a star within the Galaxy. Other quasars were then discovered to have far greater redshifts. The record is currently held by PKS 2000–330, whose redshift of 3.78 indicates that it is receding from us with over 90% of the speed of light. Astronomers today interpret the quasar redshifts as doppler shifts (*see* doppler effect) arising from the expansion of the universe, making them the most distant and hence the youngest extragalactic objects we observe.

Many quasar spectra also show absorption lines, which sometimes have a wide range of redshifts up to the value of the emission redshifts. Those close to the emission redshift presumably arise from clouds of matter ejected from the quasar, while lower-redshift systems are probably due to galaxies and clouds of intergalactic hydrogen lying in the line of sight.

To be visible at such great distances, quasars must be exceedingly luminous: many have absolute *magnitudes brighter than -27. There is however a great range in luminosity. In addition quasars themselves are often variable by a factor of two or greater, on timescales sometimes as short as a few hours. This indicates that their light-producing regions are sometimes less than a light-day across, which poses several problems in explaining their energy generation. The only process known to be efficient enough is some sort of accretion on to a supermassive *black hole (of the order of 10^9 solar masses) that is located at the nucleus of the quasar.

Some of the nearer quasars are surrounded by a 'fuzz', believed to be a surrounding galaxy too faint to be detected around the more distant objects. This gives support to the suggestion that quasars are the nuclei of *active galaxies. The fuzz often has indications of spiral structure, and sometimes a companion galaxy; the latter may be 'spilling' gas on to a black hole in the centre of a spiral galaxy. This suggests a difference from *radio galaxies, *BL Lac objects, and possibly *N galaxies, which appear to represent activity in an elliptical galaxy. Quasars are then the most violent examples in a sequence that ranges through *Seyfert galaxies to normal spiral galaxies like our own.

quasi-stellar galaxy (QSG). *See* quasar.

quasi-stellar object (QSO). *See* quasar.

quiescent prominence. *See* prominences.

quiet sun. The term applied to the sun around the minimum of the *sunspot cycle, when the absence of *active regions allows relatively 'quiet' conditions to prevail. The sun is not then free of all disturbance but there are few (if any) concentrations of intense magnetic flux to produce significant departures from a uniform distribution of radiative flux across the disc. *Compare* active sun.

R

RA. *Abbrev. for* right ascension.

radar astronomy. The study of celestial bodies (as yet) within the solar system by means of the faint reflections from them of powerful high-frequency radio transmissions aimed in their directions from the earth. Some of the very large radio *dishes such as the 305 metre dish at Arecibo, are equipped for radar work. The time interval between transmission of a signal and reception of the reflected signal is an accurate measure of intervening distance. The determination of planetary distances led to a precise value for the *astronomical unit. The transmitted frequency may be chosen to penetrate the atmosphere of a planet in order to establish a profile of the surface, which is otherwise hidden from view. In addition, rotation periods can be determined from the doppler shift produced in pulses bounced off different parts of an object. The rotation periods of Venus and Mercury were determined by radar measurements.

radar meteor. *Another name for* radio meteor. *See* meteor.

radial velocity (line-of-sight velocity). Symbol: v_r. The velocity of a star

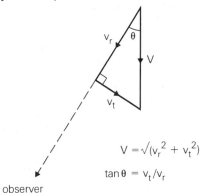

$$V = \sqrt{(v_r^2 + v_t^2)}$$

$$\tan \theta = v_t / v_r$$

observer

Space velocity and its components

along the line of sight of an observer. It is calculated directly from the doppler shift (*see* doppler effect) in the lines of the star's spectrum: if the star is receding there will be a *redshift in its spectral lines and the radial velocity will be positive; an approaching star will produce a blueshift and the velocity will be negative. About 60% of stars have values of about ±20 km/s. Radial velocity is usually given in terms of the sun's position so that the earth's orbital motion has no effect.

The velocity and direction of motion of the sun relative to the *local standard of rest can be determined from a statistical analysis of the observed radial velocities of nearby stars: stars in the region of the solar *apex have predominantly positive velocities while those around the antapex have predominantly negative ones.

A star's velocity in space relative to the sun – its *space velocity* – is a vector quantity and can be split into two components: its radial velocity and its tangential velocity, v_t, along the direction of *proper motion. Measurements of these components

give both the magnitude, V, and direction, θ, of the star's space velocity (see illustration).

radial-velocity curve. A curve showing how the radial velocity of a *spectroscopic binary changes during one revolution.

radian. A unit of angle used in plane geometry that is the angle subtended at the centre of a circle by an arc of the circle equal in length to the circle's radius. Thus 2π radians equals 360°, 1 radian equals 57°.296, and 1° equals 0.017 radians. Solid angles, used in spherical geometry, are measured in steradians.

radiant. The point on the celestial sphere, usually given in the coordinates *right ascension and *declination, from which a *meteor appears to originate. Shower meteors all seem to radiate from this point and the *trains of a series of shower meteors, extrapolated backwards, intercept at the radiant. Its position is found by calculating the intercept with the celestial sphere of the vector sum of the meteor's velocity vector and the earth's velocity vector on collision. The radiant position changes slightly with time as the earth moves through the *meteor stream. Meteor showers are usually named after the constellation that contains the radiant.

radiant flux. The total power emitted or received by a body in the form of radiation. It is measured in watts.

radiation. *See* electromagnetic radiation; energy transport.

radiation belts. One or more doughnut-shaped regions in the *magnetosphere of a planet in which energetic electrons and the ionized nuclei of atoms, chiefly protons from hydrogen atoms, are trapped by the planet's magnetic field. The particles within the belt, whose axis is the magnetic axis of the planet, spiral along magnetic field lines, travelling backwards and forwards between reflections which occur as they approach the magnetic poles. This motion produces *synchrotron emission from the particles. The particles are either captured from the *solar wind or are formed by collisions between *cosmic rays and atoms or ions in the planet's outer atmosphere.

The earth, Jupiter, Saturn, and Uranus are all known to possess radiation belts. The magnetic field of Mercury appears to be too weak to sustain belts. The radiation belts of Jupiter are many times more intense than the *Van Allen radiation belts of the earth and pose a threat to spacecraft systems in the vicinity of the planet: they would certainly be very hazardous to manned exploration of the *Galilean satellites of Jupiter.

radiation era. *See* big bang theory.

radiation pressure. The very small pressure exerted on a surface by light or other electromagnetic radiation. The radiation is considered as a stream of *photons, and the pressure is then due to the transfer of momentum from the photons to the surface. Although radiation pressure has a negligible effect on large particles or on dense matter, the effect is considerable with very small particles, which are driven away from the radiation source. Thus solar radiation pressure, aided by the *solar wind, clears tiny particles of dust ($<1\,\mu m$) from the solar system. *See also* Poynting–Robertson effect.

radiation resistance. A ficticious resistance in which the power radiated away from or collected by an *antenna is considered to be dissipated. For a perfect antenna made of superconducting material, the ohmic losses (power dissipation arising from resistance) in the elements themselves are zero; the resistive part of the impedance presented at the feed point is then equal to the radiation resistance. In any practical antenna the ohmic losses are kept as small as possible in comparison to the radiated power. Maximum power transfer between transmitter or *receiver and the antenna is effected when the *feeder has a characteristic impedance equal to the radiation resistance.

radiation temperature. *See* effective temperature.

radiative transport, zone. *See* energy transport.

radioactive decay. The spontaneous transformation of one atomic nucleus into another with the emission of energy. The energy is released in the form of an energetic particle, usually an *alpha particle, *beta particle (i.e. an electron), or positron, sometimes accompanied by a gamma-ray photon. The unstable isotopes of an element that can undergo such transformations are called radioactive isotopes, or *radioisotopes*. The emission of a particle from the nucleus of a radioisotope results in the production of an isotope of a different element, as in the beta decay of carbon–14 to nitrogen–14 or the alpha decay of radium–226 to radon–222. The isotope produced is itself often radioactive.

The average time taken for half a given number of nuclei of a particular radioisotope to decay is the *half-life of that radioisotope; values range from a fraction of a second to thousands of millions of years. *See also* radiometric dating.

radio astrometry. A branch of *astrometry in which the coordinates of celestial objects are measured with high precision using a *radio telescope. It has become more accurate than its optical counterpart in recent years with the advent of interferometers having very long and accurately determined *baselines.

radio astronomy. The study of celestial bodies by means of the *radio waves they emit. Extraterrestrial radio waves were first detected in 1932 by the American radio engineer Karl Jansky. The first *radio telescope was built and operated by Grote Reber in the late 1930s but radio astronomy only developed after the war. It has benefitted greatly from advances in electronics and solid-state physics and is now a major branch of observational astronomy. *See also* radio source.

radio background radiation. *See* radio souce; ISM.

radio galaxy. An extragalactic radio source identified with an optical galaxy whose radio-power output lies in the range 10^{35} to 10^{38} watts. These sources often show double structure (*see* radio source structure) and are to be distinguished from normal galaxies, such as M31, whose radio emissions roughly follow the optical contours of the galaxy and whose radio powers are much lower, in the range 10^{30} to 10^{32} watts. Radio galaxies with diameters greater than one megaparsec are now known. An example of such a *giant radio galaxy* is 3C 236, which is a double radio source whose radio emission spans almost six megaparsecs of space, making it one of the biggest radio galaxies in the universe.

Radio galaxies are a type of *active galaxy. At the high-luminosity end of the range, the distinction between radio galaxy and *quasar becomes hazy. There is some evidence, however, that radio galaxies are associated with elliptical galaxies, thus possibly differentiating them from quasars, which are now thought to occur in spiral galaxies. The energy mechanism of radio galaxies is not understood but (as with quasars) supermassive *black holes have been postulated as the source of the prodigious power output. One suggestion is that a toroidal cloud of matter surrounds the black hole, gradually being sucked in and liberating a large proportion of its rest-mass energy. This energy may be channelled into *beams.

radio heliograph. A *radio telescope designed for making radio maps of the sun. The radio heliograph at *Culgoora, Australia, is a particular example.

radioisotope. *See* radioactive decay.

radioisotope thermoelectric generator. *See* thermoelectric generator.

radio map. *See* radio-source structure.

radio meteor (radar meteor). *See* meteor.

radiometer. A device that measures the total energy or power received from a body in the form of radiation, especially infrared radiation. In radio astronomy, a radiometer measures the total received radio *noise power.

radiometric dating. The dating of rocks (and also fossils and archaeological remains) by the accurate determination of the quantities of a long-lived radioactive isotope and its stable decay product in a sample. Assuming that the parent radioisotope was present at the time of formation of the rock, etc., then the number of daughter isotopes produced by *radioactive decay of the parent depends only on the *half-life of the parent and the age of the sample. Half-lives must therefore be known with great accuracy for precise dating and should range from about 10^5 to about 10^{10} years. In addition, there should be no loss or gain of parent or daughter isotope during the time the 'radioactive clock' is operating; if this condition is only partially satisfied, allowances must be made. The decay of radioisotopes can be used not only to date material but also to time very slow processes, such as the evolution of the earth's atmosphere.

Pairs of isotopes used in radiometric dating include potassium–40 which decays to argon–40 with a half-life of 1.25×10^9 years, and rubidium–87 which decays to strontium–87 with a half-life of 4.88×10^{10} years.

radio receiver. *See* receiver.

radio source. A celestial object whose radio emissions may be detected with a *radio telescope. Within our solar system these sources include Jupiter and the sun, both of which emit powerful radio *bursts. Galactic radio sources include *pulsars, *supernova remnants, *HI and *HII regions, *maser sources, and the *galactic centre itself. *Synchrotron emission from cosmic rays in the Galaxy gives rise to a diffuse background radiation,

centred broadly on the galactic plane but having several spurs away from it (*see* interstellar medium). This background radiation may limit the *sensitivity of a radio telescope, especially at frequencies below about 400 megahertz.

Extragalactic radio sources include spiral galaxies (whose HI and *carbon monoxide emissions may be used to map the distribution of hydrogen within them), *radio galaxies, and *quasars. An extragalactic radio source is often classified according to the spatial distribution of the radio emission shown on a radio map of the source. *See* radio-source structure; radio-source spectrum.

radio-source catalogue. A numbered list of *radio sources detected in a survey at a given frequency down to a limiting *flux density. A well-known example is the *third Cambridge cata-*

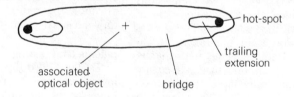

Double radio source

Bent double radio source

Radio-trail source

logue, 3C. The catalogue usually lists the source positions and flux densities together with any other information that may be available about angular sizes, etc. From one or more catalogues *source counts can be compiled, in which the total number of sources per unit solid angle (steradian) above a given flux density is plotted as a function of the flux density. These counts are important in *cosmology since different mathematical models of the universe predict different source counts: they provide one method of comparing theory with the real world.

radio-source spectrum. The variation of the *flux density of a *radio source with frequency. The emissions from most radio sources are *noise signals displaying continuous spectra that rise or fall with increasing frequency more or less smoothly. Sources of *thermal and *synchrotron emission both give spectra of this type, sometimes showing a low-frequency cut-off due to self-absorption (*see* synchrotron emission). The spectra of the radiation from *maser sources or from clouds of interstellar molecules, such as some *HI regions, have discrete lines of bright emission superimposed on a background continuum (*see* molecular-line radio astronomy).

The continuous spectra are often classified according to the *spectral index*, α, which for a source whose flux density is S is defined by

$$S \propto \nu^{\alpha} \text{ or } S \propto \nu^{-\alpha}$$

where ν is the frequency. Both definitions are in use and it is important to be clear which is intended in each case.

radio-source structure. The spatial distribution of the radio emission from a *radio source. It may be partially inferred from a *radio map* of the source, which is the *convolution of the true source distribution projected onto a plane and the *antenna pattern of the *radio telescope, and which is often plotted in contours of constant *brightness temperature or *flux density; such a map is called a *contour map*. Where the angular size of the source is smaller than the antenna pattern, the radio map reveals no details of its structure; the source is then said to be *unresolved* and is described as a *point source*. In the converse case, the source is said to be *extended*.

Maps of extragalactic radio sources often show two distinct bright features that may or may not be joined by a faint *bridge* of diffuse emission (see illustration). These are *double radio sources*. The bright features can sometimes be subdivided into distinct spots of bright emission, or *hot spots*, with trailing extensions. *Cygnus A is a fine example. Sometimes the trailing extensions and the bridge merge into one and do not lie on the straight line defined by the bright heads. The source is then described as a *bent double radio source*. The angle of bending may become so large that the trailing extensions merge into one another or become almost parallel, forming a *radio-trail source*. See also jet; beam.

radio telescope. An instrument for recording and measuring the *radio-frequency emissions from celestial *radio sources. All radio telescopes consist of an *antenna, or system of antennas, connected by *feeders to one or more *receivers. The antennas may be in the form of *dishes or simple linear *dipoles. The receiver outputs may be displayed on suitable devices such as pen-recorders or recorded on tape for further analysis by computer.

The antenna system may consist of two separated units whose electrical signals are conveyed by feeders to a common receiver, forming an *interferometer*. The antenna units are often mounted on an east-west line and are arranged to point in the same direction. The earth's rotation then causes a radio source to move through the antenna beam. Waves from the source interact when the identical signals are combined, alternately reinforcing each other (producing a signal) and cancelling out (producing no signal) during the source's passage through the beam. The amplitude of the summed signal thus changes periodically, producing *interference fringes* at the output of the receiver. The fringes are sinusoidal variations whose maximum amplitude depends on the *flux density of the source and whose period depends on both the radio wavelength and the spatial separation of the antenna units. Analysis of the changing interference fringes allows source position to be determined and source structure to be studied. An interferometer is generally used to improve the angular *resolution of the telescope: the longer the distance, or baseline, between the antennas, the finer the detail that can be resolved. As resolution increases, however, large-scale structure is lost. *See also* aperture synthesis; array; sea interferometer.

radio-trail source. *See* radio-source structure.

radio waves. Travelling electromagnetic disturbances whose frequencies lie in the radio region of the electromagnetic spectrum, extending from about 20 kilohertz to over 300 gigahertz, the highest frequencies corresponding to millimetre and submillimetre wavelengths. *Radio astronomy may in principle be carried out at any radio frequency but in practice observations are confined to certain 'radio-astronomy bands' set aside for that purpose by international agreement to minimize *interference from other services (see table). These bands lie within the *radio window.

radio window. *See* atmospheric windows.

radius of curvature. Symbol: *r*, *R*. The radius of the sphere whose surface contains the surface of a spherical mirror, lens, or wavefront. The *curvature* of the mirror, etc., is given by the reciprocal of the radius of curvature, i.e. by $1/r$. The radius of curvature of a mirror is twice its *focal length, *f*. That of a lens is given by
$$1/f = (n - 1)(1/r_1 - 1/r_2)$$
where *n* is the refractive index; it is assumed that, in the direction of light travel, the radii, r_1 and r_2, for the convex and concave lens surfaces are positive and negative, respectively.

radius vector. The line, of length *r*, joining a point on a curve to a reference point, such as the origin in spherical and cylindrical coordinate systems. In planetary motion the radius vector is the line between the position of a planet and the focus of its orbit, i.e. the sun's position. *See also* Kepler's laws.

radome. A protective covering over a radio *antenna that is transparent to radio waves at the frequency in use.

Ramsden disc (Ramsden circle). *See* exit pupil.

Ramsden eyepiece. *See* eyepiece.

Ranger. An American space programme, originally intended to hard-land instruments on the lunar surface but upgraded in 1963 into purely photographic missions. The first five Ranger craft were launched between August 1961 and October 1962. Ranger 6, launched Jan. 30 1964, reached its target but suffered camera malfunction. The last three – Rangers 7, 8, and 9 – were launched on July 28 1964, Feb. 17 1965, and Mar. 21 1965. They obtained the first close-ups of the moon, revealing the existence of boulders and metre-sized *craters in the *regolith: over 4300, 7100, and 5800 photographs were transmitted by Rangers 7, 8, and 9

Frequency Allocations for Radio Astronomy
(W.A.R.C. 1979)

37.5–38.25 MHz	secondary allocation (fixed & mobile, primary)
73–74.6 MHz	exclusive (USA only)
150.05–153 MHz	primary allocation (shared with fixed & mobile; not USA)
322–328.6 MHz	primary allocation (shared with fixed & mobile)
406.1–410 MHz	primary allocation (shared with fixed & mobile)
608–614 MHz	primary allocation (shared with satellite links; USA only)
1400–1427 MHz	worldwide exclusive
1660–1660.5 MHz	primary allocation (shared with satellite links)
1660.5–1670 MHz	primary allocation (shared with met. aids)
2655–2700 MHz	worldwide exclusive
4800–4990 MHz	secondary allocation (fixed & mobile, primary)
4990–5000 MHz	primary allocation (shared with fixed & mobile)
10.6–10.68 GHz	primary allocation (shared with fixed & mobile)
10.68–10.7 GHz	worldwide exclusive
14.47–14.5 GHz	secondary allocation (satellite links, primary)
15.35–15.4 GHz	worldwide exclusive
22.21–22.5 GHz	primary allocation (shared with fixed & mobile) [includes H_2O at 22.235 GHz]
23.6–24 GHz	worldwide exclusive [includes NH_3 at 23.7 GHz]
31.1–31.5 GHz	worldwide exclusive
31.5–31.8 GHz	exclusive (shared with fixed & mobile in USA)
42.5–43.5 GHz	primary allocation (shared with fixed & mobile)
86–92 GHz	worldwide exclusive [includes HCN at 88.6 GHz]
105–116 GHz	worldwide exclusive
182–185 GHz	worldwide exclusive
217–231 GHz	worldwide exclusive
265–267 GHz	primary allocation (shared with fixed & mobile)

respectively, leading to a great advance in knowledge of the moon. *See also* Lunar Orbiter probes; Surveyor; Luna probes; Zond probes.

Ras Algethi (α Her). A visual binary in the constellation Hercules. The primary (A) is a red supergiant that varies irregularly in magnitude from 3.0 to 4.0, the 6th-magnitude yellow giant companion (B) appearing green by contrast. B is a spectroscopic binary, period 51.6 days. Spectral type: M5 I (A), G5 III + F2 (B).

Ras Alhague. *See* Ophiuchus.

R-association. *See* association.

Rayleigh limit. *See* resolution.

rays. High *albedo (bright) streaks that surround the youngest moon *craters and help to assign them to the Copernican System – the most recent era in lunar history. Rays consist of crater *ejecta that has not yet been darkened by radiation or by admixture with local materials. The most prominent systems are associated with *Tycho and Giordano Bruno. Similar bright features have now been found around craters on other solar-system bodies.

Razin effect. The reduction of the *synchrotron emission from a relativistic electron due to the *refractive index of the surrounding medium being less than one.

R Coronae Borealis stars. A small group of *variable stars in which the brightness decreases – typically by about four *magnitudes – at irregular intervals and often quite rapidly, returning to its normal maximum often slowly and with considerable fluctuation. The frequency of occurrence of the decrease varies greatly between the stars. The prototype of the group is R Coronae Borealis, which drops from 6th magnitude usually down to 14th. They are supergiants that have carbon-rich atmospheres and an abnormally low proportion of hydrogen. The decrease in brightness apparently arises from the strong absorption of light by layers of carbon particles occasionally formed in the outer regions of the stellar atmosphere and gradually blown away by radiation pressure. This interpretation is supported by the observation that the infrared output remains nearly constant during the dramatic visual dimmings. Some R Coronae Borealis stars, including the prototype, are also *pulsating variables.

receiver. An electronic device that, in radio astronomy, detects and measures the radio-frequency signals picked up by the *antenna of a radio telescope. A receiver that measures the total *noise power from the antenna and from itself is called a *total-power receiver*. This is in distinction to other types that measure, for example, the correlated power in two antennas (*see* phase-switching interferometer; Dicke-switched receiver; correlation receiver).

Radio-astronomy receivers usually use the *superheterodyne* technique: the radio-frequency signals at frequency f_1 are combined in a *mixer* with signals from a *local oscillator* at frequency f_0 to produce a combination signal whose amplitude faithfully follows that of the radio-frequency signal but whose frequency is $f_1 - f_0$. This is the *intermediate frequency*, IF, which is amplified in the *IF amplifier* and pre-

sented to the first *detector*. The output of the nonlinear detector is a voltage that depends on the input power.

In a total-power receiver the output from the detector becomes the receiver output, having first been smoothed by a low-pass *filter to reduce the fluctuations due to the system noise (*see* sensitivity). In a switching receiver the first detector output is passed to a *phase-sensitive detector before smoothing in order to pick out the component at the switching frequency that is buried in the noise. *See also* line receiver.

recession of the galaxies. An important discovery made by Edwin Hubble in 1929 and based on the fact that the spectral lines of all galaxies beyond the *Local Group are systematically shifted towards longer wavelengths. These observed *redshifts are interpreted as being due to the doppler shift and imply that all distant galaxies are moving rapidly away from our own. The *recession velocities* of the galaxies can be determined from their redshifts and are found to be roughly proportional to their distances from us. Thus the universe as a whole appears to be expanding, and it follows that the universe was denser in the past. *See* Hubble's law; expanding universe; big bang theory.

recombination line emission. Electromagnetic radiation resulting from the *recombination* of a free electron (one not bound to a nucleus) and an ionized atom. The spectrum of the radiation consists of a continuum from the transition of free electrons to bound states and a number of distinct recombination lines, each one due to a transition between two *energy levels in the atom. The emission lines occur at radio, infrared, and optical frequencies.

The energy levels are numbered in sequence starting at one for the ground state. Recombination lines are then classified according to the number of the energy level to which the electron moves and the number of energy levels that it passes. Recombination line emission occurs, for example, in *emission nebulae.

recombination time. The period in the universe's history, according to the *big bang theory, after which the temperature was low enough for atoms to form from free nuclei and electrons. This required a temperature of less than about 3000 K, which occurred when the universe had expanded for about 300 000 years. At the time of recombination the radiation in the universe began to propagate freely and ceased to interact with free electrons. This freely propagating radiation now resides in the *microwave background.

recurrent nova. A *cataclysmic variable that suffers a series of violent nova-like outbursts at periodic intervals. The change in brightness is smaller and the decline in brightness more pronounced than with *novae. Like other cataclysmic variables, a recurrent nova is a close binary system in which one member is a *white dwarf; the other component is a *red giant, and gas is being transferred from the latter to the former. The red giant loses matter about a thousand times faster than the companion in a nova system, so that the transferred hydrogen builds up on the surface of the white dwarf at a much quicker rate. The accumulated hydrogen is hence sufficient to erupt in a thermonuclear explosion after only a few decades, and astronomers have been able to see multiple outbursts within the past century or so of systematic investigation of variable stars. Between outbursts, the system's light comprises both emission from hot gas circling the white dwarf and light from the cool red giant; as a result the spectrum shows what is apparently a star with two different temperatures (as with a *symbiotic star). T Pyxidis (1890, 1902, 1920, 1944, 1965), RS Ophiuchi (1901, 1933, 1958, 1967), and T Coronae Borealis (1866, 1946) are recurrent novae.

reddening. *See* extinction.

red dwarf. *See* dwarf star.

red giant. A *giant star with a surface temperature between 2000–3000 kelvin and a diameter 10–100 times greater than the sun. Red giants are one of the final phases in the evolution of a normal star, reached when its central hydrogen has been used up. The star resorts to burning hydrogen in a shell around its dense inert helium core, which results in a rapid inflation of the outer layers of atmosphere. During subsequent evolution, interior structural and nuclear changes cause the star's temperature, size, and luminosity to alter (*see* giant); the star may change to a more compact hotter type of giant, then swell back to a red giant, more than once as it evolves. Since red giants are so distended, gravity has only a small effect on their surface layers and many lose considerable amounts of mass into space in the form of *stellar winds. Condensing dust grains around the stars make them *infrared sources, while molecules here can emit maser radiation, making them *maser sources.

Red giants are often *variable stars since their surface layers slowly expand and contract. Because these stars are so large, their pulsations usually take about a year to complete; red giants therefore belong to the class of *Mira stars, or long-period variables. Low-mass red giants end up as *planetary nebulae, gently puffing off their distended atmospheres at low velocities (a few km s^{-1}) and leaving their dense cores exposed as *white dwarf stars, radiating away their heat into space. More massive red giants explode as Type II *supernovae. *See also* stellar evolution; supergiant.

Red Planet. *Another name for* Mars.

redshift. A displacement of spectral lines towards longer wavelength values; for an optical line, the shift would be towards the red end of the visible spectrum. The *redshift parameter, z,* is given by the ratio $\Delta\lambda/\lambda$, where $\Delta\lambda$ is the observed increase in wavelength of the radiation and λ is the wavelength of the spectral line at the time of emission from a source, i.e. the wavelength in the 'normal' terrestrial spectrum.

The redshifts of astronomical objects within the Galaxy are interpreted as doppler shifts (*see* doppler effect) caused by movement of the source away from the observer. The value of *z* is then v/c, where *v* is the relative *radial velocity and *c* is the speed of light. The redshifts of extragalactic sources, including *quasars, are also interpreted in terms of the doppler effect, which for these objects results from the expansion of the universe. The redshift parameter of a distant galaxy thus gives its velocity of recession; since recessional velocities can be very great, the relativistic expression for redshift must be used:
$$z = [(c + v)/(c - v)]^{1/2} - 1$$
From measurements of galactic redshifts it has been possible to calculate the distances of galaxies, using *Hubble's law (*see* distance determination).

The redshifts described above represent a loss of energy by the photons of radiation in overcoming the effects of recession or expansion. There is another mechanism, however, by which redshifts can be produced, i.e. by which photons can lose energy – the presence of a strong gravitational field. This *gravitational redshift was predicted by Einstein in his general theory of relativity. Although the redshifts of galaxies are often interpreted as due to the relativistic doppler effect alone, both the expansion and the gravitational field of the universe are involved. *See also* cosmic scale factor.

redshift survey. A three-dimensional mapping of galaxies in a large region of the sky, obtained by measuring the *redshifts of all the galaxies down to a certain limiting magnitude. The redshifts allow distances to be determined using *Hubble's law; the sky positions are known. Two major surveys are the Harvard–Smithsonian *CFA (Center for Astrophysics) survey* and the *Durham–Australian redshift survey, DARS.*

Red Spot. *See* Great Red Spot.

red variables. The large group of variable stars, including *Mira stars and *irregular variables, that lie in the top right of the Hertzsprung-Russell diagram of *pulsating variables.

reflectance. The ratio of the total radiant power (in watts) reflected by a body to that incident on it.

reflectance spectra. *Spectra by means of which remote chemical analyses may be made, based on the *reflectance efficiency of a planetary or satellite surface at various wavelengths, particularly near-infrared wavelengths.

reflecting telescope (reflector). A telescope employing a mirror (the *primary mirror) to bring the light rays to a focus. The various configurations include the *Cassegrain, *Newtonian, *Gregorian, and *coudé telescopes; there are also the *Maksutov and *Schmidt telescopes, which are *catadioptric. These instruments differ in the additional optical system used to bring the image to a convenient point where it can be viewed through an *eyepiece, photographed, or detected and recorded electronically, and subsequently analysed. The *aberrations in reflectors can be reduced to a very low level by suitable shaping of the primary mirror and with the possible use of a *correcting plate. In addition a mirror, unlike a lens, suffers no *chromatic aberration.

The first reflector was a Newtonian. In the period following its introduction in about 1670, reflecting telescopes with mirrors made from *speculum metal displaced the simple *Galilean and *Keplerian *refracting telescopes whose performance was limited by the chromatic aberration of their single-lens objectives. It was William Herschel who developed a special ability in making speculum-metal mirrors of really accurate figure. In 1783 he built a reflector with an aperture of 46 cm and in 1789 one of 122 cm. He used these telescopes for a wide range of observations, making many important advances in stellar astronomy.

The end of the era was marked by the construction in 1842–45 of an 183 cm diameter speculum-metal mirror by William Parsons, third Earl of Rosse, at his family seat of Birr Castle, at Parsonstown, Eire. This was the largest reflector built until the Mount Wilson 254 cm reflector was commisioned 75 years later; it was known as Leviathan. Parsons was the last of the gifted amateurs to lead the main stream of astronomical development, which in future would require the resources of large corporate bodies.

In the latter half of the 19th century, refracting telescopes with *achromatic objectives overtook reflectors in performance and held their lead until the turn of the century. By then the silvered glass mirror had been found to give a great improvement upon the reflectivity of speculum metal, which rapidly became obsolete. It was also realized that refractors had reached a real size limit at about 100 cm diameter: a lens can only be supported around its edge and larger lenses sag under their own weight to an unacceptable degree. A primary mirror can be supported over its whole rear face and thus escapes this limit.

The modern era opened with the construction of the giant reflecting telescopes at Mount Wilson Observatory and then Palomar Observatory, both in California. There are now several reflectors of *aperture exceeding 3.5 metres (see Table 11, backmatter). In the most recent telescopes, advantage has been taken of the developments in drive technology pioneered by radio astronomers to return from the *equatorial to the simple *altazimuth mounting. Telescope performance has greatly improved as a result of new techniques for detection and analysis of the image, using complex electronic equipment and computer systems (see imaging). Further improvements in instrumentation will not however greatly increase sensitivity. To observe very faint objects a new generation of ground-based telescopes with substantially larger apertures is being planned (see primary mirror). These will be augmented by orbiting telescopes, such as the *Hubble Space Telescope, which are undisturbed by the earth's atmosphere.

See also infrared telescope.

reflection. A phenomenon occurring when a beam of light or other wave motion strikes a surface separating two different media, such as air and glass or air and metal. Part of the wave has its direction changed, according to the laws of reflection, so that it does not enter the second medium. A rough surface will produce diffuse reflection. A smooth surface, as of polished metal, will reflect the radiation in a regular manner. The laws of reflection state firstly that the incident beam, the reflected beam, and the normal (the line perpendicular to the surface at the point of incidence) lie in the same plane, and secondly that the two beams are inclined at the same angle to but on either side of the normal.

reflection grating. *See* diffraction grating.

reflection nebula. A bright cloud of interstellar gas and dust that lies near and somewhat to one side of a star or stellar group, usually of *spectral type B2 or later. The starlight is scattered in all directions by the dust grains, the density of which is sufficient to produce a noticeable illumination of the cloud. The light scattered towards the observer is bluer than that of the illuminating star but the spectra of cloud and star are essentially the same – a continuous spectrum with absorption lines. (If there is sufficient dust in an *emission nebula, it will cause the stars' absorption spectra to be added to the normal emission spectrum of the nebula.) Reflection nebulosity is characteristic of young clusters and *associations; the dust is dispersed in about 100 million years. Typical reflection nebulae occur around the brighter stars of the *Pleiades. *See also* nebula.

reflector. *Short for* reflecting telescope.

refracting telescope (refractor). A telescope employing an *objective lens to bring the light rays to a focus. The first telescopes were simple refractors with single-lens objectives: these were the *Galilean and *Keplerian telescopes. They were supplanted by *reflecting telescopes because no-one could overcome their *chromatic aberrations, which introduced brilliant false colour effects in the images. Although John Dollond discovered how to make an *achromatic lens in 1756, he could only make small diameters suitable for terrestrial telescopes; glass makers did not know how to make large uniform discs of crown and flint glass. This obstacle was overcome in the early 1800s by Joseph Fraunhofer, who made the 24 cm refractor at Dorpat and the 16 cm heliometer at Konigsberg. His successor, Merz, supplied the 32.4 cm equatorial refractor

to the Royal Greenwich Observatory and the 38 cm refractor that was the first large telescope to be mounted in the USA, at Cambridge, Massachusetts, in 1847.

Perhaps because of this, American astronomers leaned strongly to the refractor. By the 1880s Alvan Clark was making relatively large refractors: 47 cm diameter for the Dearborn Observatory, 91 cm for Lick Observatory, and 102 cm for Yerkes Observatory. Other very large refractors are the 61 cm at both Lowell Observatory, Arizona, and the Pic du Midi in France, the 75 cm at Pulkovo Observatory, Leningrad, and the 83 cm at the Meudon Observatory in Paris. All were built in the late 19th century. There are great technical problems in making large lenses free of imperfections and impurities, and of supporting these lenses (only around the edge) so that distortion of the image is minimal. The desire for ever larger *apertures to reach ever further and fainter objects in space has meant that the major telescopes built in this century have been reflectors.

The surviving large refracting telescopes are mainly used in *astrometry for the measurement of stellar positions, *proper motions, and *parallax. Refractors are often preferred by visual observers because of their long focal length and closed tube; the latter avoids air currents in the tube, which often cause an unsteady image.

refraction. A phenomenon occurring when a beam of light or other wave motion crosses a boundary between two different media, such as air and glass. On passing into the second medium, the direction of motion of the wave is 'bent' towards or away from the normal (the line perpendicular to the surface at the point of incidence). The incident and refracted rays and the normal all lie in the same plane. The direction of propagation is changed in accordance with *Snell's law*:

$$n_1 \sin i = n_2 \sin r$$

i and *r* are the angles made by the incident and refracted ray to the normal; n_1 and n_2 are the *refractive indices of the two media. The change in direction of motion results from a change of wave velocity as the wave passes from the first to the second medium.

refraction, angle of. *See* zenith distance.

refraction, atmospheric. *See* atmospheric refraction.

refractive index (index of refraction). Symbol: *n*. The ratio of the *speed of light, *c*, in a vacuum to the speed of light in a given medium. This is the *absolute refractive index*; it is always greater than one. The relative refractive index is the ratio of the speeds of light in two given media. The refractive index of a medium depends on the wavelength of the refracted wave: with light waves, *n* increases as the wavelength decreases, i.e. blue light is refracted (bent) more than red light. *See also* refraction.

refractor. *Short for* refracting telescope.

regio (plural: **regiones**). A large area distinguished by shading or colour on the surface of a planet or satellite. The word is used in the approved name of such a feature.

regolith. The layer of dust and broken rock, created by meteoritic bombardment, that covers much of the surfaces of the moon and Mars and, probably, many other bodies in the solar system.

regression of nodes. *See* nodes.

Regulus (α **Leo**). A bluish-white star that is the brightest one in the con-

stellation Leo and lies at the base of the Sickle of Leo. It is a visual triple star with an orange companion (B), which itself has a 13th-magnitude companion (C). Regulus has a common proper motion with a fourth star (D). m_v: 1.34 (A), 7.6 (B); M_v: −0.8; spectral type: B7 V (A), K1 V (B); distance: 26 pc.

relative aperture. *See* aperture ratio.

relativistic astrophysics. High-energy *astrophysics, concerned with the extreme energies, velocities, densities, etc., associated with celestial objects such as *white dwarfs, *neutron stars, *black holes, and *active galaxies, and with the very early universe. It involves the theories of special and general *relativity, *quantum mechanics, and the physics of *elementary particles.

relativistic beaming. *See* synchrotron emission.

relativistic electron. An electron moving at close to the speed of light. The special theory of *relativity must be used to describe associated phenomena.

relativity, general theory. The theory put forward by Albert Einstein in 1915. It describes how the relationship between space and time, as developed in the special theory of relativity, is affected by the gravitational effects of matter. The basic conclusion is that gravitational fields change the geometry of *spacetime, causing it to become curved. It is this curvature of spacetime that controls the natural motions of bodies. Thus matter tells spacetime how to curve and spacetime tells matter how to move. General relativity (GR) may therefore be considered as a theory of *gravitation, the differences between it and Newtonian gravitation only appearing when the gravitational fields become very strong, as with black holes, neutron stars, and white dwarfs, or when very accurate measurements can be made. GR also reduces to the special theory when gravitational effects are negligible.

The basic postulate from which GR was developed is the principle of *equivalence between gravitational and inertial forces: in two laboratories, one in a uniform gravitational field and the other suitably accelerated, the same laws of nature apply and thus the same phenomena, including optical ones, will be observed. Einstein was able to show that the natural motion of a body was in a 'straight line', i.e. along the shortest path between two points (a geodesic), in the geometry required to describe the physical world (*compare* Newton's first law of motion). Three-dimensional Euclidean geometry could not be used for this purpose. The requisite geometry had to be more flexible: this was the geometry of spacetime. The presence of gravitational fields leads to the curvature of spacetime. The curvature produced by a particular gravitational field can be calculated from Einstein's *field equations*, advanced as part of GR. It is then possible to calculate the geodesics followed by particles in this curved spacetime.

Experimental tests of GR require either very strong gravitational fields or very accurate measurements. These tests must verify the predictions of GR where they deviate from the predictions of Newtonian gravitation and also where they deviate from the predictions of other relativistic theories of gravitation (including the *Brans–Dicke theory) that have been developed since 1915. The main tests involve four predicted effects.

One effect is the bending of light or other electromagnetic radiation in a gravitational field. Measurements of the positions of stars when close to the sun's limb (i.e. during total solar eclipses) and when some distance from the sun have been shown to dif-

fer: the sun's gravitational field changes the path of a photon of radiation from a straight line to a path bent towards the sun. The angle by which the photon's path is deflected is called the *deflection of light*. The measured deflection for stars on the solar limb is within 10% of Einstein's predicted value, which is 1.75 arc secs. Closer agreement (within 1%) has been found when the bending of radiation from radio sources such as quasars is determined. A related effect is the time delay in radio signals from spacecraft when they are close to the sun. The measured delays from the Viking probes and other craft agree with Einstein's prediction within 0.1%.

The predicted *gravitational redshift in the spectral lines of radiation emitted from a massive body has also been successfully demonstrated (5% accuracy), using both the sun's and also the earth's gravitational fields. A related prediction is that clocks run more slowly in a strong gravitational field than in a weaker one. An atomic clock flown at an altitude of 10 km has been shown to run faster than at sea level by a value within 0.02% of the predicted amount (47 $\times$ 10^{-9} seconds).

The effects of curved spacetime on the motions of orbiting bodies were shown convincingly when the *advance of the perihelion of Mercury was found to be accounted for by GR. Recent measurements of the very much larger periastron shift of the *binary pulsar 1913 + 16 provided much stronger evidence, being in very close agreement with prediction.

The fourth test of GR concerns the predicted existence of *gravitational waves. The emission of such waves causes a loss of energy from a system and in a binary system alters the orbital period. Recent measurements of the speed-up in the period of the binary pulsar 1913 + 16 are in almost precise agreement with the predicted value of GR. Many alternative theories of gravitation predict much higher values.

Other predictions of GR are the existence of *singularities in the structure of spacetime leading to black holes, and the constancy of the *gravitational constant. There is no direct evidence as yet for (or against) these, although indirect evidence of black holes is accumulating.

It cannot be said that GR has been conclusively proved. The experimental evidence does, however, seem to be mounting in its favour.

relativity, special theory. The theory proposed by Albert Einstein in 1905. It is concerned with the laws of physics as viewed by observers moving relative to one another at constant velocity (i.e. with observers in *inertial frames), and with how relative motion affects measurements made by these observers. At low relative velocities, special relativity (SR) predicts the same results as the classical Newtonian laws of mechanics. As the relative velocity increases beyond that encountered in everyday experience, the predictions of the two theories diverge. It is those of SR that have been conclusively verified by experiment. SR must therefore be used to describe the behaviour of bodies, such as electrons, when they are moving at velocities close to the *speed of light.

One of the basic principles of SR is that the speed of light is the same in any direction in free space, is the same for all observers, and is independent of the relative motion of the observer and the body emitting the light (or other electromagnetic radiation). This violates the classical concepts of relative motion but has been demonstrated experimentally.

Another basic principle is that all physical laws are the same for every inertial frame. Any law deduced in one inertial frame will be true in every other inertial frame. The values of the numerical constants appearing in these laws are also independent of the frame in which they are measured. The same does not, however, apply to certain variable quantities measured in different inertial frames.

Prior to SR it had been assumed that there was a universal 'absolute' time and 'absolute' space that were the same for all observers. Einstein showed that as a direct consequence of the invariance of the speed of light, time and space could not be considered as separate concepts, independent of each other and of the observer. Instead they must be regarded as a composite entity, called *spacetime. Any event in an inertial frame must therefore be described in terms of four spacetime coordinates – three spatial and one time coordinate. The spacetime coordinates of the same event measured in two different frames will differ, but can be interconverted by the *Lorentz transformation.

As the relative velocity (v) between inertial frames approaches the speed of light (c), very strange things are predicted. One is *time dilation*: two observers approaching at a relative velocity close to c will each see the clock of the other operating more slowly than their own. The time intervals will be dilated by a factor of $1/\beta$ where

$$\beta = \sqrt{(1 - v^2/c^2)}$$

Another prediction is the apparent contraction of a moving object that is observed, in the direction of motion of the object, by someone in a different inertial frame. The observed contraction, known as the *Lorentz contraction*, amounts to the factor β. The object also appears slightly rotated. No change in dimensions is observed in directions perpendicular to the direction of motion.

A further consequence of SR, when applied to the field of dynamics, is that the mass (m) of a body is not invariant but increases as the relative velocity between body and observer increases:

$$m = m_0/\beta$$

m_0 is the *rest mass of the body and is an invariant property of matter. It follows that no object with mass can reach the speed of light; only a particle with zero rest mass (such as the *photon) can travel at the speed of light. Einstein showed that rest mass is a form of energy and can be converted to other forms of energy. Mass and energy are thus not distinct physical properties but must be considered as equivalent. They are related by the equation

$$E = mc^2$$

where E is the total energy of the particle.

remote operation. The operation of a telescope and its associated instruments by off-site astronomers possibly thousands of kilometres away. Signals are passed over one or more communication links carrying voice, video, and digital information. This is a development of the current trend to transfer the operation of a telescope system to an on-site control room, which itself required development of automated equipment and communication links (*see* automation).

Remote operation by astronomers of a telescope-instrument-detector system in space was demonstrated in 1978 using the *International Ultraviolet Explorer satellite. It was achieved using a ground-based telescope in 1982 when the *UK Infrared Telescope in Hawaii was remotely operated from the Royal Observatory, Edinburgh. The UK telescopes at the *Roque de los Muchachos Observatory in the Canaries will be remotely operated from the Royal Greenwich Observatory.

réseau. A reference grid of fine lines that are photographically produced in the emulsion of a photographic plate and used in measurements of star positions.

resolution. The ability of a telescope or other instrument to distinguish fine detail, or a measure of that ability. Hence a spectrometer has a *spectral resolution*, and an electronic imaging device has a *linear resolution*. Detectors can have an *energy resolution*.

The *angular resolution* or *resolving power* of a telescope is the smallest angle between two point objects that produces distinct images. It depends on both the wavelength at which observations are made and on the diameter, or aperture, of the telescope. This minimum angle can be given by the *Rayleigh limit*:

2.52 × 10^5 λ/d arc seconds

where λ and d are the wavelength and aperture (in metres). For an optical system the Rayleigh limit is approximately

0.14/d arc seconds

The *Dawes limit*, originally determined experimentally, gives the angular resolution as

0.12/d arc seconds

When the angular separation of two stars is very small, it might be thought that the use of a large enough aperture or high enough magnification would always resolve the light into two distinct images. Because of diffraction effects however, the image of each star is not a point of light but a disc (*see* Airy disc). If the two discs substantially overlap then increased aperture or magnification merely gives a larger blur of light, and the telescope has not sufficient resolving power to separate the images. The stars will just be resolved, however, when their Airy discs touch. This gives the Dawes limit. Two stars are at a telescope's Rayleigh limit when the centre of the Airy disc of one star falls on the first dark ring of the diffraction pattern of the other.

Fig. *a* shows the images of two stars of equal magnitude separated by 0.8 seconds of arc. Viewed under perfect *seeing conditions with a 7.5 cm aperture (right) they are not resolved; with 15 cm (centre) they are just re-

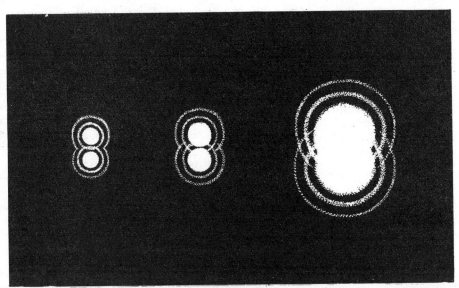

a Images of an optical double of separation 0.8″
through apertures (L to R) of 30, 15, and 7.5 cm

b Satellite of Pluto, unresolved but detected by slight
elongation of greatly enlarged image

solved and with 30 cm (left) they are clearly separated.

The discovery of Pluto's satellite in 1978 shows that the Dawes (or Rayleigh) limit cannot be too strictly applied. Although well above the limiting angle for large telescopes, Pluto's satellite has never been re- solved; it was detected as an elongat- ed photographic image (Fig. *b*). This is partly because of the difference in brightness and because Pluto is not a point source, but mainly because poor seeing prevents the larger-aperture telescopes from reaching the limiting angle. The next big step forward in optical astronomy thus depends upon using the *Hubble Space Telescope outside the earth's atmosphere.

In radio astronomy, where the same formulae for angular resolution apply but where very much greater wavelengths are studied, apertures maybe 50 times greater than those used in optical telescopes can only give poor resolution. The situation has been greatly improved by the use of *aperture synthesis and *very long baseline interferometry.

resolving power. A measure of the *resolution of a telescope, *spectrograph, *diffraction grating, or some other system.

resonance. A condition prevailing in a system when the frequency of some external stimulus coincides with a natural frequency of the system, and producing a selective or maximum response in the system.

resonance transition. A change in the energy state of an atom, involving a jump from the ground state to a higher *energy level, leading to *excitation, or a return to the ground state from an excited state. This produces a *resonant* absorption or emission line in a spectrum. *See also* ultraviolet astronomy.

rest mass. The mass of an elementary particle, etc., when at rest. According to the special theory of *relativity, a body's mass increases as its velocity increases. This is only noticeable at velocities close to the speed of light.

retardation. The difference in the moon's rising time between successive nights.

reticle. A grid or pattern of two or more fine wires set in the focal plane of a telescope eyepiece and used in determining the position and/or size of a celestial object.

Reticulum (Net). A small constellation in the southern hemisphere near the Large Magellanic Cloud, the brightest stars being one of 3rd magnitude and some of 4th magnitude. Abbrev.: Ret; genitive form: Reticuli; approx. position: RA 4h, dec $-60°$; area: 114 sq deg.

retrograde motion. *See* direct motion.

retrorocket. A small rocket motor attached to a spacecraft that, by ejecting gas in a forward direction, decelerates the craft. In this way the craft may change its orbit or, possibly with parachutes, achieve a soft landing.

reversing layer. The comparatively low-temperature stratum overlying the surface of a *main-sequence star, in which it was proposed the absorption lines in the star's spectrum were produced. It is now known that the lines originate in the *photosphere at the same levels as the *continuum (with emission predominating at low levels and absorption higher up) rather than in a well-defined layer.

revolution. 1. Orbital motion of a celestial body about a centre of gravitational attraction, such as the sun or a planet, as distinct from axial *rotation. *See also* direct motion.
2. One complete circuit of a celestial body about a gravitational centre. The earth takes one year to make one revolution around the sun.

RFT. *Abbrev. for* richest-field telescope.

RGO. *Abbrev. for* Royal Greenwich Observatory.

Rhea. The second largest of the Saturn system of satellites, with a diameter of 1530 km and a density of 1.3 g cm^{-3}. The northern hemisphere is particularly heavily cratered and resembles the rolling cratered highlands of the moon. The crater density on the surface is irregular and suggests a varied geological history. The craters vary in size up to 75 km. *See also* Table 2, backmatter.

rich cluster. *See* clusters of galaxies.

richest-field telescope (RFT). A telescope designed to show the largest possible number of stars at a single view when aimed at a dense field of faint stars, such as the Milky Way. This end is accomplished by combining a reasonably large aperture (100 to 150 mm) with a low power and the widest possible field of view. Although the true RFT is a special design, fitted with an expensive wide-field *eyepiece, binoculars and most *finder telescopes go a long way towards meeting these specifications. Apart from providing spectacular views of immense numbers of stars (which are not obtained with the high magnifications and narrow fields more usually employed), the RFT is an ideal instrument for scanning the heavens in the search for novae and comets.

ridge (scarp). *See* wrinkle ridges; lobate ridge.

Rigel (β Ori). A very massive luminous and remote bluish-white supergiant that is very conspicuous and is the brightest star in the constellation *Orion. Its luminosity is over 50 000 times that of the sun. It is a spectroscopic binary, period 9.86 days. It is also fixed at about $10''$ with a binary system (BC), separation $0''.4$. m_v: 0.11; M_v: -7.1; spectral type: B8 Ia; distance: about 250 pc.

right ascension (RA). Symbol: α. A coordinate used with *declination in the *equatorial coordinate system. The right ascension of a celestial body, etc., is its angular distance measured eastwards along the celestial equator from a *catalogue equinox to

the intersection of the hour circle passing through the body. It is generally expressed in hours (h), minutes (m), and seconds (s) from 0 to 24h: one hour equals 15° of arc.

right-ascension circle. The *setting circle on the polar axis of an *equatorial mounting that measures the *right ascension at which the telescope is set.

Rigil Kent. *See* Alpha Centauri.

rille (rill; groove; cleft). Either of two distinct types of elongated lunar depression (see table). *Sinuous rilles* are confined to the *maria, where they follow local contours, are frequently discontinuous (or consist of rows of coalescing craters), and have V-shaped cross sections. They sometimes originate in volcanic caldera and are assumed to be channels, or collapsed tubes, carved out by molten basaltic lavas. Examples exist of sinuous rilles inside sinuous rilles but tributary systems are not observed.

Faulted *linear, arcuate,* and *forked rilles* have flat floors and do not recognize highland-mare boundaries. This second type of rille is frequently associated with impact basins and craters, where they may be concentric. *See also* wrinkle ridges; crater chain.

rima. A fissure. The word is used in the approved name of such a surface feature on a planet or satellite.

ring current. An electric current flowing westwards around the earth; it arises from a net eastward flow of electrons and a net westward flow of protons trapped inside the *Van Allen radiation belts. The current produces a magnetic field that cancels part of the geomagnetic field at the earth's surface. During *geomagnetic storms, when the Van Allen belts are recharged with particles, the increased ring current results in a further decrease of the earth's surface magnetic field strength.

Lunar rilles and valleys

rille, valley	location; type
Alpine Valley	NE Imbrium; radial fracture
Aridaeus	W Mare Tranquillitatis; linear
Byrgius	W Mare Humorum; sinuous
Cauchy Fault	E Mare Tranquillitatis
Hadley	Palus Putredinus; sinuous
Hyginus	Mare Vaporum; forked rille
Lee Lincoln Scarp	Taurus Littrow area; fault
Posidonius	NE Mare Serenitatis; sinuous
Rheita Valley	S Nectaris; radial fracture
Schröter's Valley	Aristarchus Plateau; double sinuous
Schrödinger Canyon	S farside; radial fracture
Sirsalis	W Mare Humorum; sinuous
Straight Wall	Mare Nubium; linear fault

axy that has the form of an elliptical ring either with a massive nucleus, often off-centre, or with little or no luminous material visible in its interior. In many cases there is a small satellite galaxy devoid of interstellar gas. The ring configuration is thought to be unstable and it has been suggested that these systems may represent the after-effects of a collision between two galaxies.

ring mountains. *See* mountains, lunar; basin.

ring nebula. A ring or arc of nebulosity centred on and ionized by ultraviolet radiation from a *Wolf-Rayet star. The Ring nebula is, confusingly, not a ring nebula.

Ring nebula (M57; NGC 6720). A ring-shaped *planetary nebula about 700 parsecs away in the constellation Lyra. The inner regions closest to the centrally placed ionizing star are green as a result of light emitted by doubly ionized oxygen; there is an approximately symmetrical transition to the reddish outer edge where the lower-energy transitions of hydrogen

colour. See emission nebula.

ring plains. *See* maria, lunar.

ring systems. *See* Saturn's rings; Uranus; Jupiter's satellites.

rising. The daily appearance on an observer's horizon of a particular celestial body as a result of the earth's rotation, the body disappearing below the horizon, or *setting*, at some time in the following 24 hours. A star rises, on average, 2 hours earlier per month. Stars always rise and set in the same position relative to other stars whereas the sun, moon, and planets rise and set at different points on successive days. This led to the early idea of *fixed stars and wandering stars.

For an observer at the equator all stars in theory can be seen, rising and setting at right angles to the horizon. At the poles an observer will see only the stars in his hemisphere; these stars never rise nor set but move in circles parallel to the equator. At intermediate latitudes some stars (*circumpolar stars) never set and some, in a corresponding area in the sky around the opposite pole, never rise. The remaining stars rise and set at oblique angles to the horizon, the time spent above the horizon depending on the observer's latitude and the position of the star relative to the celestial pole. Stars on the celestial equator rise due east and set due west and for all observing positions except the poles remain 12 hours above the horizon.

Ritchey-Chrétien optics. An optical system that is a variation on the *Cassegrain configuration and was developed by George Ritchey and Henri Chrétien. It corrects both *spherical aberration and *coma at the Cassegrain focus, and therefore gives a high-quality image over a relatively wide field of view. A primary mirror with a hyperbolic profile is used in conjunction with an appropriate departure from classical (hyperbolic) form at the secondary mirror.

Roche limit. The minimum distance from the centre of a body at which a satellite can remain in equilibrium under the influence of its own gravitation and that of its primary. Assuming that the satellite is held together only by gravitational attraction, i.e. it has zero tensile strength, and that it has the same density as the primary, the Roche limit is about 2.4 times the radius of the primary. Inside the Roche limit a satellite would be torn apart by tidal forces.

Roche lobe. *See* equipotential surfaces.

rocket. *See* launch vehicle; sounding rocket.

rockoon. A means of exploring the upper atmosphere of the earth. It consists of a large balloon to which is attached a small rocket. The rocket is fired at a considerable altitude, thus avoiding the air drag of the lower atmosphere. The balloons are usually *Skyhooks*: large nondilatable plastic balloons that can carry a heavy load to altitudes of over 30 km.

Roque de los Muchachos Observatory. A new international observatory s. ed on La Palma in the Canary Islands at an altitude of 2400 metres (7900 feet). It was set up by Britain, Denmark, Spain (who owns the site), and Sweden; the Netherlands and the Republic of Ireland joined with the UK and share her telescopes. British participation is in the hands of the Science and Engineering Research Council (SERC), which has given the task of building and maintaining the UK telescopes to the Royal Greenwich Observatory. The site has superb observational characteristics.

The UK telescopes will be the first to be set up for *remote operation. They include the 4.2 metre *William Herschel Telescope, which should be completed in 1986, the completely refurbished 2.5 metre *Isaac Newton Telescope, and a 1 metre reflector, the *Kapteyn Telescope*, specially designed for *astrometry and also used for *photometry. The latter two began operations in 1984. The Dutch share 20% of cost plus research effort on the UK telescopes in return for 20% observational time.

The *Carlsberg Automatic Transit Circle* is a joint UK–Danish venture for the automatic precise and rapid measurement of star positions, and began work in 1984. Two Swedish telescopes – a 60 cm reflector and a 16 m solar tower – have been operating since 1982 and 1983. Additional telescopes are under consideration.

In return for necessary services, Spanish astronomers have 20% of available time on each telescope. International projects will take up 5% of each telescope's time.

ROSAT. *Abbrev. for* Röntgenstrahlen Satellit. A major West German *x-ray astronomy satellite due for launch on the space shuttle in Aug. 1987. The payload consists of a large (80 cm diameter) *grazing incidence telescope to carry out the first deep all-sky survey in the 0.2–3 keV band. Use of imaging optics will allow a survey 100–1000 times deeper than those of *Uhuru, *Ariel V, and *HEAO–1. The x-ray images will be recorded on an imaging *proportional counter or a *microchannel plate detector, the latter being provided by NASA, who will also provide the launch in exchange for a share of observing time. ROSAT will also carry a second telescope, the Wide Field Camera (WFC), being built by a consortium of UK university groups. The WFC uses grazing incidence optics at an unusually large angle and is optimized for the XUV band (0.2–0.02 keV). It is expected to provide the first all-sky survey in this energy band. *See also* XUV astronomy.

Rosette nebula (NGC 2237-38-44-46). A large *emission nebula about 15 parsecs in diameter and about 1400 parsecs distant in the direction of the constellation Monoceros. It is an *HII region heated and ionized by a centrally situated group of hot young stars.

rotating variable. A star with a hotter or cooler region on its surface that periodically rotates into our line of sight and causes slight variations in the star's brightness. Many dwarf *Me stars show this effect (prototype BY Draconis), probably because their strong magnetic fields cause a large proportion of the star's surface to be covered by dark 'starspots'. *Pulsars are related to this group of *variable stars.

rotation. 1. Spinning motion of a celestial body or a group of gravitationally bound bodies, such as a galaxy, about an axis, as distinct from orbital *revolution. Almost all celestial bodies show some degree of rotation. The hottest (O and B) stars have very great rotation rates of about 200–250

km s^{-1}. The faster the rate of rotation the broader and shallower the star's spectral lines and the stronger the magnetic field (*see* corona). *See also* direct motion; differential rotation; synchronous rotation.

2. One complete rotation of a celestial body about its axis. The earth takes one day to make one rotation. Ideally the rotation period is measured as the time interval between successive passages of a meridian line on the surface across the centre of the disc, as seen from earth. The solid surface may however be unobservable and indirect measurements, as by radar, are then employed. The rotation period of a gaseous body, such as the sun or the planet Jupiter, varies with latitude, being greatest at the equator (*see* differential rotation).

rotational parallax. A method of determining the mean distance to a star or star cluster. It is based on the *parallax resulting from the differential rotation of the Galaxy, which leads to a variation in velocity about the galactic centre. The average *radial velocity, v_r (in km s^{-1}), and the galactic longitude, l, of an object – both measurable – are related to distance d (in parsecs) by *Oort's formula:

$$v_r = dA \sin 2l$$

A is equal to about 15 km s^{-1} per kiloparsec. The object must not lie along the direction of the galactic centre.

rotation measure. *See* Faraday rotation.

Royal Greenwich Observatory (RGO). The observatory sited originally at Greenwich, London, and founded (as the Royal Observatory) in 1675 by Charles II so that astronomical measurements could be made and the data tabulated for the primary purpose of increasing the accuracy with which positions at sea could be determined. John Flamsteed was appointed first

*Astronomer Royal and took up his post in 1676. The meridian passing through the *transit circle at the Observatory was adopted internationally in 1884 as the prime meridian, i.e. as the meridian of zero longitude. Over the centuries the increase in atmospheric and light pollution necessitated the removal of the Observatory to Herstmonceux Castle in Sussex. The move was completed by 1954. The RGO was transferred from the Admiralty to the Science Research Council (now the Science and Engineering Research Council) in 1965.

In addition to studying the positions and apparent motions of stars for the determination of longitude on earth and in measuring time, the RGO's role has widened to include *astrophysics. This led to the construction of the 2.5 metre *Isaac Newton Telescope (INT) at Herstmonceux, completed 1967. The *seeing at this site averages only 3 arc seconds, and the sky is always slightly hazy with light pollution increasing. The INT has therefore been modified, reassembled at the *Roque do los Muchachos Observatory in the Canaries, and became operational again in 1984. The 4.2 metre *William Herschel Telescope, being built under the management of the RGO, should be operating by 1986 at the same site.

Royal Observatory, Edinburgh (ROE). The observatory in Edinburgh that was founded in 1818, with a Royal Charter since 1822, and is responsible for some major national facilities funded by the Science and Engineering Research Council. These include the 1.2 metre *UK Schmidt Telescope at Siding Spring, Australia and the 3.8 metre *UK Infrared Telescope in Hawaii. In addition to astrophysical observations, the Observatory is concerned with the development of advanced technologies and their applications in astronomical and space research. Its Plate Library houses several thousand photographic plates taken with the UK Schmidt.

r-process. A rapid process of *nucleosynthesis that is thought to occur when there is a very high flux of neutrons, as in certain supernova explosions. All nuclei with a mass number greater than bismuth−209 (*see* s-process) and all neutron-rich isotopes heavier than iron have been produced by the r-process. The process involves the capture by a nucleus of two or more neutrons in quick succession. The nucleus will then undergo chains of beta decay, this beta-particle (electron) emission having been suppressed during the rapid capture process. The decay product will be a stable neutron-rich nucleus. Many heavy nuclei can be produced both by the r- and the s-process.

RR Lyrae stars. A large group of *pulsating variables that are very old giant stars (halo and disc *population II stars) and are found principally in *globular clusters. They usually have periods of less than one day and median spectral types in the range A7 to F5. They were discovered in 1895 by Solon I. Bailey, the group being named after RR Lyrae, discovered in 1899. Available evidence indicates that all RR Lyrae stars have about the same mean absolute *magnitude (about +0.5); they can therefore be used as distance indicators up to about 200 kiloparsecs (*see* distance modulus).

RR Lyrae stars are of a mixed nature. They were divided by Bailey into three groups – *RRa, RRb,* and *RRc* – depending on period and the asymmetry of their light curves (see illustration and table). Groups *a* and *b* are often combined today as *RRab* stars. Other groupings have been made according to period. Many RR Lyrae stars show a periodic variation in both period and shape of light curve – the *Blazhko effect.* In addition the periods of some RR Lyrae

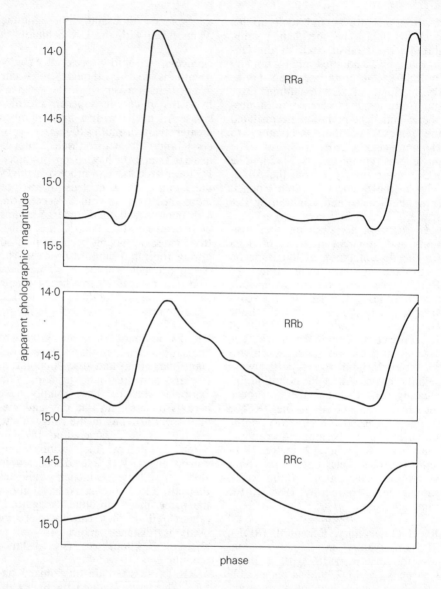

Typical light curves of RR Lyrae stars in subgroups RRa, S RRb, RRc

stars are slowly changing at a constant rate, as predicted by evolutionary theory, while others show abrupt changes in period. The pulsations of these stars are very complex and can be subdivided into one group (RRab) oscillating in fundamental mode (*see* pulsating variables) and a second group (RRc) oscillating in first harmonic mode. *See also* horizontal branch.

RRs stars. *See* dwarf Cepheids.

RS Canum Venaticorum star (RS CVn star). A short-period binary star system that contains a *subgiant of spectral type G or K exhibiting intense

Subgroups of RR Lyrae stars

type	approx. amplitude	period (days)	light curve asymmetry
RRa	1.5–2	0.35–0.55	⎰
RRb	1.2–1.7	0.5–0.8	decreasing
RRc	0.5	0.2–0.4	↓

activity at radio, ultraviolet, and x-ray wavelengths; the other component is a normal F or G main-sequence star. The orbital period can range from 1 day to 2 weeks. The two stars are not undergoing *mass transfer (they are a detached system), but *tidal forces have locked the rotation of the subgiant so that it is forced to rotate once in each orbital period, much faster than a normal subgiant. This rapid rotation apparently causes strong magnetic fields, which then produce the observed activity.

The *light curve of these stars shows a continuously changing brightness during the orbital period, believed to be due to dark 'starspots' extensively blanketing one half of the subgiant's surface. Measurements of the *Zeeman effect in light from Lambda Andromedae show that the spotted hemisphere has a field strength of over 0.1 tesla, similar to that of sunspots. Some of these stars, including the prototype RS CVn, are also eclipsing binaries; the light curve shows regular dips of about a magnitude as the subgiant eclipses the main-sequence star. The subgiant's magnetic field also causes intense radio flares (*compare* flare stars), strong ultraviolet emission lines from the *chromosphere, and powerful x-ray emission from a very hot *corona.

As the subgiant swells to giant size, it will begin to transfer mass and become a *W Serpentis star and then an *Algol variable.

R stars. *See* carbon stars.

runaway star. A young star of spectral type O or early B with a very high space velocity, examples including AE Aurigae, 53 Arietis, and Mu Columbae. It has probably been flung out from a close *binary star when its companion has exploded as a *supernova.

rupes. A scarp. The word is used in the approved name of such a surface feature on a planet or satellite.

Russell–Vogt theorem. *See* Vogt–Russell theorem.

RV Tauri stars. A small group of very luminous *pulsating variables, typified by RV Tauri, R Scuti, and R Sagittae, that are principally G and K stars with some F stars. They are yellow supergiants with extended atmospheres of gas that emit infrared radiation and have possibly been driven off by the pulsations. They have very characteristic light curves with alternating deep and shallow minima and periods ranging from 20 to 145 days. Since the luminosity fluctuations can be disturbed quite significantly in shape, period, etc., being most pronounced for longer-period stars, they are classified as *semiregular variables. RV Tauri stars can be distinguished from other similar yellow semiregular stars by the variation in their *colour index, which mimics the light curve but goes through its maximum a little before the luminosity minimum. A small group have double periodicity; DF Cygni has two separate luminosity oscillations, a rapid 50 day oscillation being superimposed on a much slower 780 day oscillation with a much greater amplitude.

S

Sagitta (Arrow). A small constellation in the northern hemisphere near

Cygnus, lying in the Milky Way, the brightest stars being one (γ) of 3rd magnitude and some of 4th magnitude. It contains the open cluster M71. Abbrev.: Sge; genitive form: Sagittae; approx. position: RA 19.7h, dec +18°; area: 80 sq deg.

sagittal plane. The plane containing all or part of the curved horizontal midline of a lens or mirror and its associated rays. The *tangential plane* contains the equivalent vertical midline and its associated rays. If the object lies off the optical axis of the lens or mirror so that light rays fall on it at an oblique angle, the sagittal and tangential rays do not converge to form a single image. *See* coma; astigmatism.

Sagittarius (Archer). A large zodiac constellation in the southern hemisphere near Scorpius, lying partly in the Milky Way. The brightest stars are the 2nd-magnitude Kaus Australis (ε) and Nunki (σ) with several of 3rd magnitude. The centre of the Galaxy lies in the direction of Sagittarius. Although obscured at optical wavelengths by considerable dark nebulosity, observations at other wavelengths have revealed very complex phenomena (*see* galactic centre). The constellation contains star clouds, many nebulae, including the *Omega, *Lagoon and *Trifid nebulae, many globular clusters, including the bright ones M22 and M54, and some open clusters. Abbrev.: Sgr; genitive form: Sagittarii; approx. position: RA 19h, dec −30°; area: 867 sq deg.

Sagittarius A. The brightest member of a complex of radio sources at the *galactic centre. It has two components. *Sagittarius A East* is a source of *nonthermal emission and is probably a *supernova remnant. *Sagittarius A West* is a source of *thermal emission from ionized gas. It also contains an intense, very compact, and variable radio source that is

probably smaller than 10 AU and lies very close to or at the Galaxy's exact centre. This 'point' source resembles a scaled-down version of the central radio sources in *active galaxies, which are believed to be accretion discs of gases around a massive *black hole. X-ray and gamm-ray observations show similar resemblances to an active galaxy's core.

Infrared and radio observations show that Sgr A West contains gas and dust with velocities up to 250 km s^{-1}; there may be a ring of material in orbit about the compact central source, and also streams of gas falling towards the central source. The velocities indicate the gravitational influence of 6×10^6 solar masses within Sgr A West. Infrared observations of a source believed to be a cluster of stars at the Galaxy's core (almost coincident with the compact radio source) are consistent with only 3×10^6 solar masses of stars. The difference, 3×10^6 solar masses, may reside in a central black hole. Many of the properties of Sgr A West could, however, result from activity caused by star formation, as occurs in a *starburst galaxy, so that the case for a black hole is not accepted by all astronomers.

Sagittarius arm. *See* Galaxy.

Sagittarius B. An immense *molecular cloud complex lying around the galactic centre, maybe forming part or all of an expanding ring of clouds. The most massive cloud in the complex, and possibly in the Galaxy, is *Sagittarius B2*. A large number of the interstellar molecules detected so far were discovered in Sgr B2.

Saha ionization equation. An equation, put forward by the Indian physicist M. N. Saha in the 1920s, that gives the ratio of the number density (number per unit volume) of ions, N^+, to the number density of neutral atoms, N_0, in the ground state in a system of atoms, ions, and free electrons at a given temperature. The system is in equilibrium so that the rate of *ionization is equal to the rate of recombination of electrons and ions. The equation can be given in the form:

$$N^+/N_0 = A(kT)^{3/2} N_e e^{-1/kT}$$

A is a constant, k is the Boltzmann constant, T the thermodynamic temperature, N_e the number of electrons per unit volume (the electron density), and I is the *ionization potential of the atom. The equation also gives the ratio N^{2+}/N^+, N^{3+}/N^{2+}, etc., if I is replaced by the second (I_2) or the third (I_3) ionization potentials, etc.

The equation shows that the higher the temperature, the more highly ionized a particular atom will be. It can therefore be used to find the relative numbers of neutral, singly ionized, doubly ionized atoms, etc., in stars of known temperature and electron density. *See also* Boltzmann equation.

Saiph. *See* Orion.

Salpeter mass function. *See* stellar mass.

Salyut. A series of Soviet space stations launched into earth orbit and visited and operated by crews of cosmonauts from the USSR and (usually) other Communist countries. The first station was launched in Apr. 1971. Other followed, Salyut–4 (1974–75), Salyut–6, and Salyut–7 being predominantly civilian in nature: in addition to medical, biological, and technological purposes, the stations have served as platforms for terrestrial and astrophysical observations. Every Salyut has been maneuvrable in orbit, either by an engine on board the station or on a craft docked to it. When unmanned they have been under automatic operation.

Salyut–6 was launched in Sept. 1977 and its programme ended in July 1982. There were crews on board for a total of about two years: the last manned mission was completed in May 1981. There was a docking port at each end of the 15 metre long 19 tonne structure. Crews were transported to and from the station by *Soyuz craft, the upgraded Soyuz T being used from 1980. Simultaneous use of the two docking ports allowed long-term and short-term visits to occur concurrently. Additional cargo and fuel was ferried to the station by unmanned *Progress craft. There were five endurance flights on Salyut–6, when crews were aboard for 96, 140, 175, 185, and 75 days (in chronological order). Post-flight medical checks showed that several months in space, although exhausting, seemed to have no permanent effect on the body. Of the experiments on Salyut–6 roughly one third were for space applications, one third for earth observations, and one sixth each for astrophysics and biomedical studies. Telescopes used included a helium-cooled 1.5 metre instrument for submillimetre observations, a radio telescope, and a gamma-ray telescope. In June 1981 an unmanned 15 tonne craft, Cosmos 1267, was docked to the unmanned Salyut–6 and the assembly studied for months. The Cosmos was described as a 'prototype space module' for a future space station.

Salyut–7 was launched in Apr. 1982. Its exterior is similar to that of Salyut–6 but its interior has been modernized both in instrumentation and living conditions. The first manned mission, May to Dec. 1982, lasted 211 days. The second mission, July to Nov. 1983, lasted 150 days. The third mission, Feb. to Oct. 1984, lasted a record 238 days.

S Andromedae. A supernova that was observed, in 1885, in the Andromeda galaxy and at maximum was just visible to the naked eye.

SAO Star Catalogue. A general whole-sky catalogue issued by the

Smithsonian Astrophysical Observatory, Washington, in 1966. It lists 258 997 stars down to a limiting magnitude of 9 and is intended as an aid in determining artificial satellite positions. An associated and highly accurate computer-plotted atlas was published in 1968.

Saros. The period of 6585.32 days (about 18 years) that elapses before a particular sequence of solar and lunar *eclipses can recur in the same order and with approximately the same time intervals. Following such a period, the earth, sun, moon, and *nodes of the moon's orbit return to about the same relative positions: the period is equal to 223 *synodic months and is almost equal to 19 *eclipse years (6585.78 days). An eclipse repeated after one Saros occurs 0.32 days later and hence 115°W of its predecessor. The Saros was known in ancient times.

SAS. *Abbrev. for* Small Astronomical Satellite. Any of a series of US *x-ray and *gamma-ray astronomy satellites. SAS–1, later named *Uhuru, was launched in Dec. 1970. It was devoted entirely to x-ray astronomy and provided a wealth of data. SAS–2, launched in Nov. 1972, carried equipment, including a *spark chamber, for gamma-ray measurements while SAS–3, launched in May 1975, carried x-ray detectors.

satellite, artificial. A man-made object that is boosted into a closed orbit around the earth, moon, or some other celestial body. It is generally unmanned. Satellites carry a variety of detectors, cameras, and measuring instruments, depending on their function, plus equipment to support these: control systems to orientate them, solar panels for power, data storage facilities, and antennae for communications with earth. The cost of the satellite and its equipment, plus the cost of launching it – e.g. from a space shuttle or on a Delta or Ariane rocket – and communicating with it, is now extremely high and countries are tending to collaborate on missions, sharing cost and expertise.

Although some malfunctioning satellites can be repaired by the crew of an orbiting shuttle, and in future may be repaired on the proposed *space stations, repairs are at present usually impossible. Each material and component part must undergo extensive testing before launch, with a carefully considered degree of redundancy in electronic components and circuitry. The problems arising from the near vacuum environment of space, from the bombardment of cosmic rays, micrometeorites, etc., and from magnetic fields and radiation belts must also be considered.

Artificial satellites have a great range of functions. Astronomical earth satellites study the radiations from space that cannot penetrate the atmosphere: *x-ray, *gamma-ray, *ultraviolet, and *infrared astronomy have been revolutionized by the recent launching of satellites specializing in these fields. Astronomical satellites can also make measurements at optical, radio, and infrared wavelengths that can penetrate, but may be severely affected by, the atmosphere. Other scientifically orientated satellites study the resources, atmosphere, and physical features of the earth.

Satellites are also used for communications (usually in geostationary orbit and allowing global long-distance live television broadcasting and telephony), for weather forecasting, and as navigational aids. In addition, the military potential of satellites has been exploited.

satellite, natural (moon). A natural body that orbits a planet. The nine major planets of the solar system have a total of at least 49 known satellites between them. In addition, the numerous small bodies that comprise the rings of Saturn, Jupiter, and Uranus may be regarded as natural satellites. A satellite may arise from one of two mechanisms, both of which probably applied in the solar system: a body may grow by accretion from material (*planetesimals) gravitationally attracted towards, and entering orbit around, a *protoplanet; secondly, bodies, such as minor planets, may be captured gravitationally by a planet. *See* Table 2, backmatter.

Saturn. The sixth major planet and, with an equatorial diameter of 120 000 km, the second largest. It orbits at a distance of between 9.0 and 10.1 AU from the sun every 29.5 years; *oppositions recur two weeks later each year. It has a polar diameter of 108 000 km so that its *flattening, 0.108, is the highest of any planet. Its mean density, 0.7 times that of water, is lower than that of any other planet. Before 1978 Saturn was thought to have 10 satellites, the three largest being Titan, Rhea, and Iapetus; the number is currently (1984) at least 22, following recent spaceprobe encounters. Orbital and physical characteristics of planet and satellites are given in Tables 1 and 2, backmatter.

The telescopic appearance of the planet is dominated by a prominent system of rings (*see* Saturn's rings) that lie in the plane of Saturn's equator, tilted by 27° with respect to its orbit. This leads first one face of the rings, and then the other, to be inclined towards the sun and the earth by up to 27°. At approximately 15 year intervals, the rings become edge-on to the earth and all but disappear: they were edge-on during 1979–80. The bright rings, about 270 000 km across, can add appreciably to Saturn's average apparent magnitude of 0.7 at opposition, when the globe and rings are 19 and 44 arc seconds wide respectively.

Saturn's disc appears similar to that of Jupiter in being crossed by yellow-ish dark and light cloud bands running parallel to the equator; these are called *belts* and *zones* respectively. They are less prominent, however, than those of Jupiter and spots within them are much less common. The spots, typically white, that have been followed indicate that Saturn, like Jupiter, rotates more rapidly near the equator (in 10 hours 14 minutes or less) than at high latitudes (about 10 hours 40 minutes).

Saturn and its environment have been studied by the *Pioneer 11 probe in 1979 and by the *Voyager probes in 1980 and 1981. These probes have provided considerable information on the meteorology and atmosphere of Saturn, which resemble those of Jupiter. Like Jupiter, Saturn emits more radiation than it absorbs from the sun and probably has an internal primordial energy reservoir. Escaping heat may drive the atmospheric convection processes that give rise to the observed cloud banks, probably of ammonia crystals at −170°C. The Saturn weather system, like those of Jupiter, are strongly zonal. At the equator the cloud top winds reach more than 500 m s^{-1}. There is evidence of a wide range of cloud systems of varied morphologies: the white spots are anticyclonically rotating systems and have features identical with Jupiter's *Great Red Spot; trains of vortex streets are seen in northern midlatitudes. Although there is a strong internal heat source, analyses of the Voyager data have shown that Saturn (and also Jupiter) and the earth drive their meteorological systems in the same way, with energy being transported from the small-scale features into the main flow. Spectroscopic studies indicate hydrogen, methane, and ethane in the upper atmosphere; hydrogen probably forms the bulk of Saturn's mass.

Models for the internal structure of Saturn suggest an earth-sized iron-rich rocky core with a 7500 km outer core of ammonia, methane, and water.

This is enclosed by about 21 000 km of liquid metallic hydrogen above which lies liquid molecular hydrogen.

Saturn's magnetic field was detected by the Pioneer 11 probe. It has the same direction as Jupiter's field but is about 20 times weaker. Surprisingly, the magnetic axis corresponds almost exactly with Saturn's rotational axis. Pioneer also discovered *radiation belts, mainly comprising energetic electrons and protons. Saturn emits radio waves that are thought to be generated as *synchrotron emission from the electrons spiralling in the magnetic field. Saturn's rings and its inner satellites sweep away these charged particles. The interactions with the rings give rise to electrostatic discharges and assist in the formation of the spoke patterns in ring B.

Saturn nebula. *See* Aquarius.

Saturn's rings. A system of coplanar rings in Saturn's equatorial plane, with an overall diameter of almost 600 000 km. The system is tilted at nearly 27° to Saturn's orbital plane. It was seen indistinctly by Galileo in 1610 and was recognized as a system of rings by Huygens in 1656. The rings are now known to consist of thousands of tiny ringlets containing numerous individual particles. The particles are up to several centimetres or, in ring B, even a few metres in size, and each is a satellite of Saturn. The rings are less than 1 km thick and almost disappear when edge-on to the earth every 15 years or so.

The *D ring* is the closest to Saturn. There is no truly well-defined inner edge so that it would appear to extend right down to the cloud tops of Saturn. The *C* (or *Crepe*) *ring* fills the gap between the D ring and the inner edge of the brightest ring, ring B, at 25 000 km above the cloud tops. It is a complicated, 'grooved' region with a large number of discrete ringlets. There are at least two particularly noticeable gaps, the outer of which is

270 km wide; these gaps contain narrow eccentric ringlets. *Ring B* is the brightest part of the entire ring system and is opaque when seen from the earth. In the central portion of the ring *spokes* have been seen, measuring 10 000 km by 2000 km wide. The spokes move radially and they are thought to be created by the interaction of the local magnetic field with the particle distributions in the rings. There are also electrostatic (lightning) discharges in this region that may also be related to the spoke formation. The particles in ring B are reddish in colour and may be as large as a few metres in diameter. These particles are quite different from those of the C and D regions.

The *Cassini Division* separates the B and A rings with inner and outer edges at 117 400 and 121 900 km above the cloud tops. It is one of the most prominent features of the ring system. The origin of the 4500 km gap may be related to the perturbation effect of the satellite Mimas and the particles in the B ring area. The *A ring* is not as bright as its inner neighbour. It contains a large number of ringlets and minor gaps. These are quite separate from the principal division, popularly known as the *Enke Divison* but now called the *A Ring Gap*. This gap also contains several ringlets, and although only 200 km wide is visible from the earth with moderate telescopes. The outer edge of ring A is a sharp boundary at a distance of 73 000 km above the cloud tops, which may be related to the presence of a tiny satellite. The rings have a thickness of 50–100 metres and are surrounded by a large rarefied cloud of neutral hydrogen extending to about 60 000 km above and below the ring plane.

Beyond the main rings are several diffuse ring systems. The braided *F ring* is not circular; on average it is about 77 000 km from Saturn but varies over a range of 400 km. Two *shepherd satellites are located on either side of the F ring. Still further out is the tenuous *G ring* at a distance of 107 000 km. It lies between the orbit of Mimas and those of its two co-orbital satellites. The distance between the F and G rings is about 30 000 km. Finally there is the *E ring*, which extends from a distance of 147 000 km from Saturn's cloud tops at the inner edge as far as 237 000 km at the outer edge. The distance of the satellite Enceladus from the planet's centre is 240 200 km so that the brightest part of Ring E lies just inside Enceladus' orbit.

scale factor. *See* cosmic scale factor.

scattering. The random deflections suffered by *waves passing through an irregular medium. If the source, medium, or observer are in relative motion, *scintillations* – random fluctuations of *amplitude – may be seen as the source is observed through the medium: scattering in the earth's atmosphere causes the stars to twinkle. Scintillations may only be seen if both the angular size of the source and the *bandwidth in which the waves are received are small enough. Otherwise, the effect of the scattering may simply be to broaden the apparent angular size of the source.

Scintillations of *radio waves are observed to occur because of irregularities in the *refractive index of the *ionosphere, the *interplanetary medium, and the *interstellar medium giving *ionospheric scintillations, interplanetary scintillations (IPS)*, and *interstellar scintillations (ISS)*, respectively. IPS may be used in determining the angular sizes of *radio sources at metre wavelengths in the range 0.1 to 2 arc seconds. ISS cause some of the random fluctuations in the intensity of pulses received from *pulsars.

Light may be deflected from its direction of travel by fine particles of solid, gaseous, or liquid matter. For very small particles (less than one wavelength in size) the effect results

from *diffraction, reflection playing a more important part with increasing size. Very small particles scatter blue light more strongly than red light. This leads to the reddening of starlight by cosmic dust and to the reddening of the sun when seen through a thick layer of atmospheric dust.

Scheat (β Peg). A red giant that is the second-brightest star in the constellation Pegasus and lies on the Great Square of Pegasus. It is an irregular variable with a magnitude range of 2.1–3.0, although the usual limits are 2.3–2.8. Spectral type: M2 II-III; distance: 60 pc.

Schedar or **Shedir (α Cas).** An orange giant that is one of the two brightest stars in the constellation Cassiopeia. The magnitude varies slowly between 2.1 and 2.6. It has a number of optical companions. M_v: −1.3; spectral type: K0 II-III; distance: 50 pc.

Schmidt telescope (Schmidt camera; Schmidt). A *catadioptric wide-field telescopic camera first built in 1930 by the Estonian Bernhard Schmidt. A short-focus reflecting telescope with a spherical mirror suffers from severe *spherical aberration. This is normally corrected by modifying the surface to a paraboloid. This method of correction is only effective, however, for a field of less than half a degree: outside this limit the star images are distorted by severe *coma. Schmidt overcame the spherical aberration of a spherical mirror, which does not suffer from coma, by placing a *correcting plate in the incoming light beam before it reached the short-focus spherical primary mirror. His correcting plate is a thin plate of complicated figure placed near the centre of curvature of the mirror (see illustration). The corrected light comes to a focus on a curved focal surface but when a photographic plate is sprung to this curve very sharp star images are formed upon it over a field of

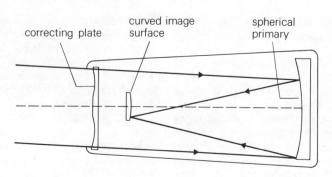

correcting plate curved image surface spherical primary

Ray path in Schmidt telescope

maybe tens of degrees: the 1.2 metre *UK Schmidt covers 40 square degrees of sky. In addition relatively short photographic exposures are required, compared to a normal telescope, because of the small *focal ratio of the Schmidt. The description '1.2 m / 1.8 m / 3.1 m Schmidt' indicates that the correcting plate has a diameter of 1.2 metres, and the spherical mirror a diameter of 1.8 m and a focal length of 3.1 m. This is normally abbreviated to '1.2 m Schmidt'.

The development of the Schmidt telescope has enabled whole-sky surveys, i.e. the *Palomar Sky Survey and *Southern Sky Survey, to include very faint (down to about 21st magnitude) stars. Information, in digital form, can now be extracted rapidly and automatically from the photographic plates using machines such as *COSMOS.

The *super Schmidt* is an extreme form of Schmidt developed by the American James Gilbert Baker in the late 1950s. By employing additional correcting plates he was able to retain the wide field while increasing the speed to focal ratios approaching 0.5. (Schmidt's original telescope was f/1.7). These extremely fast cameras were designed to record meteor and artificial satellite trails.

Schönberg–Chandrasekhar limit. An upper limit on the mass of the core of a *main-sequence star: a star will leave the main sequence and become a *red giant when it has exhausted hydrogen supplies totalling some 12% of its original mass.

Schwarzschild black hole. A *black hole that possesses mass but has zero angular momentum and zero electric charge.

Schwarzschild radius. The radius of the *event horizon of a *black hole: a critical radius that must be exceeded by a body if light from its surface is to reach an outside observer. For a body of mass M (but zero angular momentum and zero electric charge), the Schwarzschild radius, R_s, is given by

$$R_s = 2GM/c^2$$

where G is the gravitational constant and c the speed of light. If a body collapses so that its radius becomes less than this critical value, then the *escape velocity becomes equal to the speed of light and the object becomes a *black hole. The Schwarzschild radius is proportional to the mass of a body. For a star the size of the sun, the Schwarzschild radius is some three km.

Schwassmann-Wachmann 1. A *comet that moves in a near-circular orbit between the orbits of Jupiter and Saturn and suffers from random *outbursts*: increases in brightness by up to seven magnitudes that last for a few weeks.

A considerable number of comets suffer outbursts and many theories have been put forward to explain them. These include pressure release from gas pockets of methane and carbon dioxide, explosive chemical reactions involving free radicals (CH, OH, and NH), temperature-induced phase changes in amorphous ice, impact cratering by interplanetary boulders, and spin- and age-induced nuclear break-up. Impact and break-up are the most likely causes.

scintillation (twinkling). Rapid irregular variations in the brightness of light received from celestial objects, noticeably stars, produced as the light passes through the earth's atmosphere: irregularities in the atmosphere's refractive index occur in small mobile regions and can cause the direction of the light to change very slightly during its passage. In a telescope a star image will consequently wander rapidly about its mean position, producing an overall blurred enlarged image. With extended light sources, such as the planets, scintillation produces a hazy outline in a telescopic image. For stars near the horizon, where refraction effects including *dispersion are much greater, changes in colour can also be observed. *See also* scattering; speckle interferometry.

scintillation counter. A radiation detector that is used in astronomy mainly for measurements of gamma rays. It consists of a layer of scintillator crystals that emit flashes of light when radiation passes through. Each scintillation is picked up by a *photomultiplier and an amplified pulse of current is produced. These pulses can be counted electronically and the activity of the gamma-ray source can be measured. It is also possible to determine the distribution of energy of the gamma rays.

The energy of the electrons liberated from the photocathode of the photomultiplier depends on the frequency of the emitted light, which in turn depends on the energy of the incident beam. It is possible to exclude radiation of energies other than those under consideration by suitable electronic circuitry.

scopulus. Irregular scarps. The word is used in the approved name of such a surface feature on a planet or satellite.

Scorpius (Scorpion). A conspicuous zodiac constellation in the southern hemisphere, lying partly in the Milky Way. The brightest stars are *Antares (α) and the blue-white spectroscopic binary Shaula (λ), both 1st magnitude, and several of 2nd and 3rd magnitude. The area contains many star clusters, including the bright naked-eye open clusters M6 and M7 and the globular clusters M4 and M80, and the x-ray source *Scorpius X-1. Abbrev.: Sco; genitive form: Scorpii; approx. position: RA 17h, dec −30°; area: 497 sq deg.

Scorpius X-1. The first cosmic x-ray source, discovered during a sounding rocket experiment from White Sands, New Mexico in June 1962. By an order of magnitude Sco X-1 is the brightest of all nontransient cosmic x-ray sources. Optically identified in 1967 with the 13th magnitude variable star V 818 Sco, it was confirmed as a low-mass binary system with an orbital period of 0.78 days. The distance of Sco X-1 remains uncertain, probably lying in the range 300−600 parsecs.

Sculptor. An inconspicuous constellation in the southern hemisphere near Grus, the brightest stars being of 4th magnitude. R Sculptoris is an intense scarlet variable star. The south *galactic pole lies in Sculptor and many faint galaxies can be observed. Abbrev.: Scl; genitive form: Sculptoris;

approx. position: RA 0.5h, dec −30°; area: 475 sq deg.

Sculptor system. A member of the *Local Group of galaxies.

Scutum (Shield). A small constellation in the southern hemisphere near Sagittarius, lying in the Milky Way, the brightest stars being of 4th magnitude. It contains the bright *RV Tauri star R Scuti, the prototype of the *Delta Scuti stars, and many star clusters, including the just-visible fan-shaped *open cluster, the *Wild Duck* (M11). Abbrev.: Sct; genitive form: Scuti; approx. position: RA 19h, dec −10°.

S Doradus. An extremely luminous supergiant located in NGC 1910, a brilliant cluster of giant and supergiant stars lying on the line of sight of the Large Magellanic Cloud. S Dor is an *irregular variable, and may be an *eclipsing binary comprising two very massive stars. Its luminosity has faded in recent years: it is now less bright than HD33579, also in NGC 1910.

sea interferometer. An interferometer (*see* radio telescope) that uses a single *antenna mounted on a clifftop overlooking the sea. Fringes are observed on a *radio source at low elevation as a result of the *interference between radio waves from the source itself and from its image reflected in the sea.

seasons. The four approximately equal divisions of the year, usually taken as lasting from *equinox to *solstice − spring and autumn − or from solstice to equinox − summer and winter. The seasons, which are more noticeable away from the equator, result from the inclination (23.5°) of the earth's axis to the perpendicular to the earth's orbital plane so that the sun spends half the year north and the remaining year south of the celestial equator (see illustration). The varying distance of the earth from the

sun has a minor effect on the seasons. Mars, with an axial tilt of 24°, also has seasons.

Sechi classification. *See* spectral types.

second. Symbol: s. The scientific (SI) unit of time, defined since 1967 in terms of *atomic time. The second is the duration of 9 192 631 770 periods of the radiation corresponding to the transition between two hyperfine energy levels of the ground state of the caesium−133 atom.

secondary. 1. *See* primary.
2. *Short for* secondary crater.

secondary cosmic rays. *See* cosmic rays.

secondary craters. Depressions produced by the impacts of low-velocity crater *ejecta. Secondary craters are usually elongated radially with respect to the primary *crater and form loops, *crater chains, and *crater clusters. They tend to lie outside the *ejecta blanket.

second of arc. *See* arc second.

secular. Continuing or changing over a long period of time.

secular acceleration. The apparent speeding up of the moon in its orbit around the earth, caused by *tidal friction and the gravitational attraction of other planets. It amounts to 10.3 arc seconds per century. It first became apparent from records of ancient *eclipses.

secular parallax. The apparent and continuously increasing displacement of stars as a result of the sun's motion through space. This form of *parallax can be used to determine the distance of nearby groups of stars. It is assumed that the stars are at approximately the same distance from the sun and that their random indi-

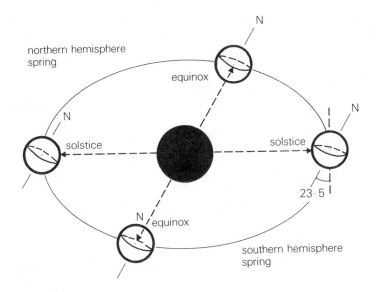

northern hemisphere spring

equinox

N

N

solstice

solstice

N

23·5

N equinox

southern hemisphere spring

Annual change in seasons

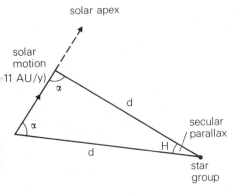

solar apex

solar motion 11 AU/y)

α

d

α

d

H

secular parallax

star group

Secular parallax

vidual proper motions tend to cancel out leaving only the parallactic motion. The baseline for this parallax measurement is the distance moved by the sun relative to the *local standard of rest and is hence increasing by the sun's velocity of 19.5 km/s, i.e. 4.11 AU per year (*see* apex). The secular parallax per annum, H (in arcsecs), is then given by:

$$\sin H = 4.11 \sin \alpha / d$$

where d is the mean distance to the group and α is the angle of the group to the solar apex (see illustration).

secular perturbation. *See* perturbation.

seeing. The quality of the observing conditions at the time of telescopic observation. The various degrees of good or bad seeing depend on the amount of turbulence in the earth's atmosphere. Turbulence distorts the plane wavefront of a beam of radiation, which has travelled more or less undisturbed through space, and leads to a perturbed 'corrugated' wavefront. On the few occasions when atmospheric turbulence is very low, a small steady disc-shaped optical image of a point source (such as a star) results. Poor seeing mainly produces small erratic movements in the image position (*see* scintillation) so that the overall image is blurred distorted and enlarged. No matter how well-designed the telescope, it is the seeing that imposes the main limitation on the instrument's angular *resolution: the minimum diameter of the optical image, resulting from diffraction (*see*

Airy disc), can be as low as 0–025 arc seconds but the angular resolution may only be one or two arc seconds, or more, as a result of poor seeing.

In amateur astronomy the seeing can be evaluated using the *Antoniadi scale*: the observer allocates a Roman numeral from I to V to conditions varying from perfect seeing, through good, moderate, poor, to appalling.

segmented mirror. *See* primary mirror.

seismology. The study of *seismic waves,* which propagate through the solid material of the earth, or other celestial bodies such as the moon, following earthquakes (or equivalent), explosions, or impact phenomena. Four types of wave are studied: *S-waves* are shear waves that vibrate at right angles to their direction of motion; *P-waves* are compressional waves that oscillate along their direction of motion; *Rayleigh waves* propagate as a rolling motion of the surface; *Love waves* produce horizontal vibrations of the surface at right angles to the direction of propagation. The S- and P-waves travel deep into the body in question where they are refracted at the boundaries between layers of differing density. S-waves can only travel through a solid; P-waves can travel through solid, liquid, or gas. The waves are measured and analysed to derive information about the events that generated them or about the internal structure of the body through which they travel.

seismometer. An instrument used in *seismology to measure the passage of seismic waves, normally by making a continuous recording of the position of a suspended mass in relation to a frame fixed to the ground.

selenography. The study of the moon's physical features. *See* moon, surface features.

selenology. The scientific study of the *moon. *See also* moonrocks.

self-absorption. *See* synchrotron emission.

self-propagating star formation. A process whereby spiral structure in galaxies is explained by waves of star formation triggered by the explosion of a previous generation of stars as *supernovae. Some of the new stars then explode in turn, and thus continue propagating the spiral pattern around the galaxy. P.E. Seiden and H. Gerola calculated in 1979 that the distribution of stars formed by supernova triggering would form spiral arms in a galaxy that was rotating. The theory gives a better fit than the rival *density-wave theory to the patchy spiral arms found in many spiral galaxies.

semidetached binary. *See* binary star; equipotential surfaces.

semidiameter. Half the *apparent diameter of a celestial body.

semimajor axis. Symbol: *a.* The distance from the centre of an ellipse to the edge, through the intervening focus. It is half the greatest distance across the ellipse, i.e. half the *major axis.* The semimajor axis of an elliptical orbit is one of the *orbital elements.

semiregular variables (SR variables). A heterogeneous group of giant and supergiant *pulsating variables showing brightness variations that do not usually exceed one or two magnitudes, that have a noticeable periodicity ranging from several days to several years, but are also disturbed at times by various irregularities. The light curves have diverse shapes. Semiregular regulars have been divided into four subgroups: *SRa* and *SRb variables* are giants of late spectral type (M, C, and S) either with relatively stable periods (SRa) or ill-defined periods (SRb) and are difficult to differentiate from long-period *Mira variables; *SRc variables* are red supergiants of late spectral type, such as Betelgeuse, Antares, and Mu Cephei; *SRd variables* are highly luminous yellow (F, G, and K) supergiants and giants.

sensitivity. The minimum signal power that can be distinguished from the random fluctuations in the output of a measuring system caused by *noise inherent in the system. In a *radio telescope contributions to the *system noise* include noise generated in the first stage of the *receiver, thermal *noise due to loss in the *antenna/ *feeder system, *synchrotron emission received by the antenna from the Galaxy (*see* radio source), and *thermal emission from the atmosphere and from the ground. If the system noise has a power P watts, then an equivalent *system temperature*, or *noise temperature*, can be ascribed to it given by

$$T = P/kB$$

where k is the Boltzmann constant and B the *bandwidth.

The *signal/noise ratio* of a radio telescope is the ratio of the power in the output that is due to the *radio source under observation to that caused by the system noise. The sensitivity is usually defined as the *flux density of a source that would produce the same signal power as the noise power, i.e. a signal/noise ratio of one. *See also* confusion.

separation. Symbol: ρ. The angular distance between the two components of a *visual binary or optical *double star. Separation can be measured with a *filar micrometer in the eyepiece of the telescope; its value is expressed in arc seconds ($''$).

Serenitatis Basin. One of the youngest lunar *basins, formed 4000 million years ago and filled with *mare lavas

from at least 3800 million years ago to produce the circular *Mare Serenitatis (Sea of Serenity)*. The Mare has striking colour contrasts and numerous *rilles and *wrinkle ridges. It was visited by *Lunokhod 2 and the margins sampled by the crew of *Apollo 17.

Serpens (Serpent). A constellation divided into two unequal parts, originally called *Serpens Caput* (Serpent's Head) and *Serpens Cauda* (Serpent's Body), that lie on either side of the constellation Ophiuchus. The larger part (the Head) lies in the northern hemisphere near Corona Borealis; the smaller section lies in the southern hemisphere near Scorpius, partly in the Milky Way. The two sections use a single set of Bayer letters (*see* stellar nomenclature) to identify the stars. The brightest stars are Alpha Serpentis, of 2nd magnitude, and several of 3rd magnitude. The globular cluster, M5, in the larger part contains over 100 RR Lyrae stars. The *Eagle nebula, M16, lies in the smaller section. Abbrev.: Ser; genitive form: Serpentis; approx. position: RA 15.5h, dec $+10°$ (Head), RA 18.3h, dec $-5°$ (Body); area: 428 sq deg (Head), 208 sq deg (Body).

service module (SM). *See* Apollo.

SETI. *Abbrev. for* search for extraterrestrial intelligence. Any of many searches in which neighbouring stars, and sometimes more distant stars, have been studied primarily by means of radio telescopes for signals indicative of life. The pioneering programme was Frank Drake's *Ozma Project of 1960. It was unsuccessful, as were subsequent projects using more sophisticated equipment and targeting more stars. Microwave frequencies (1–10 GHz, i.e. wavelengths of 3–30 cm) are thought to be the most favourable to study: natural sources of noise in our atmosphere and in the interstellar medium are at a minimum in this waveband and some common molecules have strong microwave emissions. These include neutral hydrogen and the hydroxyl radical; many searches have been conducted at the 21 cm line of neutral hydrogen.

In 1983 a long-term SETI programme began, financed by NASA. A spectral analyser has been built that electronically breaks down the input from a radio telescope into 74 000 separate but simultaneously sampled frequency channels, each 1 Hz wide. Bands of 32 and 1024 Hz can also be sampled. This device has been tested on various radio dishes. It is hoped to increase the number of channels to 8 million. Half of the observing plan will involve a 3 year sweep of the entire sky in 32 Hz steps from 1–10 GHz. The other half involves careful surveys of the roughly 800 sun-type stars within 25 parsecs of earth, scanned in 1 Hz steps over 1–3 GHz.

Other SETI programs are also underway.

setting. *See* rising

setting circles. The circular scales provided on the polar and declination axes of a nonportable *equatorial mounting. With their aid the telescope may be set to chosen values of *right ascension and *declination, respectively, so as to bring an object of known position into the field of view.

Seven Sisters. *See* Pleiades.

Sextans (Sextant). An inconspicuous equatorial constellation near Leo, the brightest star being of 4th magnitude. Abbrev.: Sex; genitive form: Sextantis; approx. position: RA 10h, dec −2°; area: 314 sq deg.

Seyfert galaxy. A galaxy with an exceptionally bright nucleus, first described by Carl Seyfert in 1943. Over 150 Seyfert galaxies have now been identified and the vast majority of

these are otherwise normal nearby spiral galaxies. Seyferts have substantial *nonthermal emission in their nuclei. The nuclei emit radiation at optical wavelengths (by definition) but most of their output is in the infrared; they are also strong x-ray and ultraviolet sources but are not often powerful at radio wavelengths. Their output can vary on a timescale of months. This indicates that the energy source in Seyfert galaxies must be a few light-months across and astronomers believe that Seyferts are a less powerful example (by a factor of 100) of the *quasar phenomenon.

A region of hot gas (at a temperature of 20 000 kelvin) surrounds Seyfert nuclei, much of it concentrated into clouds moving outwards and presumably accelerated by the central energy source. Velocities of these clouds can exceed 10 000 km s^{-1}. In *Type I Seyferts,* whose spectra closely resemble those of quasars, the Balmer lines of hydrogen emitted by the gas are broader than the emission lines of ionized metals; this indicates velocities of several thousand km s^{-1}. The widths of the two sets of lines are comparable in *Type II Seyferts,* showing lower gas velocities – less than 1000 km s^{-1}.

NGC 4151 is the nearest bright Seyfert. Observations with the *International Ultraviolet Explorer satellite have shown that the ultraviolet spectral lines in NGC 4151 tend to brighten several weeks after a brightening of the illuminating central source. This indicates that the gas responsible for the emission lies in clouds at a radius of several light-weeks from the galaxy's centre; the gas is moving at high velocity. (There are also more distant gas clouds moving at lower velocities.) A central mass of about 5×10^8 solar masses has been determined. This implies that a black hole powers the central activity.

Some 10% of giant spiral galaxies are Seyferts. As Seyfert activity is probably only a transitory period in the life of a galaxy, then it is possible that all giant spiral galaxies, including our own, spend 10% of their lives in a Seyfert phase.

S0 galaxy. *See* galaxies.

shadow bands. Elongated patches or mottled patterns of light and shadow that have irregular form and movement and are briefly observed to cross rapidly over light surfaces on the earth just before and after totality in a solar *eclipse. They probably result from irregular refraction, in the earth's atmosphere, of light from the crescent sun.

Shane telescope. *See* Lick Observatory.

Shaula. *See* Scorpius.

Shedir. *See* Schedar.

shell star. A hot main-sequence star, usually of spectral type B, whose complex spectrum shows sharp deep absorption lines with wings of bright emission lines superimposed on the normal absorption spectrum. It is now known that they are *Be stars, the differences in appearance between the two arising from the directions in which they are observed.

shepherd satellites (shepherding satellites). Eight minor satellites in the inner part of the Saturnian system that have been observed by the *Voyager spacecraft as guarding one of *Saturn's rings or sharing an orbit. They are located by the outer edge of the A ring ($\sim$ 40 km in diameter); on either side of the F ring ($\sim$ 200 km in diameter); two co-orbiting moons ($\sim$ 100 km in diameter) that may well be the fractured portions of a once single satellite located beyond the F ring; two satellites situated in the *Lagrangian points of the satellite Tethys ($\sim$ 50 km in diameter); and a

co-orbital satellite near to the satellite Dione ($\sim$ 160 km in diameter).

shock wave. A compression wave of large amplitude that is produced when there is a sharp and violent change in pressure, density, and/or temperature in a fluid. It occurs when a disturbance in the fluid cannot be dispersed quickly enough and thus banks up. A shock wave can lose energy by heating the medium through which it travels.

shooting star. *Colloquial name for* meteor.

short-period comet. *See* comet.

shower. *Short for* air shower. *See* cosmic rays.

shower meteor. *See* meteor shower.

shuttle. *Short for* space shuttle.

Sickle of Leo. *See* Leo.

side lobe. *See* antenna.

sidereal. Relating to or measured or determined with reference to the stars.

sidereal day. The interval between two successive passages of a *catalogue equinox across a given meridian. It is divided into 24 sidereal hours. The sidereal day is 3 minutes 56 seconds shorter than the 24 hour *day, i.e. the period of the earth's rotation on its axis.

sidereal hour angle. The angular distance measured westwards along the celestial equator from a *catalogue equinox to the intersection of the *hour circle passing through a celestial body. It is equal to 360° minus the *right ascension (in degrees).

sidereal month. The time of 27.321 66 days taken by the moon to complete one revolution around the earth, measured with respect to a background star or stellar group considered fixed in position.

sidereal noon. *See* sidereal time.

sidereal period. The time taken by a planet or satellite to complete one revolution about its primary, measured by reference to the background of stars. The *sidereal month and *sidereal year are the sidereal periods of the moon and earth. *See also* synodic period; Tables 1 and 2, backmatter.

sidereal rate. *See* equatorial mounting.

sidereal time. Time measured in *sidereal days and in sidereal hours, minutes, and seconds. The sidereal time at any instant is given by the *sidereal hour angle of a *catalogue equinox and ranges from 0 to 24 hours during one day. The day starts at *sidereal noon,* which is the instant at which the equinox crosses the local meridian. The hour angle at a particular location gives the *local sidereal time, the hour angle at Greenwich being the *Greenwich sidereal time. A celestial object will be on the meridian of a particular place when the local sidereal time becomes equal to the object's *right ascension.

Apparent sidereal time is measured by the hour angle of the true equinox and thus suffers from periodic inequalities, the position of the true equinox being affected by the *precession and *nutation of the earth's axis. *Mean sidereal time* relates to the motion of the mean equinox, which is only affected by long-term inequalities arising from *precession (*see* mean (and true) equator and equinox). Apparent minus mean sidereal time equals the *equation of the equinoxes.* Sidereal time is directly related to *universal time and *mean solar time and is used in their determination.

sidereal year. The time taken by the earth to complete one revolution

around the sun with reference to a background star or stellar group, which is regarded as fixed in position. It is thus the interval between two successive passages of the sun, in its apparent annual motion around the celestial sphere, through a particular point relative to the stars. It is equal to 365.256 36 days and is about 20 minutes longer than the *tropical year.

siderite. *See* iron meteorite.

siderolite. *See* stony-iron meteorite.

siderostat. A flat mirror that is driven in such a way as to reflect light or infrared radiation from a celestial body to a fixed point, such as a spectrograph slit at the coudé focus of a telescope, over the duration of an observation. A siderostat is often installed outside the main dome of an observatory and can be used with a coudé spectrograph, say, at the same time as the main telescope is being used to study a different field of view. Siderostats often have computer-driven *altazimuth mountings.

Siding Spring. Mountainous site in New South Wales of the *Anglo-Australian Observatory. *See also* Australia Telescope.

SIGMA. A *coded-mask imaging low-energy gamma-ray satellite telescope originally designed by a consortium of British, Italian, and French groups. It is now planned as a Soviet-French mission for launch in the late 1980s.

signal/noise ratio. *See* sensitivity.

signature. A group of spectal lines usually in an emission spectrum that identifies a chemical – an atom or molecule, possibly ionized and/or in a rare isotopic form – in a star, stellar environment, galaxy, etc.

signs of the zodiac. *See* zodiac.

silicon stars. *See* Ap stars.

silvering. A process whereby a reflective layer of silver is applied to a mirror base from a solution of silver salts containing reducing agents. The glass is first activated by immersion in stannous chloride solution. Silvering was first used in the 1850s but has been largely supplanted by the *aluminizing process, which provides a reflective layer of greater durability and able to reflect lower wavelengths, i.e. into the ultraviolet.

sine formula. *See* spherical triangle.

singularity. A mathematical point at which space and time are infinitely distorted. Calculations predict that matter falling into a *black hole will ultimately be compressed to infinite densities at a single point, and in such conditions our laws of physics, including quantum mechanics, must break down. One black-hole theorem – the principle of *cosmic censorship* – states that singularities are always concealed by an event horizon so that they cannot communicate their existence to an observer in our universe. However, if a *naked singularity* – a singularity without an event horizon – is found, then some of our physical concepts will need re-examination.

Sinope. A small satellite of Jupiter, discovered in 1914. *See* Jupiter's satellites; Table 2, backmatter.

sinuous rille. *See* rille.

sinus. A bay or semienclosed break in a scarp. The word is used in the approved name of such a feature on a planet or satellite.

Sirius (Dog Star; α CMa). A white star that is the brightest star in the constellation Canis Major and the brightest and one of the nearest stars in the sky. It lies in a descending line from Orion's Belt. It is a visual binary (separation 7″.6, period 49.97 years), the companion, *Sirius B,* being the first *white dwarf to be discovered. Bessel suggested (1844) that Sirius had a dark companion to account for the star's wobbling movement. With improved telescope lenses Alvan G. Clark detected (1862) a tiny companion whose spectrum, first taken (1915) by W.S. Adams Jn., identified Sirius B as a white dwarf. The spectrum demonstrated the *gravitational redshift predicted by the general theory of relativity. m_v: -1.46 (A), 8.67 (B); M_v: 1.41 (A), 11.54 (B); spectral type: A9 IV (A), wdA5 (B); mass: 2.31 (A), 0.98 (B) times solar mass; radius: 1.7 (A), 0.022 (B) times solar radius; distance: 2.67 pc.

SI units. An internationally agreed system of coherent metric units, increasingly used for all scientific and technical purposes. It was developed from the MKS system of units and replaces the CGS and Imperial systems of units. There are seven base units that are arbitrarily defined and dimensionally independent (see Table 1). These include the kelvin and the second. The metre has recently been redefined but is still a base unit. The base units can be combined by multiplication and/or division to derive other units, such as the watt or the newton (see Table 2). A set of prefixes, including kilo-and micro-, can be used to form decimal multiples or submultiples of the units (see Table 3). Any physical quantity may then be expressed in terms of a number multiplied by the appropriate SI unit for that quantity.

six-colour system. A system of stellar *magnitudes in which photoelectric magnitudes U, Vi, B, G, R, and I are measured at six wavelength bands: the measurement of U (ultraviolet) is centred on 355 nanometres (nm), for Vi (violet) on 420 nm, for B (blue) on 490 nm, for G (green) on 570 nm, for

Table 1. Base and Supplementary SI Units

physical quantity	name of SI unit	symbol for unit
length	metre	m
mass	kilogram(me)	kg
time	second	s
electric current	ampere	A
thermodynamic temperature	kelvin	K
luminous intensity	candela	cd
amount of substance	mole	mol
*plane angle	radian	rad
*solid angle	steradian	sr

*supplementary units

Table 2. Derived SI Units with Special Names

physical quantity	name	symbol	base units
frequency	hertz	Hz	s^{-1}
energy	joule	J	$kg\ m^2\ s^{-2}$
force	newton	N	$kg\ m\ s^{-2}$
power	watt	W	$kg\ m^2\ s^{-3}$
pressure	pascal	Pa	$kg\ m^{-1}\ s^{-2}$
electric charge	coulomb	C	$A\ s$
electric potential difference	volt	V	$kg\ m^2\ s^{-3}\ A^{-1}$
electric resistance	ohm	Ω	$kg\ m^2\ s^{-3}\ A^{-2}$
magnetic flux	weber	Wb	$kg\ m^2\ s^{-2}\ A^{-1}$
magnetic flux density	tesla	T	$kg\ s^{-2}\ A^{-1}$
luminous flux	lumen	lm	$cd\ sr^{-1}$
illuminance	lux	lx	$cd\ sr^{-1}\ m^{-2}$
absorbed dose	gray	Gy	$m^2\ s^2$
activity	becquerel	Bq	s^{-1}
dose equivalent	sievert	Sv	$m^2\ s^2$

Table 3. Decimal Multiples and Submultiples to be used with SI Units

submultiple	prefix	symbol	multiple	prefix	symbol
10^{-1}	deci-	d	10^1	deca-	da
10^{-2}	centi-	c	10^2	hecto-	h
10^{-3}	milli-	m	10^3	kilo-	k
10^{-6}	micro-	μ	10^6	mega-	M
10^{-9}	nano-	n	10^9	giga-	G
10^{-12}	pico-	p	10^{12}	tera-	T
10^{-15}	femto-	f	10^{15}	peta-	P
10^{-18}	atto-	a	10^{18}	exa-	E

R (red) on 720 nm, for *I* (infrared) on 1030 nm.

Skyhook. *See* rockoon.

Skylab. A large manned US space station launched on May 14 1973 into

Skylab missions

spacecraft	astronauts	launch date	flight duration (hrs)
Skylab 2	Conrad Kerwin Weitz	1973, May 25	672.8 (record)
Skylab 3	Bean Lousma Garriott	1973, July 28	1427.2 (record)
Skylab 4	Carr Gibson Pogue	1973, Nov. 16	2017.3 (record)

a 435 km circular earth orbit, inclined at about 50° to the equator. The station – Skylab 1 – provided an environment in which men could live and work under controlled but weightless conditions. During 1973 and 1974 three three-man crews were ferried to and from the station by *Apollo spacecraft consisting of a command and service module. The manned missions are known as Skylab 2, 3, and 4 (see table). Although a record flight duration of 2017 hours was achieved by the Skylab 4 crew, this has been broken by *Salyut crews. With the Apollo craft docked, the overall length was 36 metres and the weight 82 000 kg; the orbital workshop was 14.7 metres long and 6.6 metres in diameter. Power was obtained from several panels of *solar cells: the first crew had to do extensive repairs to the solar panels damaged during take-off; in addition the meteroid shield and one of the solar panels had been torn away at take-off so that the workshop was exposed to intense heat and the available power was reduced.

Physiological and psychological reactions to weightlessness were carefully monitored and studied, and ability to perform experiments was tested. In over 513 man days in space 73 experiments were conducted. The instruments of the *Apollo Telescope Mount provided the majority of the observational data, other work including technological, biological, and materials-processing experiments. Over 180 000 solar photographs and 40 000 earth-resources pictures were taken. The 740 hours spent on solar studies and 215 hours of astrophysical investigation, made above the earth's distorting and absorbing atmosphere, led to great advances in astronomy.

NASA had assumed that Skylab would stay in orbit for 10 years, with the space shuttle being used to reboost it for further use or deboost it for controlled re-entry and crash-landing in the Pacific. In March 1978, however, it was reported that the space station had dropped to an altitude of about 380 km and that uncontrolled re-entry could occur as early as mid-1979. Despite successful attempts by NASA to manoeuvre the station into a more stable attitude, and so decelerate descent, NASA revealed in December 1978 that it had abandoned the rescue mission: on July 11 1979 Skylab plunged into the Indian Ocean, with fragments landing in Western Australia.

slow nova. *See* nova.

slow pulsator. *See* x-ray pulsator.

small circle. Any circle on the surface of a sphere that is not a *great circle. It is the circular intersection on a sphere of any plane that does not pass through the centre of the sphere.

Small Magellanic Cloud (SMC; Nubecular Minor). *See* Magellanic Clouds.

SMC. *Abbrev. for* Small Magellanic Cloud. *See* Magellanic Clouds.

SNR. *Abbrev for* supernova remnant.

SO galaxy. *See* galaxies.

solar activity. The collective term for time-dependent phenomena on the sun. *Sunspots, *faculae/*plages, *filaments (or *prominences), and *flares all belong in this category, but the *granulation and *chromospheric network do not since their gross configuration is stable. The level of solar activity at present varies over an average period of approximately 11 years – the so-called solar cycle (*see* sunspot cycle).

solar antapex. *See* apex

solar apex. *See* apex.

solar calendar. Any *calendar, such as the *Gregorian calendar, that is based solely on the motions of the sun. *See also* tropical year; lunar year.

solar cell. A device by which incident solar radiation is converted directly into electrical energy. It is a semiconductor device that is identical in principle to the *photovoltaic detector and has a p-n junction with a large surface area. Various differently doped materials are used. Solar radiation falling on or near the junction produces an external voltage. The conversion efficiency can reach 20–25%.

Solar cells form the main power supply in satellites, space stations, and short-range planetary probes. The cells are arranged on flat *solar panels* outside the craft to receive the maximum amount of radiation from the sun. On probes travelling beyond Mars the radiation flux is insufficient to power the instruments: the solar constant at Jupiter's orbit is only about 4% of the value at the earth's orbit. Power must then be obtained from other sources, such as *thermoelectric generators.

solar constant. The total energy, radiated from the sun, that passes perpendicularly through unit area per unit time at a specified distance from the sun. It is given by the ratio $L/4\pi r^2$, where L is the luminosity of the sun and r the distance. The value measured at the mean sun-earth distance has recently been given as 1.367×10^3 joules per square metre per second, i.e. 1.367 kilowatts per square metre. This value is quoted as being accurate to within $\pm 5\%$. Another recent measurement gives 1.353 kW m^{-2}, with an accuracy of $\pm 1.5\%$. Ideally measurements should be made, from spacecraft, over at least one *sunspot cycle. Instruments on satellites such as the *Solar Maximum Mission are monitoring the solar surface to detect whether variations in the solar constant occur, as at present suspected.

solar cycle. *See* sunspot cycle.

solar eclipse. *See* eclipse.

solar mass. Symbol: $M_\odot$. The astronomical unit of mass. It is equal to the mass of the sun, 1.9891×10^{30} kilograms. *See also* solar units.

Solar Maximum Mission (SMM). A solar observatory of NASA that was launched into a 574 km 96 minute earth orbit in Feb. 1980 to study *flares and other phenomena on the sun during the then current maximum of *solar activity. Instruments include gamma-ray, ultraviolet, and x-ray spectrometers, a coronagraph, and a radiometer. The SMM was the first *multimission modular spacecraft. Later in 1980 a blown fuse and other control problems left it unable to point correctly. In Apr. 1984 Solar Max was manoeuvred into the bay of a space shuttle, repaired, and successfully relaunched, and should still be operating at the next solar maximum in 1990/91.

solar motion. *See* apex; galactic rotation.

solar neutrinos. *See* neutrino detection experiments.

Solar Optical Telescope. A 1.25 metre telescope of Gregorian design that is being built for NASA and is scheduled to be carried aboard the space shuttle on a series of missions around 1989. It will provide detailed observations of solar features, with simultaneous observations at different wavelengths.

solar oscillations. *See* photosphere.

solar panel. *See* solar cell.

solar parallax. The angle subtended by the earth's equatorial radius at the centre of the sun, at a distance of one astronomical unit. It is equal to 8.794 148 arc seconds.

solar spectrum. The sun's spectrum extends from gamma-ray to radio wavelengths. It has an immense range in intensity, peaking at visible wavelengths. Although the central part of the curve varies little with *solar activity, the long and short wavelength sections can be very considerably affected. The radiation intensity at visible and infrared wavelengths compares with that of a *black body at a temperature of about 6000 kelvin; the maximum intensity occurs at wavelengths of about 500 nanometres. This is the continuous spectrum of the *photosphere in which absorption lines – *Fraunhofer lines – appear. At the shortest and longest wavelengths, the solar spectrum corresponds to the radiation curve of a black body at about a million kelvin, representative of the temperatures of the *corona and of solar *flares. At ultraviolet and soft x-ray wavelengths

the spectrum does not agree with either of these black-body curves. *See also* flash spectrum.

solar system. A group of celestial bodies comprising the sun and the large number of bodies that are bound gravitationally to the sun and revolve in approximately elliptical orbits around it. The latter include the nine known major planets, Mercury, Venus, Earth, Mars, Jupiter, Saturn, Uranus, Neptune, and Pluto; at least 49 known natural satellites of the major planets; the numerous minor planets, mostly in orbits between the orbits of Mars and Jupiter; the comets, which exist in great numbers beyond the most distant planets; and countless meteoroids. The sun contains 99.86% of all the mass of the system, while most of the remaining mass is concentrated in the planet Jupiter. In extent the solar system may be considered as a sphere with a radius greater than 100 000 AU although the planets alone extend to a radius of less than 50 AU (*see* Oort cloud).

With the exception of the comets, all bodies of the solar system orbit the sun in the same direction as the earth, along orbits that lie close to the plane of the earth's orbit and the sun's equator. This direction is also close to that of the axial rotation of the sun and most planets, and of the orbital motion of most planetary satellites. This motion reflects the common origin postulated for the solar system by the contraction of a rotating cloud of interstellar gas and dust about 4600 million years ago (*see* solar system, origin).

The sun and its entire retinue are moving in a nearly circular orbit around the centre of the Galaxy, which lies about 10 000 parsecs away in the constellation Sagittarius. They have a mean velocity of about 250 (or possibly 220) km s^{-1} (*see* galactic rotation). In relation to the stars in its neighbourhood, the solar system is moving with a velocity of 19.4 km s^{-1}

towards a point in the constellation Hercules – the solar *apex.

solar system, origin. There have been various theories on the origin of the solar system, many of which have been discarded or modified as observational data has slowly accumulated. The sun and planets are now believed to have formed together 4600 million years ago by the contraction of a cloud of interstellar gas, mainly hydrogen and helium, and dust grains – the *primeval nebula. The contraction may have been triggered by shock waves from a nearby supernova, but continued under the mutual gravitational attraction of the contents of the nebula (*see* stellar birth).

Conservation of *angular momentum dictated that as the cloud contracted its rotation rate increased, causing it to become disc-shaped around the developing sun. Collisions between dust grains became frequent, leading to a rapid concentration of particles in the plane of the disc, which lay close to the present plane of the ecliptic, and a gradual *accretion of grains into pebbles, boulders, and larger bodies began. Only when these *planetesimals had reached a few kilometres across could their gravity accelerate the accretion process by attracting more solid material. Eventually the *protoplanets of the present planets and satellite systems were built up, most of them rotating and revolving in the same sense as the parent nebula. Possibly any protoplanets that formed between the orbits of Mars and Jupiter suffered gravitational perturbations by Jupiter and were fragmented by collisions to leave the minor planets of today.

Although the sun had yet to begin nuclear fusion processes, temperatures in the inner solar system were high enough to maintain volatile substances such as water, methane, and ammonia in a gaseous state; the planetesimals that gave rise to the *inner planets were therefore formed from the nonvolatile component of the nebula, such as iron and silicates. Further from the embryo sun, lower temperatures allowed volatiles to become important constituents of the *giant planets, their satellite retinues, and the ring systems of Saturn and of Jupiter and Uranus. The origin of the volatile-rich comets is more obscure, but they may have formed concurrently as planetesimals near the outer edge of the solar system's disc. The low temperatures also made it possible for the giant planets to greatly augment their masses by attracting and retaining large quantities of the light elements hydrogen and helium from the surrounding nebula.

As the sun continued to collapse, the density and temperature· in its central core rose (*see* stellar birth). Once the core temperature reached some ten million kelvin, the sun began to generate energy by the nuclear fusion of hydrogen. The onset of fusion processes in the sun set up the *solar wind, which drove uncollected gas and dust grains from the solar system. Heating of the protoplanets by gravitational contraction and interior radioactivity caused partial melting that led to the present internal differentiated structure of the planets. Impacts by remaining planetesimals scarred the surfaces of the planets and their satellites, although the earth's craters have been largely removed by erosion. Cratering events continue on a much reduced scale to the present day as errant minor planets or cometary nuclei collide with the planets or their satellites.

The suggestion that a contracting nebula formed the solar system was the basis of the *nebular hypothesis, popular during the 19th century. However this was unable to explain why most of the solar system's angular velocity resides in the revolution of the planets, rather than in the rotation of the sun, and was replaced by various *encounter theories early in the 20th century. The renaissance of a nebular theory came with the realization that *magnetohydrodynamic forces would transfer angular momentum from the early sun to its surrounding nebula, and that the sun would shed further angular momentum through the solar wind. Calculations incorporating these ideas suggest that the primeval nebula contained at least three solar masses of material, most of which was expelled again into interstellar space.

solar telescope. A reflecting (or refracting) telescope plus associated instruments used in studying the sun. The sun is the only star whose disc can be resolved and studied by a telescope. The diameter of the solar image is given by 0.0093f, where f is the focal length of the telescope primary mirror or objective. Long-focus telescopes must therefore be used in order for a large image to be formed. Such telescopes are usually fixed in position, with the sunlight being reflected on to the mirror or lens by a *heliostat. This light is then directed in a fixed direction into a spectro-

graph or some other measuring instrument.

The intense heat of solar radiation requires the measuring instruments to be cooled. In some telescopes the entire building is cooled. In most cases the primary mirror and instruments are underground to facilitate this. The heat problem can be reduced by evacuating the air from the rooms housing the instruments and in some cases from the whole telescope housing.

Large solar telescopes are often built as *tower telescopes* (or *solar towers*). The sunlight is directed down a vertical path inside a solid or girder construction and reflected into underground rooms containing the measuring instruments. The *McMath solar telescope* at the Kitt Peak National Observatory, Arizona, has a somewhat different design (see illustration). Sunlight is reflected from a 2 metre heliostat, which is 30 metres above the ground. It traverses 150 metres along the polar axis and is focused by a 1.5 metre concave mirror and a second flat mirror on to the various measuring instruments.

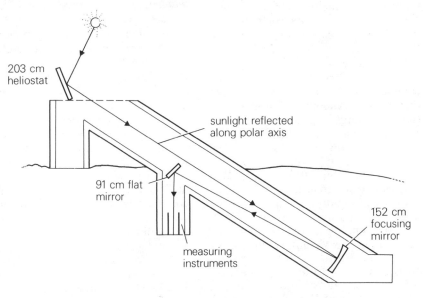

Optical path in McMath solar telescope

The best solar seeing is found at sizable lakes; in addition high altitudes are required in order to reduce the disturbing effects of the earth's atmosphere. The site of the *Big Bear Solar Observatory has both these features.

solar time. 1. *Short for* mean solar time.

2. *Short for* apparent solar time.

solar tower. *See* solar telescope.

solar units. Dimensionless units whereby the mass, luminosity, and other physical properties of a celestial body can be expressed in terms of the sun's mass, luminosity, etc. For example, a star having 10 times the mass of the sun is said to have a mass of 10 solar masses, or 10 $M_\odot$ (*see* solar mass). Likewise the luminosity of a source is frequently expressed as a multiple or fraction of the sun's luminosity, $L_\odot$, which is equal to 4×10^{26} watts.

solar wind. A flow of energetic charged particles – mainly *protons and *electrons – from the solar *corona into the *interplanetary medium. The thermal energy of the ionized coronal gas is so great that the sun's gravitational field cannot retain the gas in a confined static atmosphere. Instead, there is a continuous, near-radial, outflow of charged particles into interplanetary space, this highly tenuous plasma carrying mass and angular momentum away from the sun. The expansion is controlled by the sun's magnetic field. It is low in the inner corona but rapidly becomes supersonic and reaches anywhere between 200 and 900 km s^{-1} at one astronomical unit, i.e. in the vicinity of the earth's orbit. The density of the plasma decreases with increasing distance from the sun and at the earth's orbit is only about eight particles per centimetre cubed.

Two forms of the solar wind are readily recognizable: a slow-moving low-density flux (<400 km s^{-1}), and high-speed streams. The latter are thought to form the normal state and originate in areas of relatively weak divergent open (and therefore primarily unipolar) magnetic field. The emission is greatly increased over *coronal holes and the long-term stability of such regions may cause recurrent enhancement over several solar rotations. Similar long-lived jets are apparently emitted from the polar regions. Confinement by magnetic fields in the corona results in the slow-moving flow and this is more frequent around the maximum of the *sunspot cycle. In addition, transient mass ejections occur during *flares and in some of the more frequent but less energetic active *filaments/ *prominences.

The sun's magnetic field is transported by the expanding plasma and becomes the *interplanetary magnetic field*, the lines of which are wound into spirals by the sun's rotation. The region of expanding solar wind, or *heliosphere*, is bounded (possible at 50–100 AU) by the *heliopause*, where its pressure equals that of interstellar space. The heliosphere is probably very elongated due to the presence of an *interstellar wind* of neutral hydrogen, flowing from the direction of the galactic centre.

solstices. 1. (solstitial points). The two points that lie on the *ecliptic midway between the vernal and autumnal *equinoxes and at which the sun, in its apparent annual motion, is at its greatest angular distance north of the celestial equator – *summer solstice* – or south of the celestial equator – *winter solstice*.

2. The times at which the sun reaches these points, on about June 21 and Dec. 22; the hours of daylight or of darkness are then at a maximum.

solstitial colure. *See* colures.

solstitial points. *See* solstices.

Sombrero galaxy (M104; NGC 4594).
A spiral (Sa/Sb) galaxy in the constellation Virgo that has a relatively large bright nucleus. The spiral structure is difficult to trace because the equatorial plane is tipped only 6° from our line of sight. A dark absorption band of interstellar dust in the galactic plane is clearly visible against the bright inner regions. Distance: 11 megaparsecs; total mass: about 5×10^{11} solar masses; total magnitude: 8.1.

sonde. A device, carried by rocket or balloon, for observations and measurements at high altitudes.

sounding rocket. A rocket providing relatively inexpensive brief flights (typically 4–10 minutes) for studying the earth's upper atmosphere and for making high-altitude astronomical observations. Information collected by instruments on board is radioed or sometimes parachuted back to earth. First launched in 1946, sounding rockets are generally single-stage rockets. After completion of rocket motor firing they can coast upwards to a peak altitude of 150–200 km before dropping back to earth.

source count (number count). A compilation of the number, *N,* of sources per unit solid angle that are brighter than a flux density level *S*. In Euclidean space the radius of the sphere out to which sources of a certain luminosity are observed varies as $S^{-\frac{1}{2}}$; the number of sources contained therein thus varies as $S^{-3/2}$. A plot of log *N* versus log *S* is a straight line of slope -1.5 and is independent of source luminosity. Deviations from this value may indicate source evolution, departures from a uniform space distribution, cosmological effects, or the

effect of an intervening obscuring medium. Differential source counts (number per unit flux interval) are to be preferred when dealing with data since the errors on each point are then independent of other points. Radio source counts provided the first evidence for the *big bang theory as against *steady-state theory. *See also* radio source catalogue.

south celestial pole. *See* celestial poles.

Southern Coalsack. *See* Coalsack.

Southern Cross. Common name for *Crux.

Southern Sky Survey. A photographic star atlas of the southern sky, comprising two sets of photographic plates covering overlapping areas of the sky. The celestial objects recorded are very much fainter and more distant than those on previous southern sky maps, and about four times fainter than objects on the *Palomar Sky Survey of the northern and equatorial sky issued in the 1950s. The project has been a joint effort involving the use of the 1.0 metre *Schmidt telescope at the *European Southern Observatory, Chile, and the 1.2 metre *UK Schmidt at the Anglo-Australian Observatory, Australia.

The UK Schmidt contribution has been a set of sky-limited blue-sensitive plates; objects between a declination of −17° to the south celestial pole have been recorded, down to a magnitude of 22. This survey was virtually complete by 1983. A new survey is working northwards towards the celestial equator. The ESO Schmidt is providing a corresponding set of red-sensitive plates, covering the same area as the blue-sensitive plates. Progress has been slower but (in 1984) the survey is near completion.

south point. *See* cardinal points.

Soviet six metre. The world's largest reflecting telescope, sited at the Special Astrophysical Observatory of the Academy of Sciences of the USSR near *Zelenchukskaya in the Caucasus.

Soyuz craft. A series of Soviet manned spacecraft, the first of which – Soyuz–1 – was launched in Apr. 1967. The earliest craft were involved in docking experiments, Soyuz–4 and –5 achieving the first docking of two spacecraft and a transfer of crew, in Jan. 1969. Soyuz craft ferry cosmonauts to and from the *Salyut space stations, the first docking and transfer of crew occurring between Salyut–1 and Soyuz–11 in June 1971. An upgraded version, Soyuz T, has been used for this purpose since June 1980, when Soyuz T–2 docked with Salyut–6.

space. The near-vacuum existing beyond the atmospheres of all bodies in the universe. The extent of space, i.e. whether it is finite or infinite, is as yet unresolved. *See* interplanetary medium; interstellar medium; intergalactic medium.

Spacelab. A reusable space laboratory that is carried into earth orbit as a payload of one of NASA's *space shuttles, and in which a team of scientists, known as *payload specialists*, can oversee or take part in experiments devoted to science and technology. It was designed and constructed by the European Space Agency (ESA), West Germany bearing over half the cost. It will remain in the cargo bay of the shuttle during an orbital flight, and is thus used for short-duration experiments where it is necessary or important for people to be present. Spacelab is designed for a 10 year life or 50 seven-day flights.

There are two basic sections in Spacelab: a pressurized module in which the payload specialists can operate the experiments on board, and an unpressurized U-shaped pallet (platform) where instruments can be directly exposed to space. A pressurized tube connects the module to the shuttle's living quarters. The pressurized module is a cylindrical aluminium-alloy shell, either 4.5 or 7.2 metres in length and 4.0 metres in diameter, with a working volume of 22 cubic metres. The core segment of the module contains the basic Spacelab subsystems while the experiment segment provides space for packaged equipment that can be removed and replaced after each mission. The pallet segments are 2.9 metres long and 4.0 metres in width and may be attached individually to the module. Alternatively the pallets may be flown by themselves in the cargo bay, in a series of up to three lots; subsystems will then be carried in an attached 'igloo', pressurized and thermally controlled. The third configuration that can be used in a Spacelab mission is the module by itself.

The first mission, Spacelab–1, originally scheduled for 1980, was launched on 28 Nov. 1983 on board the shuttle Colombia. The long module and one pallet were used. The first flight was primarily to test engineering and project management, but much research was done: of the 70 or so experiments, 60 were devised by ESA countries (8 by the UK), the rest coming from the USA, Japan, and Canada. Most were devoted to human physiology and materials sciences. NASA waived the launch fee for the European experiments on Spacelab–1. Spacelab then became the property of NASA, which has definite plans for three further missions in 1984, 1985, and 1986: Spacelab–2, using pallets only and devoted to astronomy, will fly in 1985, after Spacelab–3, which is scheduled to be launched in Nov. 1984 carrying materials sciences experiments. West Germany has booked its own flight in 1985.

spaceprobe. *See* planetary probe.

space shuttle. A partially recoverable manned space transportation system developed and tested by NASA and operational by 1982: the first orbital test flight, by the shuttle *Colombia*, occurred in Apr. 1981. The shuttles *Challenger* and *Discovery* have been in service since 1983 and 1984, with *Atlantis* scheduled for launching in 1984/85. The space shuttle will eventually replace the more costly expendable launch vehicles used by NASA.

A shuttle consists of an *Orbiter* with the appearance of a delta-wing aircraft, a huge expendable external propellant tank, on which the Orbiter is mounted at launching, and two solid-fuel rocket boosters. The whole system weighs about 2000 tonnes at launch and has an overall length of about 56 metres. The Orbiter is 37 metres in length.

It is launched in a vertical position by the simultaneous firing of its two rocket boosters and three very powerful liquid hydrogen/liquid oxygen main engines. About two minutes into the flight the empty rocket boosters are detached, parachute into an ocean area, and are recovered for further use. Just before the craft reaches its orbit the propellant tank is discarded. The Orbiter can then manoeuvre by means of two on-board engines. The altitude, eccentricity, and inclination of the orbit can be varied, within limits, as can the flight duration – between about 7 to 30 days.

The Orbiter has a large cargo bay – 18.3 metres long and 4.6 metres in diameter – in which the payload is housed. *Spacelab first flew in the cargo bay in 1983. Satellites can be launched into orbit from the cargo bay, and can also be brought back into the bay for repair and redeployment or for return to earth for refurbishment (or repair) and eventual reuse. The shuttles are also used to conduct or oversee medical, scientific, and technological experiments, either by the *mission specialists* (who are in charge of flight engineering, on-board

systems, etc.) or by *payload specialists* (*see* Spacelab). It will also be possible to launch deep-space missions from shuttles. A total mass of 29.5 tonnes can be carried into a low-altitude orbit, with smaller loads for higher or less accessible orbits. Payloads to be placed in orbits above the shuttle ceiling, such as a geostationary orbit, require additional propulsion; the shuttle ceiling is about 1000 km in altitude. A payload of 11.5 tonnes can be returned to earth. On mission completion the Orbiter uses its rocket motors to put it into a re-entry path, enters the atmosphere in a shallow glide, and finally makes an unpowered landing.

space sickness (space adaptation syndrome). *See* weightlessness.

space station. An orbiting space laboratory, with a lifetime of several years or more, on which people can live and work in controlled but weightless conditions. Crews are ferried to and from the station and remain on board either for short periods or on a continuous or near continuous rota basis.

The first Soviet *Salyut space station was launched in 1971, and the one-off US *Skylab station in 1973. Both the USA and the USSR are planning low-orbit permanently manned space stations, probably of modular design. NASA's station could be assembled in orbit by 1992, while the Soviet station could be in orbit before 1990. These space stations will probably be used for scientific and technological research, as manufacturing bases, and as launching and docking platforms.

Space Telescope (ST). *See* Hubble Space Telescope.

spacetime. The single physical entity into which the concepts of space and time can be unified such that an event may be specified mathematically by four coordinates, three giving the position in space and one the time. The path of a particle in spacetime is called its *world line*. The world line links events in the history of the particle.

The concept of spacetime was used by Einstein in both the special and the general theories of *relativity. In special theory, where only *inertial frames are considered, spacetime is flat. In the presence of gravitational fields, treated by general relativity, the geometry of spacetime changes: it becomes curved. The rules of geometry in curved space are not those of three-dimensional Euclidean geometry.

Matter tells spacetime how to curve: massive objects produce distortions and ripples in the local spacetime. The question of whether the spacetime of the real universe is curved, and in what sense, has yet to be resolved. If the universe contains sufficient matter, i.e. if the *mean density of the universe is high enough, then the spacetime of the universe must be bent round on itself and closed.

space velocity. *See* radial velocity.

spark chamber. A device in which the tracks of charged particles are made visible and their location in space accurately recorded. It consists essentially of a stack of narrowly spaced thin plates or grids in a gaseous atmosphere, partially surrounded by one or more auxiliary particle detectors such as scintillation counters. Any particle detected in one of the auxiliary devices triggers the application of a high-voltage pulse to the stack of plates. The passage of the particle through the plates is then marked by a series of spark discharges along its path. The tracks are recorded by electronic or photographic means. Use of the auxiliary devices leads to a select triggering of the chamber for a particular type of energy or radiation. Spark chambers can be used to detect gamma rays following their conversion to an electron-positron pair.

special relativity. *See* relativity, special theory.

speckle interferometry. A technique whereby the limit to the *resolution of a telescope, imposed not by its design but by atmospheric turbulence (*see* seeing), may be considerably improved: factors of 50 have been reported. A typical stellar image has, at best, a diameter given by that of the diffraction-limited *Airy disc; atmospheric turbulence however causes small continuous erratic movements in image position on a long-exposure photograph, producing a final blurred image enlarged many times. In optical speckle interferometry and its more recent infrared counterpart, many short exposures (10–20 milliseconds) of the object are taken in rapid succession. These freeze the effects of turbulence so that the individual 'speckles' making up the overall image are distortion-free stellar images: those of supergiants are relatively large, those of binary stars are double. Substantial differences between the speckles of the many short exposures require the application of statistical analysis to the images. This leads to a range of information, including the separation and other properties of close binary stars, often hitherto unresolvable. In addition reconstructions have been achieved of the discs of supergiants, such as Betelgeuse; apparent diameters can thus be measured and large-scale surface details discerned. Very small areas of the sun have also been studied.

spectral classes. *See* spectral types.

spectral index. *See* radio-source spectrum.

spectral lines. *See* line spectrum; emission spectrum; absorption spectrum.

spectral resolution. *See* spectrograph.

spectral types (spectral classes). The different groups into which stars may be classified according to the characteristics of their *spectra, principally the absorption lines and bands resulting from the presence of particular atoms and molecules in the star. Spectral lines of the various elements and compounds have widely different strengths in stars of different temperatures: this is the basis of the current classification. In the *Sechi classification* made in the 1860s by the Italian Angelo Sechi, stars were divided into four groups based on the visual observation of spectra. The subsequent development of photographic spectroscopy allowed a much more precise division. The *Harvard classification* was introduced by astronomers at Harvard Observatory in the 1890s and used in subsequent volumes of the *Henry Draper Catalogue. It was developed into its present form in the 1920s by E. C. Pickering, Annie J. Cannon, and others.

The original scheme was based on the strength of the hydrogen Balmer absorption lines in stellar spectra (*see* hydrogen spectrum), the order being alphabetical from A to P: A stars had the strongest hydrogen lines. Some letters were later dropped and the ordering rearranged to correspond to a sequence of decreasing surface temperature. The majority of stars can be divided into seven spectral types – O, B, A, F, G, K, M – the order often remembered by the mnemonic 'Oh Be A Fine Girl (or Guy) Kiss Me'. The principal spectral features of these classes are shown on the graph and listed in Table *A* and also at separate entries: *O stars, *B stars, etc. Since temperature and *colour are directly linked the Harvard classification is also a sequence of colours ranging from hot blue O stars to cool red M stars. Additional classes are the *carbon stars (C stars) and the heavy-metal *S stars. The spectral types are further

subdivided into 10 subclasses denoted by digits 0 to 9 placed after the spectral type letter, as with A5 – midway between A0 and F0. These subdivisions are based on a variety of complex empirical criteria, mainly the ratios of certain sets of lines in a spectrum, and may be further subdivided, as with O9.5. Other spectral characteristics, such as the presence of emission lines, are indicated by an additional small letter placed after the spectral type, as with M5e (see Table *B*).

It was realized in the 1890s that stars of a particular spectral type could have widely differing *luminosities and several luminosity classifications were developed. The currently used *MK system* was introduced in the 1930s by W. W. Morgan and P. C. Keenan at Yerkes Observatory: stars of a given spectral type are further classified into one of six *luminosity classes*, denoted by Roman numerals placed after the spectral type, as with G2 V, and indicating whether the star is a *supergiant, *giant, *subgiant, or *main-sequence star (see Table *C*). *Subdwarfs and *white dwarfs are also sometimes considered as luminosity classes. All these classes occupy distinct positions on the *Hertzsprung-Russell diagram. Stars of a similar temperature but different luminosities must differ in surface area and consequently in *surface gravity and atmospheric density. These differences produce spectral effects that are used to differentiate between the luminosity classes. A major effect is the *pressure broadening of spectral lines, which increases as the atmospheric density and pressure increase (and radius decreases): spectral lines of a bright supergiant are much narrower than those of a main-sequence star of the same spectral type.

spectrograph. An optical instrument used in separating and recording the spectral components of light or other radiation. Spectrographs are a major

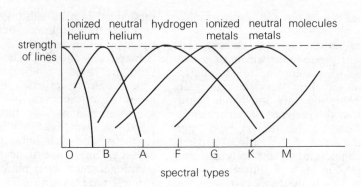

Spectral absorption features of spectral types

(A) Spectral features of star types

O	hottest blue stars	ionized helium (HeII) lines dominant strong UV continuum
B	hot blue stars	neutral helium (HeI) lines dominant no HeII
A	blue, blue-white stars	hydrogen lines dominant ionized metal lines present
F	white stars	metallic lines strengthen H lines weaken
G	yellow stars	ionized calcium dominant metallic lines strengthen
K	orange-red stars	neutral metal lines dominant molecular bands appear
M	coolest red stars	molecular bands dominant neutral metal lines strong

(B) Nonstandard spectral features

e	emission lines present
n	nebulous lines present
s	sharp lines
k	interstellar lines present
m	metallic lines present
p	peculiar spectrum
v	variable

(C) Luminosity classes

Ia	bright supergiants
Ib	supergiants
II	bright giants
III	giants
IV	subgiants
V	main sequence (dwarfs)
VI	subdwarfs
VII	white dwarfs

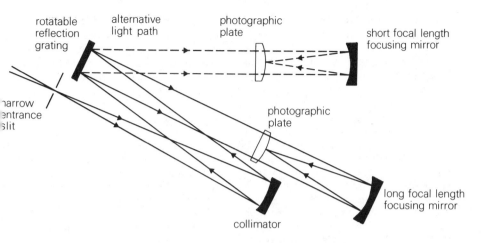

rotatable reflection grating

alternative light path

photographic plate

short focal length focusing mirror

narrow entrance slit

photographic plate

long focal length focusing mirror

collimator

Optical path in spectrograph

astronomical tool for analysing the emission and absorption spectra of stars and other celestial objects. They are used with reflecting (or refracting) telescopes, usually mounted at the Cassegrain focus. If the telescope has a coudé or Nasmyth facility a spectrograph can be permanently positioned at these foci: such spectrographs are used for high-dispersion work. The *dispersion* of a spectrograph is the linear separation of spectral lines per unit wavelength difference, often quoted in millimetres per nanometre or per angstrom. The wavelength range over which the instrument will operate depends· on the recording medium, and also on the optical elements of the device, but is generally in the region 300–1300 nanometres – i.e. light, near-infrared, and near-ultraviolet wavelengths.

The radiation is focused on and enters the instrument through a narrow rectangular slit. The diverging beam is made parallel by a collimator – a converging mirror or lens – and falls on or passes through a *diffraction grating or prism. The beam is thus split into its component wavelengths. This spectrum is focused on a photographic plate or an electronic *imaging device. Large quantities of digital information from highly sensitive elec-

tronic devices, such as *CCD detectors, can be fed into a computer for rapid analysis and manipulation. The *spectral resolution* or *resolving power* of a spectrograph is a measure of the detectable separation of wavelengths that are very nearly equal. It is given by the ratio $\lambda/\Delta\lambda$, if at a wavelength λ it is just possible to distinguish between two spectral lines of wavelength difference $\Delta\lambda$. In a large spectrograph, such as that in the illustration, there is often a choice of mirrors, of different focal lengths, for focusing the spectral image on the recording medium and also a choice of diffraction gratings.

spectroheliogram. A photograph of the sun made in *monochromatic light. It can be obtained either by means of a *spectroheliograph or by placing an *interference filter in front of a photographic plate; in the latter case it is more usually called a *filtergram*. The images are of the solar *chromosphere taken in the residual light of a strong *Fraunhofer line. The lines most commonly used are the K line of singly ionized calcium at 393.4 nanometres and the hydrogen-alpha (Hα) line at 656.3 nm. Different layers of the chromosphere can be imaged by varying the isolated wavelength about the absorption-peak wavelength.

spectroheliograph. A high-dispersion *spectrograph designed to be used, usually with a *solar telescope, for studying the sun's spectrum and for obtaining photographs of the sun at particular wavelengths, such as that of the hydrogen-alpha (Hα) line at 656.3 nm. In addition to an entrance slit, the instrument has a second narrow slit situated directly in front of the photographic plate so that only one spectral line falls on the plate. When the first slit is at the prime focus of the telescope a narrow strip of the sun's image will enter the instrument and a photograph of that narrow image, at the required wavelength, will be formed. If the first slit is moved across the sun's image, and the second slit is moved in step with it across the photographic plate, a photograph – a spectroheliogram – of the whole sun, at the wavelength under consideration, will be obtained. This photograph will be composed of tiny line elements maybe 0.03 nanometres wide.

spectrohelioscope. The optical counterpart of the *spectroheliograph, the principal difference being that both the entrance slit and secondary (viewing) slit are given a rapid oscillatory movement that enables the whole (or a part) of the sun's disc to be seen by virtue of the observer's persistence of vision. Nowadays observation of the sun in monochromatic light is routinely undertaken using automatic photographic patrol telescopes utilizing a *birefringent filter, but the spectrohelioscope is still used at some observatories and by serious amateurs, particularly for measuring the velocities of material *doppler shifted away from the centre of the hydrogen-alpha (Hα) line.

spectrometer. 1. An instrument, such as a *spectrograph, in which the spectrum of a source is recorded by electronic means. Spectrometers covering various wavelength ranges are used in ground-based and satellite and spaceprobe measurements.
2. An instrument for determining the distribution of energies in a beam of particles.

spectrophotometer. An instrument designed to measure the intensity of a particular spectral line or a series of spectral lines in an absorption or emission spectrum.

spectroscopic binary. A *binary star in which the orbital motions of the components give rise to detectable variations in *radial velocity, revealed by changes in *doppler shift in their spectral lines (see illustration). Since a binary star is more likely to be detected spectroscopically if its orbital period is relatively short and the orbital velocities are relatively high, most spectroscopic binaries are close binaries (*see* binary star). Spectroscopic detection is not possible if the orbital plane lies at right angles to the line of sight.

When the components are nearly equal in brightness both spectra are visible and the spectral lines appear double for most of each orbital period, coinciding only when the stars are moving at right angles to the line of sight. Such a system is called a *double-lined spectroscopic binary*. Comparison of the doppler shifts in the component spectra gives the relative velocities and hence the relative masses of the two stars:

$$v_1/v_2 = M_2/M_1$$

The individual masses cannot be determined unless the inclination of the orbital plane to the line of sight is known (*see* eclipsing binary).

The presence of a comparatively faint component is revealed only by its gravitational effect on the motion of the brighter star. Invisible components of low mass are rarely detected because doppler shifts corresponding

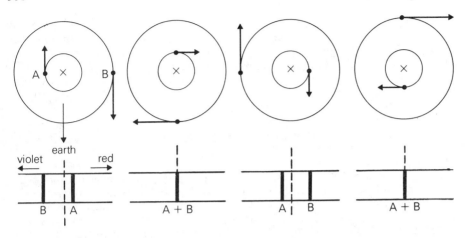

× = centre of mass

Doppler shifts of spectral lines at various orbital positions of spectroscopic binary

to orbital speeds of less than about 2 km/s are obscured by observational errors.

spectroscopic parallax. *See* parallax.

spectroscopy. In general, the production and interpretation of spectra. Astronomical spectroscopy is used over the whole range of *electromagnetic radiation from radio waves to gamma rays. It is the main source of information on the composition, temperature, and nature of celestial bodies.

The lines and bands in *emission and *absorption spectra are characteristic of the atoms, molecules, and ions producing them, and *spectral analysis* leads to the identification of these components in planetary atmospheres, comets, stars, nebulae, galaxies, and in the interstellar medium. Measurements of the intensities of spectral lines (*spectrophotometry*) can give quantitative information on the chemical composition.

In addition, spectroscopy yields information on the physical conditions and the processes occurring in celestial bodies. For instance, temperature may be measured by the vibrational excitation of molecules (if present)

through analysis of the intensities of the individual lines in their band spectra. The degree of ionization can be used to measure higher temperatures. The width and shape of spectral lines indicates temperature, movement, pressure, and the presence of magnetic fields (*see* line broadening; Zeeman effect). The processes occurring are often directly responsible for the production of the spectra, as in the recombination of ions and electrons in HII regions or synchrotron emission from electrons in magnetic fields. *See also* doppler effect; redshift.

spectrum. A display or record of the distribution of intensity of *electromagnetic radiation with wavelength or frequency. Thus, the spectrum of a celestial body is obtained by dispersing that radiation from it into its constituent wavelengths so that the wavelengths present and their intensities can be observed. A variety of techniques exist for obtaining spectra, depending on the part of the electromagnetic spectrum studied. The result may be a photographic record (as can be obtained in a *spectrograph) or a plot of intensity against wavelength or frequency (as can be produced by electronic and associated equipment).

The spectrum of a particular source of electromagnetic radiation depends on the processes producing emission of radiation (*see* emission spectrum) and/or on the way in which radiation is absorbed by intermediate material (*see* absorption spectrum). Spectra are also classified by their appearance. A *line spectrum has discrete lines caused by emission or absorption of radiation at fixed wavelengths. A *band spectrum has distinctive bands of absorption or emission. A *continuous spectrum occurs when continuous emission or absorption takes place over a wide range of wavelengths. *See also* spectroscopy.

spectrum variables (α^2 **CVn stars**). Peculiar main-sequence *Ap stars whose spectra have anomalously strong lines of metals and rare earths (Sr, Eu, Mn, Si, Cr) that vary in intensity by about 0.1 magnitudes over periods of about 1–25 days. These variations are related to changes in the stars' magnetic fields. The brighter component of the binary star *Cor Caroli (α^2 CVn) is typical.

speculum metal. An alloy of copper and tin used by Newton and his successors to make telescope mirrors because it could easily be cast, ground to shape, and took a good polish. It was made obsolete by the use of glass provided with a thin highly reflective silver coating.

speed of light. Symbol: c. The constant speed at which light and other *electromagnetic radiation travels in a vacuum. It is equal to 299 792 458 m s^{-1} (value adopted by the IAU in 1976). Since space is not a perfect vacuum, radiation travels through space at a very slightly lower speed, which decreases somewhat more when the radiation enters the earth's atmosphere. The speed of light, c, is independent of the velocity of the observer: there can therefore be no relative motion between an observer and a light beam, i.e. the speed of light cannot be exceeded in a vacuum. This radical notion was basic to Einstein's special theory of *relativity.

spherical aberration. An *aberration of a spherical lens or mirror in which light rays converge not to a single point but to a series of points whose distances from the lens or mirror decrease as the light rays fall nearer the periphery of the optical element (see illustration). It is most obvious with elements of large diameter. It can be reduced by using an *achromatic doublet rather than a single lens or by using a *correcting plate. Paraboloid surfaces are free of spherical aberration. *See also* Schmidt telescope.

spherical angle. *See* spherical triangle.

spherical astronomy. *See* astrometry.

spherical triangle. A triangle on the surface of a sphere formed by the intersections of three great circles (see illustration). The arcs BC, AC, and AB form the sides of the spherical triangle ABC. The lengths of these sides (a, b, and c), in angular measure, are equal to the angles BOC, COA, and AOB, respectively, where O is the centre of the sphere. This assumes a radius of unity. The angles between the planes are the *spherical angles A, B,* and *C* of the triangle.

The relationships between the angles and sides of spherical triangles are extensively used in *astrometry. The three basic relationships are the *sine formula:*
$$\sin A/\sin a = \sin B/\sin b = \sin C/\sin c$$
the *cosine formula:*
$$\cos a = \cos b \cos c + \sin b \sin c \cos A$$
$$\cos b = \cos a \cos c + \sin a \sin c \cos B$$
$$\cos c = \cos a \cos b + \sin a \sin b \cos C$$
and the *extended cosine formula,* which can be derived from the other two.

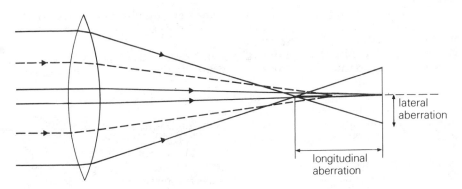

Spherical aberration in convex lens

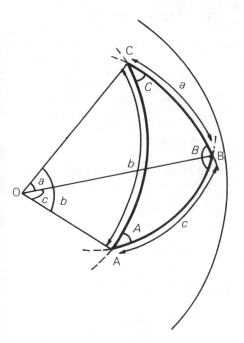

Spherical triangle

For a *spherical right-angled triangle,* in which one angle, say C, is equal to 90° so that $\sin C = 1$, $\cos C = 0$, the formulae are much simplified. They are also simplified if the sides of the triangle are small so that, say, $\sin a$ tends to a and $\cos a$ to $1 - a^2/2$. As a, b, and c approach zero the formulae reduce to those used in plane geometry. The formulae may be extended by replacing the sides and angles by the supplements of the corresponding angles and sides, respectively. Thus
$$\sin A = \sin a, \ \cos A = -\cos a.$$

spheroid. *See* ellipsoid.

spherules. Small round particles of rock and wüstite (iron oxide) formed by the solidification of molten meteoritic material that flows off a meteorite during its passage through the earth's atmosphere. Magnetic spherules can easily be recognized in deep sea sediments. Sizes range typically from 10 μm to 200 μm.

Spica (α Vir). A conspicuous blue-white star that is the brightest one in the constellation Virgo. It is an *eclipsing binary, period 4.014 days, the primary (A) being a *Beta Cephei star near to core hydrogen exhaustion. Spica, Denebola, Arcturus, and Cor Caroli form the quadrilateral shape of the *Diamond of Virgo.* m_v: 0.97; M_v: −3.1; spectral type: B1 V; distance: 65 pc.

spicules. Predominantly vertical narrow (1000 km diameter) jets of gas, visible (in the monochromatic light of certain strong *Fraunhofer lines) beyond the sun's limb. They ascend with velocities of up to 20–30 km s^{-1} from the lower *chromosphere for several thousand kilometres into the inner *corona and then fall back and/or fade away. Their average lifetime is 5–10 minutes. They are closely associated with the *chromospheric network, being located in regions of enhanced vertical magnetic field at the boundaries of supergranular cells (see supergranulation), and have temperatures of the order of 10 000 to 20 000 kelvin.

spider. The diagonal support in a *Newtonian telescope.

spider diffraction. The characteristic spikes of light in the form of a cross seen in the image formed by a Newtonian reflector. They result from diffraction effects caused by the diagonal supports – the spider – and are additional to the diffraction from the circular edge of an objective lens or a mirror, which converts a star image from a point to an *Airy disc in refractors and reflectors.

spillover. The part of the system noise (see sensitivity) of a radio telescope using a *dish antenna that results from pick-up by the feed – the secondary antenna – from directions that do not intercept the reflecting surface of the dish.

spin. A fundamental intrinsic property of *elementary particles that describes the state of rotation of the particle. This spin is always found to be in whole or half integer units of $h/2\pi$, where h is the *Planck constant. See also fermions; bosons.

spinar. A hypothetical supermassive object situated in the nuclei of *active galaxies and *quasars. Some astronomers consider that the rapid rotation of such a million solar mass object could give rise to the quasar's huge output of energy by a process akin to the *pulsar mechanism. Most now believe, however, that a supermassive black hole is responsible, partly because calculations indicate that a

spinar is unstable and likely to collapse to a black hole anyway.

spiral arms. See Galaxy; galaxies; density-wave theory; self-propagating star formation.

spiral galaxy (spiral). See galaxies; Hubble classification.

sporadic meteor. A *meteor that is not associated with a specific *meteor shower and has a random orbit.

Spörer's law. See sunspot cycle.

sprays. Impulsive events on the sun that are characterized by the ejection of chromospheric material into the inner *corona with velocities exceeding the *escape velocity (618 km s^{-1}). They invariably occur during the 'flash phase' of *flares and are less confined than *surges, their material fragmenting as it flies out. See also prominences.

spring equinox. Another name for vernal equinox. See equinoxes.

Springfield mount. See coudé telescope.

spring tide. See tides.

s-process. A slow process of *nucleosynthesis by which heavy stable nuclei are synthesized from the iron-peak elements (mass number 56) by successive captures of neutrons. The process occurs when there is a low density of neutrons in a star, the neutrons being by-products of nuclear-fusion reactions. If the nucleus produced by a neutron capture is stable it will eventually capture another neutron; if the nucleus is radioactive it will have sufficient time to emit a beta particle (i.e. an electron) to stabilize itself before further neutron capture. Thus the stable isotopes of an element are synthesized until a radioisotope is produced, at which point

365

SS433

a new element forms by beta decay. Many years or decades may elapse between successive neutron captures. The most abundant nuclei produced by the s-process will be those with a low ability to capture neutrons. The s-process cannot synthesize nuclei beyond bismuth-209 since neutron capture by this nuclei results in rapid alpha decay. *Compare* r-process.

Sputnik. Any of a series of Soviet unmanned satellites, the first of which − Sputnik 1 − was the first spacecraft to be placed in orbit. This 58 cm diameter sphere, weighing 84 kg, was launched on Oct. 4 1957; it burnt up in the earth's atmosphere 92 days later. The orbit had a period of 96 minutes, an apogee and perigee (highest and lowest altitudes) of about 950 km and 230 km, and was inclined at 65° to the equator. Its radio signals were transmitted every 0.6 seconds. It contained few instruments, being intended as a test vehicle. The launching of Sputnik 1 had a profound effect in accelerating America's space programme.

Sputnik 2, launched on Nov. 3 1957, was very much bigger. It carried about 500 kg in payload, including a live dog, Laika, which survived the launch, as well as 10 experiments. Sputnik 3, launched on May 15 1958, weighed about 1330 kg and remained in orbit for 691 days.

The word sputnik is now used to denote any Soviet satellite.

SR variables. *See* semiregular variables.

SS433. A close *binary star in Aquila that is ejecting a pair of oppositely directed jets at very extreme speed. It lies 5 kiloparsecs from the sun. The star system appears faint optically but is characterized by strong hydrogen emission lines, which led to its inclusion in the Stephenson-Sanduleak (SS) catalogue of such stars (1977); its optical variability has led to an alternative variable-star designation, V1343 Aquilae. It is an x-ray source, discovered by Ariel V in 1976. It is also a radio source, lying in the centre of a large *supernova remnant designated W50, and a gamma-ray source. The discovery that it has weak hydrogen emission lines that move considerably in wavelength led to intense investigation and analysis in 1978–81; the following picture has emerged.

SS433 comprises a fairly normal star that is transferring mass at a high rate to a compact companion; as in other *x-ray binaries, the gas forms an accretion disc around the compact star. This hot accretion disc produces most of the system's light so that the nature of the normal star is not evident. It is perhaps a star of 1 solar mass losing matter to a *neutron star; it is now thought more likely to be a star of 10 or more solar masses with a companion that is a *black hole of some 5 solar masses. These two stars have an orbital period of 13 days, and the stars and the accretion disc form a partially *eclipsing binary system.

Two jets of gas stream off the faces of the accretion disc, at a constant speed of 80 000 km s^{-1}. The disc precesses in a period of 164 days, so that the jets trace out a cone on either side with an opening angle of 40°. The component of velocity along the line of sight thus varies periodically, and the changing *doppler effect causes the emission lines from the jets to move up and down the spectrum as the wavelength varies with the 164 day cycle. Radio observations show the precessing jets out to a distance of 0.05 parsecs; the Einstein Observatory recorded their x-ray emission in a broader band at distances up to 30 pc to either side. Here they strike the surrounding supernova remnant W50.

SS433 may represent a transitory phase in the evolution of many x-ray binaries, and its apparent uniqueness may indicate that fast jets are only a short-lived phenomenon. Two other instances of stars located within supernova remnants resemble SS433 superficially, and a similar phenomenon is apparently occurring on a more moderate scale in the jet-emitting *symbiotic star R Aquarii.

SS Cygni. The brightest and most carefully observed *dwarf nova. Normally of 12th magnitude, it rises to maybe 8th magnitude every month or so: its period varies widely from the mean of 51 days; the maxima vary in shape, brightness, and duration. Like other dwarf novae it is a close binary system in which one component is a white dwarf; the other is a normal G5 star. The orbital period is 6.5 hours. SS Cygni is also an x-ray source, i.e. an *x-ray binary.

S stars (zirconium stars; heavy-metal stars). Stars of *spectral type S, about half of which are *irregular long-period variables. They are red giants similar to *M stars but distinguished spectroscopically by the presence of molecular bands of the heavy-metal oxides of zirconium (ZrO), lanthanum, yttrium, and barium rather than the oxides of titanium (TiO), scandium, and vanadium. Lines of technetium (longest half-life 2×10^6 years) can also be present. Pure S stars have very strong ZrO bands with either very weak or no TiO bands. *MS stars* are M-type stars showing ZrO bands.

ST. *Abbrev. for* Space Telescope. See Hubble Space Telescope.

standard epoch. *See* epoch.

standard time. The time in any of the 24 internationally agreed *time zones* into which the earth's surface is divided; the primary division is centred on the Greenwich (0° longitude) meridian. All locations within a single zone keep the same time. Zone times differ by a whole number of hours, or in some cases of half hours, from *Greenwich mean time. Zones west of the Greenwich zone are behind GMT, those east of it are in front of GMT.

star. A luminous gaseous body that generates energy by means of nuclear fusion reactions in its core. Below a certain mass (0.05 solar masses), the central pressures and temperatures in a body are insufficient to trigger fusion reactions; all stars must therefore exceed this mass limit. Stars of progressively higher mass are less and less common; the highest reliably determined values are about 60 solar masses, but there may be a very few more massive stars (*see* stellar mass).

Although there is only a relatively narrow range in normal stellar masses, it is the mass of a star that determines its other properties – luminosity, temperature, size – and the way in which it evolves. These quantities are related to the mass by the equations of *stellar structure. Other stellar parameters show a far greater range than stellar mass: for example, the *luminosity of a star is proportional to roughly the cube of its mass for *main-sequence stars (*see* mass-luminosity relation); stars therefore show a range in luminosity of some 10^{10}. A star's lifetime also depends on its mass: low-mass stars live considerably longer than high-mass ones.

The characteristics of a star can be determined only when its distance is known. The method of *distance determination depends on the distance itself: the shortest distances are found directly by measuring a star's *annual parallax; greater distances require indirect methods. The nearest star, Proxima Centauri, is 1.31 parsecs away. The stars in our immediate locality are found in the main to be young stars occupying the outer parts of the disc of our Galaxy, rotating around the galactic centre. Many still remain in open clusters, and over 50% of all the stars we observe are *binary or multiple stars.

See also magnitude; population I, population II; proper motion; radial velocity; spectral types; stellar birth; stellar diameter; stellar evolution; variable star.

star atlas. A collection of maps of a particular region of the sky or of the whole sky, showing the relative positions of stars and other celestial objects down to a specified *limiting magnitude. They now contain prints of photographic plates. The *Palomar and *Southern Sky Surveys are examples. *See also* star catalogue.

starburst galaxy. A galaxy in which a massive burst of star formation is currently taking place: it is characterized by an infrared luminosity that is considerably larger than its optical luminosity, sometimes by a factor of 50 or more. The starburst occurs in a region over a kiloparsec in size, thus distinguishing these galaxies from *active galaxies, which have a tiny central powerhouse. A few nearby starburst galaxies (e.g. M82) have long been known from their disturbed optical appearance; the category was only established in 1983 when the pioneering infrared satellite *IRAS revealed thousands of starburst galaxies. They are basically spirals. The reason for the burst of star formation is unclear although in some cases the gravitational effect of a companion galaxy may be responsible (*see* interacting galaxies). The new stars are still enveloped in the galaxy's dense *molecular clouds; their ultraviolet radiation is absorbed by dust in the clouds and reradiated as infrared.

star catalogue. A systematic list of stars in a specified area of the sky down to a specified *limiting magnitude. It gives stellar coordinates at a particular *epoch, apparent *magnitudes, *proper motions, *radial velocities, *spectral types, etc. In a whole-sky catalogue stars in both hemispheres will be included. Modern cat-

alogues are divided into two general types: *survey catalogues* list a large number of stars whose positions are given with only moderate accuracy; *precision catalogues,* such as *fundamental catalogues, give very precise positions for a relatively small number of stars. General catalogues include the *Bonner Durchmusterung and the *AGK. There are also catalogues specifically for double stars, variable stars, nebulae, star clusters, galaxies, radio sources, x-ray sources, etc. *See also* star atlas; catalogue equinox.

star clouds. Areas of the sky where great numbers of stars are seen so close together that they appear as continuous irregular bright clouds. Star clouds are particularly noticeable in the direction of the galactic centre in the constellation *Sagittarius.

star cluster. *See* cluster.

star density. The number of stars per unit volume of space, usually per parsec cubed. It is often given in terms of the fraction of solar mass per volume. *See also* luminosity function.

star gauges. Systematic counts of stars in different regions of the sky. They were first made in the 1780s by William Herschel who counted the total number of stars visible in each of several adjoining areas within the region under study. By averaging the sample counts he obtained the star gauge for that region. Herschel compiled star gauges for several hundred regions representative of the northern hemisphere, his work being extended to the southern hemisphere by his son John. This provided the first precise information on the local distribution of stars in the Galaxy. By comparing gauges made to different limiting magnitudes (using larger and larger telescopes) they were able to show that the star system extends to great depths in directions towards the *Milky Way. *See also* source count.

Starlab. An international mission in which it is hoped to launch a 1 metre telescope into orbit to be used for optical and ultraviolet observations in conjunction with the *Hubble Space Telescope. The angular resolution and spectral coverage of the two telescopes would be comparable but Starlab's field of view would be about a hundred times larger. Launched by the space shuttle it would periodically be taken back to earth for refurbishment and eventual relaunch.

Starlink. A network of computer systems located at six major astronomical centres in the UK. It provides image processing and data reduction facilities for astronomers. Inaugurated in 1980, it is a project of the Science and Engineering Research Council (SERC). Computer hardware and software are coordinated from the Rutherford Appleton Laboratory in Oxfordshire, the other points in the network being linked directly to this central 'node'. Most astronomical information now gathered by ground-based and orbiting telescopes is in digital form so that it can be manipulated by computer. Starlink enables astronomers to find and extract the information they require, and to evaluate it rapidly, precisely, and effectively. *See also* imaging.

star trails. Curved images on a photographic plate that record the changing positions of stars when a telescope is not driven so as to keep up with their diurnal motion. Even if such a drive is being used, comet trails will still be recorded.

static universe. A universe in which the *cosmic scale factor is independent of time. Einstein proposed a static universe in 1916 by including an ad hoc repulsion term – the *cosmological constant,* Λ – in his field

equations of general *relativity. This cancelled out the natural tendency for a gravitating universe either to expand or contract, depending on its energy content. *See* expanding universe; cosmological models.

stationary orbit. *Short for* geostationary orbit. *See* geosynchronous orbit.

stationary point. *See* direct motion.

statistical parallax. The mean parallax of a group of stars that are all at about the same distance and whose space velocities are randomly orientated. It is derived from the *radial velocities and the tau components of the stars' *proper motions: the tau component is measured perpendicular to the great circle joining a star to the solar *apex. It is thus perpendicular to and independent of the sun's motion, i.e. it results entirely from the motion of the star. The average velocity in one direction will equal that in any other direction. The average radial velocity is thus a measure of average velocity. The average distance is then found from the ratio of average radial velocity to average tau component.

steady-state theory. A cosmological theory obeying a perfect *cosmological principle in which the universe appears the same both at all points and all times: it thus had no beginning, will never end, and has a constant mean density of matter. Proposed by Hermann Bondi, Thomas Gold, and Fred Hoyle in 1948, the steady-state theory employs a flat space expanding at a constant rate. There is continuous creation of matter throughout the universe – or perhaps in galactic centres – to compensate for the expansion. The creation rate is about 10^{-10} nucleons per metre cubed per year. The steady-state theory is not readily able to account for the *microwave background radiation or the steep radio *source counts that indicate an evolving universe. In its original form it has been abandoned, even by its originators.

Stefan's law (Stefan-Boltzmann law). The law relating the total energy, E, emitted over all wavelengths, per second, per unit area of a *black body, with the temperature, T, of the body:

$$E = \sigma T^4$$

σ is *Stefan's constant*, which has the value 5.6697×10^{-8} W m^{-2} K^{-4}. The law can be derived from Planck's radiation law (*see* black body), but was first deduced by Josef Stefan in 1879 and then derived from thermodynamics by Ludwig Boltzmann in 1884. *See also* luminosity.

stellar association. *See* association.

stellar birth. Stars are formed by the *gravitational collapse of cool dense (10^9–10^{10} atoms per cubic metre) gas and dust clouds. Most are born in giant *molecular clouds; a few lower-mass stars condense from *Bok globules. There are problems, however, in initiating the collapse of a gas cloud, which resists compression because of its internal motions and the heating effects of nearby stars. In a massive dense cloud shielded by dust, it is believed that collapse can be triggered when the cloud is slowed on passing through the spiral *density-wave pattern of our Galaxy: it is noticeable that star formation preferentially occurs along spiral arms. Alternatively compression may be caused by a shock wave propagated by a nearby *supernova explosion – which may explain how stars are formed in galaxies without spiral arms or with only patchy spiral arms.

Once collapse starts, instabilities cause the cloud to divide into successively smaller fragments until these become too dense and too warm to divide further. These final fragments are *protostars, which now collapse individually. This collapse is gentle at first but after some half a million years (for a star of one solar mass) the density and temperature rise sharply at the centre, forming a hot core that will be the future nucleus of the star. Theoretical models show that the protostar will contract along the *Hayashi track in the *Hertzsprung-Russell diagram. The slow contraction heats the core to a temperature of 10 million kelvin, which is sufficient to initiate *nuclear fusion reactions in hydrogen. Once the nuclear energy source is established, the contraction is halted and the star joins the *main sequence.

Infrared and radio observations of star-formation regions are now filling out the theory. The infrared satellite *IRAS found many small infrared sources within molecular clouds and Bok globules. Most are probably young stars still cocooned in dust, but a few may be true protostars. *Molecular line radio astronomy has revealed the gas motions in regions of stellar birth. The outer parts are observed as collapsing inwards, as expected; in addition surprisingly high velocity gas has been revealed near the young stars at the centre, forming two oppositely directed beams – *bipolar outflows – from the massive young stars. These outflows must affect stellar birth in the cloud, either disrupting protostars or promoting their formation by compressing the surrounding gas. Each young massive star has been found to be surrounded by a doughnut-shaped disc of dense gas; this gas disc also contains compact clumps that emit powerful maser radiation from water molecules (*see* maser source). The gas in the disc may be spiralling inwards and increasing the star's mass substantially *after* hydrogen fusion begins. The effects of bipolar outflows and late accretion from a surrounding disc have yet to be incorporated into theoretical models; they may alter quite substantially the simple theory given above.

The young stars gradually dissipate the surrounding gas cloud by the effects of heating, *radiation pressure, and the gas outflows. The latter often break through first, indicating the presence of the star by an optical *bipolar nebula or a string of *Herbig-Haro objects. When massive O and B stars are present they light up the residual gas as an *HII region. The massive stars take the shortest time to form, 100 000 years or less, and are normal main-sequence stars by the time they become visible. Less massive stars are seen while they are still contracting and have appreciable amounts of surrounding gas and dust: these *pre-main-sequence stars* appear as *Be, *Ae and *T Tauri stars. A star like the sun takes 50 million years to reach the main sequence.

stellar diameter. The diameters of stars range from several hundreds of millions of kilometres for *supergiants through about one million km for the sun down to a few thousand km for *white dwarfs and tens of km for *pulsars. For convenience the diameter or radius is usually expressed in terms of the sun's radius, $R_\odot$. The size of a star changes drastically towards the end of its life as it progresses through the *red giant stage to its final gravitational collapse. It can also change periodically (*see* pulsating variables).

In most cases stars are too distant for their size to be determined directly. The diameter of a large star at a known distance can however be found geometrically if its *apparent diameter can be measured. Measuring techniques are based on the *interference of light. The apparent diameters of bright *supergiants and red giants were first measured in the 1920s using a stellar *interferometer: the diameters lay between 0.02 arc seconds (Aldebaran, $45R_\odot$) and $0''.047$ (Mira Ceti, $390R_\odot$). Very much smaller apparent diameters, including that of Sirius ($0''.00589$, $1.7R_\odot$) and Alpha Gruis ($0''.00102$, $2.2R_\odot$), have recently been measured with an intensity *in-

terferometer. A telescope cannot resolve the disc of a star; the disc of supergiants can however be imaged and measured using *speckle interferometry, if the apparent diameters are in the region of $0''.02$ to $0''.05$. *Occultations of stars by the moon provide a further interference method: the star does not disappear instantaneously but fades over a few seconds; the unobstructed light produces a characteristic interference pattern of intensity against time from which angular size can be determined.

The sizes of many *eclipsing binaries have been found from the shape of their *light curves. In these measurements the sizes are obtained in terms of stellar radius relative to orbital radius. If the binary is also a *spectroscopic binary the absolute values of the radii can be found. For the majority of (smaller) stars diameters can only be inferred from *Stefan's law, i.e. from known values of *effective temperature, T_{eff}, and luminosity, L:

$$L = 4\pi R^2 \sigma T_{eff}^4$$

where σ is Stefan's constant.

stellar evolution. The progressive series of changes undergone by a star as it ages. Stars similar in mass to the sun, having contracted from the *protostar phase (*see* stellar birth) stay on the *main sequence for some 10^{10} years. During this period they give out energy by converting hydrogen to helium in their cores through the *proton-proton chain reaction until the central hydrogen supplies are exhausted. Unsupported, the core collapses until sufficiently high temperatures are reached to burn hydrogen in a shell around the inert helium core. Shell-burning causes the outer envelope of the star to expand and cool so that the star evolves off the main sequence to become a *red giant.

Further contraction of the core increases the temperature to 10^8 kelvin, at which stage the star can convert its helium core into carbon through the *triple alpha process. The sudden onset of helium burning – the *helium flash – may disturb the equilibrium of a low-mass star. The triple alpha reaction gives out less energy than hydrogen fusion and soon the star finds itself once again without a nuclear energy source. Further contraction may result in helium-shell burning but it is doubtful if low-mass stars have sufficient gravitational energy to go further than this stage. The star becomes a red giant again, pulsating and varying in its brightness because of the vast extent of its atmosphere. Eventually the atmosphere gently drifts away from the compact core of the star with velocities of only a few km s^{-1}, resulting in the formation of a *planetary nebula. The collapsed core forms a *white dwarf star, which continues to radiate its heat away into space for several millions of years.

Stars more than twice as massive as the sun convert hydrogen to helium through the *carbon cycle, which makes their cores fully *convective and therefore less dense than solar-mass stars. When after only a few million years they exhaust their hydrogen, the onset of helium burning occurs only gradually: this is because the core is nondegenerate, and so no instabilities arise. Having consumed its helium, a massive star has the potential energy to contract further so that carbon – formed by the triple alpha process – can burn to oxygen, neon, and magnesium (*see* nucleosynthesis). Should the star be sufficiently massive, it will build elements up to iron in its interior. But iron is at the limit of nuclear-fusion reactions and further contraction of a massive star's core can only result in catastrophic collapse, leading to a *supernova explosion. The core collapses to become a *neutron star or *black hole.

stellar interferometer. *See* interferometer.

stellar mass. The mass of a star is usually given in terms of the sun's mass, i.e. as some number of *solar masses, $M_\odot$; it ranges from about 0.05 to 60 solar masses. Stars of higher mass are much less common that those of low mass. The symbol $\psi(M)$ is given to the number of stars with a particular mass M in a unit volume of space, and Salpeter (1955) showed that it is related to the star mass by the formula

$$\psi(M) \propto M^{-2.35}$$

Because of *mass loss and *mass transfer, this *Salpeter mass function* holds strictly only for stars at the instant of birth, and it is therefore sometimes called the *initial mass function (IMF)*. The IMF is determined by fragmentation and other little-understood processes during *stellar birth.

The lower limit on a star's mass is the minimum amount of gas whose gravitational compression will raise the central temperature high enough for nuclear fusion to occur; less massive fragments from the initial cloud may contract directly to become a degenerate star known as a *brown dwarf. It is still uncertain whether there is a theoretical upper limit to star masses, or whether the rarity of very massive stars means that a galaxy is unlikely to contain a star over about 100 $M_\odot$. The highest masses reliably measured are about 60 $M_\odot$ – for each star in a *spectroscopic binary called *Plaskett's star*. Indirect methods indicate that *Eta Carinae and some *Of stars may be as massive as 120 $M_\odot$. More speculatively, there may be very rare *supermassive stars, up to 2500 $M_\odot$.

Most stars lie on the main sequence of the *Hertzsprung-Russell diagram. Their position on the main sequence, i.e. their *spectral type, depends on their mass, which varies from 0.1 $M_\odot$ (M stars) to over 20 $M_\odot$ (O stars). *Supergiants are generally 10 to 20 $M_\odot$ but young O and B supergiants are much more massive.

The mass of a star can be determined directly if it has a significant gravitational effect on a neighbouring star. Thus the combined mass and in some cases the individual masses of *visual and *spectroscopic binary stars have been found (*see also* dynamical parallax). Mass can be estimated using the *mass-luminosity relation, or from a detailed study of the spectrum that indicates the star's surface gravity.

The evolutionary pattern and lifetime of a star depend on its mass. The least massive stars have the longest lifetimes of thousands of millions of years; the most massive exist only a few million years before exploding as *supernovae. At the end of its life, following mass loss in a *planetary nebula, etc., a star will become a *white dwarf if its mass is less than about 1.5 M_Θ; if the mass exceeds about 3 M_Θ it is likely that the star will become a *black hole; a star with intermediate mass will end up a *neutron star.

stellar motion. *See* proper motion; peculiar motion.

stellar nomenclature. Although most of the brighter stars have special names, as with Sirius, the general and most convenient method of naming stars was introduced (1603) by Johann Bayer. In the system of *Bayer letters,* letters of the Greek alphabet – alpha (α), beta (β), gamma (γ), delta (δ), epsilon (ε), zeta (ζ), eta (η), etc. – are allotted to stars in each constellation, usually in order of brightness. The Greek letter is followed by the genitive of the Latin name of the constellation, as with Alpha Orionis or in abbreviated form α Ori. When the 24 Greek letters have been assigned, small Roman letters – a, b, c, etc. – and then capitals – A, B, C, etc. – are used. The letters R–Z were first used by Argelander (1862) to denote *variable stars in a particular constellation, and when those have been as-

signed the lettering RR–RZ, SS–SZ, etc., up to ZZ is employed. If that proves insufficient AA–AZ, BB–BZ, up to QZ (the 343rd variable in the constellation) must be used (omitting Js), followed by V344, etc.

The fainter stars, usually invisible to the naked eye, are designated by their number in a star catalogue. The numbers in Flamsteed's *Historia Coelestis Britannica* (1725) are often adopted when a star has no Greek letter, as with 28 Tauri (*Pleione in the Pleiades), the *Flamsteed number* generally being followed by the genitive of the constellation name. For stars unlisted in Flamsteed's catalogue the number in some other catalogue is used, the number being preceded by the abbreviation of the name – BD (*Bonner Durchmusterung), HD (*Henry Draper Catalogue), etc. Flamsteed numbered the stars in a constellation in order of *right ascension. Most modern catalogues ignore the constellation and number purely by right ascension.

stellar planetaries. *See* planetary nebula.

stellar populations. *See* population I, population II.

stellar statistics. The use of statistical methods in the study of spatial and temporal distributions and motions of celestial objects within a galaxy, region of the universe, etc., and of spatial and temporal distributions of certain features or of certain categories of these objects. The number of stars in a galaxy can be of the order of 10^{11} and the number of galaxies runs into countless millions. Only a very small number of stars, galaxies, radio sources, etc., can be examined in any detail and from this sample, analysed statistically, a more general picture emerges. Much of the data that is used comes from *star catalogues, *source counts, or detailed studies of selected areas. Almost all the information concerning the shape and structure of our Galaxy and the spatial distribution of other galaxies and of clusters of galaxies has been determined statistically.

stellar structure. The interior constitution of a star, defining the run of temperature, pressure, density, chemical composition, energy flow, and energy production from the centre to the surface. The structure of *main-sequence stars, initially of uniform composition, is relatively simple: only very near the centre are the temperature and pressure high enough for nuclear reactions to occur. In stars more than about twice the sun's mass, convection currents in the core enlarge this zone and keep it mixed; less massive stars in contrast have a convective layer near the surface and an unmixed core.

Changes in structure as stars evolve can be calculated by following changes in chemical composition resulting from nuclear reactions, and recalculating the structure for the new composition. After the star has passed the *Schönberg-Chandrasekhar limit, the structure changes to that of a *giant, with an inert helium core surrounded by a hydrogen fusion shell and an extended envelope. Further core reactions in a very massive star will give it an 'onion-shell' structure, culminating in an iron core surrounded by successive shells of silicon, neon and oxygen, carbon, helium, and outermost, the hydrogen-rich envelope.

In principle, from an assumed composition, structure, and total mass, the other parameters of a stellar interior are derived by solving four differential equations:

$$\text{(1)} \quad dP/dr = -GM\rho/r^2$$
$$\text{(2)} \quad dM/dr = 4\pi r^2 \rho$$
$$\text{(3)} \quad dL/dr = 4\pi r^2 \rho \varepsilon$$
$$\text{(4)} \quad dT/dr = 3\kappa L\rho / 16\pi a c r^2 T^3$$

Equation 1 is that of *hydrostatic equilibrium, 2 is that of continuity of mass, 3 is that of energy generation, and 4 is that of radiative transport (*see* energy transport). Accurate solutions require a large computer since the pressure (P), opacity (κ), and energy generation rate (ε), also depend on the density (ρ), the temperature (T), and the chemical composition; in addition in some parts of the star energy may be transported by convection rather than by radiation. Of the other symbols, r is the radius, M the mass within that radius, G the gravitational constant, L the luminosity at radius r, a the radiation density constant, and c the speed of light.

stellar temperature. *See* surface temperature.

stellar wind. The steady stream of matter ejected from many types of stars, including the sun (*see* solar wind). Although the mass loss is small in sunlike stars, it has been found to be very considerable in *red giants and in hot luminous *ultraviolet stars, particularly in *Of stars and *Wolf-Rayet stars. This large mass loss could have a substantial effect on the final evolution of the stars into either white dwarfs or into more massive neutron stars (or black holes). The mechanisms by which stellar winds are produced is still not clear, but in red giants with hot coronae and low gravity it may simply be the expansion of the coronal gases into space. *See also* T Tauri wind.

Stephan's Quintet. A close group of five galaxies whose four brightest members, first observed by M. E. Stephan in 1877, are very distorted and thus possibly interacting. The observed spread in *radial velocities (and hence *redshifts) suggests however that the group is either rapidly dispersing or that the objects might not be associated but be at very different distances.

stereocomparator. *See* comparator.

Steward Observatory. An observatory of the University of Arizona at Tucson that owns several instruments at the *Kitt Peak National Observatory, principally a 2.3 metre (90 inch) reflector. It operates the *Multiple Mirror Telescope jointly with the Smithsonian Institute.

Stickney. See Phobos.

Stokes' parameters. See polarization.

stone. Short for stony meteorite.

stony-iron meteorite (siderolite). Any of a relatively small group of *meteorites containing on average 50% nickel-iron and 50% silicates.

stony meteorite (aerolite). A type of *meteorite that consists of silicate minerals generally with some nickel-iron. The two main subgroups are termed *chondrites and *achondrites, depending on the presence or absence of *chondrules.

stop (diaphragm). A circular opening that sets the effective aperture (diameter) of a lens, mirror, *eyepiece, etc., and also reduces stray light in an optical system.

stratosphere, stratopause. See atmospheric layers.

strewn field. See tektites.

Strömgren sphere. An approximately spherical region of ionized gas, mainly ionized hydrogen (*HII), that surrounds a hot O or B star. If a hot star were embedded in very low-density gas of uniform distribution, the ultraviolet radiation from the star would almost completely ionize the gas out to a particular radius – the *Strömgren radius.* Equilibrium is established within this volume when the rate of ionization is balanced by the rate of recombination of ions with electrons. Outside this volume lies a thin transition zone between ionized gas and much cooler neutral gas. If the gas cloud is relatively small and of low density, the entire cloud can be ionized. The Strömgren radius depends on the gas density and on the flux of ionizing radiation, i.e. of ultraviolet photons. The flux is dependent mainly on the star's temperature. *See also* emission nebula.

strong force, interaction. See fundamental forces.

subdwarf. A star that is smaller and about 1.5–2 magnitudes fainter than normal dwarf (main-sequence) stars of the same *spectral type. The subdwarfs are mainly old (halo population II) objects that lie just beneath the main sequence on the *Hertzsprung-Russell diagram. Their anomolous position is due to their low metal content, which affects their apparent temperature and hence spectral type. They are usually denoted by the prefix *sd*, as in sdK4 or given the luminosity class VI.

subgiant. A *giant star of smaller size and lower luminosity than normal giants of the same *spectral type, Alpha Crucis being an example. Subgiants form luminosity class IV, lying between the main-sequence stars and the giants on the *Hertzsprung-Russell diagram.

subluminous stars. Types of stars, such as *white dwarfs, *subdwarfs, *high-velocity stars, and nuclei of *planetary nebulae, that are fainter than main-sequence stars. Most are old (population II) stars.

sublunar point. See substellar point.

submillimetre astronomy. A branch of radio astronomy covering the wavelength range from 0.1 to 1 millimetre approximately. It is the highest frequency range, 300–3000 gigahertz, in which radio astronomy can be carried out, and is particularly important because of the large number of molecular emission lines to be found in the range (*see* molecular-line radio astronomy). These originate primarily from giant *molecular clouds.

A parabolic radio *dish plus a *line receiver is normally used for submillimetre observations. The receiver can be at the prime focus of the dish but is more usually at the Cassegrain focus behind the dish, to which radio waves are reflected by a small *subreflector* surface. The receiver is cooled to a very low temperature using liquid helium. To maintain *sensitivity the dish must be more accurately shaped and smooth than dishes for longer-wavelength studies, and must keep its shape under different orientations and weather conditions. Observations have to be made at high-altitude sites where there is very little atmospheric attenuation and absorption by water vapour. A 15 metre submillimetre telescope is being built by the UK and the Netherlands on *Mauna Kea mountain in Hawaii. *See also* millimetre astronomy.

subsolar point. *See* substellar point.

substellar point. The point on the earth's surface that lies directly beneath a star on the line connecting the centre of earth and star. The star would thus lie at the zenith of an observer at that point, the terrestrial latitude of which would equal the star's *declination. The *sublunar point* and *subsolar point* are the equivalent points for the moon and sun, respectively.

sulcus. An area of subparallel furrows and ridges on the surface of a planet or satellite. The word is used in the approved name of such a feature.

summer solstice. *See* solstices.

summer triangle. The large distinctive triangle formed by the three bright stars *Vega, *Altair, and *Deneb. The stars are the first ones visible in northern summer evenings.

sun (or **Sun**). The central body of the solar system and nearest star to earth, situated at an average distance of 149 600 000 km. A *main-sequence star of *spectral type G2 V, diameter 1 392 000 km, mass 1.9891×10^{30} kg, luminosity 3.83×10^{26} watts, and absolute visual magnitude $+4.8$, the sun is a representative yellow dwarf. It is the only star whose surface and outer layers can be examined in detail.

The sun is composed predominantly of hydrogen and helium (relative abundance by mass approximately 3:1), with about 1% of heavier elements. Its mean density is 1.41 g cm^{-3}. It generates its energy by nuclear fusion processes, the most important of which is the *proton-proton chain reaction. These take place in a central core, which is thought to be about 400 000 km in diameter and to contain about 60% of the sun's mass (in barely 2% of its volume). Outside the core is an envelope of unevolved material, through which energy from the core is transported by successive emission and absorption of radiation (*see* energy transport). This envelope extends to within about 100 000 km of the surface, where convection becomes the more important mode of energy transport (*see* convective zone). From the centre to the surface of the sun the temperature falls from around 15 000 000 K to 6000 K.

The surface of the sun, i.e. the *photosphere, represents the boundary between the opaque convective zone and the transparent solar atmosphere. It is a stratum several hundred kilometres thick, from which almost all the energy emitted by the sun is radiated into space. A permanent feature of the photosphere is the *granulation, which gives it a mottled appearance. More striking are the *sunspots and their associated *faculae, the numbers of which fluctuate over an average period of approximately 11 years – the so-called *sunspot cycle.

Observation of the transit of sunspots across the sun's disc discloses a *differential rotation, the *synodic period of the rotation increasing with *heliographic latitude from 26.87 days at the equator to 29.65 days at $\pm 40°$ (beyond which sunspots are seldom seen). The mean synodic period is taken to be 27.2753 days, which is equivalent to a *sidereal period of approximately 25.38 days (and corresponds to the actual period at around $\pm 15°$ latitude). Spectroscopic measurements show that the rotation period continues to increase right up to the polar regions and that at any given latitude it decreases with height above the photosphere, except at the equator where the periods are approximately equal. The reason for this differential rotation is unknown, though it has been suggested that it may be caused by rapid rotation of the sun's core. From the base to the top of the photosphere the temperature falls from 6000 K to 4000 K approximately.

Immediately above the photosphere is the *chromosphere, a stratum a few thousand kilometres thick, in which the temperature rises from about 4000 K to around 50 000 K as the density decreases exponentially with height. Between this and the exceedingly rarified *corona is the *transition region*, a stratum several hundred kilometres thick, in which the temperature rises further to around 500 000 K. The corona itself attains a temperature of around 2 000 000 K at a height of about 75 000 km. It extends for many million kilometres into the *interplanetary medium, where the *solar wind carries a stream of atomic particles to the depths of the solar system.

The sun is thought to possess a weak general magnetic field, though this has yet to be distinguished from the transient polar fields resulting from the dispersal of the intense localized fields of sunspots (*see* sunspot cycle).

See also quiet sun; active sun; solar activity.

sunrise, sunset. The times at which the apparent upper limb of the sun is on the astronomical *horizon. The true *zenith distance (referred to the earth's centre) of the centre of the disc is then 90°50′, the sun's semidiameter being 16′ and horizontal *atmospheric refraction being 34′. *See also* twilight.

sunrise, sunset terminator. *See* terminator.

sun's longitude. A measure of the earth's position around its orbit and more specifically the point in this orbit at which a *meteor shower has its maximum intensity. It is the angle between the earth–sun line when the sun is at the vernal *equinox and the earth–sun line at any other time, measured in the direction of the earth's motion.

sunspot cycle. The semiregular fluctuation of the number of *sunspots over an average period of approximately 11 years – or 22 years if the respective magnetic polarities of the p- and f-spots are considered, since these reverse at sunspot minimum for the groups of the new cycle (*see* sunspots). The nature of the cycle is shown in graph a, where the index of sunspot activity is the *Zurich relative sunspot number. It can be seen that the rise to *sunspot maximum* usually occupies a shorter time than the fall to *minimum* and that the amount of activity may vary considerably between two consecutive cycles. There is some evidence for a modulation of the amplitude of the cycle over a pe-

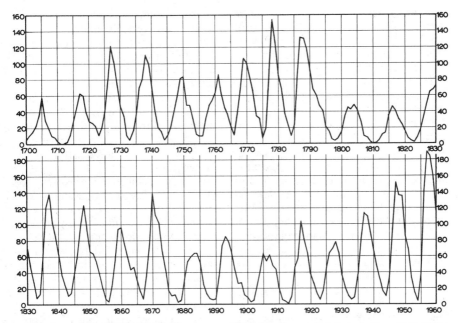

a Yearly means of Zürich relative
     sunspot numbers 1700–1960

(From 'The Sunspot-Activity in the Years 1610-1960' by
M. Waldmeier, Schulthess & Co., Zürich (1961); courtesy
of M. Waldmeier.)

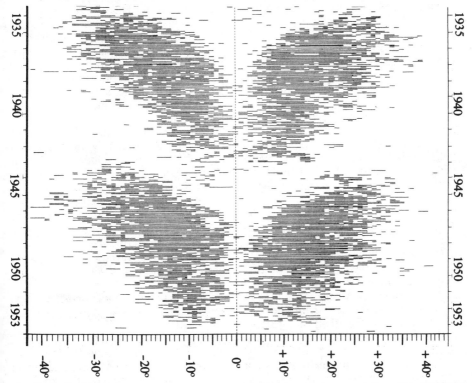

b Butterfly diagram

(From 'Sunspot and Geomagnetic Storm Data 1874-1954',
HMSO, London (1955); courtesy of the Royal Greenwich
Observatory.)

riod of around 80 years but more data are required before the reality of this can be established. The last sunspot maximum occurred in Dec. 1979; the next is predicted for 1990/91.

The phase of the sunspot cycle also determines the mean *heliographic latitude of all groups. At minimum the first groups of the new cycle appear at $\pm 30°-35°$. Thereafter the latitude range moves progressively towards the equator, until by the next minimum the mean latitude is around $\pm 7°$. Then, while the equatorial groups are petering out, those of the succeeding cycle begin to appear in their characteristic higher latitudes. This latitudinal progression is known as *Spörer's law*. At any one time there may be a considerable spread in latitude, but groups are seldom seen farther than 35° or closer than 5° from the equator. The *butterfly diagram* is a graphical representation of Spörer's law obtained by plotting the mean *heliographic latitude of individual groups of sunspots against time (see graph *b*). Its appearance has been likened to successive pairs of butterfly wings, hence the name.

The underlying cause of the sunspot cycle is thought to be the interplay between a large-scale relatively weak poloidal magnetic field beneath the *photosphere, *differential rotation, and *convection. The poloidal field, which is constrained to move with the ionized solar material, becomes increasingly distorted by differential rotation until an intense toroidal field is produced. The strength of this field is further enhanced by the perturbing effect of convection, which twists the field lines into rope-like configurations that may penetrate through the surface to form sunspots. This will occur first in intermediate latitudes, where the rate of shearing of the field is greatest, and thereafter in increasingly low latitudes.

The inclination to the equator of the fields of opposite polarity associated with the *p*- and *f*-spots is such that they may drift apart in both longitude and latitude, as a result of differential rotation and the cyclonic rotation of individual supergranular cells (*see* supergranulation). The latitudinal drift is responsible for an accumulation in the polar regions of magnetic flux of the same polarity as the *f*-spots in the respective hemispheres. Thus the intense (0.2–0.4 tesla) localized fields of sunspots are gradually dispersed to form weak ($1-2 \times 10^{-4}$ tesla) polar fields, which reverse polarity (not necessarily synchronously) around sunspot maximum. When this happens differential rotation no longer intensifies the subphotospheric toroidal field but rather weakens it and re-establishes a poloidal field of opposite direction to its predecessor.

The sunspot cycle may therefore be regarded (if this model is correct) as a relaxation process that is continually repeating itself. There is reason to believe, however, that at least some of the features of the present cycle may be transitory. In particular, a prolonged minimum, termed the *Maunder minimum*, from about 1645 to 1715, suggests that there is more than one circulatory mode available to the solar dynamo.

Sunspots are the most obvious but by no means the only manifestation of *solar activity to undergo a cyclical change over a period of around 11 years. It is therefore proper to restrict the use of the term *sunspot cycle* to consideration of the fluctuation of the number of sunspots and to use the more general term *solar cycle* when considering the variation in the level of solar activity as a whole.

sunspot maximum, minimum. *See* sunspot cycle.

sunspots. Comparatively dark markings in the *granulation of the solar *photosphere, ranging in size from *pores,* which are no larger than individual granules, to complex structures covering several thousand million square kilometres. All but the smallest spots consist of two distinct regions: a dark core or *umbra* and a lighter periphery or *penumbra*; the latter, which can be resolved into delicate radial filaments, accounts for as much as 0.8 of the total area of the spot. Most spots do not occur singly but tend to cluster in groups. The formation, development, and decay of large groups may occupy several weeks, or even months in exceptional cases, but the majority of clusters are small, having lifetimes of not more than a couple of weeks.

Sunspots are the centres of intense localized magnetic fields, which are thought to suppress the currents bringing hot gases from the *convective zone before they reach the photosphere. In the umbra, where the field is strongest (0.2–0.4 tesla), convection is almost completely inhibited; in the penumbra, where the field is more horizontal, a radial flow of material takes place (*see* Evershed effect). The umbra is therefore much cooler than the penumbra, which is in turn cooler than the surrounding granulation, their temperatures being of the order of 4000, 5600, and 6000 kelvin respectively. An alternative explanation for the lower temperature of sunspots suggests that far from inhibiting the flow of heat the magnetic field actually enhances it, converting at least three-quarters of the flux into magnetohydrodynamic waves (Alfvén waves) that propagate rapidly along the field lines without dissipation.

Sunspot groups exhibit a great diversity of structure and are most conveniently classified according to the configuration of their magnetic field as unipolar, bipolar, or complex. *Bipolar groups* are by far the most common. In their simplest form they consist of two main spots of opposite magnetic polarity, termed (with respect to the direction of the sun's rotation) the preceding (*p*-) and following (*f*-) spots, of which the *p*-spot usually lies slightly closer to the equator. The *line of inversion* separating the regions of opposite polarity (where the vertical component of the magnetic field is zero) is often the location of a relatively stable *filament and the scene of violent *flares, both of which occur above the sunspot group in the upper *chromosphere/inner *corona.

The number of sunspots fluctuates over an average period of approximately 11 years – the so-called *sunspot cycle – during which time there is a corresponding variation in the mean heliographic latitude of all groups. Moreover the respective polarities of the *p*- and *f*-spots, which are consistent within a particular hemisphere but opposite on the other side of the equator, reverse at sunspot minimum for the groups of the new cycle.

Sunspots are the most obvious manifestation of *solar activity. Together with their associated *faculae/*plages, filaments (or *prominences), and flares, they constitute *active regions, whose influence extends from the photosphere through the chromosphere to the corona.

supercluster. A loosely bound aggregate of several *clusters of galaxies, about 100 megaparsecs in extent. As seen in the sky, different superclusters overlap; their members can be distinguished only after the distances of the constituent galaxies have been determined from a *redshift survey. Over a dozen superclusters have been identified; one is the *Local Supercluster. Superclusters contain between 2 and 15 rich clusters. Although the galaxies within each cluster are tightly bound, the clusters' gravitational effect on each other is not sufficient to overcome the *expansion of the universe. Each supercluster is thus growing in extent, although more slowly than the space between the superclusters. The larger superclusters appear to be elongated, and form parts of filaments that surround empty voids (*see* filamentary structure).

supergiant. The largest and most luminous type of star, lying above both the *main sequence and the *giant region in the *Hertzsprung–Russell diagram. They are grouped in luminosity classes Ia and Ib (*see* spectral types) and generally have absolute magnitudes between −5 and −9. Only the most massive stars can become supergiants and consequently they are very rare. They are so bright, however, that they stand out in external galaxies and can be used as approximate distance indicators. Examples of supergiant stars are Rigel and Betelgeuse in Orion, Antares in Scorpius (which would engulf the orbit of Mars if placed in the solar system), and the brightest *Cepheid variable stars.

supergiant elliptical. *See* galaxies; clusters of galaxies.

supergranulation. A network of large-scale (30 000 km diameter) convective cells in the solar *photosphere. The individual cells have an upward velocity of about 0.1 km s^{-1} and exhibit a horizontal flow of material, outward from the centre, with a velocity of about 0.4 km s^{-1}; their average lifetime is about one day (*compare* granulation). The space between adjacent cells is the preferred birthplace of *sunspots, whose intense localized magnetic fields are dispersed by differential rotation and a cyclonic rotation of the cells, to form a weak polar field in both hemispheres.

Supergranulation is not readily visible but may be detected in velocity-cancelled *spectroheliograms. (These are obtained by the photographic cancellation of two spectroheliograms taken simultaneously in the wings, i.e. on either side of the core, of a strong *Fraunhofer line.) It is seen as alternate bright-dark elements away from the centre of the sun's disc, corresponding to a predominantly horizontal flow of material in the line of sight that has been rendered visible by its doubled *doppler shift. *See also* chromospheric network; spicules.

supergravity. Any of a number of theories seeking to unify the gravitational force with the three other *fundamental forces of nature, the electromagnetic, strong, and weak forces.

superheterodyne receiver. *See* receiver.

superior conjunction. *See* conjunction.

superior planets. The planets Mars, Jupiter, Saturn, Uranus, Neptune, and Pluto, which orbit the sun at distances greater than that of the earth.

superlight source (superluminal source). A *radio source in which components appear to be moving at velocities greater than the speed of light. Physical motion faster than light is impossible according to the special theory of *relativity. It is possible, however, to explain such apparent superlight velocities in terms of geometrical effects involving objects moving more slowly than light.

superluminal source. *See* superlight source.

supermassive black hole. *See* black hole.

supermassive stars. Hypothetical very rare stars with masses of up to 1000 solar masses or more. The most massive of normal stars, such as *Eta Carinae and *Of stars, are about 100 solar masses, but it is not certain whether there is an actual upper limit to *stellar mass. Optical and ultraviolet observations of an object (R136a) at the centre of the 30 *Doradus nebula in the Large Magellanic Cloud have indicated that it is a star maybe 1000 times the sun's mass;

more recent results favour the idea that R136a is a compact cluster of stars. Analyses of certain slow *supernovae have suggested that they result from explosion of supermassive stars. To date, however, the case for these objects is unproven.

supernova. A star that temporarily brightens to an absolute *magnitude brighter than about −15, over a hundred times more luminous than an ordinary *nova. A supernova explosion blows off all or most of the star's material at high velocity, as a result of the final uncontrolled nuclear reactions in the small proportion of stars that reach an unstable state late in their evolution (see stellar evolution). The debris consists of an expanding gas shell (the *supernova remnant), and possibly a compact stellar object (a *neutron star or a *black hole), which is the original star's collapsed core. Supernovae are important in the *nucleosynthesis of heavy elements.

Supernova searches, organized on a regular basis since 1936, have discovered more than 400 extragalactic supernovae. In addition, past supernovae in our *Galaxy have been recorded in Western Europe (*Tycho's star of 1572 and *Kepler's star of 1604) and in China and Korea (AD 185, 386?, 393, 1006, 1054, and 1181). The total supernova production rate in the Galaxy is currently estimated as two or three per century.

On observational grounds supernovae have been divided into five types, although the vast majority fall into Types I and II. *Type I supernovae* occur in both elliptical and spiral galaxies; they all have remarkably similar light curves and absolute magnitudes (−19) at maximum. Their spectra were for long an enigma but are now interpreted as very broad absorption lines. Surprisingly there is little evidence for hydrogen in the ejected gas, which has a mass roughly equal to the sun's and a velocity around 11 000 km s^{-1}: the total kinetic energy is about 10^{44} joules. The cause of Type I supernovae is still uncertain. The pre-supernova may be a *white dwarf that accretes enough mass from a close companion to exceed the *Chandrasekhar limit. The *degenerate matter containing carbon and oxygen nuclei will explode spontaneously at this point, and probably disrupt the star entirely. The ejecta would contain radioactive nickel, whose decay to cobalt and then iron could explain the accurately exponential drop in the supernova's light. On this theory Type I supernovae are an important source of iron in the Galaxy.

Type II supernovae occur only in the spiral arms of spiral galaxies. They show more diversity in light curves and absolute magnitude than Type I, though most reach a maximum of around −17. Their spectra show a normal abundance of the elements and indicate that several solar masses are ejected at around 5000 km s^{-1}. The pre-supernova is almost certainly a supergiant of more than eight solar masses, with a diameter of about 10 AU, and surrounded by a shell of gas lost from the star, about 10 times larger. Its iron core is incapable of generating energy by nuclear fusion and collapses catastrophically to become a *neutron star; if the star is extremely massive it may collapse completely as a *black hole. The link between the core's implosion and the explosion of the outer layers is still uncertain. It is thought most likely that the mantle of gas immediately surrounding the core falls inwards, following the collapsing core; when the collapse stops, the neutron star core suddenly presents a rigid surface to the infalling gases and the mantle bounces back, creating a shock wave that blows off the star's outer layers. The high pressures and temperatures suddenly created in the mantle cause a spate of nuclear reactions; neutrons from these reactions build up very heavy nuclei by the *r-process.

supernova remnant (SNR). The expanding shell of gas from a supernova explosion, consisting of the supernova ejecta and 'swept-up' interstellar gas. Young (<1000 year old) supernova remnants are generally optically faint but are fairly strong radio and x-ray sources; the *Crab nebula is exceptionally bright because it is energized by a central *pulsar. Older supernova remnants appear as rings of bright filaments, again with associated radio and x-ray emission. Compression by an expanding supernova remnant can trigger *stellar birth in interstellar clouds, and possibly initiated the formation of the *solar system. See also emission nebula.

super Schmidt. See Schmidt telescope.

supersynthesis. Another name for earth-rotation synthesis. See aperture synthesis.

surface gravity. Symbol: g. The *acceleration of gravity experienced by an object when it falls freely towards the surface of a planet or star.

surface temperature. The temperature of the radiating layers of a star at which its continuous spectrum is produced. The star is assumed to be a perfect radiator, i.e. a *black body, and its temperature is usually expressed as its *effective temperature or its *colour temperature. Surface temperature is very much lower than the temperature at the centre of a star.

surges. Impulsive events on the sun that are characterized by the ejection of chromospheric material into the inner *corona with velocities of $100-200$ km s^{-1}. They occur in *active regions, where they often accompany *flares but more generally stem from small flare-like brightenings (termed *moustaches* or *Ellerman bombs*) at the outer edge of the penumbra of *sunspots. The surge material, which is usually directed radially away from the spot, may reach a height of a couple of hundred thousand kilometres before fading or returning to the *chromosphere along either a curved trajectory or, more frequently, the trajectory of its ascent. Surges typically have a lifetime of $10-20$ minutes and show a strong tendency to recur. See also prominences.

survey catalogue. See star catalogue.

Surveyor. An American unmanned space programme designed to investigate the bearing strength, physical structure, and chemistry of the lunar *regolith by means of trenching devices and alpha-scattering analysis (see table). The spacecraft were soft-landed by retrorockets. Each was equipped with a television camera, powered by solar cells, for panoramic photography. The *maria were shown to have a basaltic composition, whereas the *highlands were found to be richer in calcium and aluminium. See also Ranger; Lunar Orbiter probes; Luna probes; Zond probes.

Swan nebula. See Omega nebula.

Sword-Handle. See h and Chi Persei.

symbiotic star. A variable 'star' whose spectrum shows lines characteristic of gases at two very different temperatures, typically of an M star (3500 K) and a B star (20 000 K) superimposed. It is in fact a semidetached close *binary star: a *red giant component produces the low-temperature spectrum while the higher-temperature spectrum comes from gas streams that are falling on to a companion star, usually a *white dwarf. The mass loss is due to the giant's *stellar wind, and so is much slower than the gravitational transfer in the otherwise-similar *recurrent novae. Symbiotic stars suffer smaller and more irregular outbursts than other *cataclysmic vari-

Surveyor probes

Spacecraft	Launch date	Comments*
Surveyor 1	1966, May 30	Soft-landed in Oc. Procellarum (11 150)
Surveyor 2	1966, Sept. 20	Struck moon in Oc. Procellarum
Surveyor 3	1967, Apr. 17	Tested soil in Oc. Procellarum (6315); later visited by Apollo 12
Surveyor 4	1967, July 14	Landed in Sinus Medii but radio contact lost
Surveyor 5	1967, Sept. 8	Analysed soil in M. Tranquillitatis (18 006)
Surveyor 6	1967, Nov. 7	Analysed soil in Sinus Medii (30 000)
Surveyor 7	1968, Jan. 7	Analysed/tested soil near Tycho (21 000)

*Figures in brackets indicate the number of photographs transmitted

ables. An outburst in the R Aquarii system has produced a narrow jet some 1500 AU in length, visible to optical and radio telescopes. The gas in the jet is travelling at 2000 km s^{-1}, and is apparently a milder version of the ejection found in *SS433.

synchronous orbit. An orbit made by an artificial satellite that has a period equal to the period of rotation of the orbited body. *See also* geosynchronous orbit.

synchronous rotation (captured rotation). The rotation of a natural satellite about its primary in which the period of rotation of the satellite is equal to its orbital period. The same hemisphere thus always faces the primary. The moon is in synchronous rotation, although *libration allows slightly more than one hemisphere to be seen from earth. There are good dynamical reasons for satellites fairly close to their planet being locked in synchronous rotation. *See* tidal force.

synchrotron emission (synchrotron radiation; magnetobremsstrahlung). *Electromagnetic radiation from very high energy *electrons moving in a magnetic field. It is thus *nonthermal emission. It is the mechanism most often invoked to explain the radio emission from extragalactic *radio sources and the emission from *super-

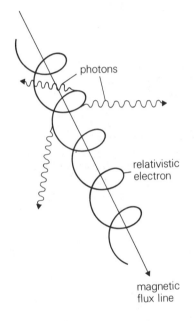

a Synchrotron emission

nova remnants. Synchrotron emission is strongly polarized.

When an electron moves in a magnetic field, it follows a circular path making $v = eB/2\pi m$ revolutions per second, where e and m are the electronic charge and mass and B is the magnetic flux density; v is the *gyrofrequency*. The acceleration of the electron in its circular path causes it to radiate electromagnetic waves at

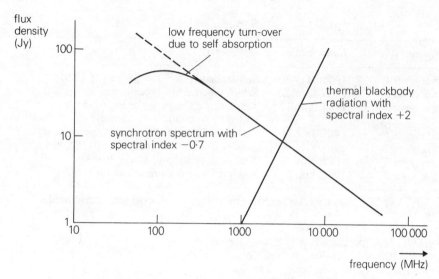

b Spectrum of synchrotron emission

frequency v. This emission is called *cyclotron radiation*.

If the velocity of the electrons becomes comparable with the speed of light, the theory of *relativity must be used to explain the observed phenomena. One effect is *relativistic beaming* in which the radiation is beamed forward in the direction of motion of the relativistic electron (see illustration *a*). The nature of the radiation changes from the simple cyclotron case and it becomes synchrotron emission. At low frequencies, the conditions in the source may cause the synchrotron emission to be reabsorbed before it can escape. The *radio-source spectrum then exhibits a turnover, as is often seen in extragalactic radio sources (see illustration *b*). This is called *synchrotron self-absorption*. See also Razin effect.

syndynames. *See* comet tails.

synodic month (lunation; lunar month). The interval of 29.530 59 days between two successive new moons.

synodic period. The average time between successive *conjunctions of two planets as seen from the earth, or between successive conjunctions of a satellite with the sun as seen from the satellite's primary. Synodic period, P_s, and *sidereal period, P_1, of an inferior or superior planet are related, respectively, by the equations
$$1/P_s = 1/P_1 - 1/P_2$$
$$1/P_s = 1/P_2 - 1/P_1$$
P_2 is the sidereal period of the earth, i.e. 365.256 days or 1 year. For a satellite the first equation applies, P_2 being the sidereal period of the primary.

synthesis telescope. A system of radio telescopes that is used for *aperture synthesis.

Syrtis Major. A dark triangular plateau near the Martian equator and prominent to earth-based observers. It slopes downwards to the east and is prone to strong winds that move and remove light-coloured dust to expose the darker rocks below. *See also* Mars, surface features.

system noise. *See* sensitivity.

system temperature (noise temperature). *See* sensitivity.

syzygy. The configuration arising when the sun, earth, and either the moon or a planet lie approximately in line, i.e. when the moon or planet is at *opposition or *conjunction.

T

TAI. *Abbrev.* (*French*) *for* International-al Atomic Time.

tail. *Short for* comet tail.

tangential velocity (transverse velocity). *See* proper motion.

T-antenna. *See* Mills cross antenna.

Tarantula nebula. *See* 30 Doradus.

T-association. *See* association.

Tau Ceti. *See* Table 7 (nearest stars), backmatter; Ozma project.

Taurids. A minor *meteor shower with a double radiant, RA 56°, dec +14° and +22°, that maximizes on Nov. 8.

Taurus (Bull). A large zodiac constellation in the northern hemisphere near Orion, lying partly in the Milky Way. The brightest stars are the 1st-magnitude *Aldebaran, the 2nd-magnitude *El Nath, which was once part of the constellation Auriga, and several of 3rd magnitude. The area contains the prototypes of the *RV Tauri and *T Tauri stars, the *Hyades and *Pleiades, both beautiful open clusters, and the supernova remnant the *Crab nebula with its associated *optical pulsar. Abbrev.: Tau; genitive form: Tauri; approx. position: RA 4.5h, dec +15°; area: 314 sq deg.

Taurus A. The *radio source in the constellation Taurus that is identified with the *Crab nebula. It has a *flux density at 178 megahertz of 1420 *jansky, most of which comes from *synchrotron emission in the nebula, which is a *supernova remnant. A small fraction of the radio flux is due to a rapidly rotating *pulsar in the nebula, the remains of the original star that exploded. Unlike other supernova remnants, such as *Cassiopeia A, the Crab nebula does not have a regular ring-like radio structure, probably because of its internal energy source – the pulsar.

Taurus-Littrow. A *highland region on the moon on the eastern borders of Mare *Serenitatis. It was the target for *Apollo 17.

Taurus X-1. The second x-ray source to be discovered (1963) and the first to be optically identified. It was found to coincide with the *Crab nebula supernova remnant.

TD-1. A satellite developed by the European Space Research Organization and launched in March 1972. It made a systematic ultraviolet survey of the whole sky, with one experiment on board recording spectrophotometric data at 3.5 nanometre resolution on thousands of stars in the wavelength range 135–274 nm. All hot O and B stars were recorded down to 9th magnitude in addition to observations of many of the brighter cooler stars of later spectral types. A catalogue of more than 30 000 ultraviolet sources has been compiled from the data. A second UV instrument detected UV spectra, at about 0.2 nm resolution, of brighter preselected stars and a catalogue of these data has also been prepared.

TDT. *Abbrev. for* terrestrial dynamical time. *See* dynamical time.

tectonics. Large-scale movements of the crust of a planet, such as those that give rise to mountain building, faulting, and, on the *earth, continental drift.

tektites. Small glassy rounded objects, two to three cm in size, found only in certain parts of the world, often in vast *strewn fields*. About 650 000 had been discovered by 1960. They have shapes varying from spheroidal to pear-, lens-, and disc-shaped with surface-flow structures indicating that they were formed by rapid cooling of molten material. The chemical composition is that of silicic igneous rock with high SiO_2 content (70–80%) and moderate Al_2O_3 content (11–15%). They are named according to the place where they are found, e.g. *moldavites* (Czechoslovakia), *australites* (Australia), etc. They were probably formed by the fusion of terrestrial material during the impact of giant *meteorites; terrestrial and lunar volcanic origins have also been proposed. They are between 0.75–35 million years old. Small tektite-like objects have been found in the lunar regolith.

telemetry. The transmission of data to or from a satellite, space station, spaceprobe, etc., by means of radio waves.

telescope. 1. An optical device for collecting light so as to form an image of a distant object. The light is collected and focused by an object lens in a *refracting telescope or by a primary mirror in a *reflecting telescope. A telescope is usually described in terms of its *aperture: *light-gathering power, angular *resolution, and *limiting magnitude all increase as the aperture increases.
 2. Any device by which radiation from a particular spectral region can be collected and focused on or brought to a suitable recording system for analysis. *See* radio telescope; infrared telescope; grazing incidence; coded-mask telescope.

telescopic object. A star or other celestial object that is less than 6th apparent *magnitude and cannot be seen without a telescope.

Telescopium (Telescope). An inconspicuous constellation in the southern hemisphere near Scorpius, the brightest stars being of 4th magnitude. Abbrev.: Tel; genitive form: Telescopii; approx. position: RA 19.5h, dec −50°; area: 252 sq deg.

television camera. An electronic instrument that converts the optical image of a scene into an electrical signal, termed a *video signal*. This signal may then be transmitted by means of radio waves or by cable to a distant receiver. *CCD cameras and *vidicons are small highly sensitive TV cameras that are used in astronomy both in space probes and satellites and also in combination with ground-based telescopes.

telluric lines. Absorption lines seen in the spectrum of a celestial object but due to components, such as water vapour, oxygen, or carbon dioxide, in the earth's atmosphere. Telluric lines can be distinguished because they are generally narrower than solar or stellar lines, etc. In addition, they vary in strength with the position of the source and are strongest when the source is on the horizon.

Tellus. *Latin name for* the *earth.

Tempel's comets. Two of the comets discovered by Ernst W. L. Tempel were named Tempel 1 (1879 III) and Tempel 2 (1957 II). The first disappeared after 1879 but reappeared in 1967. His most famous comet, 1866 I, named *Tempel-Tuttle*, has a period of 33.176 years and an orbit that is practically identical to that of the *Leonid meteor stream.

temperature. Symbol: T. A measure of the heat energy possessed by a body, i.e. of its degree of hotness. Astronomical temperatures are usually determined from spectroscopic measurements. Since astronomical bodies do not generally have a uniform distribution of heat energy and do not obey exactly the laws determining temperature, there are several different types of temperature; each characterizes a particular property of the body and has a slightly different value. They include *effective, *colour, *brightness, ionization, and excitation temperature. The latter two are determined from the *Saha ionization equation and the *Boltzmann equation. *See also* thermodynamic temperature scale; kelvin.

temperature minimum. *See* photosphere.

Tenma. A Japanese *x-ray astronomy satellite launched in Feb. 1983. Like its predecessor, the satellite *Hakucho, it was designed to study time variability and spectra of the brighter galactic and extragalactic x-ray sources. A feature of the satellite payload is a large area ($\sim$800 cm^2) gas scintillation *proportional counter (SPC). This SPC has an energy resolution a factor of two better than conventional proportional counters in the 1–50 keV band and is providing impressive spectral data, particularly of the iron-line complex and absorption edges seen in many cosmic sources at 6–8 keV.

terminator. The boundary between the dark and sunlit hemispheres of the moon or a planet, the rotation of which gives rise to a *sunrise terminator* and a *sunset terminator*. The appearance of the terminator is affected by topographic irregularities, such as mountains, and by the presence of an atmosphere. *See also* phases, lunar.

terra. An upland area on the surface of a planet or satellite. The word is used in the approved name of such a feature.

terraced walls. *See* craters.

terrestrial. Of, relating to, or similar to the planet earth.

terrestrial dynamical time (TDT). *See* dynamical time.

terrestrial planets. The planets Mercury, Venus, Earth, and Mars. They are small rocky worlds with shallow atmospheres in comparison with the mainly gaseous *giant planets.

tesla. Symbol: T. The SI unit of *magnetic flux density, defined as one weber of magnetic flux per metre squared. One tesla is equal to 10^4 gauss.

Tethys. The largest of the inner satellites of Saturn, having a diameter of 1050 km. It was discovered in 1684. In size it is virtually the twin of the satellite Dione. Its low density of 1.2 g cm^{-3} indicates that it is probably made up mainly of water ice. Tethys has a feature currently unique in the solar system: a huge trench extending around the globe from near the north pole down to the equator and extending further to the south pole; its average width is 100 km and its depth is 4–5 km. There is also an enormous crater 400 km in diameter at 30°N, 130° longitude: its diameter is 40% that of the satellite and greater than the diameter of Mimas. The remainder of the surface is covered in craters of a range of sizes up to 20–50 km, indicating that Tethys has suffered intense bombardment in its history. *See also* shepherd satellites; Table 2, backmatter.

Tharsis Ridge. The principal area of volcanic activity on Mars, measuring 5000 km in diameter and elevated by up to 7 km from the average level of the planet's surface. It includes the four largest Martian volcanoes: *Olympus Mons, *Ascraeus Mons, *Pavonis Mons, and *Arsia Mons. *See also* Mars, volcanoes.

thassaloid. *See* basins, lunar.

thermal black-body radiation. The electromagnetic emission from a perfect radiatior – a *black body – whose brightness follows Planck's radiation law.

thermal emission (thermal radiation; thermal bremsstrahlung). Electromagnetic radiation resulting from interactions between electrons and atoms or molecules in a hot dense medium. The term usually refers to thermal emission from ionized gas (but *see also* thermal black-body radiation). Free electrons in the ionized gas radiate when they are deflected in passing near *protons. The electrons are unbound to a nucleus before and after each encounter and the radiation is therefore also known as *free-free emission*. The radiation has a continuous spectrum, which at low frequencies has the spectral index, α, of a thermal black-body radiator, i.e.
$$S \propto \nu^{-\alpha}; \; \alpha = -2$$
where S is the flux density, and ν the frequency. At higher frequencies the spectrum flattens to a spectral index of $\alpha = 0$. The *emission measure* of an ionized region of length l containing a density n_e of free electrons is given by
$$EM = \int_0^l n_e^2 ds$$
where ds is an element of the path through the region. *Compare* nonthermal emission.

thermodynamic equilibrium. A condition existing in a system when all the atoms and molecules have an equal share in the available heat energy. The temperature is then the same in all parts of the system by whatever method of measurement. A system that is not in thermodynamic equilibrium is unstable and the state of the system will change until equilibrium is reached.

thermodynamic temperature scale (absolute temperature scale). A temperature scale in which the temperature, T, is a function of the energy possessed by matter. Thermodynamic temperature is therefore a physical quantity that can be expressed in units, termed *kelvin. The zero of the scale is *absolute zero. The temperature of the ice point (the zero of the Celsius temperature scale) is 273.15 kelvin. Thermodynamic temperature can be converted to Celsius temperature by subtracting 273.15 from the thermodynamic temperature.

thermoelectric generator. A device that converts heat energy directly into electrical energy. It consists basically of two dissimilar metals or semiconductors joined at two junctions: if the junctions are at different temperatures an electromotive force (e.m.f.) develops between the junctions. An e.m.f. can also be generated between two regions on one piece of metal when a temperature difference exists between the regions. In both cases the e.m.f. is proportional to the temperature difference and is also dependent on the materials used. The e.m.f. can be used to power an external electric circuit. Thermoelectric generators are used in long-range planetary probes when *solar cells are unable to produce sufficient power. The hot junction or region of the device is then often heated by means of the decay of radioisotopes.

thermonuclear reaction. *See* nuclear fusion.

thermosphere. *See* atmospheric layers.

third quarter. *Another name for* last quarter. *See* phases, lunar.

tholus. A hill or dome on the surface of a planet or satellite. The word is used in the approved name of such a feature.

three-body problem. A specific case of the *n-body problem in which the trajectories of three mutually interacting bodies are considered. There is no general solution for the problem although solutions exist for a few special instances. Thus the orbits can be determined if one of the bodies has negligible mass, as in the case of a planetary satellite, such as the moon, subject to perturbations by the sun or an asteroid whose motion is perturbed by Jupiter.

three kiloparsec arm. *See* galactic centre.

thrust. *See* launch vehicle.

Thuban (α Dra). A white giant in the constellation Draco that was the *pole star about 2700 BC when it lay 10′ from the celestial pole. Its brightness has probably decreased in the past few hundred years. It is a spectroscopic binary, period 51.38 days. m_v: 3.64; spectral type: A0 III; distance: 92 pc.

tidal bulge. *See* tides.

tidal force. A force arising in a system of one or more bodies as a result of differential gravitation: different parts of the system experience different accelerations. This can result in the production of *tides and in general terms elongates a body in the direction of a nearby massive body. The force can alter a body's rotation rate until it is equal to the revolution period: this is true of most natural satellites of the planets, including the moon, and is the case in some close binary stars, e.g. *RS CVn stars. In extreme cases tidal forces can lead to the disruption of a body.

tidal friction. Energy dissipated by the raising of *tides. As with tidal heights, the total energy dissipated on the earth by tidal friction depends on the topography of coastlines and on the areal extent of adjacent continental shelves. Tidal friction is currently slowing down the earth's rotation rate (i.e. the length of the day) by 16 seconds every million years and is causing the moon to recede from the earth by about 5 cm every year at present. *See also* secular acceleration.

tidal height. *See* tides.

tidal lag. *See* tides.

tidal theories. *See* encounter theories.

tides. Distortions of a planet, star, etc., produced by the differential gravitational attraction of other astronomical bodies on parts of that planet, star, etc. The sun and moon combine to generate two *tidal bulges* in the earth's oceans, one directed towards the moon and the other diametrically opposite. If the earth is assumed to be spherical and to be covered with water, then the gravitational pull of the moon will have a different force at different points on the earth's surface (*see* illustration *a* (i)). The gravitational acceleration at the earth's centre is equal to the centripetal acceleration of the earth-moon system. If this acceleration (which is a vector quantity) is subtracted from the surface accelerations, the differential (tidal) acceleration is found (illustration *a*(ii)), showing the two tidal bulges on the earth-moon line.

The daily rotation of the earth and the slower eastward revolution of the moon in its orbit produce (usually) two *high tides* and two *low tides* every 24 hours 50 minutes. The rotation and lunar revolution are also responsible for a *tidal lag* between the time when the moon crosses the local *meridian and the time of high tide.

The tidal pull of the sun is less than half that of the moon but reinforces it at full moon and at new

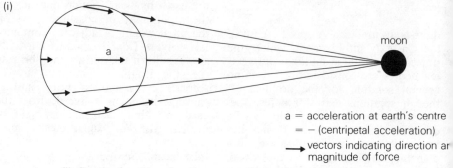

a (i) Gravitational acceleration on earth due to moon

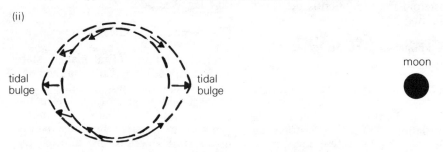

a (ii) Differential (tidal) acceleration found by subtracting *a* from other vectors on Fig. (i)

moon to produce very high, or *spring tides;* these tides are exceptionally high when the earth is close to *perihelion and the moon is close to *perigee (*see* illustration *b*). The lowest high tides, or *neap tides,* occur when the moon is at *quadrature. The *tidal height* depends on the topography of the coastline and on the area of the adjacent continental shelf. Tides are also generated in the earth's atmosphere and in the solid earth.

Lunar tides are raised by the earth as a consequence of the moon's eccentric orbit. They trigger *moonquakes and *transient lunar phenomena. *See also* tidal friction.

time. The continuous and nonreversible passage of existence, or some interval in this continuum. It is also a quantity measuring the duration of such an interval or of some process, etc. The unit of time in the SI system is the *second. The astronomical unit of time is often the interval of one day of 86 400 seconds. The time systems used in astronomy are *International Atomic Time, *universal time, *sidereal time, and *dynamical time (which has replaced *ephemeris time). The divergence of International Atomic Time and universal time is minimized by the intermediary of coordinated *universal time, which is the basis for civil timekeeping and for recording observations. *See also* spacetime.

time constant. Symbol: τ. A measure of the speed with which an electronic system can respond to a sudden change at its input. In particular, the time constant of a simple resistance-capacitance circuit is the product of the resistance and the capacitance.

time dilation. *See* relativity, special theory.

time zone. *See* standard time.

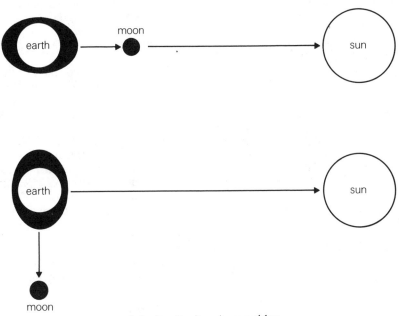

b Spring (top) and neap tides

Titan. The largest satellite of Saturn, discovered in 1655. It orbits the planet at a mean distance of 1 222 000 km in a period of 15.95 days. With a diameter of 5150 km it is the second largest satellite in the solar system, after Jupiter's Ganymede. The satellite appears reddish or orange in colour but the two hemispheres are not identical. The atmosphere is made up almost entirely of nitrogen, with traces of methane, ethane, acetylene, propane, diacetylene, methylacetylene, hydrogen cyanide, cyanoacetylene, cyanogen, carbon dioxide, and carbon monoxide. The surface pressure is 1.6 times that of the earth and the surface temperature is 92 K, so that it is warmed by a small *greenhouse effect. The earth and Titan are currently the only bodies known to have nitrogen atmospheres. The colouration of the atmosphere is due to photochemical reactions and the production of fogs and aerosol particles, which probably extend nearly to the surface. Methane on Titan plays a similar role in the satellite's atmosphere as water on earth, occurring in the three states of solid, liquid, and gas. The atoms and molecules of hydrogen that are produced by the photochemical reactions can easily escape from Titan and form a massive doughnut-shaped torus in the orbit of Titan. The material diffuses inwards but is prevented from extending much beyond the orbit of Titan through the location of the boundary between the Saturn magnetosphere and the solar wind. The density of Titan is approximately 1.88 $g\,cm^{-3}$, nearly twice that of water. This suggests that the interior is composed roughly of half rock and half water ice. It has a surface that could be either liquid or solid in part, with oceans of methane. With no evidence of an intrinsic magnetic field, there would not seem to be an electrically conducting core. *See also* Table 2, backmatter.

Titania. The largest satellite of Uranus, discovered in 1787. *See* Table 2, backmatter.

Titius—Bode law. *See* Bode's law.

tonne (metric ton). A unit of mass equal to 1000 kilograms and close to the ton (1016 kg, 2240 lb).

topocentric coordinates. The coordinates of a celestial body measured from the surface of the earth. For a star there is little or no detectable difference between topocentric coordinates and those referring to the earth's centre in a *geocentric coordinate system. For a member of the solar system the difference can be significant and a correction must be made.

topocentric zenith distance. *See* zenith distance.

total eclipse. *See* eclipse.

totality. *See* eclipse.

total magnitude (integrated magnitude). A measure of the brightness of an extended object, such as a galaxy or nebula, obtained by summing the surface brightness over the entire image area or by measuring the amount of light received through a series of apertures of increasing diameter with a photoelectric photometer. The readings are found gradually to approach an upper limiting magnitude, which, corrected for night-sky brightness, gives the total magnitude.

total-power receiver. *See* receiver.

tower telescope (solar tower). *See* solar telescope.

train. 1. The faint glow sometimes left by a brighter *meteor along its path. It is much less conspicuous than the visible radiation that comes from the immediate vicinity of the ablating meteoroid. It is known as a *wake* if it glows for less than a second and a *persistent train* if the duration is longer. Sudden and brief enhancements of light during the meteor's passage are termed *flares* (or *bursts*).
2. The ionization and dust left along the meteoroid path.

trajectory. The curved path followed by a moving spacecraft, rocket, etc., under the forces of gravity, lift, and/ or drag.

Tranquillitatis, Mare (Sea of Tranquillity). An ancient irregular *mare on the eastern *nearside of the moon. It was the landing area for *Ranger 8, *Surveyor 5, and Apollo 11 (*Tranquillity Base* was the first manned landing point). The basaltic lavas in the mare are 3900 to 3600 million years old.

transfer orbit. The trajectory followed by a spacecraft moving from one orbit to another, usually to the orbit of another body, such as another planet. The trajectory is generally part of an elongated ellipse – a *transfer ellipse* – that intersects the new orbit. The spacecraft would have to manoeuvre into an orbit around the planet by firing its rocket motors.

The transfer orbit requiring the minimum expenditure of energy is an ellipse that just touches the original orbit and the new orbit. This is called a *Hohmann transfer,* after the German engineer who described it in 1925. Enough velocity is put in at the *perigee (or equivalent point) of the Hohmann transfer for the craft to reach the new orbit at *apogee; at apogee an additional velocity input injects the craft into the desired orbit. Although requiring the lowest possible energy, Hohmann transfers between planets involve a long flight time. The flight time can be much reduced by using a somewhat greater velocity at perigee than that needed for a Hohmann transfer.

transient lunar phenomena (TLP). Short-lived changes in the appearance of lunar features, including reddish glows and obscurations of surface detail, that are triggered by lunar *tides. They often occur within lunar *craters and around the perimeters of *basins. A carbon spectrum was obtained from a TLP in the crater Alphonsus.

Transit dates

Mercury		Venus	
November	*May*	*June*	*December*
1960	1970	1761 ⎫	1874 ⎫
1973	2003	1769 ⎭	1882 ⎭
1986	2016		
1993		2004 ⎫	2117 ⎫
2006		2012 ⎭	2125 ⎭

transit. 1. The passage of an inferior planet across the sun's disc. Normally when Venus or Mercury move between the earth and sun, at inferior *conjunction, they pass above or below the sun. A transit can only occur when the planet is at or near one of its *nodes at inferior conjunction. Transits of Mercury take place either about May 7 (descending node) or Nov. 9 (ascending node) (see table). Transits occur at intervals of 7 or 13 years or in combinations of these figures, e.g. after 33 or 46 years. The much rarer transits of Venus take place either about June 7 (descending node) or Dec. 8 (ascending node). They occur in pairs with a separation of 8 years, the pairs being separated by over 100 years.

2. The passage of a planetary satellite or its shadow across the face of the planet.

3. (upper culmination). The passage of a celestial body, during its daily path, across an observer's meridian through the point closer to the observer's zenith. *See also* culmination.

transit circle. An optical telescope mounted in such a way that it can swing only in a vertical north-south plane. It is used to measure the positions of celestial bodies with great accuracy as they cross the *meridian of the telescope site (*see* culmination). The *altitude of a body as it crosses the meridian is now usually measured very precisely by electronic devices; earlier instruments used an accurately graded circular scale set in the vertical plane. The *declination of the body can then be determined using the latitude of the observatory. Measurement of the precise *local sidereal time at the meridian passage determines the *right ascension of the body.

transition region. *See* sun. *See also* chromosphere.

transit telescope. Any telescope that can swing in only one direction, up and down the north-south line in the sky, i.e. the observer's meridian. In optical astronomy a *transit circle can be used to determine the positions of celestial bodies. In radio astronomy a transit telescope can be used to build up two-dimensional maps of radio sources: the earth's daily rotation causes a radio source to move through the beam of the stationary telescope's *antenna, providing information in the east-west direction, the direction in which the telescope points being altered slightly each day.

transmission grating. *See* diffraction grating.

transmission line. *See* feeder.

transverse velocity. *Another name for* tangential velocity. *See* proper motion.

Trapezium. A cluster of four young stars that lies in the *Orion nebula and is part of the *Orion association. They are main-sequence stars, of spectral class O6, B1, B1, and B3, and they excite and ionize the nebula. They form a trapezium and are visible with a small telescope.

Triangulum (Triangle). A small constellation in the northern hemisphere near Perseus, the brightest stars being of 2nd magnitude. It contains the *Triangulum spiral. Abbrev.: Tri; genitive form: Trianguli; approx. position: RA 2h, dec $+30°$; area: 132 sq deg.

Triangulum Australe (Southern Triangle). A small constellation in the southern hemisphere near Crux, lying partly in the Milky Way. The brightest stars are the 2nd-magnitude orange giant Alpha Trianguli Australis and 3rd-magnitude Beta and Gamma, which form a triangle. The area contains the bright open cluster NGC 6025. Abbrev.: TrA; genitive form: Trianguli Australis; approx. position: RA 16h, dec $-65°$; area: 110 sq deg.

Triangulum spiral (M33; NGC 598). A spiral (Sc) galaxy in the constellation Triangulum that is the third largest of the confirmed members of the *Local Group and is about 720 kiloparsecs distant. Although it is both smaller and more distant than the *Andromeda galaxy it can appear larger because it is orientated face-on to us. The spiral arms are more open than those in our own Galaxy and the nucleus is relatively small and faint. It has a large number of young *population I stars.

Trifid nebula (M20; NGC 6514). An *emission nebula about 2000 parsecs away in the direction of Sagittarius. There is dark matter associated with the nebula.

trigonometric parallax. *See* annual parallax.

triple alpha process (3α process). A nuclear reaction occurring in an evolved star's interior when all the hydrogen has been used up and the core temperature has reached about 10^8 kelvin. Three helium nuclei (alpha particles) fuse together to form carbon by the reaction
$$3\,^4He \rightarrow \,^{12}C + \gamma$$
This reaction gives out only 10% as much energy per unit mass as hydrogen burning.

triple star. A *multiple star with three components. *See also* binary star.

Triton. The larger of the two satellites of Neptune. Discovered in 1846, it has a diameter of about 3800 km, a little larger than the moon. It is the only large satellite to follow a retrograde orbit (*see* direct motion) around its planet. Spectroscopic measurements have detected methane and nitrogen on Triton. It is possible that the satellite may possess an atmosphere of nitrogen with methane ice on its icy surface. Triton's reddish colour may be due to photochemical reactions with the atmospheric constituents. It is going to be an interesting object to observe from *Voyager 2 in 1989. *See also* Table 2, backmatter.

Trojan group. Either of two groups of *minor planets that lie in the vicinity of one of the two *Lagrangian points in the orbit of Jupiter around the sun: each group occupies one corner of an equilateral triangle with the sun and Jupiter at the other corners. Beginning with the first Trojan to be discovered, *Achilles in 1906, most have been named after heroes from the two sides of the Trojan Wars. In fact, perturbations by other planets mean that each member of the group oscillates considerably in position (mainly in longitude along the orbit of Jupiter) from its stable Lagrangian point. It is likely that perturbations and collisions between them cause the total number of members of the group to change with time. Recent surveys suggest that more than 1000 Trojan group members may be bright enough to be seen from the earth but less than 2% of these have yet received official minor planet designations. They are all exceptionally dark bodies.

tropical year. The interval between two successive passages of the sun through the vernal *equinox, i.e. the time taken by the sun, in its apparent annual motion, to complete one revolution around the celestial sphere relative to the vernal equinox. It is equal to 365.242 19 days. If the vernal equinox were a fixed point on the celestial sphere the tropical year would be equal to the *sidereal year. *Precession however produces an annual retrograde motion of the equinoxes amounting to 50.27 seconds of arc relative to the fixed stars. The tropical year is thus about 20 minutes shorter than the sidereal year. Since the seasons recur after one tropical year, the average length of a *calendar year should be as close as is practicable to the tropical year.

The tropical year of 1900 was the primary unit of *ephemeris time. Use of ephemeris time, and hence of the tropical year, was discontinued in 1984.

troposphere, tropopause. See atmospheric layers.

true anomaly. See anomaly.

true equator and equinox. The coordinate system determined by instantaneous positions of *celestial equator and *ecliptic. This reference system slowly changes as a result of the progressive effects of *precession and the short-term variations of *nutation. See also mean equator and equinox.

Trumpler's classification. A method of classifying *open clusters according to their appearance, using the three criteria of stellar concentration towards the centre, range in brightness of the member stars, and the total number of stars in the cluster. The method was devised by the Swiss-American astronomer Robert Trumpler who also found a correlation between this appearance classification and the cluster diameter.

T Tauri stars. *Irregular variable stars of late spectral type whose spectra is dominated by strong emission lines, usually attributed to chromospheric activity and stellar winds in these stars. They are invariably embedded in dense patches of gas and dust that may require observations in the infrared: the dust absorbs the visible light of the star and reradiates it at longer wavelengths. The prototype is T Tauri. T Tauri stars are usually found together in groups (T associations) and are the youngest optically observable stages in the life of a star of about the sun's mass; more massive counterparts are observed as *Ae and *Be stars. They are frequently associated with *Herbig–Haro objects.

There is much evidence that T Tauri stars are young objects, for instance they have a high abundance of lithium, an element destroyed fairly early in a star's life, and they are surrounded by gas and dust. They are thought to be young *protostars that have only recently contracted out of the interstellar medium (see stellar birth). Lying above the main sequence in the *Hertzsprung-Russell diagram, they are still contracting and losing mass (see T Tauri wind). Their spectral lines reveal that some are extremely rapid rotators, throwing off material at speeds of up to 300 km s^{-1}. The irregular light variations are believed to arise partly from activity in the chromospheres of these young stars, and partly by the obscuring effect of the patchy dust in the cocoon as it moves in front of the star.

T Tauri wind. A continuous flow of material from the surface of a *T Tauri star. Qualitatively this flux resembles the *solar wind but the amount of material lost is far greater, approaching 10^{-7} solar masses per year in some cases. It is possible that this outflow acts as a braking mechanism for fast-spinning young stars. In some T Tauri stars this wind is collimated into a *bipolar outflow.

Tucana (Toucan). A constellation in the southern hemisphere near Grus, the brightest stars being of 3rd magnitude. It contains the Small *Magellanic Cloud and the globular clusters 47 Tucanae, easily visible, and NGC 362, just visible. Abbrev.: Tuc; genitive form: Tucanae; approx. position: RA 0h, dec −65°; area: 295 sq deg.

Tully–Fisher method. The most accurate method for measuring the distances to distant spiral galaxies; it uses a relation between a galaxy's absolute magnitude and its rotation velocity that was discovered by R. B. Tully and J. R. Fisher in 1977. Radio observations at 21 cm wavelength give the rotation velocity; the Tully–Fisher relation then indicates the absolute magnitude, and a comparison with the apparent magnitude gives the distance. The Tully–Fisher relation must be calibrated by observations of nearby spirals, whose distances are known from *Cepheid variables, etc. (see distance determination).

Tunguska event. A gigantic explosion that occurred at about 7.17 a.m. on June 30, 1908 in the basin of the River Podkamennaya Tunguska, Central Siberia. Devastation rained over an area 80 km in diameter and eye witnesses up to 500 km away saw in a cloudless sky the flight and explosion of a blindingly bright pale blue bolide. The sound of the explosion reverberated thousands of kilometres away, the explosion air wave recorded on microbarographs going twice round the world. The main explosion had an energy of 5×10^{16} joules and occurred at an altitude of 8.5 km.

Many theories have been put forward to explain the explosion, the less conventional involving black holes, critical masses of extraterrestrial fissionable material, antimatter bodies, and alien spacecraft. In fact it was probably caused by the impact of a small comet or a very fragile *Apollo asteroid. When the object encountered the earth it would have been coming from a point in the dawn sky comparatively close to the sun and would thus have been most difficult to detect and observe.

turnoff point. The point on the *Hertzsprung–Russell diagram of a *globular or *open cluster at which stars are leaving the *main sequence and entering a more evolved phase. Assuming that all the stars in a cluster contract out of a single gas cloud at approximately the same time, the locations of these stars on the H–R diagram depend on the time elapsed since the initial contraction. The length of time spent on the main sequence depends on stellar mass: stars of about solar mass will remain there for maybe 10^{10} years while more massive, and hence more rapidly evolving stars will turn off earlier. Since position on the main sequence also depends on mass, the lower the turnoff point, the older the cluster. The turnoff point of a cluster is the most accurate means of determining its age, and hence provides astronomers with a sample of stars of known age.

twenty-one cm line (hydrogen line). See HI region.

twilight. The time preceding *sunrise and following sunset when the sky is partly illuminated. *Civil twilight* is the interval when the true *zenith distance (referred to the earth's centre) of the centre of the sun's disc is between 90°50′ and 96°, *nautical twilight* is the interval between 96° and 102°, and *astronomical twilight* is that between 102° and 108°.

twinkling. See scintillation.

twin quasar. *See* gravitational lens.

two-body problem. A special case of the *n-body problem in which a general solution can be found for the orbits of two bodies under the influence of their mutual gravitational attraction. The motion of a planet around the sun is an example as long as the attractive forces of other planets are assumed to be negligible. A solution of the two-body problem is often acceptably realistic. Most theories of celestial motion thus use functions and principles, such as *orbital elements and *Kepler's laws, that have been derived by consideration of a two-body problem.

two-colour diagram. A plot of the *colour indices $B-V$ against $U-B$. Stars of different luminosity classes (*see* spectral types) lie on slightly different curves.

Tycho. A very recent lunar *crater, possibly only about 100 million years old; it is 100 km in diameter. *Rays from the crater traverse much of the nearside of the moon and are particularly striking at full moon. The *ejecta blanket was sampled by *Surveyor 7, the only Surveyor to land in the *highlands; some craters in the highland region of *Taurus-Littrow may be *secondaries of Tycho.

Tychonic system. A model of the solar system proposed by Tycho Brahe in the 1580s and based on observations he made in order to test the *Copernican system. He concluded that the sun and moon revolved around the earth, which he maintained lay at the centre of the solar system, but suggested that the other planets orbited the sun.

Tycho's star. A type I *supernova that was observed in 1572 in the constellation Cassiopeia and was studied by astronomers in Europe, China, and Korea. Tycho Brahe's careful observa-

tions of its position have identified it with a known radio and x-ray source, which is the expanding supernova remnant. It was considerably brighter than Kepler's star of 1604, reaching a magnitude of -4 and remaining visible for about 18 months.

Type I, II Cepheids. *See* Cepheid variables.

Type I, II Seyferts. *See* Seyfert galaxy.

Type I, II supernovae. *See* supernova.

U

U–B. *See* colour index.

UBV, UBVRI systems. *See* magnitude.

U Geminorum stars. *See* dwarf novae.

Uhuru. ('Freedom' in Swahili). The first artificial earth satellite for *x-ray astronomy (*SAS–1), launched into a 500 km circular equatorial orbit in December 1970 from the San Marco platform off the coast of Kenya. Two sets of proportional gas counters, each 840 cm^2 in area and mounted on opposite sides of the slowly spinning spacecraft, scanned the sky for x-ray sources. Uhuru provided the first detailed x-ray sky map, leading to the *3U catalogue* of 161 sources, published in 1974. Its most important discoveries were of *x-ray binaries (Cen X-3, Her X-1, etc.) and of extended regions of x-ray emission associated with *clusters of galaxies. Uhuru operated successfully until July 1971, when its transmitter failed. The 3U catalogue is based on the data to that time. Subsequently the star sensors, used to determine the precise attitude of the spacecraft and hence the sky coordinates of each x-ray source, also degraded; surveying recommenced in Dec. 1971, however, after partial transmitter recovery. These later observations, with the aid of a magnetometer attitude solution, continued until Apr. 1973, when the Uhuru battery failed. All the Uhuru observations were finally put together to form the *4U catalogue,* issued in 1978 and containing 339 x-ray sources.

UK Infrared Telescope (UKIRT).

The British *infrared telescope sited on *Mauna Kea, Hawaii, at the very high altitude of 4200 metres. The site is well above the main weather layers and the bulk of atmospheric water vapour, opaque to most infrared wavelengths; in addition there is excellent *seeing for work at visible wavelengths. The telescope is funded by the Science and Engineering Research Council and is run by the Royal Observatory, Edinburgh. It became operational in 1979. The mirror is 3.8 metres (150 inches) in diameter, making the instrument the world's largest infrared telescope. It is unusually thin (29 cm) and weighs only 6.6 tonnes, allowing an extremely lightweight supporting structure to be used. The mirror surface has been ground sufficiently accurately that it can also be used at visible wavelengths for short exposures.

A full range of cryogenic (very low temperature) facilities is provided for cooling the detectors in order to reduce spurious signals generated by the warmth of the detectors. There are also means for removing spurious signals generated by the telescope itself. The instruments mounted at UKIRT's Cassegrain focus include infrared spectrometers and photometers; the latter work at various wavebands – optical, infrared (1–5 μm, 4–35 μm), and submillimetre (0.3–1 mm). Heavier instruments are mounted at the coudé focus.

UK Schmidt Telescope (UKST).

The British 1.2 metre *Schmidt telescope at the Anglo-Australian Observatory, Australia. It is funded by the Science and Engineering Research Council and is operated as a field station of the Royal Observatory, Edinburgh. It began regular observation in 1973. Originally planned to work in conjunction with the *Anglo-Australian Telescope on the same site, it has proved a major research tool in its own right. Its highly rated performance stems from recent improvements in telescope technology and photographic emulsions and also from experience gained from the use of the 48 inch Schmidt at the Palomar Observatory. It has been involved primarily in the production of photographic plates for the ESO/SERC *Southern Sky Survey. It can reach limiting magnitudes in exposures of about an hour as a result of its low focal ratio of f/2.5. Each photograph covers a very wide area of the sky, amounting to 40 square degrees, and gives images of arc-second resolution over a wavelength range of 340–1000 nanometres. *See also* COSMOS.

ultraviolet astronomy.

Astronomical observations in the waveband between the Lyman limit of atomic hydrogen at 91.2 nm and the earth's atmospheric absorption cut-off at about 320 nm (*see* hydrogen spectrum; atmospheric windows). The waveband between the Lyman limit and the x-ray band starting at about 12 nm is called the extreme ultraviolet (EUV or XUV – *see* XUV astronomy). Strong atomic and molecular absorption in the earth's atmosphere requires UV observations to be carried out at balloon altitudes (about 30 km) for the 250–320 nm waveband and on rockets or satellites, at higher altitudes, for wavelengths below 250 nm. The UV band includes many of the resonant absorption and emission lines of most of the more abundant chemical elements, including hydrogen in both its atomic and molecular form. Such *resonance transitions (to and from atomic ground states) are very important for the determination of the physical and chemical properties of astronomical sources, such as the interstellar gas, the outer atmospheres of stars, and the gaseous regions of active galaxies, in which the excitation is such that the electrons in atoms or molecules are mainly in the ground state.

Most ultraviolet astronomy has been carried out using satellite instrumentation either to survey the sky at low spatial and spectral resolution or to observe individual preselected sources at high resolution. The *OAO–2 satellite, launched in 1968, carried a telescope that made a partial sky survey at 100–300 nm, while the *TD–1 satellite carried an experiment that surveyed the whole sky between 135 and 274 nm, observing over 30 000 *ultraviolet stars. OAO–3, renamed *Copernicus* after launch in 1972, carried a high-resolution reflecting telescope and grating spectrometer that obtained spectra of many stars and interstellar regions in the wavelength range 90–300 nm over the following 8 years. Copernicus was restricted to observations of hot stars brighter than about 6th magnitude and concentrated on studies of the interstellar gas and the *stellar winds of hot luminous stars. The *Astronomical Netherlands Satellite (ANS), launched in 1974, provided UV photometric observations of many thousands of stars in the waveband 155–320 nm.

In Jan. 1978 the *International Ultraviolet Explorer (IUE) satellite was launched into a geosynchronous orbit over the mid-Atlantic. This satellite is operated like a ground-based observatory with astronomers carrying out observations at two stations in the USA and in Spain. IUE carries a 45 cm aperture ultraviolet telescope feeding two *echelle grating spectrographs, which cover the waveband 115–320 nm at about 0.01 nm resolution. IUE is much more sensitive than any previous UV satellite and has allowed observations of stars and active galaxies as faint as 17th magnitude. It has provided important data on a wide variety of astronomical fields: there have been studies of the solar system, of stars and the interstellar medium in our own Galaxy and the Magellanic Clouds, and of external galaxies, Seyfert galaxies, and quasars.

It has been found that an extensive hot gaseous halo surrounds the Galaxy and the Magellanic Clouds, and that the sun is located in a warm low-density 'hole' in the interstellar medium. Extensive *mass loss from hot stars, first studied in high-luminosity objects with Copernicus, has been discovered to exist also in low-luminosity hot stars such as O and B subdwarfs and the central stars of planetary nebulae; in many cases the mass loss rates are too high to be accounted for by current theory. Many stars have been found to show considerable variations in their mass loss. The outer chromospheres and coronae of the cool stars have been observed in the UV with IUE for the first time, allowing the densities and temperature distributions in these atmospheric regions to be mapped and shedding light on their energy balance. Because of the observatory nature of IUE it has been possible to respond quickly to the occurrence of transient events and the UV spectra of several novae and supernovae have been obtained. It is anticipated that IUE will have an operational lifetime of at least 10 years. NASA's *Hubble Space Telescope, due for launch in 1987, will carry several cameras and spectrographs for UV observations in the waveband 115–320 nm and will be several orders of magnitude more sensitive than IUE.

ultraviolet radiation. *Electromagnetic radiation lying in the wavelength range beyond the earth's atmospheric absorption at about 320 nm to the hydrogen Lyman limit at 91.2 nm (*see* hydrogen spectrum). The gap between the x-ray and UV wavebands is the XUV region. Long-wavelength ultraviolet radiation, i.e. with wavelengths up to 350 nm, near that of *light, is often called *near ultraviolet* with *far ultraviolet* being applied to short wavelengths.

ultraviolet stars. Hot stars that have considerable emission at ultraviolet and XUV wavelengths, i.e. between about 12–320 nm; these include early-type *O and *B stars, *subdwarf O and B stars, hot *white dwarfs, and the central stars of most *planetary nebulae. For an O star with a surface temperature of about 30 000 K, over 80% of its radiation lies in the ultraviolet. In comparison the sun, with a surface temperature of about 5800 K, emits most of its radiation in the visible region of the spectrum (380–750 nm). Space observations have allowed the UV energy distribution of such stars to be measured, providing a more precise determination of their total radiated power. In addition emission and absorption line spectra have given information on the structure and element abundances in the stellar atmospheres, and have shown for both luminous and subluminous hot stars the presence of strong *stellar winds. In many cases such winds involve a rate of mass loss that will substantially affect the evolution of the star. Extensive regions of ionized interstellar gas – the *HII regions – surround many such hot luminous stars. *See also* ultraviolet astronomy.

umbra. 1. The dark inner region of a shadow cast by an object illuminated by a light of finite size rather than a pinpoint of light. The less dark outer region of the shadow is the *penumbra* (see illustration). The source of light is totally obscured to someone in the umbral region but only partly obscured in the penumbral region. *See* eclipse.

2. *See* sunspots.

Umbriel. A satellite of Uranus, discovered in 1851. *See* Table 2, backmatter.

unit distance. The astronomical unit of length, i.e. the length for which the *Gaussian gravitational constant takes its exact value when the units of measurement are the astronomical units of length, mass, and time. It is equal to the *speed of light, c, times the *light-time for unit distance, τ_A, i.e. to $1.495\ 978\ 70 \times 10^{11}$ metres

United States Naval Observatory. The observatory in Washington D.C. concerned mainly with astrometric measurments for purposes of timekeeping and navigation: the observatory broadcast the world's first radio time signals, in 1904. The original observatory was built in 1833. The site has since moved and expanded several times.

universal time (UT). The precise measure of time that forms the basis for all civil timekeeping. It conforms closely to the mean diurnal motion of the sun.

UT is defined by a mathematical formula relating UT to *Greenwich mean sidereal time. It is thus determined from observations of the diurnal motions of stars. It is not a uniform timescale, however, due to variations in the earth's rotation. The timescale determined directly from stellar observations is dependent on the place of observation, and is designated UT0. The timescale independent of location is obtained by correcting UT0 for the variation in the observer's meridian that arises from the irregular varying motion of the earth's geographical poles; it is designated UT1.

Coordinated universal time (UTC) is based on *International Atomic Time

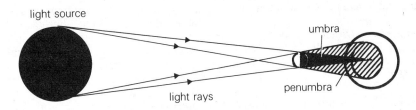

light source

umbra

penumbra

light rays

Umbral and penumbral regions of a shadow

(TAI) and was introduced in order to minimize the divergence of universal time from the uniform timescale of atomic time. Since Jan. 1 1972 the timescale distributed by most broadcast time services has been UTC. UTC differs from TAI by an integral number of seconds. UTC is kept within 0.90 seconds of UT1 by the insertion (or deletion) of exactly one second when necessary, usually at the end of December or June; these step adjustments are termed *leap seconds*. An approximation (DUT1) to the value of UT1 minus UTC is transmitted in code on broadcast time signals. *See also* dynamical time.

universe. The sum total of potentially knowable objects. The study of the universe on a grand scale is called *cosmology. Barriers between potentially knowable and unknowable objects exist in even a simple expanding Euclidian universe with flat space (i.e. in which the surface area of a sphere is $4\pi r^2$). More and more distant objects are seen to recede at faster and faster velocities. At a certain distance objects are receding at the speed of light and suffer an infinite *redshift; they are then potentially unobservable.

Within the framework of general *relativity the space of the universe can be curved, and depending on the nature of the curvature the universe may be *closed* or *open*. In a closed universe gravitational distortion of light would cause space to be spherically curved: space would curve back on itself to form a finite volume with

no boundary. The universe would then be finite and the gravitational attraction among galaxies would be sufficient eventually to halt and reverse the expansion of the universe so that it would begin to contract. In an open universe space would still be curved but would not turn back on itself: the curvature would be hyperbolic rather than spherical. The universe would then be infinite, with gravity too weak to halt the universe's expansion. In a *flat universe* space is not curved: it can be described by Euclidean geometry. The universe would then be infinite and would expand forever, but less rapidly than an open universe.

The dynamic behaviour of the universe, i.e. whether it is closed, open, or flat, depends on the value of the *mean density of matter, and hence on the *deceleration parameter q_0:

if $q_0 > \frac{1}{2}$, universe is closed

if $q_0 < \frac{1}{2}$, universe is open

if $q_0 = \frac{1}{2}$, universe is flat

The universe exists close to the dividing line between the ever-expanding model and the eventually contracting model, but it is not yet certain which of these radically different prospects will occur. *See also* age of the universe; cosmological models; big bang theory.

unresolved source. A source of radiation whose angular size is too small for details of its structure to be revealed.

unseen component. *See* binary star.

upper culmination. *See* culmination.

uranometry. *Obsolete name for* star catalogue or star atlas. The *Uranometria* of Johannes Bayer was published in 1603.

Uranus. The seventh major planet of the solar system and the first to be discovered telescopically – by William Herschel in 1781. One of the *giant planets and closely resembling *Neptune, it has an equatorial diameter of about 50 800 km, a mass 14.5 times that of the earth, and a density 1.3 times that of water. Uranus orbits the sun every 84 years, varying in distance between 18.3 and 20.1 AU. At *oppositions, which recur four days later each year, it has a mean angular diameter of 3.7 arc seconds and a magnitude of 5.7, near the limit of naked eye visibility. Telescopically Uranus shows a featureless greenish disc making visual determination of its rotation period impossible. The period is probably about 16 hours. Orbital and physical characteristics are given in Table 1, backmatter.

Uranus' equator is tilted by 98° with respect to its orbit, making it unique in having its rotation axis close to its orbital plane. This means that each pole is presented almost face-on to the sun and earth during part of each orbit. In the equatorial plane are five satellites (*see* Table 2, backmatter) and a modest *ring system*. This system consists of nine rings, between 3 and 100 km broad and lying 42 000 to 51 000 km from the centre of the planet, within the *Roche limit for Uranus. The material comprising the rings may be water-ice chunks, like those postulated for *Saturn's rings, although they have a much lower *albedo (about 0.05). The rings were discovered only in 1977 when they occulted a star. Perturbations by Uranus' satellites may govern the position and thickness of each ring.

It is thought that the visible surface of Uranus corresponds to a methane cloud layer seen through a haze of molecular hydrogen: both methane and hydrogen are detected spectroscopically in the upper atmosphere. The temperature of the cloud tops, −210°C, is close to what is expected if Uranus does not have a net heat outflow (as do Jupiter and Saturn). Ammonia, which provides the visible clouds of Jupiter, freezes at a higher temperature than methane and probably exists as clouds below the methane layer where the pressure and temperatures are higher. Lower still, models imply a 10 000 km layer of molecular hydrogen followed by an 8000 km thick ice mantle and a 16 000 km diameter rocky core. This core may be partly metallic and molten and thus able to generate a magnetic field; this would account for observations of auroral activity that confirm that Uranus has a *magnetosphere.

The planetary probe *Voyager 2 should encounter Uranus in Jan. 1986 and should greatly amplify our knowledge of the planet and its satellites and rings.

Ursa Major (Great Bear). An extensive very conspicuous constellation in the northern hemisphere that can be used to indicate the positions of brighter neighbouring stars (see illustration). The brightest stars are *Alioth, *Alcaid, and *Dubhe, all of 2nd magnitude, the seven brightest stars forming the *Plough. The area contains several double stars including the naked-eye double *Mizar and Alcor, the planetary *Owl nebula, and the close spiral galaxies M81 and M82. M82 is a *starburst galaxy. Abbrev.: UMa; genitive form: Ursae Majoris; approx. position: RA 11.5h, dec +55°; area: 1280 sq deg.

Ursa Major cluster. An *open cluster containing over 100 stars that are scattered over an area of sky more than 1000 minutes of arc in diameter. It is an example of a *moving cluster

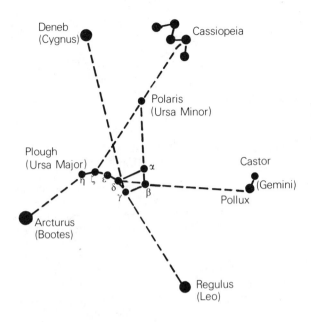

Ursa Major and bright neighbouring stars

and its wide angular diameter is due to its proximity to earth. It includes the stars Beta, Gamma, Delta, Epsilon, and Zeta Ursae Majoris in the *Plough, Zeta Leonis, Beta Aurigae, Beta Eridani, Gamma Coronae Borealis, and *Sirius.

Ursa Minor (Little Bear). A constellation in the northern hemisphere that includes and surrounds the north celestial pole. The two brightest stars, both of 2nd magnitude, are *Polaris (α) (the present *pole star) and the orange subgiant Kochab (β). Abbrev.: UMi; genitive form: Ursae Minoris; approx. position: RA 0 – 24h, dec +80°; area: 256 sq deg.

Ursa Minor system. A member of the *Local Group of galaxies.

Ursids. A minor *meteor shower, radiant: RA 217°, dec +78°, maximizing on Dec. 22.

UT, UT0, UT1, UTC. *See* universal time.

Utopia Planitia. A smooth Martian plain located northwest of the *Elysium Planitia volcanoes and chosen as the landing site of the *Viking 2 Lander in 1976.

UV. *Abbrev. for* ultraviolet.

uvby system. A system of stellar *magnitudes in which photoelectric *magnitudes u, v, b, and y are measured at four wavelength bands: measurements of u (ultraviolet) are centred on 350 nanometres (nm), v (violet) on 410 nm, b (blue) on 470 nm, and y (yellow) on 550 nm. The bandwidth (20–30 nm) is much narrower than for the UBV system.

UV Ceti stars. *See* flare stars.

V

Valles Marineris. A vast system of Martian canyons photographed by *Mariner 9. Long recognized as a feature of changing brightness, known as *Coprates canyon,* it was renamed in honour of the Mariner project. *See* Mars, surface features.

valley, lunar (canyon). A radial fracture on the moon associated with an impact basin or crater. *See table at* rille.

vallis (plural: **valles**). Valley. The word is used in the approved name of such a surface feature on a planet or satellite.

Van Allen radiation belts. Two regions within the earth's *magnetosphere in which charged particles become trapped and oscillate backwards and forwards between the magnetic poles as they spiral around magnetic field lines. The lower belt, from 1000 to 5000 km above the equator, contains protons and electrons either captured from the *solar wind or derived from collisions between upper atmosphere atoms and high-energy *cosmic rays. The upper belt, between 15 000 and 25 000 km above the equator but curving downwards towards the magnetic poles, contains mainly electrons from the solar wind. The belts were discovered by James Van Allen in the course of his analysis of observations by early *Explorer satellites in 1958.

Vanguard project. Part of the US programme for the *International Geophysical Year, in which three small satellites were orbited. Vanguard 1, launched on Mar. 17 1958, was the first satellite powered by *solar cells: it continued transmitting for six years and long-term tracking of it later revealed the slight pear shape of the

earth. Two other Vanguards launched in 1959 completed the programme, which provided information not only on the earth's shape but on its surface and interior, its magnetic field and radiation belts, and its space environment.

variable. *Short for* variable star.

variable radio source. A *radio source whose measured parameters change with time, especially the *flux density. Many extragalactic radio sources have been observed at frequencies above 400 megahertz to vary over periods of weeks or months. The time-scale of the variations imposes an upper limit on the linear size of the varying region since the conditions are unlikely to change more quickly than the time it takes for light to travel across the region.

variable star. A star whose physical properties, such as brightness, radial velocity, and spectral type, vary with time: about 30 000 have been catalogued so far. The variation in brightness is the easiest to detect, occurring in almost all variables. It can be seen on the *light curve of the star and may be regular – with a period given by one complete cycle in brightness on the light curve – or may be irregular or semiregular. The periods range from minutes to years. The range between maximum and minimum brightness – the *amplitude – is also very wide.

There are three major groups of variables: eclipsing, pulsating, and cataclysmic variables. The eclipsing or geometric variables are systems of two or more stars whose light varies as one star is periodically eclipsed by a companion star. These *eclipsing binaries can be subdivided into *Algol variables, *W Serpentis stars, and *W Ursae Majoris stars. *Pulsating variables, such as the *Cepheid variables, periodically brighten and fade as their surface layers expand and contract. *Cataclysmic variables are close *binary stars where one component is a white dwarf to which mass is being transferred from the other component. This group includes *novae, *recurrent novae, and *dwarf novae. Minor groups of variables include *flare stars, *rotating variables, *spectrum variables; and *R Coronae Borealis stars.

These many types also fall into three other main categories: with *extrinsic variables* the brightness variation results from some process external to the star, as with some eclipsing binaries; with *intrinsic variables* the variation results from some change in the star itself, as with the pulsating variables; in other cases, such as W Ursae Majoris stars, the variation can result from a combination of intrinsic and extrinsic agencies.

Different types of variable can be distinguished by period and also by light-curve shape: by amplitude of the variation, depth of the minima, size and shape of the maxima, characteristic irregularities, and by the continuous or step-like changes involved. Spectroscopic and photometric measurements can be used to differentiate between borderline cases and to investigate changes in temperature, radial velocity, and spectral type. *See also* stellar nomenclature.

variation. The periodic *inequality in the moon's motion that results from the combined gravitational attraction of the earth and sun. Its period is half the *synodic month, i.e. 14.77 days, and the maximum longitude displacement is $39'29''.9$. *See also* evection; parallactic inequality; annual equation; secular acceleration.

vastitas. An extensive plain of lowlands. The word is used in the approved name of such a surface feature on a planet or satellite.

Vega (α Lyr). A conspicuous white star, apparently blue in colour, that is the brightest star in the constellation Lyra. It was the *pole star about 14 000 years ago. It is the standard A0 V star ($B-V = U-B = 0$) in the UBV system. Observations of Vega by the satellite *IRAS reveal an *infrared excess beyond 10 μm. The excess is attributed to thermal radiation from particles more than 1 mm in radius located about 85 AU from Vega and heated by the star to a temperature of 85 K. It appears that the large particles have arisen from the residual prenatal cloud of gas and dust and may perhaps represent an embryonic and developing planetary system. As Vega is a bright A0 star with a short lifetime (3×10^8 years) a 'solar system' will not have sufficient time to form around Vega. (The age of our solar system is 4.6×10^9 y.) m_v: 0.03; M_v: 0.5; distance: 8.1 pc.

Veha. A spaceprobe that is to be launched in Dec. 1984 to study both Venus and *Halley's comet. The mission is sponsored by the USSR, France, East and West Germany, Austria, Bulgaria, Hungary, Poland, and Czechoslovakia. The probe will reach Venus in June 1985 and send a module to the surface of the planet. It will then be set on a new course to intercept Halley in Mar. 1986. It will return pictures and investigate the structure and chemical composition of the comet and its dust tail. It will also assist ESA's *Giotto, arriving in the comet's vicinity slightly later, to achieve its close approach.

Veil nebula. *See* Cygnus Loop.

Vela (Sails). A constellation in the southern hemisphere near Crux, lying partly in the Milky Way. It was once part of the large constellation *Argo. The four brightest stars, Gamma, Delta, Kappa, and Lambda Velorum, are of 2nd magnitude and form a quadrilateral. Gamma, a remote luminous triple star, is the only conspicuous *Wolf-Rayet star. The area contains the *Vela pulsar. Abbrev.: Vel; genitive form: Velorum; approx. position: RA 9.5h, dec −50°; area: 500 sq deg.

Vela pulsar. The second *optical pulsar to be discovered, the optical flashes being detected in 1977. With a brightness of only about 26 magnitudes − one of the faintest optical bodies to be detected − it is considerably fainter than the optical pulsar in the *Crab nebula, although at a distance of about 500 parsecs it is four times nearer. It was already known to emit pulses of radio emission in a period of 0.089 seconds and pulses of gamma rays twice every revolution. At gamma-ray wavelengths it is the brightest object in the sky. The optical pulses were found by sampling the light from a field only five arc seconds across, centred on the accurate radio position of the pulsar, and folding the data to the precise radio period.

Like other pulsars, the Vela pulsar is gradually slowing down in its rotation rate. Several brief and temporary increases in rotation rate − glitches − have however been observed since 1969 (*see* pulsar).

velocity dispersion. A measure of the spread of velocities of members within a *cluster of galaxies. If both radius, R, and velocity dispersion, V, of a cluster are measured, an estimate of mass, $V^2 R/G$, is obtained.

velocity of light. *See* speed of light.

Venera probes (Venus probes). A series of successful Soviet probes to Venus. Venera 1, launched on Feb. 12 1961, passed Venus although radio contact had been lost after seven million km. Venera 2 passed Venus on Feb. 27 1966 but, like Venera 3 which plunged into the Venusian atmosphere two days later, it failed to return data.

Veneras 4−8 ejected capsules into Venus' atmosphere to measure the atmospheric temperature, pressure, and composition. The Venera 4 capsule transmitted for 94 minutes on Oct. 18 1967 during its parachute descent, while the capsules from Veneras 5 and 6 returned data for 53 and 51 minutes on May 16 and 17, 1969, respectively; all were crushed by the atmospheric pressure before they reached the surface. The capsule from Venera 7 did survive to reach the surface on Dec. 15 1970, returning data for 23 minutes and indicating a temperature of 475°C and a pressure of 90 atmospheres. Venera 8's capsule returned data from the surface for 50 minutes of July 22 1972.

Veneras 9 and 10 each consisted of a lander and an orbiter through which signals from the lander were relayed to earth. On Oct. 22 1975, Venera 9's lander sent the first photograph from the surface of another planet, showing sharp angular rocks near the probe. Another panorama one day later from the Venera 10 lander revealed an apparently older more eroded landscape. The landers operated for 53 and 65 minutes respectively on the Venusian surface.

Veneras 11 and 12 reached Venus on Dec. 25 and Dec. 21 1978 and landers were sent to the surface. As they descended they transmitted information concerning the atmosphere to the two spacecraft, and continued relaying for about 100 minutes from the surface.

Veneras 13 and 14 reached Venus on Mar. 2 and Mar. 5 1982 and landers were sent to the surface. Coloured panoramas were taken of the landing sites and soil analyses made, and this information, together with atmospheric data, was relayed back to the spacecraft.

Veneras 15 and 16 went into orbit around Venus on Oct. 11 and Oct. 16 1983. They studied the atmosphere and using radar mapped the mountains, craters, and other features on the N. hemisphere of the planet's surface, to within about 20° of the equator and with a resolution of a few kilometres. *See also* Venus.

Venus. The second major planet in order from the sun and the one that comes closest to the earth. It orbits the sun every 225 days at a distance varying between 0.72 and 0.73 AU. It has a size (12 104 km equatorial diameter), density, and, probably, internal constitution comparable with those of the earth but measurements by *Mariner, *Pioneer Venus, and *Venera probes have revealed extremely hostile surface and atmospheric conditions. Orbital and physical characteristics are given in Table 1, backmatter.

Venus is never far from the sun in the sky, reaching its greatest *elongation of 45° to 47° about 72 days before and after inferior conjunctions, which recur at intervals of about 584 days. At greatest brilliancy the planet is near magnitude −4.4 and brighter than every other celestial object except the sun and the moon; this occurs at an elongation of 39° about 36 days before and after inferior conjunction, while Venus is an *evening star or a *morning star.

Telescopically the planet shows a brilliant yellowish-white cloud-covered disc. Because of its orbital position between sun and earth, Venus exhibits *phases: gibbous and crescent phases occur between full at superior conjunction, when the planet's apparent diameter is 10 arc seconds, and new at inferior conjunction, when it has a diameter of about 61 arc seconds. *Transits of Venus across the sun's disc at inferior conjunction are rare: the next will occur in 2004.

The clouds never permit views of the surface; it was not until 1961 that radar established that Venus rotated in a retrograde direction 243 days relative to the stars, or 117 days in relation to the sun (the Venusian day). Features in the clouds, seen indistinctly from the earth but observed in detail by ultraviolet photography from Pioneer Venus and *Mariner 10, indicate a four day east to west rotation period for the upper atmosphere, corresponding to 350 km/hour winds. The atmosphere thus rotates in the same direction but about 60 times faster than the solid planet. The cloud patterns swirl from the equator towards the poles and appear to be driven by solar heating and convection near the subsolar point in the middle of the sunlit hemisphere.

Possibly because of the slow rotation of its nickel-iron core, Venus has little or no intrinsic magnetic field. A magnetic field is however induced in the planet's ionosphere by the *solar wind. The resulting bow shock wave was found by one of the Pioneer Venus probes to be very strong and can therefore buffer the outer atmosphere of Venus from the solar wind.

The atmosphere extends to about 250 km above the surface. Pioneer Venus measurements showed there to be a layered structure to the Venusian clouds, which lie between altitudes of roughly 45–60 km within the troposphere of Venus (i.e. the region in which the atmospheric temperature decreases with altitude). There is haze above and below the clouds. Three clouds layers have been distinguished, having different concentrations and sizes of suspended particles; the particles are identified as liquid droplets of sulphuric acid with an admixture of water and solid and liquid particles of sulphur.

The temperature increases from about 300 K at the cloud tops to about 750 K at Venus's surface, which experiences a crushing atmospheric pressure 90 times that of the earth. The surface temperature, higher than that of any other planet, is the result of a *greenhouse effect involving the clouds and the large proportion of carbon dioxide in the atmosphere. Pioneer Venus showed that the atmosphere as a whole consists of about 98% carbon dioxide, 1–3% nitrogen, with a few parts per million (ppm) of helium, neon, krypton, and argon. Although the amounts of neon, krypton, and argon are small, they indicate very much greater amounts of primordial neon, krypton, and argon than those found in the earth's atmosphere. This is currently raising problems concerning the established view of the origin of the solar system. In the lower atmosphere, below the clouds, trace amounts of water vapour (0.1–0.4%) and free oxygen (60 ppm) were detected. The high surface temperature may have prevented any atmospheric water vapour from condensing into oceans, as occurred on earth; it may also have prevented carbon dioxide from combining with the crustal material to form carbonaceous rocks. The upper atmospheric winds moderate to only 10 km/hour or so at the surface.

Photographs of the surface returned by the Venera 9 and 10 landers in 1975 showed a stony desert landscape, while tests for radioactive elements in the crust implied a composition similar to that of basalt or granite. Radar mapping of the surface of Venus by the Pioneer Venus and Venera 15 and 16 spacecraft have provided a clear picture of the varied geological processes that have shaped the surface hidden beneath the ubiquitous clouds. We have now found evidence of huge rolling plains that cover 70% of the surface, depressions covering 20%, and highland areas covering the remaining 10% and concentrated in two main areas. Basins are comparatively rare, with the most extensive being *Atlanta Planitia*, which is about the size of the Gulf of Mexico. Some craters, about 20–50 km in diameter, have been observed and must be related to impacts before the atmosphere developed to its current opacity. The highland areas include towering volcanic regions. The two main areas are *Ishtar Terra* and *Aphrodite Terra*. Ishtar Terra, the more

northerly, is a flattish plateau bounded on three sides by mountains. *Maxwell Montes*, one of these mountainous regions, is the highest region on Venus, rising to more than 10 km above the surface. A huge rift valley with a depth of 2 km below the surface and width of nearly 300 km (comparable to Valles Marineris on Mars) is probably of tectonic origin. The surface with its huge plateaus greater than those seen on earth are examples of active geological history.

Venus probes. *See* Venera probes; Mariner 2; Mariner 5; Mariner 10; Pioneer Venus probes.

Venus Radar Mapper (VRM). A proposed NASA mission to Venus, scheduled for launch in 1988, that will produce a definitive map of Venus (i.e. with better resolution than that of the *Pioneer Venus Orbiter) in the following year. The VRM mission is a cut-price version of the cancelled Venus Orbiting Imaging Radar (VOIR) mission.

Venus, transits. *See* transit.

vernal equinox (spring equinox). *See* equinoxes.

vernier. A short scale used to increase the accuracy of the graduated scale to which it is attached. If nine divisions of the vernier correspond to 10 divisions of the instrument scale, the latter can be read to a further decimal place.

vertical circle. A great circle on the celestial sphere that passes through an observer's zenith and cuts the *horizon at right angles.

Very Large Array (VLA). *See* aperture synthesis.

very long baseline interferometry (VLBI). Radio observations of the same cosmic source made with very widely separated *radio telescopes, i.e. with an *interferometer whose longest baseline exceeds about 100 km and can be as great as the earth's diameter. The very long baseline greatly increases the angular *resolution of the interferometer. This allows the fine structure of very distant *radio sources, such as *quasars, to be studied.

It is usually not possible to combine the signals from the two arms of the interferometer directly so that tape recorders attached to high-precision clocks have to be used; the recorded signals are later processed in a correlating *VLBI terminal*.

Networks of telescopes in the USA and in Europe are diverted from their normal uses in order to make VLBI observations, and occasionally the networks are combined to give baselines of over 10 000 km. The USA plans to build an array of telescopes distributed over the whole country and dedicated solely to VLBI.

Vesta. The fourth *minor planet to be discovered, found in 1807. It is the third largest minor planet but has a higher *albedo (0.23) than any other large minor planet and is the only one bright enough to be visible to the naked eye at favourable *oppositions. Its unusual spectrum suggests that it has a surface composed of basaltic minerals. *See* Table 3, backmatter.

vidicon. A sensitive semiconductor-based instrument derived from television technology and used in astronomy for detecting and measuring light, ultraviolet, and near-infrared radiation. The target area is photosensitive, responding to the radiation falling on it by producing an electronic signal that varies linearly with the intensity of the incident radiation. The vidicon has been superseded by *CCD detectors.

Viking probes. Two identical American spacecraft, each comprising an Orbiter and a Lander, that were launched in 1975 on a mission to search for life on *Mars. Each Orbiter weighed 2325 kg and carried two television cameras to photograph the surfaces of Mars and its satellites, and instruments to map atmospheric water vapour and surface temperature variations; 52 000 pictures were relayed to earth. After detaching from its Orbiter, each Lander, weighing 1067 kg, made a successful landing using a combination of aerodynamic braking, parachute descent, and retro-assisted touchdown. The Viking 1 Lander set down on July 20 1976 in *Chryse Planitia after a delay while Orbiter photographs were searched for a smooth landing area. Viking 2 landed on Sept. 3 1976 in *Utopia Planitia, 7420 km northeast of the Viking 1 Lander.

Each Lander carried two television cameras, a meteorological station, a seismometer, and a set of soil-analysis experiments serviced by a sampling arm. Only the seismometer carried by Viking 1 failed to function; that on Viking 2 registered mainly wind vibrations and a few minor Martian 'earth' tremors. The television cameras returned views of a red rock-strewn surface under a pink dusty sky. Sand or dust dunes were evident at the Viking 1 site but no life forms were seen. The meteorology instruments reported temperatures varying between −85°C and −30°C with mainly light winds gusting to 50 km per hour.

X-ray analysis of soil samples showed a high proportion of silicon and iron, with smaller amounts of magnesium, aluminium, sulphur, cerium, calcium, and titanium. Gas chromatograph mass spectrometers carried by the Landers failed to detect any biological compounds in the Martian soil, but more ambiguous results were returned by three experiments designed to test for microorganisms in various soil samples.

A *labelled–release experiment* tested for radioactive gases that might have been expelled from organisms in a soil sample fed with a radioactive nutrient. In a *pyrolytic release experiment* soil was placed with gases containing the radioactive isotope carbon-14. The idea was that Martian organisms might assimilate the isotope into their cell structure in a process similar to earthly photosynthesis. Later the soil was baked and a test made for the carbon isotope in gases driven off. Finally, a *gas exchange experiment* monitored the composition of gases above a soil sample, which might or might not have been fed with a nutrient. If Martian organisms were breathing the gases, a change in composition might be detected.

In practice, the pyrolytic release experiment produced negative results, while the gas exchange experiment gave a marginally positive response, which could be explained in terms of a chemical reaction involving the soil and nutrient. The positive response of the labelled release experiment was marked but disappeared when the soil was sterilised by preheating.

Detailed analysis of the Viking Lander experiments has led to the conclusion that there is no form of life on Mars.

Virgo (Virgin). An extensive zodiac constellation straddling the equator between Bootes and Centaurus. The brightest stars are the 1st-magnitude *Spica and several of 3rd magnitude, including the binary *Porrima. The area contains the prototype of the W Virginis stars, a class of *Cepheid variables. The area is rich in faint galaxies, which are members of the *Virgo cluster, and contains the first detected *quasar 3C–273. Abbrev.: Vir; genitive form; Virginis; approx. position; RA 13.5h, dec −5°; area: 1294 sq deg.

Virgo A. An intense *radio source in the *Virgo cluster of galaxies that is associated with the brightest galaxy, M87. It has a *flux density of 970 *jansky. M87 is a giant elliptical galaxy with a *jet from which a considerable fraction of the radio radiation comes by *synchrotron emission. The close packing and high speeds of stars in the galaxy's centre suggest that it may contain a supermassive black hole of several thousand million solar masses.

Virgo cluster. A giant irregular *cluster of galaxies lying near the north galactic pole in the constellation Virgo. It is the nearest large cluster: its distance is currently estimated as 15 megaparsecs. Some 2500 galaxies have been observed of which about 75% are spirals, the remainder being mainly ellipticals with a few irregulars. One of its brightest members, the giant elliptical galaxy M87, is both a radio source (*Virgo A) and an x-ray source (Virgo X-1). The Virgo cluster is the centre of the *Local Supercluster; measurements of the *microwave background radiation indicate that the Virgo cluster has a significant gravitational pull on the *Local Group of galaxies, reducing the measured recession velocity of the Virgo cluster by some 250 km s^{-1} (to about 1000 km s^{-1}).

visible radiation, spectrum. *See* light.

visual binary. A *binary star, such as *Sirius or Gamma Centauri, whose components are sufficiently far apart to be seen separately, either with the naked eye or with a telescope. In most cases the components differ in brightness: the brighter star is designated the *primary* and the other is the *companion*. Observations over a period of time reveal the apparent orbit of the companion relative to the primary, from which the true orbit is derived by adjusting for the inclination of the orbital plane. If the distance is known the total mass of the binary can be obtained from Kepler's third law:

$$M_1 + M_2 = a^3/p^3P^2$$

a is the semimajor axis of the true orbit (in arc seconds), p is the parallax, and P is the orbital period; the total mass is obtained in terms of the solar mass. It is sometimes possible to measure the absolute orbit of each component against the background of more distant stars. Then, if a_1, a_2 are the semimajor axes of these orbits, the ratio of masses is given by

$$a_1/a_2 = M_2/M_1$$

and by applying both equations the individual masses can be obtained. If the distance is unknown, both it and the total mass can be found using the method of *dynamical parallax.

Visual binaries generally have long orbital periods and in some cases the period is too long for any orbital motion to be detected. Such pairs are called *common proper motion stars* because, unlike *optical double stars, they appear to stay together as they move across the sky.

visual magnitude. *See* magnitude.

visual meteor. *See* meteor.

VLA. *Abbrev. for* Very Large Array. *See* aperture synthesis.

VLBI. *Abbrev. for* very long baseline interferometry.

Vogt–Russell theorem. A theorem proposed independently by H. N. Russell and H. Vogt in 1926 stating that, if the age, mass, and chemical composition of a star are specified, then its temperature and luminosity are (in almost all cases) uniquely determined. Hence for a star whose age, mass, and chemical composition are known, the equations of *stellar structure will have only one solution.

void. *See* filamentary structure.

Volans or **Piscis Volans (Flying Fish).** A small inconspicuous constellation in the southern hemisphere near the Large Magellanic Cloud, the brightest stars being of 4th magnitude. Abbrev.: Vol; genitive form: Volantis; approx. position: RA 8h, dec −70°; area: 141 sq deg.

Voskhod. Two Soviet manned spacecraft that followed the *Vostok series. Voskhod 1 was launched on Oct. 12 1964 and was the first multi-crewed spacecraft: three cosmonauts were on board. Voskhod 2 was launched on Mar. 18 1965 and during its flight one of the two-man crew, Alexsei Leonov, made the first spacewalk.

Vostok. A series of six Soviet manned spacecraft that were able to carry one person into orbit. Vostok 1 was launched on Apr. 12 1961. Its crew member, Yuri Gagarin, became the first man in space, the flight lasting 1.8 hours. In subsequent Vostok missions the flight duration was extended up to 119 hours so that the effects of prolonged weightlessness could be studied. The double launching of Vostoks 3 and 4 in August 1962 and their close approach in orbit led to an appreciation of rendezvous techniques.

Voyager probes. Two 825 kg US probes launched in 1977 towards the planets of the outer solar system and making use of the unique *planetary alignment that occurred at the beginning of the 1980s. Voyager 1, launched on Sept. 5 1977, reached *Jupiter in Mar. 1979 moving in a path that took it close to the satellites *Callisto and *Ganymede, then quite near *Europa, and close to *Io, before it made its closest approach (280 000 km) to Jupiter on Mar. 5; it then passed *Amalthea. Using the slingshot action of Jupiter, Voyager 1 was set on a course for *Saturn, which it reached in Nov. 1980. It approached within 125 000 km on Nov. 12, dipping below Saturn's rings just after passing close to the satellite *Titan. The spacecraft is now heading out of the solar system with no further planetary encounters.

Voyager 2, launched on Aug. 20 1977, reached Jupiter in July 1979. It passed close to Callisto, Ganymede, and Europa, then passed Amalthea before its closest approach (650 000 km) to the planet on July 9. Since Voyager 1's encounter with Titan was successful, Voyager 2 was set on a path that took it past Saturn in Aug. 1981 and on towards Uranus and Neptune, which it should encounter in Jan. 1986 and Aug. 1989. Voyager 2 passed closer to several of Saturn's satellites (in particular *Iapetus, *Hyperion, *Tethys, and *Enceladus) than Voyager 1, but further from Titan. Moving above the plane of Saturn's rings it made its closest approach (101 000 km) to the planet on Aug. 25 1981.

The two spacecraft are carrying a variety of instruments for scientific investigations, and three engineering subsystems for on-board control of spacecraft functions. Trajectory correction manoeuvres can only be enabled from earth. The science instruments for viewing the planets and their satellites and ring systems are mounted on a platform that can be pointed very precisely. The instruments include a narrow-angle and a wide-angle TV camera for high-resolution imaging of satellite surfaces, etc., an infrared spectrometer, interferometer, and radiometer, an ultraviolet spectrometer, and instruments for studying plasmas, low-energy charged particles, and cosmic rays. In addition there are magnetometers mounted on an extendable boom and two planetary radio astronomy and plasma wave antennae. A 3.7 metre aperture high-gain radio dish points continuously towards earth, allowing two-way communications of commands and information. Power is provided by three radioisotope thermoelectric generators, and hot gas jets provide thrust for at-

titude stability and trajectory corrections.

A wealth of information has been returned to earth from the two Voyager craft, and has revealed many unexpected aspects concerning the satellites, in particular the volcanic activity on Io, the fine details of Saturn's rings, and the existence of several new satellites and a Jovian ring system. (Details are given at individual entries in the dictionary.)

Vulcan. A hypothetical major planet, thought during the 19th century to orbit the sun within the orbit of Mercury. Searches for it during total solar eclipses and at suggested times of *transit across the sun were all unsuccessful. It is now known not to exist.

Vulpecula (Fox). A constellation in the northern hemisphere near Cygnus, lying in the Milky Way, the brightest stars being of 4th magnitude. It contains the *Dumbbell nebula. Abbrev.: Vul; genitive form: Vulpeculae; approx. position: RA 20h, dec +25°; area: 268 sq deg.

V/V$_m$ test. *See* luminosity-volume test.

W

wake. *See* train.

walled plains. *See* craters.

wandering stars. *See* fixed stars.

watt. Symbol: W. The SI unit of power, defined as the power resulting from the dissipation of one joule of energy in one second.

waveband (band). A region of the electromagnetic spectrum lying between frequency (or wavelength) limits that are defined according to some property of the radiation, or some requirement or functional aspect of a detecting device or system or transmission channel.

wavefront. The imaginary surface of a wave of light or other radiation, connecting points of the same phase. The advancing wavefront is normally perpendicular to the direction of travel and may be plane, spherical, cylindrical, or more complex.

waveguide. A metal tube, usually of rectangular cross section, down which travelling electromagnetic waves may be propagated. In a more general sense it is any system of material boundaries that fulfils the same purpose, such as layers of plasma in the *ionosphere. Waveguides are used as transmission lines at high frequencies where dielectric losses in other types of radio cables become excessive; they are therefore used as *feeders in radio telescopes. The guided waves may be radiated away at the end of the waveguide by a *horn antenna*, which is a flared metal device having the dimensions of the waveguide at one end and opening out to a large aperture at the other end.

wavelength. Symbol: λ. The distance over which a periodic wave motion goes through one complete cycle of oscillation, i.e. the distance travelled during one period. Thus for a sinusoidal wave motion, such as *electromagnetic radiation, it is the distance between two successive peaks or troughs. For electromagnetic radiation, wavelength is related to frequency, ν, by $\nu\lambda = c$, where c is the speed of light. Wavelength is measured in metres or in multiples or submultiples of metres; for example, the wavelength of light is usually given in nanometres while that of infrared radiation is usually quoted in micrometres.

WC stars. *See* Wolf-Rayet stars.

weak force. *See* fundamental forces.

weber. Symbol: Wb. The SI unit of magnetic flux.

weight. The force experienced by a body on the surface of a planet, natural satellite, etc., that results from the gravitational force (directed towards the centre of the planet, satellite, etc.) acting on the body. A body of mass m has a weight mg, where g is the acceleration of gravity.

weightlessness. The condition associated with *free fall, i.e. the motion of an unpropelled body in a gravitational field. The acceleration of a person in a freely falling object, such as a spacecraft, is equal to that of the freely falling object. The person thus experiences no sensation of weight and floats freely. Many crew members on the US space shuttles and the USSR and US space stations have suffered a combination of symptoms, akin to those of motion sickness, as they adapt to conditions of weightlessness. It usually takes 1–3 days to overcome the problem (known as *space adaption syndrome* or *space sickness*), duties being performed more slowly than scheduled. This can be serious on short-term missions. The adaption to conditions of normal earth gravity following periods of weightlessness have been investigated by both the USSR and the US after spaceflights of ever-increasing duration. No long-term effects have been observed to date.

Weizsäcker's theory. A theory, proposed by C.–F. von Weizsäcker in 1944, for the *accretion of the planets at the points of contact between turbulent eddies in the nebula that contracted to form the solar system.

West. A brilliant comet that passed within 30 million km of the sun in Feb. 1976. The cometary nucleus broke into at least four fragments, accompanied by massive outbursts of gas and dust.

Westerbork Radio Observatory. An observatory at Westerbork, near Groningen in the northern Netherlands. It has a highly sensitive *aperture synthesis system, which went into operation in 1970 and is run by the Netherlands Foundation for Radio Astronomy. Fourteen 25 metre radio dishes lie on a line running east-west and 2.7 kilometres in length. (Originally there were 12 dishes on a 1.6 km baseline.) Four of the dishes are moveable: two at one end of the line and a further two nearer the centre of the line and the ten fixed dishes.

western elongation. *See* elongation.

west point. *See* cardinal points.

Wezen. *See* Canis Major.

Whipple Observatory. *See* Fred Lawrence Whipple Observatory.

Whipple's comet. A comet that started life with a near-circular orbit, eccentricity 0.167, period 10.3 years, and has been perturbed by Jupiter so that it now has an eccentricity of 0.348 and a period of 7.5 years. The comet seems to decrease in brightness by one magnitude at each return.

Whirlpool galaxy (M51; NGC 5194). A well-defined type Sc spiral galaxy (*see* Hubble classification) that is face-on to us. It lies at a distance of 6 megaparsecs in the constellation Canes Venatici, the nearest bright star being Eta Ursae Majoris in the *Plough. A small companion galaxy, NGC 5195, appears to be connected to it by an extension of one of the spiral arms. Total mass: about 8×10^{10} solar masses; total magnitude: 8.3.

white dwarf. A star that has a mass below the *Chandrasekhar limit (about 1.4 solar masses) and has undergone *gravitational collapse. Having diameters only 1% that of the sun these stars are consequently very faint, with absolute *magnitudes ranging from $+10$ to $+15$. The description 'white' is misleading since they display a range in colour as they cool down, through white, yellow, and red, until finally ending up as cold black globes – called *black dwarfs*. However, those of *spectral types A and F (i.e. white stars) are the brightest and were the first to be discovered.

White dwarfs are the final phase in the evolution of a low-mass star. Their progenitors are stars of up to 8 solar masses, which lose up to 90% of their matter in the form of *planetary nebulae. The core shrinks to become a white dwarf following the exhaustion of its nuclear fuel. Since most of the matter in the core of the collapsing star is in a *degenerate state – with the electrons stripped from their nuclei and packed tightly together – the star contracts until its gravity is balanced by the degeneracy pressure of the electrons, and the density rises to $10^7 - 10^{11}$ kg m^{-3}. Because of the peculiar behaviour of degenerate material (which is subject to the quantum mechanical uncertainty principle), the most massive white dwarfs collapse to the smallest diameters and highest densities. Stars above the Chandrasekhar limit are too massive to be supported in this way and must collapse further to become *neutron stars or *black holes. In practice, the addition of extra matter to a carbon-oxygen white dwarf at the Chandrasekhar limit may alternatively produce a runaway explosion as a Type I *supernova.

The light from white dwarfs does not arise from internal nuclear reactions but in a thin gaseous atmosphere that slowly leaks away the star's heat into space. The spectral lines arising in this atmosphere are broadened by the extremely high surface gravity, and in extreme cases the light loses enough energy to suffer a measurable *gravitational redshift. Some white dwarfs show no hydrogen at all in their spectra, while others are enriched in helium, carbon, and calcium. A few have strong magnetic fields (10^4 tesla) and several have high rotational velocities. Some white dwarfs, known as ZZ Ceti stars, are *pulsating variables, fluctuating by 0.01 to 0.3 magnitudes with periods of 2 to 20 minutes. *Novae, *recurrent novae, and *dwarf novae are close binary stars in which one component is a white dwarf (see cataclysmic variable).

It has been estimated that there may be 10^{10} white dwarfs in our Galaxy, many having by now cooled to become black dwarfs. Best known of all white dwarfs is Sirius B, companion to *Sirius (αCMa), which was discovered in 1862 by Alvan Clark after F. W. Bessel had predicted its existence (in 1844) from the unusual motion of Sirius. Sirius B has a radius of only 10^4 km, about twice that of the earth, but a mass similar to that of the sun.

white hole. The reverse of a *black hole: a region where matter spontaneously appears. Early calculations on black holes indicated that the extreme distortion of space and time inside the *event horizon should connect our universe with another through an Einstein–Rosen bridge (or wormhole). Matter falling into the black hole should then correspondingly appear in the other universe as outpouring of material through a white hole. Certain highly energetic objects, such as quasars, were proposed as examples of white holes in our own universe. With better calculations however, it now seems likely that such links, and hence white holes, do not exist.

white-light corona. See corona.

white noise. See noise.

WHT. *Abbrev.* *for* William Herschel Telescope.

wide-field eyepiece. See eyepiece.

Widmanstätten figures. Patterns of bands crossing one another in two, three, or four directions that are revealed when polished surfaces of most *iron meteorites are etched. They are lamellar intergrowths of kamacite, taenite, and plessite.

Wien displacement law. See black body.

Wild Duck. See Scutum.

Wild's Trio. A group of three irregular barred spiral galaxies that lies in the constellation Virgo. The galaxies appear to be connected by bridges of luminous material. Their differing radial velocities, derived from the *redshifts in their spectra, suggest that the group is rapidly disintegrating.

William Herschel Telescope (WHT). The 4.2 metre (165 inch) reflecting telescope with an *altazimuth mounting that is to be sited at the new *Roque de los Muchachos Observatory in the Canaries. It is largely a UK telescope in which the Dutch have a 20% share. It should be completed in 1986. It has a glass-ceramic (Cer-Vit) primary mirror of great thermal stability and a focal ratio of f/2.4. The telescope can be used in a *Cassegrain configuration (there is also an off-axis 'broken Cassegrain' focus). In addition, light from the secondary mirror can be diverted through the altitude bearings of the Y-shaped fork mounting to two *Nasmyth foci, where heavy instruments can be mounted. Sophisticated electronic equipment will detect and analyse the light. Astronomers will be able to use the WHT by *remote operation from the *Royal Greenwich Observatory in the UK.

Wilson-Bappu effect. An experimentally derived linear relation between certain properties of the calcium K line of late-type F, G, K, and M stars (the widths of their double reversals) and their absolute magnitude.

Wilson effect. The apparent displacement of the umbra of a *sunspot relative to the penumbra as the spot is carried by the sun's rotation from the east to the west limb of the disc. For a symmetrical sunspot, the side of the penumbra closest to the centre of the disc is more foreshortened than that towards the limb. This was first interpreted as implying that the spot is a depression, but it is now known that the opacity of the umbra is less than that of the penumbra and that this also contributes to the effect.

window. *See* atmospheric windows; launch window.

wing. A broad feature occurring on one side of the position of a spectral line or band. The matter producing this feature is travelling at much greater speeds than that producing the main peak and thus has a greater doppler shift. A redshifted wing (i.e. shifted to longer wavelengths) is produced by gas receding from us, a blueshifted (shorter-wavelength) wing by gas moving towards us.

winter solstice. *See* solstices.

WN stars. *See* Wolf-Rayet stars.

Wolf number. *See* Zürich relative sunspot number.

Wolf-Rayet stars (WR stars; W stars). A small group of very luminous very hot stars, with temperatures possibly as high as 50 000 kelvin, that have anomalously strong and broad emission lines of ionized helium, carbon, oxygen, and nitrogen but few absorption lines. Since their discovery (by C. J. E. Wolf and G. Rayet, 1867), over 300 have been found in our Galaxy and its neighbours. The majority either have lines of He, C, and O – *WC stars* – or He and N – *WN stars*; both types have anomalously low abundances of hydrogen. The emission lines are thought to arise in an expanding stellar atmosphere moving at very high speeds of up to 2000 km s^{-1} so that the star is continuously and rapidly losing mass. The average mass for Wolf-Rayets is 10 solar masses. About half are known to be binary stars, usually with O or B stars as companions, an example being Gamma Velorum (WC8 + O7).

These unusual properties provide clues that Wolf-Rayets are the centres of very massive *Of stars stripped of their outer envelopes. Since the Wolf-Rayets are the more evolved members of each binary, they must originally have been the more massive partner, a star of at least 20 solar masses. Half that mass has thus been lost in their stellar winds, whose high outflow rate would strip this mass off the star in only 100 000 years. This gas is in fact often visible as a ring nebula surrounding the Wolf-Rayet star. In a close binary the companion's gravity may assist the stripping, but for single stars the cause of the high mass loss is still uncertain.

world line. *See* spacetime.

world models. *See* cosmological models.

wrinkle ridges. Linear deformed and folded lunar features that are confined to the *maria and may be elevated by several hundred metres; they may however continue into the *highlands as scarps. Wrinkle ridges are frequently compound, discontinuous, and associated with *rilles. They often reflect underlying topography, including *basin ring structures beneath the maria. They may result from fissure eruptions, faulting, or the contraction of a surface crust following the draining away of subsurface lavas. *Surveyor 6 landed close to a wrinkle ridge in Sinus Medii. *See also* lobate ridge.

WR stars (W stars). *See* Wolf-Rayet stars.

W Serpentis star (Beta Lyrae star). A close *binary star system where matter is being transferred very rapidly from one star to the other. This occurs when the more massive member of a close binary evolves to become a red giant (the previous stage may be an *RS CVn star); the *mass transfer can be so efficient that 85% of the red giant's mass is transferred to the other star. The system ends up as an *Algol variable.

Observationally, W Ser stars are *spectroscopic binaries characterized by emission lines from an extended gas envelope that is hotter than either of the stars in the system; the emission is thousands of times brighter than the lines from a star's *chromosphere. This gas represents part of the giant star's mass that is lost to the system during the rapid transfer. The bulk of the gas forms an accretion disc around the giant's companion, and as this gas spirals inwards the inner regions can heat up to 100 000 K and supply the radiation that ionizes the extensive tenuous gas producing the emission lines.

The outer cooler regions of the accretion disc can camouflage the accreting star and make it look larger and cooler than it actually is. This star will thus often have the appearance of a giant rather than that of the underlying main-sequence star. When mass transfer is most rapid, as in *Beta Lyrae*, the disc conceals this component entirely. Thus in Beta Lyrae we detect only the expanding giant star, a supergiant of spectral type B8.5 that is about 13 times the diameter of the sun and elongated towards its companion because it fills its Roche lobe (*see* equipotential surfaces). The companion is hidden by a disc about twice as wide but only half as thick as the supergiant's diameter. The system is an *eclipsing binary, with disc and star alternately eclipsing one another; the ellipsoidal shape of the supergiant causes the light curve to peak between eclipses, when the maximum extent of the star is seen. The B8.5 star is losing mass at a rate of about 10^{-5} solar masses per year, and this causes an increase in the orbital period (12.9 days) at a rate of 19 seconds per year.

W Ursae Majoris stars. A class of contact *eclipsing binaries that have very short periods amounting to only a few hours and components that are so close that they are grossly distorted by tidal forces into ellipsoidal shapes. Mass transfer occurs between the stars (*see* equipotential surfaces). The components are approximately equal in brightness, as seen by the equal depths of the minima of the light curves. As with *W Serpentis stars the light curves show continuous variation resulting from the distorted stellar shapes and variable surface intensity. The two components are more similar in luminosity than in mass (the ratio of masses can be 12:1), so energy from the more massive star's core is being fed to the companion's photosphere. Resulting chromospheric activity is thought to be responsible for the unusually high x-ray output of these stars.

About 0.1% of all main-sequence stars should be born in close enough doubles to become contact systems: the lower mass F, G, and K stars are the W UMa class while the rare more massive analogues are called *SV Centauri stars*. In both cases as the more massive component swells to become a giant, its outer layers will surround both stars to create a *common envelope star; the final outcome will be a coalesced star or a *cataclysmic variable system.

W Virginis stars. *See* Cepheid variables.

WW Aurigae. An eclipsing binary (period 2.52 days) in the constellation Auriga. The size and magnitude of the two components are similar so that the primary and secondary minima on the light curve are approximately equal. m_v: 5.7; M_v: 1.2; M_B: 1.7 (A), 2.0 (B); mass, radius: 1.92 (A), 1.90 (B) times solar mass, radius; spectral type: A7 (A), F0 (B); distance: 77 pc.

X

x-ray astronomy. The study of objects lying beyond the solar system in the photon energy band 100 to 500 000 electronvolts (corresponding to wavelengths from 12 nanometres (nm) to 0.0024 nm). X-ray observations are now an integral part of astronomy, relating particularly to galactic and extragalactic systems where violent or energetic phenomena give rise to copious high-energy particles or super-hot gas.

Opacity of the terrestrial atmosphere throughout the x-ray band requires that observations be made above about 150 km and hence x-ray astronomy could begin only after high-altitude rockets became available. The first detection of a cosmic x-ray source was made by Riccardo Giacconi and colleagues in July 1962, during an exploratory rocket launch equipped with an x-ray detector to search for lunar fluorescence. Confirmation of this source (*Scorpius X-1) and the discovery of a second source (Taurus X-1) in 1963 began an active period of rocket and balloon observations that, by 1970, had yielded 25–30 sources spread throughout the Galaxy; there was also one likely extragalactic source, apparently associated with the powerful radio galaxy *Virgo A (M87). The first optical identification was of Taurus X-1, which was found to coincide with the *Crab nebula supernova remnant in a classical lunar occultation observation by Herbert Friedman and his colleagues in the USA.

The launch of the first x-ray astronomy satellite, *Uhuru, in Dec. 1970 accelerated the development of the subject, yielding many new sources including a large number at high galactic latitude. In particular, Uhuru discovered *x-ray binaries and showed these to be the most common form of galactic x-ray source. A second major discovery was of powerful x-ray emission from *clusters of galaxies, with evidence that this emission arose in an extended region comparable in size (about 0.5 megaparsecs) to the cluster.

The launch of other x-ray astronomy satellite experiments continued the rapid expansion of the subject. The *Ariel V sky survey extended the Uhuru catalogue and led to the establishment of a second major class of extragalactic source: the x-ray *Seyfert galaxies. Observations by *Copernicus and Ariel V, followed by *SAS-3, found the slow *x-ray pulsators, periodically variable x-ray sources with periods of a few minutes, substantially longer than those of *Centaurus X-3, *Hercules X-1, etc. Ariel V, followed by the satellite OSO-8, discovered an emission line near 7 keV in the spectra of several *supernova remnants and many of the rich *clusters of galaxies, showing the x-rays to be of thermal origin, arising from hot gas at temperatures of 10^6 to 10^8 kelvin and containing an abundance of iron similar to that found in the solar system. *X-ray transients and *x-ray bursters were detected by several of the satellites. Optical identifications of a few of these have shown them to be most probably x-ray binaries in which the *mass transfer or accretion rate is highly variable. Further major advances were made by the *Einstein Observatory, whose great sensitivity revealed substantial x-ray emission from a wide range of normal stars. This emission comes from the stars' hot *coronae.

In addition to the discrete x-ray sources an apparently diffuse sky *x-ray background radiation* is observed. It seems likely that this is due mainly to unresolved and distant sources, such as clusters and *active galaxies.

x-ray background radiation. *See* x-ray astronomy.

x-ray binary. The most common type of luminous galactic x-ray source, involving a close binary system in which gas flows (via the *inner Lagrangian point) or blows (by a strong *stellar wind) from a normal nondegenerate star on to a compact companion. For the most luminous sources, radiating x-rays at $10^{29}-10^{31}$ watts, this companion is probably a *neutron star or *black hole; in less luminous cases, such as SS Cygni and AM Hercules, it is more likely to be a *white dwarf. Gravitational energy powers these sources and both the luminosity (L) and temperature (T) are proportional to the mass-to-radius ratio (M/R) of the accreting star:

$$L \propto (M/R)Gm'$$

where G is the gravitational constant and m' is the rate of mass accretion, and

$$T \propto (M/R)G\varepsilon$$

where ε depends on the efficiency of the gas heating, being high for shocks and low for viscous heating.

Two main types of x-ray binary are distinguished: in high-mass binaries, such as *Centaruus X-3 and *Cygnus X-1, the nondegenerate star is a giant or supergiant of early spectral type (O, B, or A); in low-mass binaries the visible star is a middle or late main-sequence star of near solar mass. Several binaries contain a pulsating x-ray source, probably involving a rotating magnetized neutron star; these bina-

ries, which include Centaurus X-3, *Hercules X-1, and *Cygnus X-3, are among the best-determined of all binary systems.

x-ray burst sources (x-ray bursters). Sources of intense flashes of cosmic x-rays, discovered in 1975. The bursts, seen in the 1–30 keV energy band, are characterized by a rapid onset, often less than one second, followed by an exponential decay with a time constant of a few seconds to a minute. They are located in our Galaxy, mostly within 40° of the galactic centre. Two types have been distinguished: *type I* repeat on timescales of hours or days; *type II* emit bursts every few minutes or less for a period of several days. It is widely believed that x-ray bursts arise in *x-ray binary systems and are due to thermonuclear flashes when material accreted on to the surface of a neutron star exceeds a critical temperature and pressure (type I) and to spasmodic mass accretion on to a neutron star (type II).

x-ray mirror. *See* grazing incidence.

x-ray pulsars. *See* x-ray pulsators.

x-ray pulsators. Regularly variable *x-ray binaries that have periods of a few seconds or in the case of the *slow pulsators* of a few minutes. This pulsation is widely interpreted as being associated with the rotation of a magnetized neutron star, so that these objects may be regarded as *x-ray pulsars*. The x-ray pulsations are believed to arise from channelling of the accreting gas onto the magnetic poles of the neutron star. The gas flow affects the neutron star's spin and as a result all x-ray pulsars (unlike other *pulsars) are gradually speeding up. Examples include *Centaurus X-3, *Cygnus X-3, and *Hercules X-1.

x-rays. High-energy *electromagnetic radiation lying between gamma rays and ultraviolet radiation in the electromagnetic spectrum. The XUV region bridges the gap between the x-ray and ultraviolet bands. X-rays, unlike light and radio waves, are usually considered in terms of photon energy, $h\nu$, where ν is the frequency of the radiation and h is the Planck constant. X-ray energies range from about 100 electronvolts (eV) up to about 500 000 eV, corresponding to a wavelength range of about 12 nanometres (nm) to about 0.0024 nm. Low-energy x-rays are sometimes called *soft x-rays* to distinguish them from high-energy or *hard x-rays*. In astronomy, thermal x-rays are produced from very high temperature gas ($\sim 10^6 - 10^8$ K) with nonthermal x-rays arising from the interaction of high-energy electrons with a magnetic field (*synchrotron emission) or with low-energy photons (inverse Compton emission – *see* Compton effect). *See also* thermal emission; nonthermal emission.

x-ray satellites. Artificial earth satellites devoted to cosmic x-ray observations, the first of which was NASA's *Uhuru, launched in Dec. 1970. Up to this date, i.e. throughout the period 1962–70, *x-ray astronomy was carried out exclusively with sounding rockets and balloon experiments. Uhuru produced the first all-sky survey for cosmic x-ray sources and increased the catalogue of known sources to 161 by 1974. The second x-ray astronomy satellite was the British satellite *Ariel V, which was launched in Oct. 1974 and occupied a similar 500 km circular equatorial orbit to Uhuru. It successfully extended the Uhuru sky map and made detailed spectral and temporal studies of individual sources.

Although x-ray experiments were also carried on other satellites principally devoted to the research fields (Copernicus, OSO-7, ANS, OSO-8), the next two x-ray astronomy satellites were *SAS-3 and *HEAO-1, launched respectively in 1975 and 1977. Both included modulation collimator experiments in their payloads and these provided the first accurate positions of many x-ray sources, leading to additional and more reliable optical identifications. In Nov. 1978 HEAO-2 (the *Einstein Observatory) was launched; its large *grazing-incidence telescope produced major advances. After the Einstein Observatory ceased operation in Apr. 1981 only two small satellites, *Ariel VI and *Hakucho, remained before the launch in 1983 of the Japanese *Tenma and the sophisticated European *EXOSAT spacecraft. Future x-ray satellites now in preparation are *ROSAT and *ASTRO–C, with the next major US mission, *AXAF, not due until the 1990s.

x-ray sources. Luminous sources of x-ray emission lying well beyond the solar system. Several thousand are known and are identified with a wide range of astronomical objects, from high-redshift quasars to nearby stars. The brightest sources lie in our Galaxy and are identified principally as *x-ray binaries (e.g. *Scorpius X-1, *Cygnus X-1, *Cygnus X-3, *Centaurus X-3, *Hercules X-1), many having an x-ray power 100 to 1000 times the power of the sun. A minority of galactic x-ray sources are identified with supernova remnants, such as the *Crab nebula and *Tycho's star. Fainter but intrinsically much more powerful x-ray emission is detected from many extragalactic objects, especially rich *clusters of galaxies and *active galaxies – such as *Seyfert galaxies, *quasars, and powerful *radio galaxies, including Virgo A (M87) and Centaurus A.

x-ray telescope. An instrument carried above the earth's atmosphere by an *x-ray satellite, etc., and by means of which x-rays from space can be detected and recorded. *See also* grazing incidence; proportional counter; microchannel plate detector; CCD.

x-ray transients. Bright nova-like cosmic x-ray sources that develop rapidly over a few days and remain visible for several weeks or months. *Ariel V observations, supported by simultaneous ground-based studies, led to the first optical identification: A 0620−00 (Nova Mon 1975), for a time in 1975 the brightest cosmic x-ray source ever seen, was found in the visible as a *nova of about 11th magnitude. Both x-ray and optical stars then faded rapidly, to disappear by mid-1976.

The sky distribution of x-ray transients shows them to be galactic, typically at distances of 1–10 kiloparsecs. Several have been found to have recurrent outbursts, usually on a timescale of several months or years. It is believed that x-ray transients represent close binary systems in which the *mass transfer is highly variable. The qualitative difference from most optical novae (e.g. Nova Cygni 1975, which had no detectable x-ray emission) may be due to the compact star being a *neutron star or *black hole, rather than a *white dwarf as in the optical novae.

XUV astronomy. The study of objects beyond the solar system that emit radiation bridging the gap between x-rays and ultraviolet radiation (the *XUV region*), i.e. from about 12 nm to the hydrogen Lyman limit at 91.2 nm (*see* hydrogen spectrum). The high opacity of the interstellar gas at these wavelengths prevents observations beyond about 10 parsecs from 91.2 to about 50 nm, extending to about 200 parsecs at 12 nm. The first partial sky survey in the XUV was carried out in the *Apollo-Soyuz mission in 1975, in which were found a number of intense sources, including HZ43 and Feige 24 (both hot *white dwarf stars) and *Proxima Centauri. The first deep all-sky survey in the XUV band will be carried out by the British Wide Field Camera, a 50 cm diameter *grazing incidence telescope of short focal length, to be flown in 1987 on the German satellite *ROSAT.

Y

Yagi. A directional *antenna comprising three or more parallel elements mounted on a straight beam and at right angles to it. The elements are a *dipole, to which the *feeder is attached, behind which is a reflector and in front of which is one or more directors. Named after Hidetsugu Yagi, this type of aerial is commonly used for television reception although it is also used as an element in some radio telescopes.

year. 1. The period of the earth's revolution around the sun or the period of the sun's apparent motion around the celestial sphere, both periods being measured relative to a given point of reference. The choice of reference point determines the exact length of the year (see table). *See also* Julian year.

2. *Short for* calendar year.

Yerkes Observatory. An astronomical station of the University of Chicago at Williams Bay, Wisconsin, endowed by Charles Yerkes at the suggestion of George Ellery Hale. It holds the

year	reference point	length (days)
tropical	equinox to equinox	365.242 19
sidereal	fixed star to fixed star	365.256 36
anomalistic	apse to apse	365.259 64
eclipse	moon's node to moon's node	346.620 07

world's largest refracting telescope, 40 inches (102 cm) in diameter, completed in 1897.

Yerkes system. *Another name for* MK system. *See* spectral types.

ylem (Greek: that on which form has yet to be imposed). *See* primordial fireball.

yoke mounting (English mounting). *See* equatorial mounting.

YY Orionis stars. *Irregular variable stars that appear to form a subclass of *T Tauri stars. However, their spectra reveal that matter is falling on to them rather than being ejected and so it seems likely that they represent an earlier stage of evolution.

Z

Z Camelopardalis stars. *See* dwarf novae.

Zeeman effect. An effect that occurs when atoms emit or absorb radiation in the presence of a magnetic field: the field modifies the energy configuration of the atom with the result that a spectral line is split into two, three, or more closely spaced components, each of which is polarized. (In some cases the components cannot be resolved and the effect appears as a broadening of the spectral line.) This Zeeman splitting can be used to study the magnetism of stars, including the sun, since the spacing of the components is a measure of the magnetic field strength.

Zelenchukskaya Observatory. The Special Astrophysical Observatory of the Academy of Sciences of the USSR, sited near Zelenchukskaya in the N Caucasus at an altitude of 2100 metres. The chief instruments are a 6

metre (236 inch) telescope (at present the world's largest reflector *but see* primary mirror) and a radio telescope, RATAN 600.

The 6 metre telescope was the first large optical telescope to have an *altazimuth mounting. The telescope is carried in a fork located in the vertical (altazimuth) axis of rotation, and is supported on oil 'cushions'. Instruments can be placed at the prime focus and the *Nasmyth foci of the primary mirror. The original 43 tonne 6 metre Pyrex primary had a focal length of 24 metres. A new glass-ceramic mirror of lower coefficient of expansion should reduce distortion of the image due to temperature fluctuations. Stress-relieving systems have reduced changes in mirror shape arising from weight. The telescope was operational in 1977 after a long development period.

RATAN 600 went into use in 1976. It consists of 895 vertical metal panels, each 2 metres wide by 7.4 metres high, arranged in a circle 576 metres across. The panels are tilted so that radio waves are reflected horizontally to a smaller secondary mirror and antenna within the circle. Different sections of the ring can be used to·image separate sources simultaneously. The system operates at wavelengths of 8–300 mm.

zenith. The point on the *celestial sphere that lies vertically above an observer and 90° from all points on his horizon. It lies on the observer's celestial meridian. The *geocentric zenith* lies at the point where the line connecting the earth's centre with the observer's position would meet the celestial sphere. Since the earth is not spherical the two points rarely coincide. *Compare* nadir.

zenithal hourly rate (ZHR). The probable number of *meteors observed per hour from a *meteor shower that has its *radiant in the observer's zenith. Shower rates vary as the zenith dis-

Normalized zenithal hourly rate

A	90°	52°	35°	27°	8.6°	2.6°
F	1	1.25	1.67	2	5	10

tance of the radiant changes. To obtain the normalized ZHR the observed rate must be multiplied by a factor, F, which varies for different values, A, of radiant altitude (see table).

zenith attraction. The corrections that must be subtracted from the observed velocity and direction of a meteoric body to give the true values. The gravitational attraction of the earth on the body will increase the *meteoroid's impact velocity, the effect being greatest for bodies with low initial values of geocentric velocity (i.e. meteoroids just catching up with the earth). The path of the meteoroid will also change, the new *radiant appearing to be closer to the observer's zenith than the original true radiant.

zenith distance (coaltitude). Symbol ζ. The angular distance of a celestial body from an observer's zenith measured along the vertical circle through the body. It is therefore the complement of the altitude $(90° - h)$ of the body. The *topocentric zenith distance* is the zenith distance measured at the observation point; it differs from the true value due to *atmospheric refraction, the difference being termed *angle of refraction*. See also diurnal parallax.

zenith stars. Stars that come to *culmination on the observer's zenith.

zenith tube. A telescope mounted in a vertical plane so that stars at or near the zenith are observed. Atmospheric disturbance is minimal in this direction. See also photographic zenith tube.

zero-age main-sequence (ZAMS). *See* main sequence.

Zerodur. *Trademark* A glass-ceramic material that alters very little in size or shape when heated over normal temperature ranges. It has thus been used in telescope mirrors, including that of the *Isaac Newton Telescope.

zeta (ζ). The sixth letter in the Greek alphabet, used in *stellar nomenclature usually to designate the sixth-brightest star in a constellation.

Zeta Aurigae (ζ Aur). An *eclipsing binary in the constellation Auriga (period 972 days, magnitude range 5.0–5.7). The brighter component is a hot blue B star and the secondary is a cool orange K5 II giant over 50 times the size of its companion. During the partial eclipse of the brighter star its light shines through the atmospheric layers of the giant and the spectral changes give information about the atmosphere's composition.
Zeta Aurigae stars are eclipsing binaries composed of a hotter smaller star of early spectral type and a cooler K-type giant or more usually a supergiant. The period often amounts to several years.

ZHR. *Abbrev. for* zenithal hourly rate.

zirconium stars. *See* S stars.

zodiac. A band passing around the *celestial sphere; extending about 9° on either side of the *ecliptic. It thus includes the apparent annual path of the sun as seen from earth and also the orbit of the moon and principal planets apart from Pluto. The band was divided by the ancient Greeks into 12 parts, each 30° wide, known as the *signs of the zodiac*. The signs originally corresponded in position to 12 constellations bearing the same name – the *zodiac constellations*. *Precession of the equinoxes has since shifted the zodiac constellations eastwards by

zodiac sign	symbol	dates of sun's passage
Aries	♈	21 Mar.–19 Apr.
Taurus	♉	20 Apr.–20 May
Gemini	♊	21 May–21 June
Cancer	♋	22 June–22 July
Leo	♌	23 July–22 Aug.
Virgo	♍	23 Aug.–22 Sept.
Libra	♎	23 Sept.–23 Oct.
Scorpius	♏	24 Oct.–21 Nov.
Sagittarius	♐	22 Nov.–21 Dec.
Capricornus	♑	22 Dec.–19 Jan.
Aquarius	♒	20 Jan.–18 Feb.
Pisces	♓	19 Feb.–20 Mar.

over 30°; the constellations through which the sun now passes include Ophiuchus, which is not considered a zodiac constellation. The table shows the zodiac signs and symbols with the dates of the sun's passage through them when their boundaries were fixed over 2000 years ago. On Jan. 1 the sun is now in Sagittarius rather than Capricornus.

zodiacal light. A permanent phenomenon that can be seen as a faint glow, especially at tropical latitudes, on a clear moonless night in the west after sunset and in the east before sunrise. It is shaped like a slanting cone and extends from the horizon, tapering along the direction of the *ecliptic, and visible for maybe 20°. The zodiacal light has been shown, spectroscopically, to be sunlight scattered by minute dust particles in the interplanetary medium in the inner solar system. The particles are concentrated into a belt lying close to the plane of the ecliptic. The zodiacal light can be detected right round the zodiac, there being a slight increase in size and brightness – the *gegenschein – directly opposite the sun's position. *See also* infrared background radiation.

zodiac constellations. *See* zodiac.

Zond probes. A series of Soviet interplanetary and lunar probes. Zonds 1 and 2 passed Venus and Mars in 1964 and 1965 respectively but did not return data. Zond 3 photographed the farside of the moon in 1965, obtaining the first high-quality pictures, while Zonds 5 and 6 (in 1968), 7 (1969), and 8 (1970) made circumlunar flights and were recovered on earth. Zond 4, in 1968, may have been an unsuccessful attempt at a circumlunar flight.

zone. A portion of a spherical surface lying between two parallel circles.

zone of avoidance. A region of sky, which roughly coincides with the belt of the *Milky Way, where no galaxies are seen. The apparent absence of galaxies is due to the presence of light-absorbing clouds of dust in the galactic plane.

zone of totality. Another name for path of totality. *See* eclipse.

Zürich relative sunspot number (Wolf number). Symbol: R. A somewhat arbitrary index of the level of sunspot activity, derived from the formula
$$R = k(f + 10g)$$
g represents the number of sunspot groups, f the total number of their component spots, and k is a constant dependent on the estimated efficiency of, and also the equipment used by, a particular observer. Though a more objective assessment of the level of sunspot activity relies on the measurement of sunspot areas, the Zürich relative numbers provide an almost continuous record back to the mid-18th century and have well withstood the test of time: they yield graphs similar (at least so far as the yearly means are concerned) to those based on other indices. The Zürich tradition came to an end in Jan. 1981, when the Sunspot Index Data Centre in Brussels took over responsiblity for the determination of sunspot relative numbers, which were henceforth to be known as International Sunspot Num-

bers (symbol: *RI*). The International Sunspot Number is determined from the formula for *R* given above. *See* *also* sunspot cycle.

ZZ Ceti stars. *See* white dwarf.

Table 1 Planets

Planet	Orbit				Equatorial Diameter (km)	Globe					Axis Period* (d h m)			Satellites
	Sidereal Period	Perihelion (AU)	Aphelion (AU)	Inclination (deg)		Oblateness	Mass (Earth=1)	Density (water=1)	Albedo	Tilt (deg)	d	h	m	
Mercury	87.97 d	0.31	0.47	7.0	4878	0.0	0.06	5.4	0.06	0	58	15	30.2	0
Venus	224.70 d	0.72	0.73	3.4	12 104	0.0	0.82	5.2	0.72	177	243	0	R	0
Earth	365.26 d	0.98	1.02	0.0	12 756	0.0034	1.00	5.5	0.39	23.4		23	56.1	1
Mars	686.98 d	1.38	1.67	1.8	6787	0.005	0.11	3.9	0.16	25.0	1	0	37.4	2
Jupiter	11.86 y	4.95	5.45	1.3	142 800	0.065	317.83	1.3	0.70	3.1		9	50.5	16 & rings
Saturn	29.46 y	9.01	10.07	2.5	120 000	0.108	95.17	0.7	0.75	26.7		10	14	+22 & rings
Uranus	84.01 y	18.28	20.09	0.8	50 800	0.030	14.50	1.3	0.90	97.9		~16	R	5 & rings
Neptune	164.79 y	29.80	30.32	1.8	48 600	0.026	17.20	1.8	0.82	29.6		~18		2
Pluto	247.7 y	29.6	49.3	17.2	3000?	0	0.002	1.1?	0.9?	118?	6	9	17	1

*R indicates retrograde axial rotation.

The rotation periods for Jupiter and Saturn refer to equatorial regions; the periods exceed 9h 55m and 10h 38m at higher latitudes, respectively.

Table 2 Planetary Satellites

PLANET & Satellite	Year of Discovery	Diameter (km)	Orbital Radius (10^3 km)	Eccentricity	Orbital Period (days)	Inclination (°)
EARTH						
moon	–	3476	384.4	0.055	27.32	23.4
MARS						
Phobos	1877	27 × 22 × 19	9.38	0.015	0.32	1.0
Deimos	1877	15 × 12 × 11	23.46	0.001	1.26	1.8
JUPITER						
Metis	1979	40	128		0.29	
Adrastea	1979	24 × 20 × 16	129		0.30	
Amalthea	1892	270 × 170 × 150	181	0.003	0.50	0.5
Thebe	1979	100	222	0.013	0.67	0.9
Io	1610	3630	422	0.004	1.77	0.0
Europa	1610	3050	671	0.009	3.55	0.5
Ganymede	1610	5260	1070	0.002	7.15	0.2
Callisto	1610	4800	1883	0.007	16.69	0.5
Leda	1974	20	11 094	0.148	239	26
Himalia	1904	180	11 480	0.158	250.6	28
Lysithea	1938	40	11 720	0.107	259.2	29
Elara	1905	80	11 740	0.207	259.7	25
Ananke	1951	30	21 200	0.17	631 R	147
Carme	1938	40	22 600	0.21	692 R	164
Pasiphae	1908	50	23 500	0.38	735 R	145
Sinope	1914	40	23 700	0.28	758 R	153
SATURN						
1980 S28	1980	40 × 20	138		0.61	0
1980 S27	1980	100	139		0.61	0
1980 S26	1980	100	142		0.63	0
1980 S3	1978	90 × 40	151		0.69	0
1980 S1	1978	100	151		0.69	0
Mimas c-o	1982	10	186		0.94	1.5
Mimas	1789	390	186	0.020	0.94	1.5
Enceladus	1789	500	238	0.004	1.37	0.0
Tethys	1684	1050	295	0.000	1.89	1.1
1980 S25	1980	35	295		1.9	
1980 S13	1980	35	295		1.9	
Tethys c-o	1982	15	295		1.9	0
–	1982	15	350		2.4	
Dione	1684	1120	377	0.002	2.74	0.0
1980 S6	1980	160	378		2.7	
Dione c-o	1982	15	378			0.3
–	1982		470			
Rhea	1672	1530	527	0.001	4.52	0.3
Titan	1655	5150	1222	0.029	15.95	0.3
Hyperion	1848	400 × 250 × 240	1481	0.104	21.28	0.4
Iapetus	1671	1440	3560	0.028	79.33	14.7
Phoebe	1898	160	12950	0.163	550.34 R	150

Table 2 Planetary Satellites

PLANET & Satellite	Year of Discovery	Diameter (km)	Orbital Radius (10^3 km)	Eccentricity	Orbital Period (days)	Inclination (°)
URANUS						
Miranda	1948	400	130	0.027	1.41	4
Ariel	1851	1300	191	0.003	2.52	0.3
Umbriel	1851	1100	266	0.004	4.14	0.4
Titania	1787	1600	436	0.002	8.71	0.1
Oberon	1787	1600	583	0.001	13.46	0.1
NEPTUNE						
Triton	1846	3800	355	0.000	5.88 R	160
Nereid	1949	300	5510	0.75	360.2	28
PLUTO						
Charon	1978	1000?	20	0?	6.39	0?

Note: The diameters given for the smaller satellites of Jupiter, and for many of the satellites of Saturn, Uranus, Neptune, and Pluto, are uncertain.

The orbits of the outer satellites of Jupiter vary considerably with time.

The distances refer to the centre of the parent planet and the inclinations to the plane of the equator of the planet.

R indicates retrograde motion.

c-o in a satellite name indicates co-orbital.

Table 3 Minor planets

No.	Name	Discovery Year	Period (years)	Perihelion (AU)	Aphelion (AU)	Inc (°)	Diameter (km)	Rotation (hours)
1	Ceres	1801	4.60	2.55	2.98	10.6	1003	9.1
2	Pallas	1802	4.61	2.11	3.42	34.8	540	10.5
3	Juno	1804	4.36	1.98	3.35	13.0	247	7.2
4	Vesta	1807	3.63	2.15	2.57	7.1	5|55	10.7
5	Astraea	1845	4.14	2.10	3.06	5.3	117	16.8
6	Hebe	1847	3.78	1.93	2.92	14.8	201	7.3
7	Iris	1847	3.69	1.84	2.94	5.5	209	7.1
10	Hygiea	1849	5.59	2.84	3.46	3.8	450	18.0
15	Eunomia	1851	4.30	2.14	3.14	11.7	272	6.1
31	Euphrosyne	1854	5.61	2.45	3.86	26.3	370	?
433	Eros	1898	1.76	1.13	1.78	10.8	23	5.3
588	Achilles	1906	11.90	4.44	5.98	10.3	50	?
944	Hidalgo	1920	14.04	2.00	9.64	42.5	15	?
1221	Amor	1932	2.66	1.08	2.76	11.9	0.5	?
1566	Icarus	1949	1.12	0.19	1.97	23.0	1	2.3
1862	Apollo	1932	1.78	0.65	2.29	6.3	1	?
2060	Chiron	1977	50.68	8.51	18.88	6.9	300?	?
2062	Aten	1976	0.95	0.79	1.14	18.9	1	?
–	Adonis	1936	2.56	0.44	3.30	1.4	0.3	?
–	Hermes	1937	2.10	0.62	2.66	6.2	0.5	?

Table 4 Meteor Showers

Shower	Normal Limits	ZHR at Max.
Quadrantids	Jan. 1–6	110
Corona Australids	Mar. 14–18	5
April Lyrids	Apr. 19–24	12
η -Aquarids	May 1–8	20
June Lyrids	June 10–21	8
Ophiuchids	June 17–26	6
Capricornids	July 10–Aug. 15	6
δ -Aquarids	July 15–Aug. 15	35
Pisces Australids	July 15–Aug. 20	8
α -Capricornids	July 15–Aug. 25	8
Perseids	July 25–Aug. 18	68
Cygnids	Aug. 19–22	4
Orionids	Oct. 16–26	30
Taurids	Oct. 20–Nov. 30	12
Cepheids	Nov. 7–11	8
Leonids	Nov. 15–19	10
Phoenicids	Dec. 4–5	5
Geminids	Dec. 7–15	58
Ursids	Dec. 17–24	5

Table 5 Constellations

Andromeda	And	Delphinus	Del	Perseus	Per
Antlia	Ant	Dorado	Dor	Phoenix	Phe
Apus	Aps	Draco	Dra	Pictor	Pic
Aquarius	Aqr	Equuleus	Equ	Pisces	Psc
Aquila	Aql	Eridanus	Eri	Piscis Austrinus	PsA
Ara	Ara	Fornax	For	Puppis	Pup
Aries	Ari	Gemini	Gem	Pyxis	Pyx
Auriga	Aur	Grus	Gru	Reticulum	Ret
Boötes	Boo	Hercules	Her	Sagitta	Sge
Caelum	Cae	Horologium	Hor	Sagittarius	Sgr
Camelopardalis	Cam	Hydra	Hya	Scorpius	Sco
Cancer	Cnc	Hydrus	Hyi	Sculptor	Scl
Canes Venatici	CVn	Indus	Ind	Scutum	Sct
Canis Major	CMa	Lacerta	Lac	Serpens	Ser
Canis Minor	CMi	Leo	Leo	Sextans	Sex
Capricornus	Cap	Leo Minor	LMi	Taurus	Tau
Carina	Car	Lepus	Lep	Telescopium	Tel
Cassiopeia	Cas	Libra	Lib	Triangulum	Tri
Centaurus	Cen	Lupus	Lup	Triangulum Australe	TrA
Cepheus	Cep	Lynx	Lyn	Tucana	Tuc
Cetus	Cet	Lyra	Lyr	Ursa Major	UMa
Chameleon	Cha	Mensa	Men	Ursa Minor	UMi
Circinus	Cir	Microscopium	Mic	Vela	Vel
Columba	Col	Monoceros	Mon	Virgo	Vir
Coma Berenices	Com	Musca	Mus	Volans	Vol
Corona Australis	CrA	Norma	Nor	Vulpecula	Vul
Corona Borealis	CrB	Octans	Oct		
Corvus	Crv	Ophiuchus	Oph		
Crater	Crt	Orion	Ori		
Crux	Cru	Pavo	Pav		
Cygnus	Cyg	Pegasus	Peg		

Table 6 The Twenty-Five Brightest Stars

Star	Position (1975)		Apparent magnitude (m_v)	Distance (parsecs)	Spectral Type	Absolute Magnitude (M_v)
	α	δ				
Sirius, α CMa	6^h 44.0^m	$-16°41'$	-1.5*	2.7	A1 V	$+1.4$
Canopus, α Car	6 23.6	-52 41	-0.7	60	F0 II	-4.6
α Centauri	14 38.0	-60 44	-0.3*	1.31	G2 V	$+4.2$
Arcturus, α Boo	14 14.5	$+19$ 19	-0.1	11	K2 III	-0.3
Vega, α Lyr	18 36.0	$+38$ 46	0.0	8.1	A0 V	$+0.5$
Capella, α Aur	5 14.8	$+45$ 52	0.1*	14	G8 III	-0.6
Rigel, β Ori	5 13.3	-8 14	0.1*	250	B8 Ia	-7.1
Procyon, α CMi	7 38.0	$+5$ 17	0.4*	3.5	F5 IV–V	$+2.7$
Achernar, α Eri	1 37.8	-57 22	0.5	35	B5 V	-2.2
β Centauri	14 02.1	-60 15	0.6*	120	B1 III	-4.8
Altair, α Aql	19 49.5	$+8$ 48	0.8	5.1	A7 IV–V	$+2.3$
Betelgeuse, α Ori	5 53.8	$+7$ 24	0.8†	200	M2 Iab	-5.7
Aldebaran, α Tau	4 34.0	$+16$ 28	0.8*	21	K5 III	-0.8
α Crucis	12 25.2	-63 00	0.9*	80	B1 IV	-3.7
Spica, α Vir	13 23.9	-11 01	1.0†	65	B1 V	-3.1
Antares, α Sco	16 27.8	-26 22	1.0*†	130	M1 Ib	-4.7
Pollux, β Gem	7 43.8	$+28$ 05	1.2	11	K0 III	$+1.0$
Fomalhaut, α PsA	22 56.2	-29 45	1.2	7	A3 V	$+1.9$
Deneb, α Cyg	20 40.6	$+45$ 11	1.3	430	A2 Ia	-7.2
β Crucis	12 46.2	-59 33	1.3	130	B0 III	-4.3
Regulus, α Leo	10 7.0	$+12$ 5	1.3*	26	B7 V	-0.8
Adhara, ϵ CMa	6 57.7	-28 56	1.5	200	B2 II	-5.0
Castor, α Gem	7 33.0	$+31$ 56	1.6	14	A1 V	$+0.8$
Shaula, λ Sco	17 31.8	-37 5	1.6	96	B2 IV	-3.3
Bellatrix, γ Ori	5 23.8	$+6$ 20	1.6	140	B2 III	-4.1

*Multiple star: m_v is integrated magnitude
†Variable star

Table 7 Nearest Stars

Star	Position (1975) α	δ	Distance (parsecs)	Prop. Mot. (''/yr)	App. mag. (m_v)	Abs. mag. (M_v)	Spectral Type
Proxima Centauri (C)	$14^h\ 28^m$	−62°34′	1.31	3.68	10.7	15.1	M5e V
Alpha Centauri (A) (B)	14 38	−60 44	1.31	3.68	0.01 / 1.4	4.4 / 5.8	G2 V / K5 V
Barnard's star	17 57	+4 37	1.83	10.34	9.53	13.2	M5 V
Wolf 359	10 55	+7 10	2.32	4.71	13.66	16.8	M6e V
Lalande 21185 (A) (BD 36°2147)(B)	11 2	+36 8	2.49	4.78	7.47	10.5 unseen component	M2 V
Sirus (A) (B)	6 44	−16 41	2.67	1.32	−1.46 / 8.7	1.4 / 11.5	A1 V / white dwarf
Luyten 726-8 (A) (UV Ceti) (B)	1 38	−18 5	2.74	3.35	12.5 / 12.9	15.4 / 15.8	M5.5e V / M6e V
Ross 154	18 48	−23 52	2.90	0.72	10.6	13.3	M4.5e V
Ross 248	23 41	+44 3	3.16	1.60	12.24	14.7	M5.5e V
Epsilon Eridani	3 32	−9 33	3.28	0.97	3.73	6.1	K2 V
Luyten 789-6	22 37	−15 27	3.31	3.25	12.58	14.9	M5.5e V
Ross 128	11 46	+0 57	3.32	1.40	11.13	13.5	M5 V
61 Cygni (A) (B)	21 6	+38 37	3.43	5.22	5.19 / 6.02	7.5 / 8.3	K5 V / K7 V
Epsilon Indi	22 1	−56 53	3.44	4.69	4.73	7.0	K5 V
Procyon (A) (B)	7 38	+5 17	3.48	1.25	0.34 / 10.7	2.7 / 13.0	F5 IV–V / white dwarf
BD 59° 1915 (A) (B)	18 42	+59 35	3.52	2.29	8.90 / 9.69	11.1 / 11.9	M4 V / M5 V
BD 43° 44 (A) (B)	0 17	+43 53	3.55	2.91	8.07 / 11.04	10.3 / 13.2	M2.5e V / M4e V
CD −36° 15693	23 4	−36 0	3.59	6.90	7.39	9.6	M2 V
Tau Ceti (A) (B)	1 43	−16 4	3.67	1.92	3.5	5.7 unseen component	G8 Vp
BD +5° 1668	7 26	+5 18	3.76	3.73	9.82	11.9	M4 V
CD −39° 14192	21 16	−38 58	3.85	3.46	6.72	8.7	M0 V
Kapteyn's star	5 11	−45 0	3.91	8.72	8.81	10.8	M0
Kruger 60 (A) (B)	22 27	+57 34	3.94	0.87	9.77 / 11.43	11.8 / 13.4	M3 V / M4.5e V

A, B, C refer to brightest, second brightest, third brightest components of binary or triple star

BD: Bonner Durchmusterung; CD: Cordoba Durchmusterung

Table 8 Messier numbers plus equivalent NGC, IC numbers

Messier no.	NGC IC	Type*	Const.	Messier no.	NGC IC	Type*	Const.
M1	1952	n	Tau	M41	2287	o.c.	CMa
2	7089	g.c.	Aqr	42	1976	n	Ori
3	5272	g.c.	CVn	43	1982	n	Ori
4	6121	g.c.	Sco	44	2632	o.c.	Cnc
5	5901	g.c.	Ser	45	—	o.c.	Tau
6	6405	o.c.	Sco	46	2437	o.c.	Pup
7	6475	o.c.	Sco	47	2422	o.c.	Pup
8	6523	n	Sgr	48	2548	o.c.	Hya
9	6333	g.c.	Oph	49	4472	G E	Vir
10	6254	g.c.	Oph	50	2323	o.c.	Mon
11	6705	o.c.	Sct	51	5194	G Sc	CVn
12	6218	g.c.	Oph	52	7654	o.c.	Cas
13	6205	g.c.	Her	53	5024	g.c.	Com
14	6402	g.c.	Oph	54	6715	g.c.	Sgr
15	7078	g.c.	Peg	55	6809	g.c.	Sgr
16	6611	o.c.	Ser	56	6779	g.c.	Lyr
17	6618	n	Sgr	57	6720	p.n.	Lyr
18	6613	o.c.	Sgr	58	4579	G SBb	Vir
19	6273	g.c.	Oph	59	4621	G E	Vir
20	6514	n	Sgr	60	4649	G E	Vir
21	6531	o.c.	Sgr	61	4303	G Sc	Vir
22	6656	g.c.	Sgr	62	6266	g.c.	Oph
23	6191	o.c.	Sgr	63	5055	G Sb	CVn
24	6603	o.c.	Sgr	64	4826	G Sb	Com
25	14725	o.c.	Sgr	65	3623	G Sa	Leo
26	6694	o.c.	Sct	66	3627	G Sb	Leo
27	6853	p.n.	Vul	67	2682	o.c.	Cnc
28	6626	g.c.	Sgr	68	4590	g.c.	Hya
29	6913	o.c.	Cyg	69	6637	g.c.	Sgr
30	7099	g.c.	Cap	70	6681	g.c.	Sgr
31	224	G Sb	And	71	6838	g.c.	Sge
32	221	G E	And	72	6981	g.c.	Agr
33	598	G Sc	Tri	73	6994	o.c.	Aqr
34	1039	o.c.	Per	74	628	G Sc	Psc
35	2168	o.c.	Gem	75	6864	g.c.	Sgr
36	1960	o.c.	Aur	76	650	p.n.	Per
37	2099	o.c.	Aur	77	1068	G Sb	Cet
38	1912	o.c.	Aur	78	2068	n	Ori
39	7092	o.c.	Cyg	79	1904	g.c.	Lep
40	—	d	UMa	80	6093	g.c.	Sco

*n: nebula; p.n.: planetary nebula; o.c.: open cluster; g.c.: globular cluster; G: galaxy & classification

from C. W. Allen: Astrophysical Quantities

Table 8 (continued)

Messier no.	NGC IC	Type*	Const.	Messier no.	NGC IC	Type*	Const.
M81	3031	G Sb	UMa	M96	3368	G Sa	Leo
82	3034	G Irr	UMa	97	3587	p.n.	UMa
83	5236	G Sc	Hya	98	4192	G Sb	Com
84	4374	G E	Vir	99	4254	G Sc	Com
85	4382	G SO	Com	100	4321	G Sc	Com
86	4406	G E	Vir	101	5457	G Sc	UMa
87	4486	G Ep	Vir	102	5866	G Sa	Dra
88	4501	G Ep	Com	103	581	o.c.	Cas
89	4552	G E	Vir	104	4594	G Sa	Vir
90	4569	G Sb	Vir	105	3379	G E	Leo
91	4567	G S	Com	106	4258	G Sb	CVn
92	6341	g.c.	Her	107	6171	g.c.	Oph
93	2447	o.c.	Pup	108	3556	G Sb	UMa
94	4736	G Sb	CVn	109	3992	G SBc	UMa
95	3351	G SBb	Leo				

*n: nebula; p.n.: planetary nebula; o.c.: open cluster; g.c.: globular cluster; G: galaxy & classification

from C. W. Allen: Astrophysical Quantities

Table 9 Astronomical and Physical Constants

astronomical unit, AU	$149.597\,870 \times 10^6$ km
speed of light in vacuum, c	$2.997\,924\,58 \times 10^5$ km s^{-1}
light-time for 1 AU	499.012 s
gravitational constant, G	6.672×10^{-11} N m^2 kg^{-2}
standard acceleration of free fall, g	$9.806\,65$ m s^{-2}
mass of earth	5.9742×10^{24} kg
earth's equatorial radius	6378.140 km
polar radius	6356.775 km
mean density of earth	5.517 g cm^{-3}
mean distance to moon	384 403 km
lunar mass	7.35×10^{22} kg
lunar radius	1738 km
solar mass, $M_\odot$	1.9891×10^{30} kg
solar radius, $R_\odot$	696 000 km
solar luminosity, $L_\odot$	4×10^{26} W
solar parallax	$8''.794\,148$
constant of sine parallax for moon	$3422''.451$
constant of aberration	$20''.496$
obliquity of ecliptic (1900)	$23°27'8''.26$
(1980)	$23°26'30''.78$
general precession in longitude per tropical century (1900)	$5025''.64$
constant of nutation (1900)	$9''.210$
Gaussian gravitational constant, k	$0.017\,202\,098\,95$ N$^{1/2}$ m kg^{-1}
Planck constant, h	$6.626\,196 \times 10^{-34}$ J s
Boltzmann constant, k	$1.380\,622 \times 10^{-23}$ J K^{-1}
Stefan's constant, σ	5.6696×10^{-8} W m^{-2} K^{-4}

Table 10 Famous People in the Field of Astronomy

Adams, John Couch	1819–92	English
Airy, Sir George Biddell	1801–92	English
Albategnius (or Al-Battani)	858–929 A.D.	Arabian
Aristarchus of Samos	~310–~230 B.C.	Greek
Baade, Walter	1893–1960	German-American
Bessel, Friedrich Wilhelm	1784–1846	German
Bradley, James	1693–1762	English
Brahe, Tycho	1546–1601	Danish
Cassini, Giovanni Domenico	1625–1712	Italian-French
Chandrasekhar, Subrahmanyan	1910–	Indian-American
Clark, Alvan Graham	1832–97	American
Copernicus, Nicolas	1473–1543	Polish
Eddington, Sir Arthur Stanley	1882–1944	English
Einstein, Albert	1879–1955	German-American
Eratosthenes	~276–~196 B.C.	Greek
Flamsteed, John	1646–1719	English
Fraunhofer, Joseph von	1787–1826	German
Galileo Galilei	1564–1642	Italian
Goodricke, John	1764–86	Dutch-English
Hale, George Ellery	1868–1938	American
Halley, Edmund	1656–1742	English
Herschel, Sir William	1738–1822	German-English
Herschel, Sir John	1792–1871	English
Hipparchus	fl. ~146–127 B.C.	Greek
Hertzsprung, Ejnar	1873–1967	Danish
Hubble, Edwin Powell	1889–1953	American
Huggins, Sir William	1824–1910	English
Huygens, Christiann	1629–95	Dutch
Kepler, Johann	1571–1630	German
Laplace, Pierre Simon, Marquis de	1749–1827	French
Lowell, Percival	1855–1916	American
Newton, Sir Isaac	1642–1727	English
Pickering, Edward Charles	1846–1919	American
Ptolemy (Claudius Ptolemaeus)	~100–~170 A.D.	Greek
Russell, Henry Norris	1877–1957	American
Schwarzschild, Karl	1873–1916	German
Shapley, Harlow	1885–1972	American

Table 11 Major Observatories

Optical & Infrared Observatories	Location	Principal Instruments
Anglo-Australian Obs.	Siding Spring, NSW	3.9 m reflector 1.2 m Schmidt
Cerro Tololo Interamerican Obs.	La Serena, Chile	4 m reflector
Crimean Astrophysical Obs.	Simeiz, Ukraine	2.6 m reflector
European Southern Obs.	Cerro La Silla, Chile	3.6, 3.5 (u.c.), 2.2 m reflectors 1 m Schmidt
Fred Lawrence Whipple Obs.	Mt Hopkins, Arizona	4.5 m multiple mirror
German-Spanish Astrophy. Obs.	Calar Alto, Spain	3.5, 2.2 m reflectors
Kitt Peak National Obs.	Kitt Peak, Arizona	4, 2.3, 2.1 m reflectors 1.5 m solar
Las Campanas Obs.	La Serena, Chile	2.5 m reflector
Lick Obs.	Mt Hamilton, California	3 m reflector
Mauna Kea Obs.	Hawaii	3.8, 3.6, 3.0 m IR reflectors 10 m reflector (u.d.)
McDonald Obs.	Mt Licke, Texas	2.7, 2.1 m reflectors
Mount Wilson Obs.	Mt Wilson, California	2.5 m reflector
Palomar Obs.	Palomar Mountain, California	5 m reflector 1.2 m Schmidt
Roque de los Muchachos Obs.	La Palma, Canaries	4.2 m (u.c.), 2.5 m reflectors
Zelenchukskaya Obs.	Zelenchukskaya, Caucasus	6 m reflector

u.c.: under construction; u.d.: under design

Radio Observatories	Location	Principal Instruments
Arecibo Radio Obs.	Arecibo, Puerto Rico	305 m dish
Australian Nat. Radio Astronomy Obs.	Parkes, NSW	64 m dish
CSIRO Solar Obs.	Culgoora, NSW	3 km heliograph Australia Telescope
Max Planck Inst. for Radio Astronomy	Bonn, West Germany	100 m dish
Molonglo Radio Obs.	Hoskinstown, NSW	1.6 km array
Mullard Radio Astronomy Obs.	Cambridge, England	5 km, 1.6 km arrays
National Radio Astronomy Obs.	Green Bank, W Virginia Socorro, NM	91 m, 43 m dishes VLA
Nobeyama Radio Obs.	Nobeyama Highland, Japan	45 m dish (mm waves)
Nuffield Radio Astronomy Labs.	Jodrell Bank, Cheshire	76 m, 38 m dishes MERLIN
Westerbork Radio Obs.	Westerbork, Netherlands	2.7 km array
Zelenchukskaya Obs.	Zelenchukskaya, Caucasus	600 m array

Table 12   Recent Successful Planetary Probes

Probe	Country	Planet(s)	Launch Date	Arrival Date
Mariner 9	US	Mars	30.5.71	13.11.71
Pioneer 10	US	Jupiter	3.3.72	4.12.73
Venera 8	USSR	Venus	26.3.72	22.7.72
Pioneer 11	US	Jupiter, Saturn	6.4.73	3.12.74 9.79
Mariner 10	US	Venus, Mercury	3.11.73	5.2.74 29.3.74
Venera 9 Venera 10	USSR	Venus	8.6.75 14.6.75	22.10.75 25.10.75
Viking 1 Viking 2	US	Mars	20.8.75 9.9.75	20.7.76 3.9.76
Voyager 2	US	Jupiter Saturn Uranus Neptune	20.8.77	7.79 8.81 1.86 8.89
Voyager 1	US	Jupiter Saturn	5.9.77	3.79 11.80
Pioneer Venus 1, 2	US	Venus	20.5.78 8.8.78	4.12.78 9.12.78
Venera 11 Venera 12	USSR	Venus	9.9.78 14.9.78	21.12.78 25.12.78
Venera 13 Venera 14	USSR	Venus	30.10.81 4.11.81	2.3.82 5.3.82
Venera 15 Venera 16	USSR	Venus	2.6.83 7.6.83	11.10.83 16.10.83